Fuel Economy

in Road Vehicles Powered by Spark Ignition Engines

Fuel Economy

in Road Vehicles Powered by Spark Ignition Engines

Edited by

John C. Hilliard
Combustion and Fuel Research, Inc.
Ann Arbor, Michigan

and

George S. Springer
Stanford University
Stanford, California

PLENUM PRESS • NEW YORK AND LONDON

Library of Congress Cataloging in Publication Data

Main entry under title:

Fuel economy in road vehicles powered by spark ignition engines.

Includes bibliographical references and index.
1. Automobiles—Fuel consumption—Addresses, essays, lectures. 2. Internal combustion engines, Spark ignition—Fuel consumption—Addresses, essays, lectures. I. Hilliard, John C., 1951- . II. Springer, George S.
TL151.6.F85 1984 629.2′53 84-1932
ISBN 0-306-41438-4

A Division of Plenum Publishing Corporation
233 Spring Street, New York, N.Y. 10013

Printed in the United States of America

Contributors

T. W. ASMUS • *Advanced Power Plant Engineering, Chrysler Corporation, Highland Park, Michigan*

CLAUS BORGNAKKE • *Automotive Laboratory, University of Michigan, Ann Arbor, Michigan*

S. K. CLARK • *Department of Mechanical Engineering and Applied Mechanics, University of Michigan, Ann Arbor, Michigan*

DAVID COLE • *Office for the Study of Automotive Transportation, Transportation Research Institute, University of Michigan, Ann Arbor, Michigan*

J. T. KUMMER • *Chemical Research Laboratory, Ford Motor Company, Dearborn, Michigan*

KENNETH C. LUDEMA • *Department of Mechanical Engineering and Applied Mechanics, University of Michigan, Ann Arbor, Michigan*

RUDOLF MALY • *Institut für Physikalische Elektronik, University of Stuttgart, Federal Republic of Germany*

THOMAS MOREL • *Integral Technologies Incorporated, Westmont, Illinois*

W. B. RIBBENS • *Electrical and Computer Engineering Department, University of Michigan, Ann Arbor, Michigan*

GEORGE S. SPRINGER • *Department of Aeronautics and Astronautics, Stanford University, Stanford, California*

THOMAS R. STOCKTON • *Scientific Research Laboratories, Ford Motor Company, Dearborn, Michigan*

R. W. WHEELER • *Atlantic Research Associates, Tunbridge Wells, Kent, England*

Preface

Concern about the reduced availability and the increased cost of petroleum fuels prompted great efforts in recent years to reduce the fuel consumption of automobiles. The ongoing efforts to reduce fuel consumption have addressed many relevant factors, including increased engine performance, reduced friction, use of lightweight materials, and reduced aerodynamic drag. The results of the investigations assessing the various factors affecting fuel economy have been published in journals, conference proceedings, and in company and government reports. This proliferation of technical information makes it difficult for workers to keep abreast of all developments. The material presented in this book brings together in a single volume much of the relevant materials, summarizes many of the state-of-the-art theories and data, and provides extensive lists of references. Thus, it is hoped that this book will be a useful reference for specialists and practicing engineers interested in the fuel economy of automobiles.

J. C. HILLIARD
G. S. SPRINGER

CONTENTS

4. EFFECTS OF VALVE EVENTS ON ENGINE OPERATION

T. W. Asmus

5. FLAME PROPAGATION AND HEAT-TRANSFER EFFECTS IN SPARK IGNITION ENGINES

Claus Borgnakke

6. ABNORMAL COMBUSTION EFFECTS ON ECONOMY

R. W. Wheeler

7. MECHANICAL FRICTION AND LUBRICATION IN AUTOMOBILES

Kenneth C. Ludema

8. TIRE ROLLING RESISTANCE AND VEHICLE FUEL ECONOMY

S. K. Clark

9. PROPERTIES OF SHEET MOLDING COMPOUNDS

George S. Springer

10. AERODYNAMICS OF ROAD VEHICLES

Thomas Morel

11. POWER-TRAIN MATCHING AND FUEL ECONOMY PROJECTION METHODS

Thomas R. Stockton

12. ELECTRONIC ENGINE CONTROL

W. B. Ribbens

Automotive Fuel Economy

DAVID COLE

I. INTRODUCTION AND BACKGROUND

In this day of energy uncertainty our attention is sharply focused on every major energy user in the United States, and for that matter, the entire world. No energy consumer is immune from consideration, whether it be production of goods, home heating and cooling, or transportation. Clearly, one of the most important factors in the energy use equation is automotive transportation and in particular the automobile. The incentives to improve automotive vehicle efficiency are substantial and growing year by year. It can be reliably predicted that the cost of petroleum-based energy will continue to increase and the supply will remain uncertain since many of the principal sources of this energy form are in relatively unstable areas of the world. The ongoing political uncertainties in the Middle East and other developing areas that are net exporters of petroleum are before us daily in the news media. Even with the appearance of petroleumlike synthetic gasoline, diesel fuel, or methanol derived from coal or oil shale, little impact can be made on the basic supply structure in the next 10–20 years. It is in this period of transition that the free world is most vulnerable to dislocations in petroleum availability.

In the United States it is becoming increasingly clear that the individual automobile will remain the principal mode of personal transportation and this extends to Canada, many areas of Western Europe, and for that matter, much of the developed world. In the past few years we have come to recognize this fact, and even the most ardent supporters of public or mass transit recognize that these transit forms are only going to expand at a modest rate in the domestic transportation system. One can expect, of course, to see some increased use of

---

DAVID COLE • Office for the Study of Automotive Transportation, Transportation Research Institute, University of Michigan, Ann Arbor, Michigan 48109.

buses and fixed-rail systems and perhaps other forms of public transit, but with the advent of promised highly fuel efficient automobiles and trucks the incentives are expected to be reduced dramatically for these alternative transportation modes. And, of course, the individual freedom afforded by the personal automobile is likely to continue as an important attribute of life in our society. Often this degree of freedom is taken for granted and only when we are threatened with the removal of this flexibility does it become an important and even overriding issue.

The impact of personal transportation on total energy use is very significant. It is even more significant when one relates it to total petroleum use. Table 1 shows a rather simple breakdown of the major energy consumers in the United States as well as the breakdown in the use of petroleum. At the present time the United States is consuming about 17 million barrels per day of petroleum of which approximately 6 million barrels per day are imported.

Figure 1, based on a forecast by Dr. William Agnew at General Motors Research Laboratories,[1] shows the past and expected future total national aggregate trends in automobile and light truck fuel consumption through 1990. Even with the expected increase in the number of vehicles, the trend in fleet fuel consumption for automobiles is downward because of an expected decrease in the number of miles driven per vehicle and, very importantly, the major individual vehicle fuel economy improvements expected in this time period. Light truck fuel consumption is expected to increase slightly because of an expected increase in total miles driven. Another projection of future fuel economy trends was developed in a University of Michigan Delphi study.[2] The engineering panelists in this study forecast that the average fuel economy of U.S.-produced vehicles would be 31 mpg in 1985 and 35 mpg in 1990.

Despite the despondent cries of amateur historians, those that live in the past, the "good old days" of automotive transportation were not filled with cars of excellent mileage, and the statement that "they don't build them like they

TABLE 1. Distribution of Total Energy and Petroleum Consumption in the United States

Total energy (1979) (%)		Petroleum (1977) (%)	
Industrial	28	Industrial	18.7
Electric utility losses	22	Electric utilities	9.3
Residential, commercial	25	Residental, commercial	17.2
Transportation	25	Transportation[a]	53.3
	100.0%	Other	1.1
			100.0%

[a] Approximately 50% of transportation fuel is used in passenger cars.

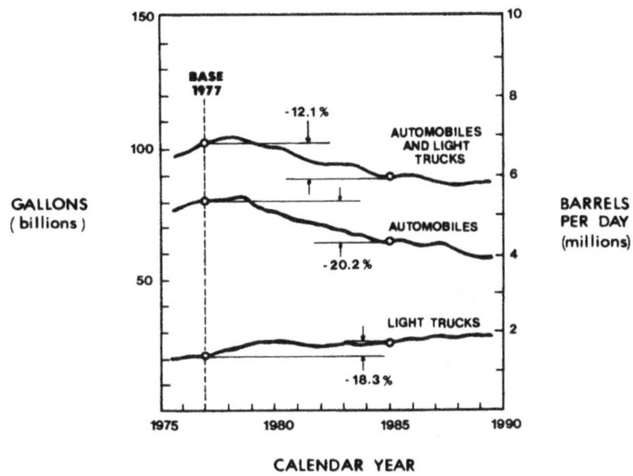

FIGURE 1. Estimated aggregate U.S. fuel consumption for automobiles and light trucks.

used to" is simply a myth. The automobiles of today are considerably better than those of the past, particularly when one considers the total range of consumer value factors embodied in modern automobiles. These include safety, ride, handling, interior noise level, comfort, emission performance, and of course fuel economy. On essentially every count, except price, the cars of today easily surpass the cars of yesteryear. Because of speculation as to past automobile fuel economy and emission performance, a mint condition 1930 Model A Ford was recently tested. It is compared to a 1981 Chrysler "K" car in Table 2. The rather dramatic differences between these vehicles are evident, and when one considers the obviously superior creature comfort aspects of the modern vehicle, far greater durability, and reduced maintenance requirements, the automobile of today is a clear winner.

TABLE 2. Exhaust Emission and Fuel Economy Comparison between a 1930 Ford Model A and a 1981 Dodge Aries (K Car)[a]

	1930 Ford Model A	1981 Dodge Aries
Fuel economy—EPA urban	14.1 mpg	25.0 mpg
Exhaust emissions		
Hydrocarbons	5.6 g/mile	<0.4 g/mile
Carbon monoxide	83.6	<7.0
Nitrogen oxides	1.54	<1.0

[a] Both equipped with manual transmission.

Because of the wealth of material and energy resources found in the United States, fuel economy has only recently become a matter of real concern. We were essentially energy independent in petroleum through the early 1970s. With this in mind it is clear that the automotive industry in the United States has made dramatic progress toward improved fuel economy in a rather short time. General Motors, for example, has improved its fleet average fuel economy from 12 mpg in 1974 to more than 22 mpg in 1981.

Many factors are responsible for the recent rapid improvement in fuel economy, and these same factors will continue to be important in the dramatic fuel economy improvements expected in the years just ahead. These factors include vehicle downsizing, tire improvements, reduced aerodynamic drag, the application of new lightweight materials, more efficient automotive structures, and greatly improved engines and total drive trains. Even with the progress made to date, there is still much to be done, as is clearly evidenced by the popularity of small U.S. and imported automobiles. There are many that believe the imports invented fuel economy and therefore are the leaders in automotive technology. Of course, in most areas of the world fuel economy has historically been an important issue because most of the petroleum-based fuel had to be imported at rather significant costs. Also, policy decisions within foreign governments often led to high taxes on motor vehicle fuel.

Most foreign-produced automobiles exhibit excellent fuel economy. However, if economy is measured on the basis of total volume (passengers plus luggage) the relative efficiency of vehicles from around the world in comparison to the U.S.-produced vehicles is very enlightening. The average new Japanese or European car is considerably better than the average new U.S. car, although on a percentage basis U.S. cars have been greatly improved by the downsizing and redesign efforts of the past several years. The major factors in the foreign car efficiency lead are their smaller size and lighter weight plus the availability of features such as five-speed manual transmissions.

To develop a valid yardstick for comparing foreign and U.S. technology, it is necessary to normalize with respect to key parameters. An appropriate factor is total interior volume. For our purposes the Environmental Protection Agency (EPA) combined passenger and luggage volume and urban fuel economy data has been selected.[(3)] Comparison should be with an equivalent transmission and engine type. It would also be useful to generate an equal performance normalization, but this is very difficult with the data at hand. For these calculations fuel economy performance has been converted to gallons/mile rather than miles/gallon. The normalized data is thus expressed as gallons/mile per cubic feet of total interior volume. Results for representative passenger cars are shown in Table 3. As in golf, the lower the score the better the performance.

Note, in the second to last column, that the best rating of 3.7 is shared by a relatively large, midsized hatchback GM "X" car, and the two classes smaller subcompact Toyota Tercel. The Chrysler "K" car at 3.8 is almost equal to the 3.7 Tercel and better than the other seven Japanese cars listed. The Ford Fairmont, at 4.0 is equal to or better than all the Japanese cars except the Tercel. The Chevrolet Caprice, classified by the EPA as a large car, rates better than most Japanese cars.

Consumption data is also presented on a gallon/mile per passenger basis in the last column of Table 3. On this basis as well it is clear that U.S. passenger cars do very well.

These combined fuel economy–volume ratings are not an attempt to downgrade the importance of fuel economy alone but to put it into perspective and to demonstrate that U.S. automotive technology is at least equivalent to the international competition.

One key point to bear in mind with regard to the future of the automotive industry and its products is the market sensitivity to fuel pricing. As the fuel economy of vehicles improve, the incremental savings associated with each added improvement in fuel economy decreases. Figure 2 provides an illustration of this point in which the cost of fuel per year is plotted as a function of miles per gallon with the assumption that the average motorist drives approximately 10,000

TABLE 3. Fuel Consumption Comparison between U.S., Japanese, and European Passenger Cars[a]

Car—Model	No. of pass.	Total EPA int. volume	EPA urban mpg	Combined fuel economy volume ratings: gal/mile/ cu. ft. $\times 10^{-4}$	Combined fuel economy volume ratings: gal/mile/ pass. $\times 10^{-3}$
G.M.—Chevette (Hatchback)	4	89	26	4.3	9.6
G.M.—"X" Car (Hatchback)	5	115	23	3.7	8.6
G.M.—Chev. Malibu (4 dr.)	6	119	19	4.4	8.7
G.M.—Chev. Caprice (4 dr.)	6	131	19	4.0	8.7
Ford—Fairmont (4 dr.)	6	113	22	4.0	7.6
Ford—LTD (4 dr.)	6	133	16	4.6	10.4
Chrysler—"K" Car	5/6	110	24	3.8	6.9
G.M. Olds—Cutlass Diesel	6	118	23	3.7	7.2
VW Rabbit—Gasoline	4	91	25	4.3	10.0
Honda—Civic (Hatchback)	4	82	29	4.2	8.6
Honda—Accord (4 dr.)	4	91	24	4.5	10.4
Datsun—210 (Hatchback)	4	85	29	4.0	8.6
Datsun—510 (Hatchback)	4	87	27	4.2	9.3
Toyota—Celica (Hatchback)	4	89	25	4.5	10.0
Toyota—Tercel (Hatchback)	4	93	29	3.7	8.6

[a] Automatic transmission, smallest engine available.

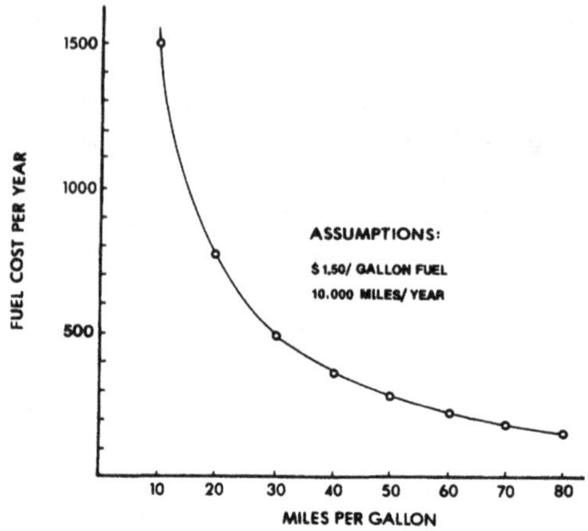

FIGURE 2. Yearly fuel cost as a function of vehicle fuel economy in miles per gallon.

miles per year and fuel is priced at $1.50 per gallon. The fuel cost per year for a 10-mpg car is $1,500. At a 20-mpg rating the yearly fuel cost drops to $750, and one can readily see the decreased sensitivity to vehicle fuel economy with the increasing miles per gallon performance of the vehicle. As we look to the future when all vehicles, even the larger ones, will be quite fuel efficient, the possibility that this parameter will decrease in importance to the consumer is very real. Even with higher fuel prices the same trend is evident, although the fuel cost difference associated with different mpg ratings is proportionally greater.

Perhaps a word of caution is warranted on the issue of fuel consumption and fuel economy. Often these terms are used interchangeably but of course one is really the inverse of the other. That is, good fuel economy means low fuel consumption and vice versa. Generally, fuel economy is reported in miles per gallon or kilometers per liter. However, in more technical circles fuel consumption is often discussed rather than fuel economy. Fewer gallons per mile or liters per kilometer is a mark of superiority. It is particularly important to use fuel consumption when dealing with the aggregate performance of vehicles. For example, if you consider both a 10-mpg and a 20-mpg vehicle, the average of the two would suggest that the fuel economy would be 15 mpg, but this is incorrect. When averaging properly, using a consumption base, a 0.1 gal/mile (10 mpg) car and another of 0.05 (20 mpg) gal/mile car averages 0.15 gallons per 2 miles of travel or .075 gal/mile the inverse of which is 13.3 mpg. Throughout this book fuel consumption and fuel economy will be used so the reader is cautioned to carefully note the distinction between the two terms.

Clearly, size and weight are extraordinarily important variables in the overall fuel economy equation of modern vehicles. Smaller and lighter vehicles generally exhibit better fuel economy than their larger and heavier counterparts. Unfortunately, downsizing is ultimately limited as a means to obtain improved fuel economy unless, of course, it is possible to decrease the size of people, families, and their packages. I think we can all agree that this is a challenging task and well beyond the ability of the average engineer. Another consideration is that the smaller vehicles are not as crashworthy as larger vehicles. Recent data shows rather dramatically the relative value of size in an accident. The small car loses in either a small car/big car confrontation or even a small car/small car collision. Consequently, other factors in the fuel economy equation in addition to downsizing are being attacked with vigor.

In each component area, whether engines, transmissions, or tires the rate of progress is slowing as we approach the ultimate limits for a given technology. Very early on in the development of engines, for example, rather rapid progress was made in both power and fuel economy performance. However, with maturing technology the incremental gains have slowed dramatically as the fundamental limits are approached. With a significant technological breakthrough, which would cause a shift to a higher limit, the same process is repeated. In most terms of the automotive fuel economy equation, we are approaching the limit of potential for efficiency improvement.

In many instances the theoretical limits of performance are difficult to define. Certainly, this is true of aerodynamic losses. In the engine, however, for a given set of operating parameters such as compression ratio and fuel–air ratio, the laws of thermodynamics place a rather specific limit on thermal efficiency.

Before beginning an assessment of individual fuel economy factors, it is useful to analyze all the forces resisting movement of the vehicle. A free-body diagram showing the various longitudinal axis forces that must be overcome by the powerplant is shown in Fig. 3. Clearly, anything that can be done to reduce

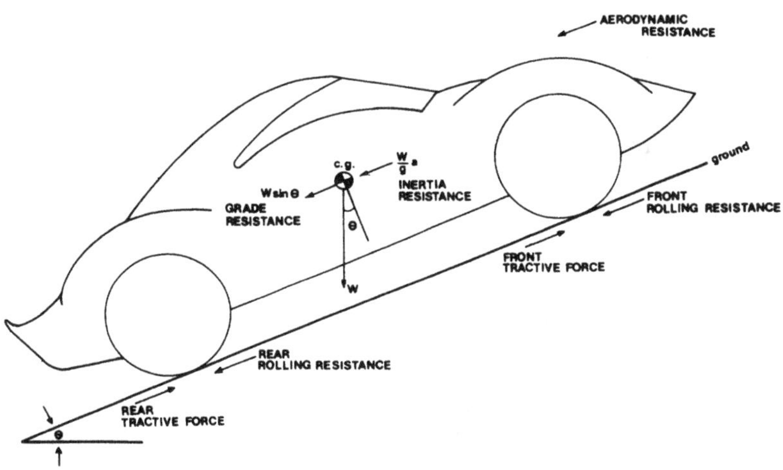

FIGURE 3. Longitudinal force free-body diagram for a ground vehicle.

the magnitude of these forces in total or in part implies that the engine power can be reduced leading to reduced fuel consumption. A major part of the subsequent discussion on fuel economy will be focused on these various motion resistance forces. These forces are qualitatively expressed graphically as a function of vehicle speed in Fig. 4. Curve A represents the rolling resistance which appears at the tire–road interface. This force is nearly independent of speed. Curve B is the aerodynamic resistance, which varies as the square of velocity. The sum of these curves, Curve C, is defined as the road load resistance of the vehicle; i.e., the force required to maintain a constant speed on a level road. The grade resistance force, which is equal to the weight times the sine of the slope, is also shown on this diagram for several values of inclination. Thus the road load resistance on a given grade is easily determined by summing the grade resistance force and the road load resistance for a 0% grade. Incidentally, the percent slope, or grade, typically used as a measure of roadway inclination is actually the tangent of the grade angle. For small angles typical of the normal paved highway system, the sine and tangent are approximately equal. Also shown in this diagram is a schematic representation of the tractive force developed at the rear wheels for an internal combustion engine powered vehicle with a three-speed standard shift transmission. As the transmission is shifted to a higher gear, the speed reduction ratio is decreased and the transmission output torque and

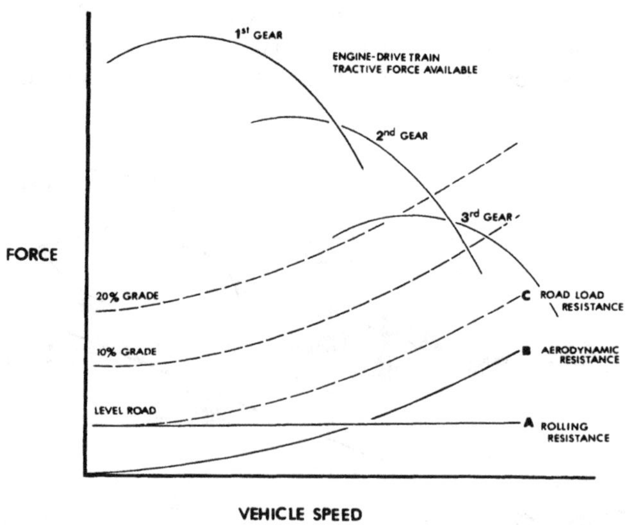

FIGURE 4. Rolling, aerodynamic, and road load resistances and tractive forces available with an internal combustion engine connected to a three-speed standard shift transmission as a function of vehicle speed.

therefore the tractive force available at the drive wheel are reduced. The difference between the tractive force available and the road load resistance on a given highway grade is known as the free-tractive effort and is the force available for acceleration or pulling a trailer. Considering all of the factors in the fuel economy equation, the ultimate fuel economy potential of an internal combustion engine-powered vehicle is a combination of the vehicle resistance forces and the combined performance of the engine and drive train. At any operating point the tractive force delivered by the drive train must match the total of the resistance forces.

II. FUEL ECONOMY FACTORS

The important parameters influencing fuel consumption can be disaggregated into the following list:

1. Engine characteristics
2. Drive-train characteristics
3. Weight
4. Aerodynamics
5. Rolling resistance
6. Driving cycle
7. Driver habits

The potential fuel economy of a four-passenger enclosed vehicle powered by an open-chamber, supercharged diesel engine appears to be in the area of 100 mpg on the EPA combined urban–suburban cycle. At least one vehicle manufacturer has demonstrated this performance level in a highly sophisticated car using the most advanced technology. However, this technology is presently not commercially applicable. We certainly would not expect to achieve 200 mpg on a similar driving cycle with this type of vehicle because of the barriers afforded by the laws of nature. A word of caution is in order when dealing with any fuel economy data that concerns the impact of the operating cycle on fuel economy. Without knowledge of the operating cycle, miles per gallon data comparison is meaningless. The cycle must be well defined. Currently, the manufacturers are required to conduct fuel economy tests according to the EPA urban driving cycle, which will be discussed later in this chapter. Fuel economy performance measured on this cycle is the published number used by the various manufacturers. The EPA also has a suburban test cycle that includes a significant fraction of simulated highway driving without the "stop and go" of the urban cycle. In order to determine compliance with the Corporate Average Fuel Economy (CAFE) standards established by the federal government, the urban and suburban cycle fuel economies are combined according to the following formula:

$$\text{mpg}_{\text{composite}} = \frac{1}{\dfrac{0.55}{\text{urban mpg}} + \dfrac{0.45}{\text{highway mpg}}}$$

To provide some perspective on fuel economy expectations, we will for a moment violate the foregoing rule related to test cycle definition and examine the qualitative economy performance of several vehicles. Recent advertisements for a small four-stroke cycle motorcycle have promoted its excellent fuel economy of about 125 mpg (cycle uncertain). When one considers the limited capacity of this motorcycle and the lack of many of the creature comforts associated with enclosed vehicles, it is clear that today's passenger car fuel economy levels represent a rather substantial achievement. Other examples of vehicle economy range from the heavy highway tractor–trailer with an open-chamber diesel engine which might deliver 6 mpg on a highway cycle to highly sophisticated and specialized vehicles that have been built specifically to obtain extremely high mileage. A number of special diesel and gasoline engine-powered vehicles have obtained more than 3000 mpg but while being driven on specialized and not very realistic driving cycles.

When looking to the future and to the efficiency improvements that can be anticipated in production vehicles, there is considerable uncertainty because of the difficulty in projecting the consumer demand for fuel economy and vehicle size as well as the potential technological gains and the trends in government regulation. There is no such thing as perfect future vision. And as the French philosopher, Paul Valery, stated, "The trouble with these times is that the future isn't what it used to be."

TABLE 4. Projected Relative Contribution of Vehicle Factors to Fuel Economy Improvement through 1985 and 1990

Factor	% Share of improvement	
	1985	1990
Downsizing	35%	30%
Use of light material	15	20
Engine efficiency	15	15
Aerodynamics	10	10
Drive-train efficiency, including transmission improvements	10	10
Tire rolling resistance	5	5
Other	10	10
	100%	100%

In the University of Michigan Delphi study referred to earlier, industry experts were asked to estimate the likely relative contribution of various technological factors to the expected future improvements in fuel economy. Table 4 summarizes their expectations. This table can be interpreted as to suggest the relative importance of technical factors that will lead to economy gains predicted for 1985 compared to 1980 and 1990 compared to 1985.

In the following sections the various parameters influencing fuel economy will be considered briefly and in subsequent chapters will be examined in detail.

A. Engine

From a pedestrian viewpoint it is common when thinking of fuel economy to focus only on automotive powerplants. While this is incorrect in total, the engine is still an extremely important element and will continue as such in the future. Basically, fuel energy is delivered to the engine in chemical form and a transformation (combustion) process occurs within the engine and torque is delivered to the drive train. The sequence of events associated with this process is rather simple in concept but is enormously complex in terms of the physical and chemical details. Essentially, fuel and air (air only in the diesel) are inducted into the engine and compressed by movement of the piston. The charge is ignited either by an external source such as a spark plug or through self-ignition as in the diesel when the fuel is injected. The chemical energy of the working fluid is converted into internal energy or the energy of molecular motion in the combustion chamber. This increases the internal gas pressure significantly and creates an added force on the piston face in the case of the reciprocating internal combustion engine and useful power is produced. Following the power stroke, the exhaust valve is opened and the products of combustion are purged from the chamber followed by the opening of the intake valve and the introduction again of a fresh flow of fuel and air and the cycle is repeated.

At the present time it is impossible to consider just the power and efficiency of the engine. Exhaust emission performance is intimately associated with the efficient production of power. Emission considerations add substantial complexity to the overall problem and consequently have been under intense study for the past 20 years. In general, the present automotive emission requirements result in a modest fuel economy penalty, but there is considerable disagreement among experts as to the magnitude of this penalty and in general the trade-off relationship between fuel economy and emission performance. An historical summary of the federal emission standards as well as those for California which have often differed from the rest of the states is shown in Table 5. The tests associated with these standards are conducted on a chassis dynamometer following a special urban driving cycle discussed later in this chapter. For further details on the emissions test procedures, the EPA should be consulted. EPA fuel

TABLE 5. Summary of Emission Control Standards for Light-Duty Vehicles in the United States

	Model year			
	1977–1979[a]		1980	1981–1985
Emissions, g/mile				
Hydrocarbons	1.5	0.41	0.41	0.41
Carbon monoxide	15.0	9.0	7.0[c]	3.4
Nitrogen oxides	2.0	1.5	2.0[c]	1.0[b]

[a] For 49 states and California.
[b] Possible exemption for alternative engines, including diesels: 1.5 g/mile.
[c] California set slightly different standards.

economy data are generated at least in part on the same test with the emission measurement. Fuel economy is determined from a carbon balance technique based on the emission output and not from a direct measurement of fuel consumption. This procedure over the years has been controversial, but the current state of the technology is such that the carbon balance calculated fuel economy and measured fuel economy are approximately the same.

To gain a better understanding of the potential for the engine to produce power efficiently, it is necessary to turn to the science of thermodynamics which is essentially the science of energy conversion. In engines we typically look to the first and second law of thermodynamics to establish our boundary conditions. No device or system has ever been found to operate in violation of either of these laws. All engines operate on thermodynamic cycles consisting of a sequence of processes which include compression, heat addition, expansion, and heat rejection. The same processes that were embodied in the elementary discussion of the simple engine. The general statement of the second law of thermodynamics states rather clearly that it is impossible to construct an engine that operates in a cycle and will produce no effect other than the extraction of heat from a reservoir in the performance of an equivalent amount of work. Thus the process of energy conversion within the engine will always result in less work output than the equivalent chemical energy delivered to the engine in the fuel. It is important to establish a distinction between energy or heat losses and heat that must be rejected from the cycle to adhere to the second law. Commonly, there is a tendency to refer to any heat which an engine system transfers to the atmosphere as a heat loss. It has been shown rather clearly, however, that there is a fundamental requirement that some of the heat supplied to the system must be rejected to a colder region. Rejection of this required amount of heat is no less necessary to the production of work than is the initial supply of heat to the system. Engines operating with a hydrocarbon fuel and air mixture even with ideal processes have an efficiency potential considerably less than 100%. Thermal efficiency is strongly a function of fuel–air ratio and compression ratio in addition to the nature of the operating cycle.

A clue to the distribution of energy after it has entered the engine with the fuel can be gained with an understanding of the overall energy balance. If we consider the engine to be enclosed in a control volume as shown in Fig. 5, it is possible to consider the various energy flows across the boundaries. Energy supplied to the system is indicated as items 1, 2, and 3 and energy removed as items 4 through 8. An excellent discussion of an energy balance is contained in Refs. 4 and 5. Items 1 and 2 indicate the energy contained in the air and fuel entering the engine, and items 3 and 4 show the liquid coolant entering and leaving at their respective temperatures. Items 5 and 6 represent energy transfers to the ambient by radiation and convection. The useful work leaving the control volume is indicated at item 7. Finally, item 8 is the collective energy represented by the exhaust products. An energy balance for a 1975 Chevrolet 5.7-liter engine with an 8.2 to 1 compression ratio is shown in Table 6. These data were taken at 1800 rpm and at the 50 mph road load power for the engine installed in a 4780-lb vehicle. In addition, these data were with optimum spark advance and a 0.9 fuel–air equivalence ratio and with no exhaust gas recirculation (EGR). The energy balance will differ slightly for this engine under other conditions and other engines will exhibit some differences, but in general the trends evident are expected for most spark-ignition engines and most operating conditions.

For analysis purposes it is convenient to divide the energy losses according to the categories suggested in Table 7. An ideal engine with a compression ratio and equivalence ratio of the Chevrolet engine should have a thermal efficiency of 0.44 with a heat rejection of 0.56 based on the fuel–air cycle. The real engine

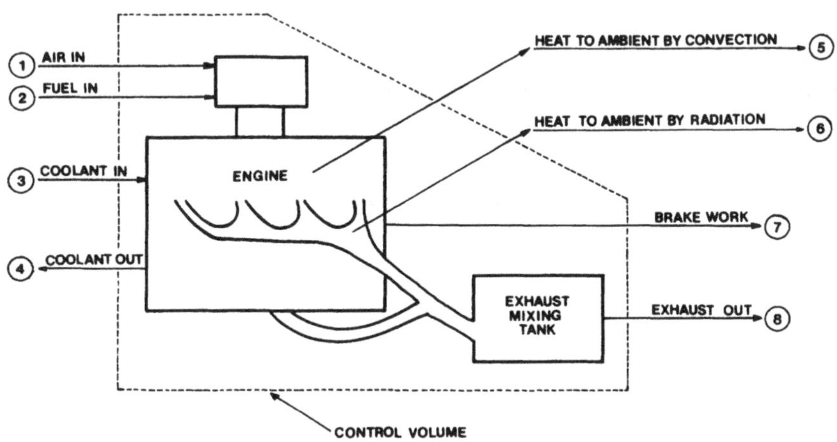

FIGURE 5. Control volume of the engine system energy flows.

TABLE 6. Energy Balance on a 1975 Chevrolet 350-cu.-in. V-8 Engine

Energy item	Fuel energy fraction
Shaft power	0.20 (19.9 hp)
Liquid coolant	0.48
Exhaust gases	0.29
Convection and radiation	0.03
Total fuel energy of combustion	1.00

produced only about half this power with a heat rejection of 0.8. Therefore, the second law heat rejection must be 0.56 of the fuel supplied in this example. The above heat rejection is necessary. The thermodynamic laws do not preclude that the efficiency of the real engine could be raised from the test value of 0.2 to the ideal value of 0.44 with an attendant decrease in heat rejection from 0.8 to the ideal level of 0.56. The real cycle and real engine effects, listed as potentially recoverable losses, are the sources of the 0.24 difference between the ideal engine heat rejection and the measured heat rejection of Table 6.

In the following discussion the potentially recoverable losses are considered. The losses are termed as cycle losses, pumping losses, and mechanical friction.

The cycle losses can be further segregated into (1) time loss, (2) heat loss, and (3) exhaust blowdown loss. The time loss is associated with the inability to burn instantaneously, the heat loss is related to the heat transfer primarily during expansion, and the exhaust blowdown loss is the small loss in useful work because of early opening of the exhaust valve. These various losses are illustrated schematically in Fig. 6 and show the differences between the ideal fuel/air cycle and the real cycle for the compression, combustion, and expansion processes. All of these losses can be potentially reduced. However, the prospect for achieving important gains here is limited. Experiments have shown that the total effect of eliminating these three cycle losses is an efficiency gain of less than 20%.

TABLE 7. Engine Energy Loss Categories

Net shaft power
Second law heat rejection
Potentially recoverable losses
Cycle losses
Pumping loss
Mechanical friction losses

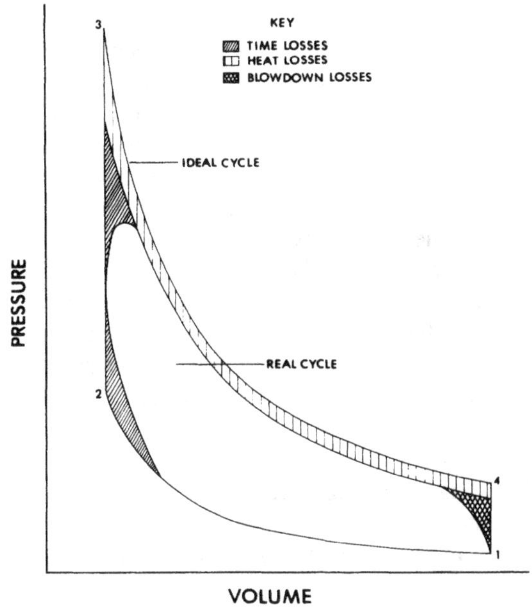

FIGURE 6. Simplified pressure– volume diagram for a gasoline engine showing various cycle losses.

The total engine cycle of Fig. 7 shows an important feature not evident in the previous cycle. This is the so-called "pumping loop" associated with the exhaust and intake processes. The piston work associated with the induction of the fresh fuel–air mixture and discharging the exhaust gases against the various flow resistances in the exhaust system is the pumping work. In real engines this loss is typically noted as the "throttling loss." At part load in a spark-ignited engine the pumping loop is a rather significant fraction of the total work produced whereas at a wide-open throttle (WOT) or maximum power this area is relatively

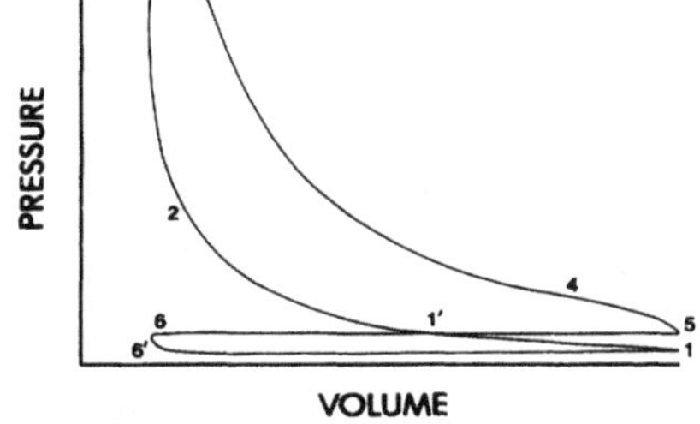

FIGURE 7. Pressure–volume diagram for a spark-ignited engine showing the compression, combustion, expansion, exhaust, and intake processes.

small in proportion to the total work area. It should be observed that in the diesel engine where no throttling valve is used to limit or control airflow, the throttling losses are relatively low. The impact of throttling loss can be illustrated by examination of the engine data cited earlier. At the conditions of the test the inlet manifold vacuum was at 17.4 in. Hg. Assuming that the exhaust manifold pressure is about 1 in. Hg above atmospheric pressure, an intake and exhaust pumping loss of approximately 7 hp can be calculated, which equals 0.072 of the total fuel supply. While this number is not absolutely accurate it does illustrate the importance of this loss. At 50 mph road load the pumping loss is almost 40% of the engine brake power.

The foregoing discussion has emphasized the work generated at the piston face whereas useful work is derived from an output shaft. The difference between the work output and the work delivered to the piston face is called the "mechanical friction." Engine engineers typically call the work at the face of the piston indicated work and the output work as the brake work. The area enclosed in the P–V diagram of Fig. 7 for a real engine is the indicated work of the cycle. Some indication of the importance of engine friction can be seen in Fig. 8, which shows a plot of road load power of a typical vehicle as a function of vehicle

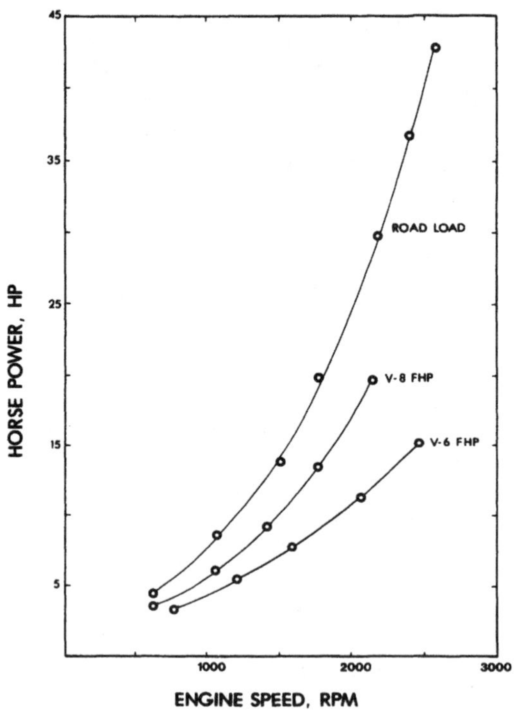

FIGURE 8. Friction horsepower as a function of engine speed for a V-6 and V-8 engine, and approximate road load horsepower for a full-sized passenger car.

speed as well as engine friction and pumping losses at WOT for typical V-8 and V-6 engines, all as a function of speed. Of course, the actual rubbing losses will result in a curve slightly below the friction curves shown. Engine friction data is very limited in the literature. However, Bishop[6] developed generalized formulas for calculating various components of friction losses. Using this procedure and applying it to the 1975 Chevrolet engine, the major friction losses are calculated as shown in Table 8. It is evident that the pistons and piston rings are the major contributors to friction.

A detailed breakdown of the energy losses associated with the Chevrolet engine used in the prior examples is shown in Table 9. Data are presented in terms of horsepower, fractions of the initially supplied fuels energy, and fractions of the engine net shaft power. The latter column is included to emphasize that if energy amounting to 1% of the fuel supply could be removed from the loss and added to the shaft power an approximate 5% increase in power and brake efficiency would result. It is evident that the second law heat rejection is predominant, and the only way it can be reduced is by increasing compression ratio or reducing the fuel–air ratio to increase the basic cycle efficiency.

All factors are being studied by engine designers in their attempt to improve efficiency and lower emissions, and one of their most important new tools in this quest is the growing availability of electronic controls. This technology is permitting a greater degree of system optimization than in the past, particularly fuel–air ratio, ignition timing, and scheduling of EGR which is an important part of the emission control strategy. It must be emphasized, however, that these sophisticated controls are not a panacea for all fuel economy ills, but rather they help minimize the fuel economy penalty associated with rigid emission requirements. A host of other technologies related to fuel preparation, valve timing, combustion chamber shape, intake and exhaust manifold design, as well as many other considerations are under continuous investigation to provide fuel economy improvements.

It is important to note that at best for a given set of key operating parameters the expected fuel economy improvements associated with these technologies are rather modest. The problems associated with the fuel economy–emissions trade-

TABLE 8. Estimated Mechanical Losses in a 1975 350-cu.-in. Chevrolet V-8 Engine

Pistons and rings	3.81
Valve gear	1.04
Bearings	1.40
Pumps and miscellany	0.75
Total	7.00 hp

TABLE 9. Energy Balance Summary for a 1975 Chevrolet 350-cu.-in. V-8 Engine

	Fraction of total fuel	Horsepower	Fraction of net shaft power
Net shaft power	0.204	19.90	1.000
Second law heat rejection	0.560	54.63	2.745
Potentially recoverable losses			
Heat loss	0.052	5.07	0.255
Time loss	0.026	2.54	0.128
Exhaust blowdown loss	0.008	0.78	0.039
Pumping loss	0.072	7.0	0.352
Friction loss	0.072	7.0	0.352
Subtotal	0.994	96.92	4.871
Leakage, incomplete combustion, errors of approximations	0.006	0.63	0.031
Total fuel supplied	1.000	97.55	4.902

off cannot be overemphasized. This issue is an underlying consideration to every aspect of engine design and development. One example of this trade-off is apparent in Fig. 9, which shows various emissions and fuel economy plotted as a function of air–fuel ratio. From this rather simple set of curves it is evident that there is no single optimum air–fuel ratio for all factors.

Another issue is the power to drive various accessories. This power is not an engine loss in the same sense as those considered earlier; however, engine power is required to drive these devices so they have an appreciable impact on performance and economy. Published data are rare; however, C. Coon *et al.*[7] have reviewed a number of these losses. Figure 10 is reproduced from this report for accessories of a 350-cu.-in. V-8 engine mounted in a 4400-lb vehicle. This engine is similar to the engine discussed earlier. Of course, with modern smaller cars and smaller engines some of these losses could be expected to be lower but others may even increase. For example, with the expectation for increased automotive electronics, alternators will likely have greater capacity in future vehicles. Similarly, the air-conditioning load of the small vehicle is not significantly different from the large vehicle, thus, its loss or power requirement will probably not change dramatically. In any event the accessory impact on fuel economy is important, as shown in Table 10.[8]

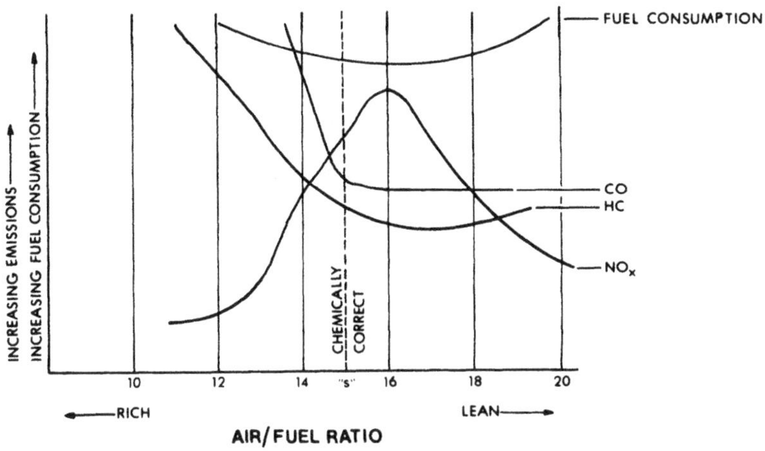

FIGURE 9. The effect of fuel–air ratio on fuel consumption, hydrocarbon, CO, and NO_x emissions in a representative spark-ignited engine.

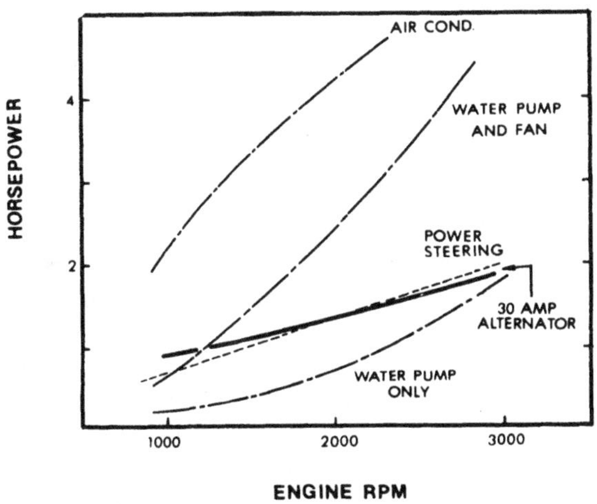

FIGURE 10. Accessory power requirements for a full-sized U.S. passenger car.

TABLE 10. Effect of Engine Accessories and Convenience Devices on the Fuel Economy of an Intermediate Car with Automatic Transmission

Type of accessory	Urban cycle (mpg)	70 mph road load (mpg)
Air conditioning	1.5	1.0
Alternator	0.9	0.5
Fan	0.1	0.5
Power steering	0.1	0.4

B. Drive Train

Various drive-train elements in a modern vehicle, excluding the engine, are the transmission (including the clutch or torque converter), the drive shaft (including universal joints), the differential and final drive assembly, and axle shafts.

The transmission and final drive/differential are the two most important drive-train components in terms of fuel consumption. Various losses incurred in the transmission of power can be substantial particularly with the automatic transmission with its relatively inefficient torque converter and in the hypoid gears of a rear drive differential assembly. Losses in the bearings, universal joints, and other elements are less important.

In current automobile transmissions either automatic or manual shift, a finite number of gear ratios are available. Therefore, it is impossible to perfectly match the engine and the vehicle across the operating range of the engine. Because of this imperfect match, the engine generally operates at less than optimun efficiency at a given speed and load. Recent developments in continuously variable transmissions (CVT) hold promise, and if they prove acceptable in terms of cost, durability, manufacturability, and other considerations it is projected that they would yield a 10% fuel economy improvement over the best of current designs. Unfortunately, the technology of the CVT is by no means certain.

Axle ratio is also an important consideration in the matching of the engine to the vehicle. The overall drive-train ratio is often spoken of as the N/V ratio where N refers to engine rpm and V, vehicle velocity in mph, with the transmission in its highest gear. The trend in recent years has been to reduce axle ratios, or lower N/V ratios. In general, a decrease in axle ratio will result in reduced fuel consumption; however, the acceleration capability is reduced as well. The essential trade-offs between performance and economy are shown in Fig. 11.[9] The lower axle ratio causes the engine to operate at a higher load factor but at a reduced rpm with an attendant reduction in engine throttling losses and friction.

Aside from the engine–drive train matchup question, the various component efficiencies are important. The torque converter in an automatic transmission in particular has been a serious problem over the years because of rather significant slippage between the input and output shafts. This device is, despite its inefficiencies, one of the keys to the success of modern automatic transmissions. Under stall conditions the torque converter efficiency is zero, and even under optimum conditions torque converter efficiency rarely exceeds 95%. Of course, some of the inherent disadvantages of the simple torque converter are being ameliorated with the advent of lockup or divided path torque converters in which some or all of the power is transmitted through a mechanical system under certain operating conditions. In the future most automatic transmissions will employ some version of a lockup mechanism.

The clutch in the manual transmission is a much more efficient device, on the order of 99%, and is not as degrading to fuel economy as the torque converter in the automatic transmission. The gear sets in both the manual shift and automatic transmission are quite efficient and the losses associated with internal friction, windage, etc., are low and little gain is anticipated in the future. The propeller and axle shafts with their various U-joints are also efficient parts of the current drive train, and major gains in overall efficiency are unlikely with improvements in these components.

The front-drive vehicle poses a particularly challenging problem because true constant velocity universal joints are required. These are substantially more

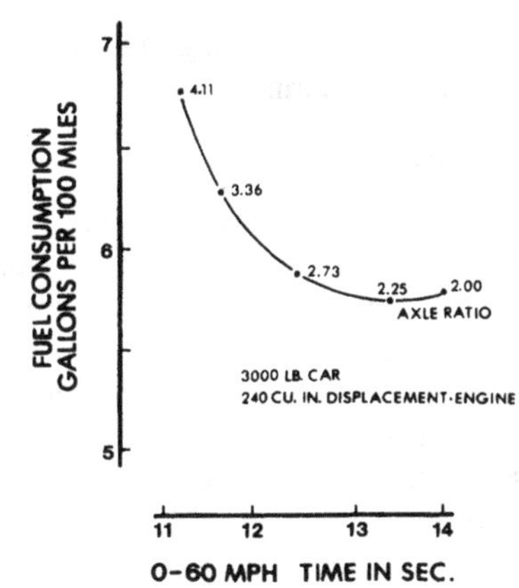

FIGURE 11. The effect of vehicle axle ratio on fuel consumption and performance.

complex and expensive than the conventional Cardan-type joint, but still their power transmission efficiencies are high.

An extremely important part of the drive train is the differential assembly. In modern vehicles a torque-equalizing differential is used which provides equal torque to the wheels on a given axle and makes dissimilar rotational speeds possible. Obviously, when turning a corner the inside and outside wheels traverse different radii and were it not for the differential, skidding would result. In the rear-drive system a hypoid gear set is used to provide a rather low position for the drive shaft entering the nose of the differential. Unfortunately, the friction in hypoid gears is significant and special viscous lubricants are required. In contrast, front-drive vehicles do not use hypoid gears but more efficient chain or gear drives.

C. Vehicle Factors

Numerous vehicle factors aside from the power train have an important impact on overall fuel economy. Just as in the case of the engine, the relative impact of these vehicle factors is often dependent on the driving cycle. For example, aerodynamics are obviously more important at high speed than at low speed. It is generally observed that any change in a parameter which results in lower fuel consumption will also result in a reduction in emissions. This is contrary to what often occurs in the engine where complex trade-offs are common.

1. Aerodynamics

Aerodynamic drag has always been an important factor in some kinds of transportation vehicles, particularly aircraft. Today aerodynamic drag or wind resistance is playing an increasingly important role in land vehicle design. It is important to note that good aerodynamic design is not just related to fuel economy but also to emission performance, stability, noise, surface dirt control, and component and interior cooling.

With the complex and irregular shapes of ground vehicles and their close proximity to the ground plane, vehicle aerodynamics is an extremely complex field even with the advanced state of modern technology. There is still little science and much art to the subject. Good aerodynamic design is generally the product of experience and experimental testing of scaled models and full-sized vehicles. The initial aerodynamic design is often evaluated on a three-eighths scale model tested in small wind tunnels. With the increased emphasis on fuel economy and reducing aerodynamic drag more and more, experimental testing is being conducted in full-scale wind tunnels which provide better resolution as to the impact of detailed design features on aerodynamic performance. Unfortunately, there are only a rather limited number of full-scale wind tunnels available for automotive research in this country. Test facilities such as at Lockheed in Marietta, Georgia, are being used 24 hours a day, seven days a week. Full-scale tunnels are enormously expensive as evidenced by the multimillion dollar cost of a facility recently built at the General Motors Technical Center in Warren, Michigan.

As a result of the increased demand for good aerodynamic and structural design, the traditional automotive stylist's role is being shifted to that of a true total system designer. A measure of this shift in roles is evident from the recent change in the name of General Motors Styling to GM Design Staff.

Aerodynamic drag is the only vehicle motion resistance factor not directly related to mass but is indirectly related to mass through the relationship between size and mass. Aerodynamic losses are primarily due to two factors in the flow of air through and over the vehicle. One is the friction loss associated with the small-scale turbulence in the boundary layer of the vehicle and the other is termed "form drag," which is related to the large-scale turbulent eddies created when the boundary layer separates from the surface. Form drag is typically responsible for about 80% of the overall aerodynamic drag. Thus one of the principal objectives of the ground vehicle aerodynamicist is to reduce turbulence creating shapes on surfaces of the vehicle that could lead to boundary layer separation.

Aerodynamic drag is proportional to a size factor, an aerodynamic efficiency factor, and the velocity squared. Speed is a particularly important consideration because of the square relationship. Further, when one considers the horsepower required to overcome aerodynamic resistance, it looms as an even more important factor since it now is proportional to the velocity cubed. Because fuel rate is essentially proportional to the horsepower rather than force or torque, the important impact of aerodynamic forces on high-speed fuel economy is apparent.

The drag coefficient, C_d, in the drag equation is a measure of the degree of slipperiness or efficiency of the aerodynamic design of the vehicle. Typical modern passenger cars have drag coefficients in the area of 0.5 and some of the newly designed vehicles are approaching 0.3. Unless rather significant breakthroughs are achieved, it will probably be difficult to attain a drag coefficient significantly less than 0.3 in conventional passenger vehicles.

The area or size term is obviously important and is related to the vehicle package requirements and, therefore, to the side-by-side seating and attitude positions of the passengers. In all vehicle aerodynamic calculations including those for lift and lateral forces as well as straight line wind resistance, the area term is taken as the projected frontal area of the vehicle.

Even cooling air is an important consideration in aerodynamic design. Underhood cooling airflow necessary for combustion and engine compartment cooling increases the drag by approximately 10%. The trailing region of the vehicle is particularly significant in creating drag because large-scale turbulent eddies are easily created. Such small considerations as a lip or midget spoiler on the rear trunk can have a profound impact on the overall drag of the vehicle and can influence other factors such as wind stability. Vehicles of the future will be characterized by much more attention to detail in the exterior surface in order to achieve optimum aerodynamic performance.

2. Automotive Structures

The structural design and weight management of an automobile are intimately entwined with one another and are the subject of intense research and development. Clearly, weight reduction is one of the most important methods of achieving improved fuel economy, but downsizing alone as a means to achieve weight reduction is faced with several key limitations. Numerous criteria are important in the design of the package, including providing adequate passenger and luggage capacity at a reasonable cost, as well as insuring corrosion resistance, good ride and handling, and crashworthiness.

Within a given set of constraints on vehicle size the modern designer's objective has been to minimize weight with a given set of materials consistent with the other design criteria. Passenger cars must be structurally rigid, which is more than just being sufficiently strong to carry the load. Computer-aided design techniques have played a crucial role in the great strides that have been made in the past few years. One example of this progress was embodied in the redesign of the GM full-sized cars from 1976 to 1977. These cars were downsized, but the internal passenger and luggage volume remained about the same with approximately an 800-lb overall weight reduction. This gain was achieved primarily through more efficient design procedures. Finite element models and the use of fast Fourier transform technology have enabled the designer at both the design and test stage to iterate much more rapidly through various configurations to approach more nearly the optimum prior to design release. Probably no single factor has played as important a role in this overall process of design and development as the modern digital computer with its peripherals and attendant software.

Several major changes are being made in the basic structural design of American automobiles. One of these is the increased use of the unibody or integral body–frame construction rather than the previous separate body–frame concept. In the past a number of advantages of the body–frame design dictated its use. These included good ride due to the added layer of isolation between the body and frame and the ease with which changes could be made in the exterior shell to meet the requirement of yearly model changes. With emergence of the overwhelming interest in weight reduction and structural efficiency, the integral body–frame design is a natural outgrowth. Almost all future passenger cars and a significant fraction of light trucks will use this configuration. Fortunately, some of the same modern technology that has brought us more efficient structures has also helped minimize some of the ride problems that were traditionally associated with this concept.

Another important fundamental design change sweeping the industry is the reconfiguration of the vehicle from the front engine-rear drive to a front engine-front drive in most cases with the engine located transversely between the front wheels. This technology has led to an important decrease in weight (approximately 125 lb in a subcompact vehicle and 250 lb in a full-sized vehicle). All this is being achieved at a cost penalty of approximately $125 per car measured in 1980 dollars. The decision to reconfigure future vehicles has placed enormous burdens on the technical and financial staffs at both the manufacturers and their suppliers. The bodies, chassis, engines, and drive train for each vehicle line are being redesigned and tooled concurrently instead of the more orderly and less expensive sequential component changes possible with previous large vehicle designs. Even the suspensions are substantially different than those used on older style vehicles. The MacPherson strut front suspension is emerging as the preferred design because of the much greater space available between the front wheels for the transversely mounted engine. A number of other advantages in addition to weight reduction are provided by front drive including:

1. Less intrusion into the passenger compartment because of the almost total elimination of the driveshaft tunnel;
2. More usable trunk space because less room is required for the rear axle and driveshaft; and
3. More protected gas tank location in front of the rear axle.

In general, this technology provides a smaller, lighter weight vehicle with a minimum of compromise in interior passenger and luggage space.

One measure of the value associated with weight saving was determined in the recent University of Michigan Delphi study of the automotive industry in which engineering executives estimated that the value of a pound of weight saved through the 1980s would exceed $1.00.

Lightweight materials offer interesting and important weight-saving potential in future vehicles. Unfortunately, it is not a simple issue and many of the choices embody rather complex trade-offs. On the surface material substitution appears to be highly attractive. It is one way to preserve large passenger volumes but with most of the fuel economy benefits of decreased weight. One of the major problems is often basic material cost, but a true assessment requires a total systems analysis from design to production to service in the field. Manufacturing considerations which include such factors as capital investment, total manufacturing energy requirements, changes in material handling, speed of processing, surface finishes, and in-plant scrap and repair are critical as are corrosion properties in the final product, field repair, scrap recycling, and many more factors.

There are literally a host of candidate materials for use in motor vehicles to reduce weight, including reinforced and nonreinforced plastics, aluminum, high-strength low-alloy (HSLA) steels, and metal–plastic laminates. There are many functional classifications for vehicle components, including interior and exterior sheet, engine/drive train and chassis structural.

For structural components there is significant interest in HSLA steel and high performance composites such as the graphitic type used in some sporting equipment and aerospace applications. Unfortunately, the cost of most substitute materials is rather high and the technology, particularly related to high production manufacturing of polymer-based composites, is still rather new. In the engine/drive-train area aluminum, magnesium, some plastics, and HSLA steels are all being seriously considered for weight reduction. Aluminum is already the state of the art in certain components like pistons and transmission cases. Novel approaches such as stamped steel manifolds that are substantially lighter than their cast-iron predecessors are being investigated. Cylinder heads, blocks, inlet manifolds, and other parts are increasingly being fabricated from aluminum. Magnesium is just beginning to appear and is a likely candidate in a few select engine/drive train components in the next few years.

An exciting competition is underway between steel, wrought aluminum, and a host of different plastics for interior and exterior sheet applications. At this point in time it is impossible to forecast a winner, but all of these materials can be found in existing vehicles and there is a strong indication that much more prominent use of lightweight sheet materials will be forthcoming in the years ahead.

An important benefit of primary weight reduction (weight reduced with the direct material substitution) is the secondary weight savings available when the primary reduction is propagated through the vehicle. With a given level of primary weight reduction, less support structure is required. The tires, brakes, even engine size can potentially be reduced for a given level of performance. Unfortunately, the secondary weight benefit is very difficult to precisely quantify. Certainly, it is highly dependent on the location of primary weight reduction. A primary weight saving of 1 lb may yield as much as 1 lb of secondary savings for a net weight reduction of 2 lb. A maximum propagation of secondary weight reduction must be thought of in the context of a complete vehicle redesign. Where a substitution is made in an existing vehicle the secondary weight savings are generally small unless major redesign of other components is considered.

Because of the complex interrelationship between economy parameters it is difficult to quantify the impact of mass on fuel economy. Numerous investigators have performed regression analyses in an attempt to quantify this relationship and spread of projections is reasonably broad. However, one reasonable rule of thumb is that each pound of weight saved is equivalent to a reduction in fuel consumption of 1.4×10^{-5} gal/mile.

3. *Tires and Rolling Resistance*

One of the most important automotive innovations of the century has been the development of the modern pneumatic tire. Today's tires are durable, contribute to an excellent ride, are quiet, and provide excellent traction on a host of different road surfaces. The tire is the key factor in rolling resistance on hard pavement. This resistance arises primarily from hysteresis in the tire carcass. Other factors such as windage and tire slip are small by comparison. Incidentally, most people do not realize that tires actually exhibit a relative slip with respect to the road surface when transmitting braking or acceleration forces. Actually, the maximum tractive forces are achieved with approximately a 15% relative slip between the tire and the ground plane.

On soft surfaces, such as sand or loose dirt, the hysteresis factor is small and tire work on the ground dominates in the rolling resistance relationship. In urban driving the rolling resistance is more important than aerodynamic resistance as shown in Fig. 4 at low speeds. At higher speeds aerodynamic resistance becomes increasingly significant.

Dramatic improvements have been made in tire rolling resistance in the past few years with the introduction of radial-ply tires in place of the older bias-ply designs. The radial configuration of the carcass plies, which are held in place by a rigid belt, results in a more efficient elastic structure and therefore lower hysteresis. This shift has led to approximately a 50% reduction in rolling resistance. Tire rolling resistance is roughly 1% of the vertical load on the tire. It is primarily a function of vertical load but it increases modestly with an increase in speed. Tire pressure has an important impact. With higher pressures there is less carcass deflection and therefore reduced hysteresis losses.

In addition to lower rolling drag the radial tire offers substantially great tire life, improved wet surface and snow traction, and improved penetration resistance. Unfortunately, some problems are also evident with this design, including reduced self-aligning torque and somewhat harsher ride, which must be compensated for with careful attention to detail in the suspension design. In the future further small improvements in rolling resistance are expected from ever-increasing tire pressure and new construction types including the so-called elliptical design and even a cast tire.

4. *Miscellaneous Factors*

In the preceding discussion the major "hardware" factors in the fuel economy equation have been reviewed. Several other vehicle factors have a modest impact, including a slight drag with most disc brake systems which employ a light rubbing action to keep the disc-pad interface dry. In addition, minor fuel economy losses are associated with such factors as wheel bearings but the impact is almost immeasurable.

D. Operating Factors

There are a number of vehicle operating factors that have a considerable influence on the fuel economy performance. These include:

1. Characteristics of the driving cycle (speed, time, path);
2. Start-up and warm-up behavior;
3. New vehicle performance;
4. Ambient conditions;
5. Roadway factors;
6. Maintenance; and
7. Driver habits.

1. Driving Cycle

The driving cycle impact is very significant. In the city or urban environment with slow speeds and stop and go driving, fuel economy is substantially lower than for the typically more steady but higher speed country, suburban, or freeway driving. Strong evidence of this driving cycle influence can be found in the EPA mileage data for both the urban and suburban fuel test cycles. These data show that the urban fuel economy is about 50% less than the suburban cycle data. Currently, EPA only publishes urban fuel economy results but suburban or highway fuel economy data is available on request. The disparity between cycles is strong evidence supporting the concern expressed earlier with regard to fuel economy reporting unless the test cycle is clearly defined. There are several reasons for the inferior fuel economy of the urban cycle, but by far the predominant factor is related to the important impact of vehicle mass. As the vehicle is accelerated the kinetic energy of the vehicle increases. During brake application, which occurs regularly during this cycle, a substantial fraction of the kinetic energy is converted to heat in the brake system and thereby lost. In the EPA urban cycle a surprising fraction of the total fuel energy is dissipated in the brakes, and this clearly is one of the incentives for exploring the concept of regenerative braking via such techniques as flywheel energy storage. In viewing the aggregate fuel consumption in the United States, it is important to consider the impact of such factors as average trip length, average speed, etc., in projecting total fuel usage. Statistical data such as that developed by Austin and Hellman[10] show several interesting trends, as shown in Fig. 12. One clear revelation from this study is the great importance of short-trip driving in the United States.

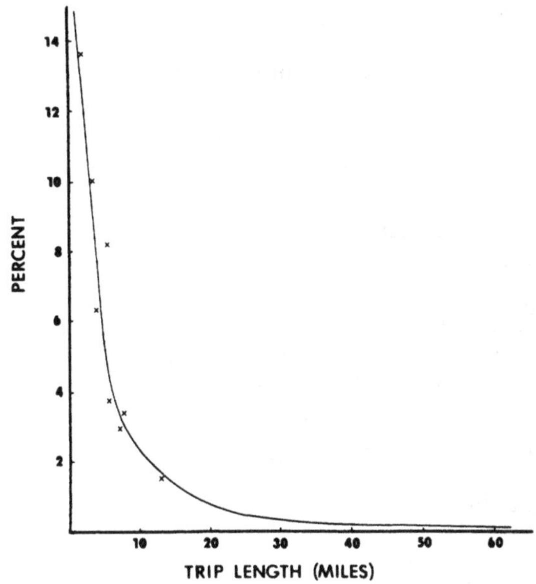

FIGURE 12. Trip frequency as a function of trip length in the United States.

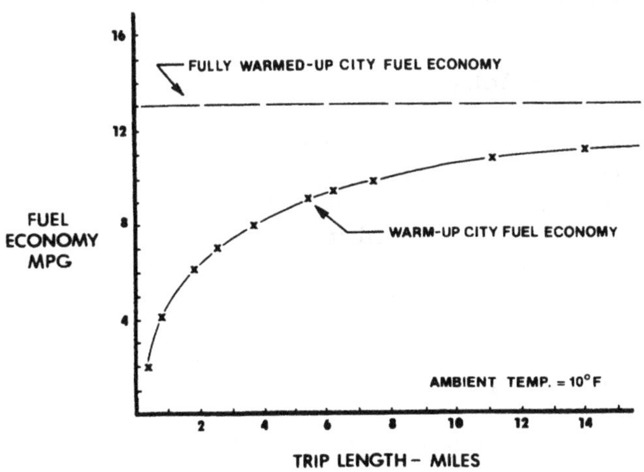

FIGURE 13. The effect of trip length on fuel economy.

2. Cold Start/Warm-Up

Cold start and warm-up has a surprisingly important impact on overall fuel economy. At the start of a cold vehicle friction in both the engine and drive train is high, and in the spark-ignited engine very rich fuel–air mixtures are required. Both of these factors significantly degrade fuel economy in relation to the fully warmed-up vehicle. Figure 13, which was obtained from a paper by Sheffler and Niepoth[11] graphically illustrates the important impact of warm-up on economy. These data show that at roughly 2.5 miles into a city driving cycle the fuel economy is only 50% that of the fully warmed-up vehicle. This area holds great research promise related to redesign of components for low friction and the use of less viscous lubricants as well as the application of advanced fuel management systems that can minimize problems with rich mixtures during cold operation. Of course, the diesel engine, while experiencing friction problems during start up, does not require excessively rich mixtures during this phase.

3. New Vehicle Performance

In a new or rebuilt vehicle engine frictional losses can be high depending upon the tolerance and fit of the various moving parts. The piston–cylinder bore interface is particularly important. As break-in occurs, friction is steadily reduced. In current vehicles a full break-in may require more than 12,000 miles of operation. The fuel economy difference between new or "green" condition in the fully warmed-up vehicle may be as much as 10%. This is also an important area for study of new methods that can accelerate break-in of major engine/drive-train components without contributing to excessive wear and long-term durability problems.

4. Ambient Conditions

The ambient temperature or environment also influences fuel economy through a host of factors ranging from temperature effects on tire hysteresis, engine friction, and engine mixture requirements to the atmospheric humidity impact on combustion. Temperature is perhaps the most important factor as evidenced by the data shown in Fig. 14 from Ref. 12. This confirms the qualitative findings of most drivers that it is more economical to drive in the summer than in the winter.

5. Roadway Factors

Modest variations in roadway or highway surfaces have little impact on fuel economy at least for dry pavement. Of course, totally different pavement surfaces such as unpaved gravel or sand or wet and snowy surfaces can result in substantial fuel economy degradation.

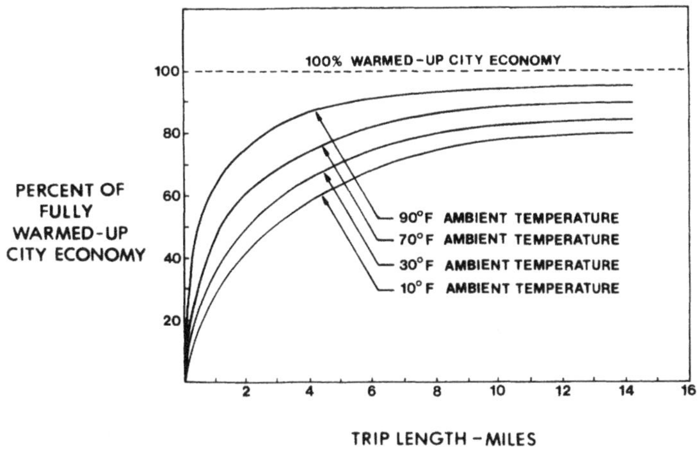

FIGURE 14. The effect of ambient temperature on fuel economy for a typical U.S. vehicle.

6. *Maintenance*

The state of maintenance is another important factor, particularly if a vehicle is improperly maintained. Most vehicle subsystems require some periodic inspection and maintenance, although in recent years the interval between scheduled maintenance has increased considerably. Today, 8,000 to 10,000 miles maintenance intervals are not unusual. Clearly, such factors as low tire pressure and poor engine tune are important. For example, in a six-cylinder engine just one misfiring spark plug can result in a 20% decrease in fuel economy. In addition, almost an order of magnitude increase in hydrocarbon emissions results from this maintenance-related factor. In one sense this might be termed a driver factor because each driver has the responsibility to ensure that proper maintenance is done on a vehicle per the manufacturer's specifications. In the past few years some states have developed rather significant vehicle inspection programs that help ensure high operating efficiency for vehicles. The primary focus of this has been on the safety and emission performance, but clearly fuel economy benefits can be a spin-off of emission inspection.

7. *Driving Habits*

One of the more intangible factors in the fuel economy equation is the driver. Drivers through their skill, or lack of skill, can have a profound impact in many ways on fuel economy ranging from maintenance to the selection of the driving cycle to the consistency of behavior on a given driving cycle. We are all well aware of the impact of "jack rabbit" starts or high speeds. In addition, such factors as the excessive use of brakes on a given driving cycle is equally detrimental. A good driver, in terms of obtaining the best possible fuel economy, is sensible, steady, and drives at modest speeds.

E. Test Cycles

Throughout this chapter we have voiced caution concerning the driving cycle and its impact on the fuel economy performance. It is perhaps appropriate to discuss this subject in somewhat greater depth to provide added perspective. Over the years the manufacturers and other various organizations have developed their own standard driving cycles that they believed were representative of consumer behavior. Generally, multiple cycles were used, at least one representing city driving and another suburban or highway driving. Unfortunately, all of these cycles were slightly different, and therefore the results from tests performed on these various cycles were not generally comparable. Of all the standard driving cycles the most discussed and significant today are those that have been developed by the EPA. All vehicles sold in the United States are tested on these cycles to determine compliance to exhaust emission and fuel economy standards. The EPA driving cycles are based on observations and measurements of actual drivers in Los Angeles. Both the urban and suburban cycles are shown schematically in Fig. 15. (Note: the cycle begins at the left side of the figure in this presentation.) The important differences between cycles are evident, and it is little wonder that the fuel economy data difference between these various cycles is large. In recent years the SAE has attempted to further add to the understanding of fuel economy measurement as has the American Society for Testing and Materials (ASTM).

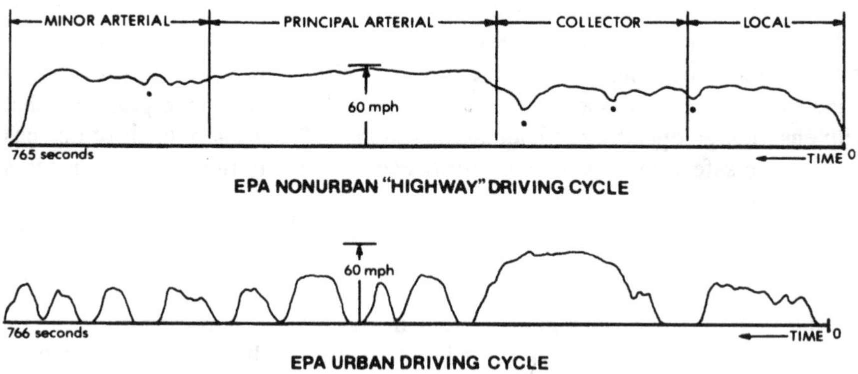

FIGURE 15. EPA urban and nonurban, or highway, driving cycles.

REFERENCES

1. Agnew, W. G., "Automotive Fuel Economy Improvement," General Motors Research Publication, GMR-3493, November (1980).
2. Cole, D. E., Harbeck, L. T., and Smith, D. N., "Delphi Forecast and Analysis of the U.S. Automotive Industry in the 1980's," Office for the Study of Automotive Transportation and Industrial Development Div. of the Institute of Science and Tech. (1981).
3. 1982 Gas Mileage Guide, EPA Fuel Economy Estimate, U.S. Dept. of Energy and Transportation, DOE/CE-0019, September (1981).
4. Ament, F., Cole, D. E., and Patterson, D. J., "Heat Balance and Comparison at Part Load of a 1975 Chevrolet 350-Cubic-Inch V-8 and an Experimental 222-Cubic-Inch V-6 Engine," Final Report Project 320445, Dept. of Mech. Eng., University of Michigan, December (1974).
5. *Guidelines for Evaluating Energy-Related Automotive Engine Inventions,* Society of Automotive Engineers, under U.S. Dept. of Energy, Contract EU-78-C-01-6572 (1978).
6. Bishop, I. N., "Effects of Design Variables on Friction and Economy," SAE Paper 812 A, January (1964).
7. Coon, C. W., and Wood, C. D., "Improvement of Automobile Fuel Economy," SAE paper 740969, October (1974).
8. Gasser, D. J., and Huebner, G. J., "Energy and the Automobile—General Factors Affecting Vehicle Fuel Consumption," SAE paper 730518 (1973).
9. Marks, C., and Niepoth, G., "Car Design for Economy and Emissions," SAE paper 750954, October (1975).
10. Austin, T. C., and Hellman, K. H., "Passenger Car Fuel Economy—Trends and Influencing Factors," SAE paper 730790, September (1973).
11. Niepoth, G. W., and Scheffler, C. E., "Customer Fuel Economy Estimated from Engineering Tests," SAE paper 650861, November (1965).
12. Hwang, D. H., prepared discussion of "Customer Fuel Economy Estimated from Engineering Tests," SAE paper 650861, November (1965).

2

Fuel Economy and Emissions

J. T. KUMMER

I. INTRODUCTION

The relationship between vehicle fuel economy and government-regulated exhaust emissions is very complex. It involves a number of factors subject to personal bias. For example, one must to some extent subjectively judge what is acceptable vehicle performance during cold start and normal driving because performance in these operating modes can, as will be shown, be related to both emission levels and fuel economy, either directly or indirectly. It is important to decide what economic penalty should be involved when one attempts to maximize fuel economy at a given emission level. One must also decide what degree of sophistication the emission control system can have and still function well in the hands of a nontechnical owner with minimal maintenance.

The relationship involves quantitative factors set by the government such as an exact description of the vehicle tests used to measure fuel economy, vehicle emission levels, and system durability. It requires a knowledge of the fuel to be used, its contaminant level (Pb, P, S), its octane number, its volatility, and a knowledge of the economic trade-off between its octane number and cost of production. And finally, the fuel economy–emission relationship will of course be a function of the specific vehicle under consideration: its weight, power-to-weight ratio, aerodynamic design, the type of emission control system used, the engine design, etc.

In this chapter I would like to consider the emission regulations and tests, the general ways these emission regulations can affect fuel economy, fuel octane effects, the specific engine sources of pollutants and how their control by engine modifications can effect fuel economy, the use of aftertreatment devices to

---

J. T. KUMMER • Chemical Research Laboratory, Ford Motor Company, Dearborn, Michigan 48121.

improve fuel economy at the required emission levels, and a brief look at today's vehicle performance.

A methodology for the optimization of fuel economy with a given engine–vehicle combination will be the subject of a later chapter.

II. EMISSION REGULATIONS

United States government regulations require that automobile manufacturers control the amount of carbon monoxide (CO), hydrocarbons (HC), and nitrogen oxides (NO_x) in the exhaust of gasoline and diesel-fueled passenger vehicles and light and heavy trucks (see Tables 1 and 2). In Table 2 is shown the gasoline consumption in gal/mile, fuel economy in miles per gallon (mpg), and the *average* pollutant concentration in the exhaust which will meet the 1981 emission standards (assuming an air-to-fuel ratio of 15) in order to orient one to the magnitudes involved.

In Table 1 the emissions are expressed in terms of grams of each pollutant

TABLE 1. Passenger Car Emission Standards in Grams per Mile (Results from 1975 Test[a])

Model year	Hydrocarbons	CO	NO
Uncontrolled	10.6	84	4.1
1970–71(F)	4.1	34	NR
1971(C)	4.1	34	5.8
1972(F)	3	28	NR
1972(C)	3	28	3.2
1973	3	28	3.1
1974(F)	3	28	3.1
1974(C)	3	28	2.1
1975–76(F)	1.5	15	3.1
(C)	0.9	9	2.0
1977–79(F)	1.5	15	2.0
(C)	0.41	9	1.5
1980(F)	0.41	7	2.0
(C)	0.39[c]	9	1.0
1981(F)	0.41	3.4[b]	1.0[d]
(C)	0.39[c]	7	0.7
1982(F)	0.41	3.4[b]	1.0[d]
(C)	0.39[c]	7	0.7

[a] Measured by the CVS-CH Federal Test Procedure. (F) = Federal regulations; (C) = California regulations.
[b] Certain engine families have received a 7-g/mile waiver.
[c] Nonmethane hydrocarbon standard.
[d] Certain diesel engine families have received a 1.5-g NO_x waiver.

TABLE 2. Fuel Usage

			Average exhaust concentration during FTP[c]		
mpg	g/mile[a]	% Fuel[b]	Mole % CO	ppm NO	ppm hydrocarbons as C_6
10	279	0.15	0.084	94	34
15	186	0.22	0.13	142	51
20	140	0.29	0.17	188	67
25	112	0.37	0.21	235	84
30	93	0.44	0.25	283	101
35	80	0.51	0.29	329	118
40	70	0.59	0.34	376	134

[a] Gasoline 6.15 lb/gal

[b] Maximum percent of fuel consumed that can be emitted as hydrocarbons in order to meet the 0.41 g/mile standard.

[c] Maximum *average* exhaust concentrations during the FTP if emission levels of 3.4 g/mile CO, 0.4 g/mile NO, and 0.41 g/mile of hydrocarbon are to be met.

emitted per mile of vehicle travel over a prescribed test driving cycle simulating an average trip in an urban area (see Fig. 1). The engine predominantly emits NO; the grams per mile are calculated as NO_2 and designated as NO_x. Larger and heavier vehicles will have more difficulty in meeting these emission requirements than smaller and lighter vehicles.

The U.S. federal emission test procedure (FTP) requires that the vehicle under test be at a temperature of 16–30 °C for a period of 12 h before the test and that the total pollutant emissions be measured from the moment that the ignition key is turned on until the end of the test. A conventional carburetted engine requires the use of a choke in order to start the engine at ambient temperatures. This produces a perturbation such that for a short period of time the vehicle runs rich and emits high concentrations of CO and hydrocarbons. If a catalyst is used, it is cold at the beginning of this cold start and until it reaches the temperature of 250–300 °C necessary for the oxidation of CO and hydrocarbons; these are emitted from the tail pipe. The time required for this to occur is a function of catalyst design and position but can be from 20 to 120 sec, and

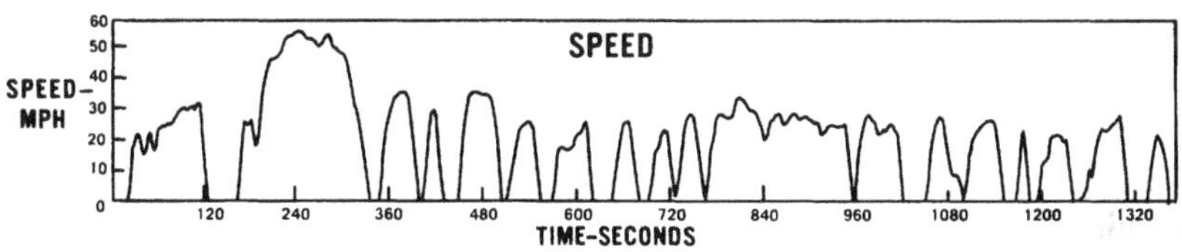

FIGURE 1. Vehicle driving schedule during the FTP. The procedure is used for urban fuel economy measurements and to determine compliance with emission standards.

the hydrocarbons emitted during this period can be from $\sim^1/_4$ to $\sim^3/_4$ of the allowable limit.

The amount of NO emitted during the cold start is low (10% of the total) so that when relating NO emissions and fuel economy, the NO emitted during the cold start portion of the test is not as important as in the case of carbon monoxide and hydrocarbons unless the cold start extends longer than ~160 sec which is the beginning of the acceleration to 55 mph cruise in the FTP (see Fig. 1).

The FTP was conceived to represent a typical urban driving pattern. The fuel consumed during this test is used to calculate the vehicle fuel economy for city driving. The emissions of CO, hydrocarbons, and NO are used to certify that the vehicle under test has met the government-mandated emission standards.

In basic concept the FTP can be divided into two parts: an engine start and run part of 505 sec and a following stabilized part of 867 sec for a distance of 7.5 miles.[(1)] Two engine start and run tests are made, one with the vehicle at an ambient temperature of 16–30 °C before the start, and one with the engine and catalyst hot (after a shutdown and 10 min wait following the stabilized part of the test). A small portion of the exhaust flow is collected during the FTP by a method (constant volume sampling) that results in the concentration of the pollutants in the collected gas being proportional to their mass emissions during the time of collection. A sample of the exhaust emitted in the cold start and run, the stabilized part, and the hot start and run part, are collected separately in three bags and referred to as bag 1, bag 2, and bag 3 emissions, respectively. The emissions in grams during the two start and run tests are averaged with a weighting factor of 0.57 for the hot start and 0.43 for the cold start, added to the emissions in grams during the stabilized part, and divided by 7.5 to obtain the final test emissions in g/mile.

The U.S. law requires that the vehicle comply with the emission standards for 5 years or 50,000 miles, whichever occurs first. The vehicle durability tests consists of 50,000 miles of use in a prescribed durability driving schedule which consists of continuous repetition of a 40.7-mile test cycle (AMA cycle). The test cycle consists of numerous stops, accelerations, and high (55 mph max.) and medium speed driving.[(1)] Emission standards must be met during this durability test.

It is often possible to prepare catalysts or thermactors or other emission control hardware that enable certain emission standards to be met at zero mile usage. It is much more difficult to meet these emission standards during 50,000 miles of durability testing in which the engine, catalyst, and other emission hardware have seen both high temperatures, carbon deposits, and inorganic contaminants such as lead, phosphorus, sulfur, and zinc which can be present in the motor oil or in commercial unleaded gasoline. These can degrade the cold-start and steady-state performance of the catalyst, the performance of air–fuel ratio sensors, the durability of thermactors, and the operation of exhaust gas recirculation (EGR) regulation valves.

As stated before, the FTP is used to obtain a value for fuel economy in city driving.[2] The federal highway fuel economy test[3] is used to obtain the fuel economy with highway driving. This test is 10.2 mi long; the time is 12.75 min (average speed 49.4 mph). During this test the vehicle is accelerated to between 40 and 50 mph and run at this approximate steady speed for ~4 min, decelerated to 30 mph, accelerated to between 50 and 60 mph, and run at this approximate steady speed for the remainder of the test.

III. GENERAL CONSIDERATIONS

Under urban driving conditions our present vehicles, which are powered by spark-ignited IC engines calibrated for high fuel economy but with careful attention to emission control (for example, with a cold-start air–fuel ratio of stoichiometry or slightly lean, low choke time, no rich transient excursions of the air–fuel ratio during acceleration or deceleration modes, etc.) will have *engine-out* emission levels higher than the present legal limits. This is illustrated in Table 3. Compare to Table 1.

The data in Table 3 are representative of today's vehicles, but it should be realized that although the vehicles represented were calibrated for high fuel economy, most were designed for use with aftertreatment devices and with economic considerations and do not necessarily represent a design for the very maximum fuel economy without any emission considerations.

It is evident, however, that the engine-out emission levels will have to be lowered in order to meet the legal emission requirements. A number of things can be done to accomplish this.

One approach that has been taken is to attempt to reduce the amount of unburned CO and hydrocarbons and of NO produced, by means of changes in conventional engine design. This approach has given a small but significant lowering of emissions from spark-ignited engines and is to some degree reflected in the data of Table 3. The sources of CO and hydrocarbons have been reduced, but not to the extent necessary to meet emissions standard, by improvements in air–fuel ratio control, cylinder-to-cylinder distribution of air and fuel, choke operation, if carburetors are used, etc. These engine design changes can lead to small improvements in fuel economy in addition to reducing emissions but can also, in some cases, increase costs. This subject will be covered in more detail later. There should always be optimism that control of pollutants at their source can be further improved.

The principal approaches taken to control emissions in the 1973–74 model year vehicles, where the required emission levels were reduced from 1972 (see

TABLE 3. Some Engine-Out Emissions

Vehicle wt. (lb)	Engine size (liters)	CR	FTP mpg	Emissions g/mile HC	CO	NO_x	Comments
	Emission Control by Methods with Undesirable Emission Constraints on Fuel Economy						
5000	6.9	8	8.8	0.62	18	1.7	Rich + spark retard + air injection at outlet valves
5000	5.7	8	12.1	3.6	53	2.6	Rich + valve overlap + spark retard air at exhaust valves
5000	6.9	8	8.8	4.2	63	0.95	Rich + valve overlap + spark retard air at exhaust valves
4500	6.5	8	11.0	1.55	27	2.8	EGR + spark retard, air at exhaust valves
4000	4.0	8	12.2	2.32	71	2.0	Rich + spark retard + air injection at exhaust valves
	Vehicles with Minimum Emission Constraints on Fuel Economy						
2250	1.3	8.2	26.9	2.4	27	2.0	No emission constraints, near stoichiometry (Ref. 4)
3000	1.6	8.0	21.8	3.1	29	4.0	No emission constraints, near stoichiometry (Ref. 4)
4500	5.8	8.0	13.1	1.7	6	6	Near MBT, lean operation (Ref. 56)
4500	5.7	8.3	14.0	3.0	7.4	1.4	Lean operation + EGR; 3% fuel economy loss from MBT (Ref. 9)
3000	2.4	8.3	21.5	2.3	8.0	1.6	Lean operation + EGR; 3% fuel economy loss from MBT (Ref. 9)
3750	3.8	8	19.3	2.9	12.5	2.8	EGR; near max fuel economy
3500	3.8	8	19	3.3	7.3	1.4	EGR; near max fuel economy

Table 1), were to use leaner air–fuel ratios, control deceleration hydrocarbons, retard the spark from the optimum setting, and use valve overlap[8,53] to provide internal port extracted EGR for NO and, to some extent, hydrocarbon control. These latter changes very significantly reduce NO and hydrocarbon emissions but also cause losses in fuel economy when large reductions of these pollutants are required (see Figs. 2 and 3). The change in weight-adjusted fuel economy in these years was small as seen in Fig. 4. The fuel economy was increased in 1975–76 by the use of catalysts and external EGR, even though emission standards were reduced further. This increase, in part, reflects the decrease in spark retard, valve overlap, etc., which was used before 1975–76 and which allowed the expected increase in fuel economy due to leaner operation and other changes to be observed.

The inability of engine redesign and changes of engine calibrations, either alone or together, to control emissions with or without an undesirable loss in fuel economy (even in very small vehicles) has led to the use of aftertreatment devices,[4] particularly catalysts.

An ideal aftertreatment system would remove essentially *all* of the three pollutants over all operating modes of the engine and possess excellent durability. Such a system could of course enable the vehicle to operate at its maximum fuel economy and meet emission standards. Of the two aftertreatment devices that have been proposed, the thermal homogeneous reactor (thermactor) and the catalytic converter, the latter has proven to be effective over a wider temperature range than the former and to be capable of NO removal. As a consequence, it has been extensively used. However, it is important to realize that catalytic devices, even though they can be highly effective when they are operated above their catalytic ignition temperature, may still not allow one to obtain optimum fuel economy. There are a number of reasons for this:

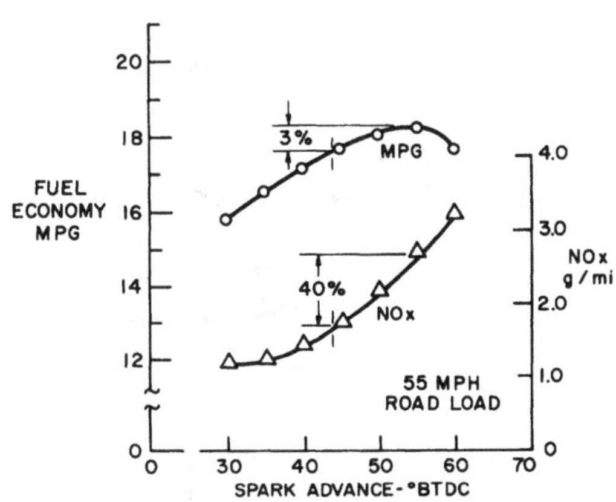

FIGURE 2. Relation between ignition timing, NO_x emissions, and fuel economy. 4500-lb vehicle, 350-CID engine, 55-mph road load. From Ref. 9. Reprinted with permission.

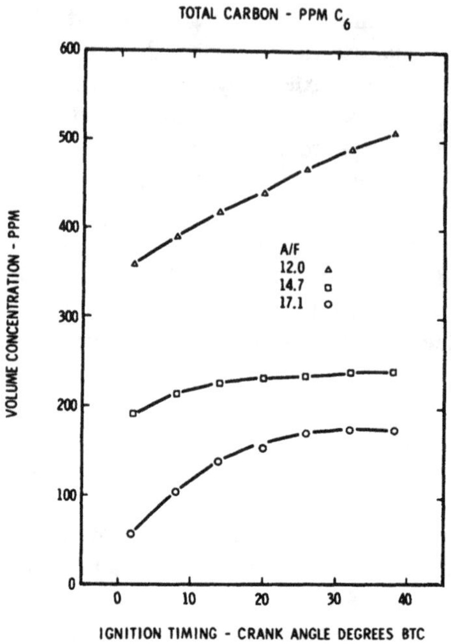

FIGURE 3. Effect of ignition timing on total hydrocarbon concentration in the exhaust. From Daniel, W., "Engine Variable Effects on Exhaust Hydrocarbon Concentration (A Single-Cylinder Engine Study with Propane as the Fuel)." SAE paper 670124. Reprinted with permission, © 1967, Society of Automotive Engineers, Inc.

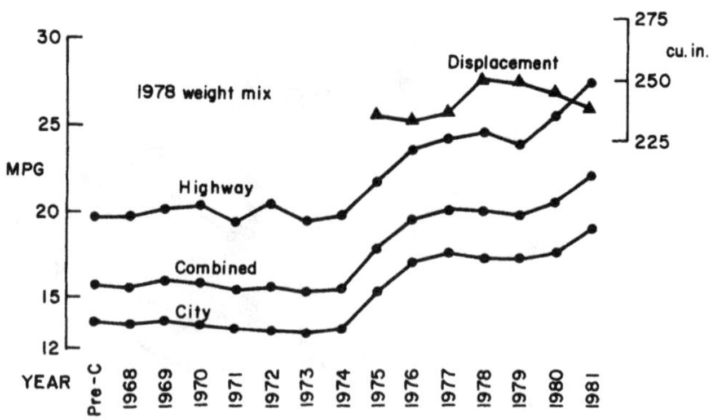

FIGURE 4. Fuel economy (mpg), after 4000 mile use, for each model year (from Ref. 77). Reprinted with permission, © 1981, Society of Automotive Engineers, Inc.

1. Catalysts require the use of unleaded fuel in order to maintain high efficiency during 50,000 miles use, particularly when used for NO removal. This requires the use of lower compression ratio engines than might otherwise be used if fuel of octane equivalent to leaded fuel were employed, and as a consequence, the engines operate at lower efficiency. This will be discussed later.
2. NO can be removed catalytically only by reduction to nitrogen.[29] When catalytic NO control is needed, the engine must be operated either at a stoichiometric air–fuel ratio or slightly rich of this. Because the maximum fuel economy of an internal combustion (IC) engine occurs at an air–fuel ratio lean of stoichiometry (see, for example, Fig. 5), operation of a catalyst-equipped vehicle for NO removal will entail some fuel economy loss from the maximum possible without emission constraints.[22] Without the use of EGR, operation at stoichiometry entails a fuel economy penalty of ~5% from that which can be obtained at a leaner air–fuel ratio of 17 with correct burning rate.
3. As stated before, during the vehicle cold start, a considerable fraction of the allowable emissions can be emitted before the catalyst becomes active. The unburned CO and hydrocarbons do not themselves represent any large loss in fuel economy during the FTP, but large cold-start emissions place a larger emission reduction burden on the catalyst system during the remainder of the FTP. If the catalyst system available cannot handle this load, a reduction of engine-out emissions can be required which may include some spark retard for hydrocarbon reduction and consequently a fuel economy penalty.

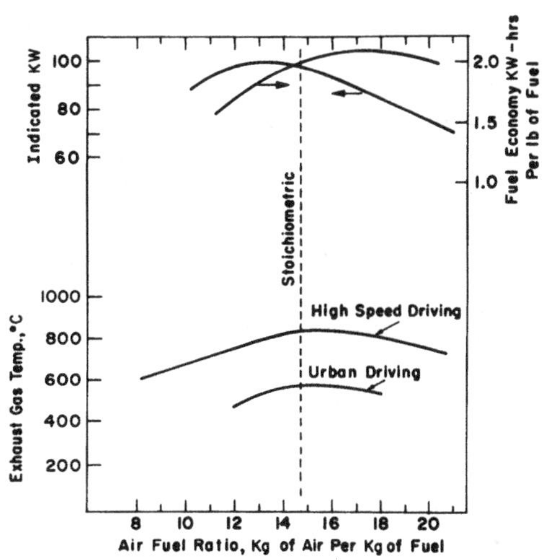

FIGURE 5. Fuel consumption and power at full load, and average exhaust gas temperatures of a spark-ignited IC engine as a function of the inlet air–fuel ratio.

4. The hydrocarbon emissions of an engine contain many molecular species. Some of these, such as methane and saturated hydrocarbons, react catalytically with oxygen slowly enough after 50,000 miles durability so that their catalytic removal can be very incomplete. If the engine-out hydrocarbons of these species are sufficiently high (large vehicle), with a high fuel economy calibration so that they cannot be removed catalytically by catalysts aged for 50,000 miles to give the required hydrocarbon emission levels, then these engine-out emissions may have to be reduced by spark retard with a concomitant fuel economy loss.
5. If the catalytic system employed requires the use of an air pump for supplying secondary air (for oxidation catalysts, for example), then the power requirement of this pump will reduce the vehicle fuel economy. The pump power requirements can be up to several horsepower. A fuel economy loss as high as 1 mpg has been observed.[5] For today's small vehicles the loss is estimated at from 0.3 to 0.5 mpg. The use of a system that uses the cylinder blowdown to introduce secondary air into the exhaust manifold[6] can prevent this loss.

IV. COMPRESSION RATIO, ECONOMY, AND EMISSIONS

The well-known benefit of increasing engine compression ratio (CR) is illustrated in Fig. 6. In order to realize this fuel economy benefit in practice one must consider (1) the effect of increasing the CR on the emissions of CO, NO, and hydrocarbons and (2) the need for a higher octane number fuel required for the higher CR. These problems have been investigated by a number of groups.[7–10]

Engine mapping data of Kirklin and Wise[8] at 9.3 and 8 CR has shown that the brake specific (BS) NO and hydrocarbon emissions (gram per horsepower-hour) increase as the CR increases (see Figs. 7 and 8). These data are representative of what one would expect, but it should be remembered when looking at the absolute numbers that the engine is a 400-CID engine and only one speed-load point is illustrated. The increased emissions found for the CR of 9 over that found for a CR of 8 would place an additional burden on the aftertreatment device if the higher CR were used. This is particularly burdensome for hydrocarbon emissions. Simulating the performance of a *5000-lb* vehicle, Kirklin and Wise[8] found that the lowest engine-out hydrocarbons obtainable were 1.4 g/mile at 8 CR and 3.6 g/mile at 9 CR with the NO constrained by EGR and some spark retard to 2 g/mile (although the fuel consumption was 7% lower for the 9.3 CR engine).

Gumbleton *et al.*[9] have presented data for a 3000-lb vehicle powered with two engines (151 CID) of different CR, one of 8.3 CR and one of 9.2 CR. The axle ratio used with the 9.2 CR engine was changed to provide the same per-

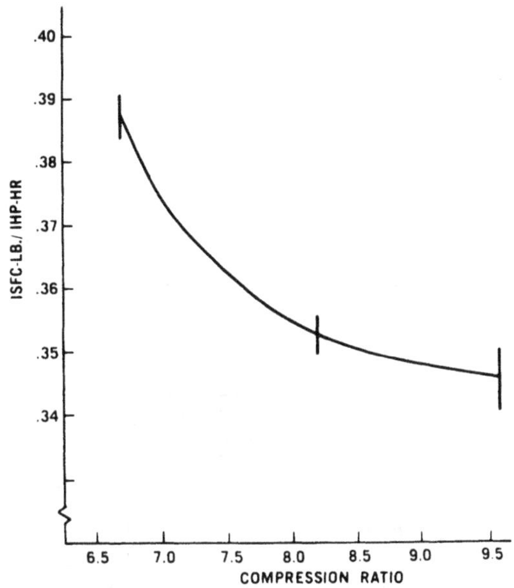

FIGURE 6. Simulated highway *indicated* fuel economy versus compression ratio. 5000-lb vehicle, 2000-rpm, 17 A–F, no EGR, 400-CID engine, 17.5 ft lb torque. From Baker, R., Daby, E., Pratt, J., "Selecting Compression Ratios for Optimum Fuel Economy with Emission Constraints," SAE paper 770191. Reprinted with permission, © 1977, Society of Automotive Engineers, Inc.

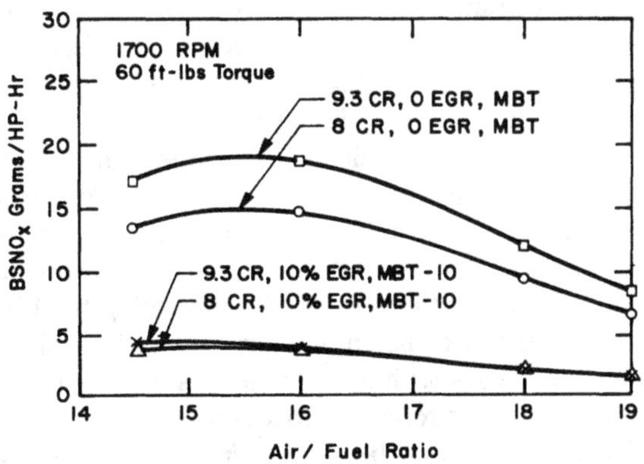

FIGURE 7. Effect of compression ratio on brake-specific NO_x emissions at 1700 rpm, and 60 ft lb torque; 400-CID engine. Data at two spark setting and various A–F's (from Ref. 8). Reprinted with permission, © 1978, Society of Automotive Engineers, Inc.

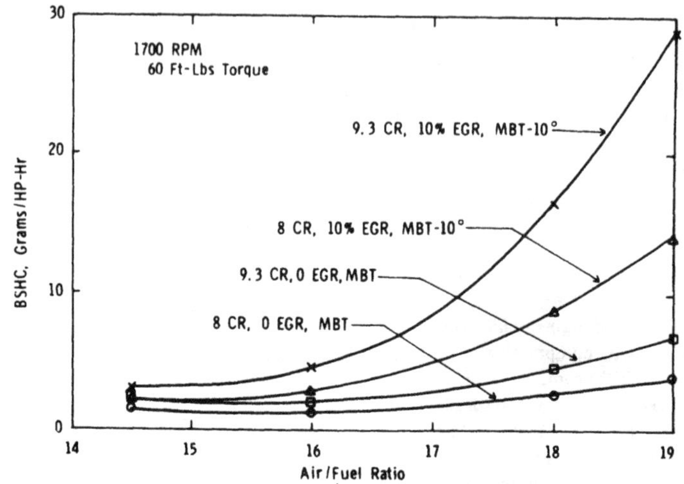

FIGURE 8. Effect of compression ratio and EGR on brake-specific hydrocarbon emissions at 1700 rpm, and 60 ft lb torque; 400-CID engine. Spark retarded 10° when 10% EGR used (from Ref. 8). The effect of EGR is more pronounced at lean operation. Reprinted with permission, © 1978, Society of Automotive Engineers, Inc.

formance as with the 8.3 CR engine. A 160 in.^{3} oxidation catalytic converter was used as an aftertreatment device. The pollutant emissions and fuel economy evaluated for the vehicle with either the high or low CR engine are shown in Table 4. As can be seen, there is an increase in fuel economy but at the same time an increase in hydrocarbon emissions. The viability of using a higher CR engine to improve fuel economy (if the higher octane fuel is available) will depend upon how difficult it is to meet the hydrocarbon emission levels with the engine–vehicle system under consideration.[(57)]

Production of higher octane unleaded gasoline required[(11)] by the higher compression ratio engine exacts an energy penalty.[(12)] This problem has been addressed by Gumbleton *et al.*[(9)] The energy gain (increased fuel economy at equal performance) in going from an 8.3 to a 9.2 CR is estimated to be 5.8%, whereas the energy penalty in the refinery for producing the higher octane unleaded fuel required by high octane engine has been estimated to be 9.5%. This value is an estimate that varies with the method of summing up the total refinery energy balance and the type of refining process used. See, for example, Ref. 81 in which this value is estimated to be 4–5%. From a total energy point of view there may be a loss in going to the high CR. This is shown in Fig. 9. The required octane can also be obtained in the refinery by the addition of tetraethyl-

TABLE 4. Economy and Emissions During FTP; 3000-lb IW Vehicle[a]

Compression ratio	g/mile			Economy EPA		
	HC	CO	NO	City	Highway	Composite
8.3	0.35	4.6	1.6	21.5	25.5	23.1
9.2	0.46	4.6	1.6	22	27.4	24.1

Compression ratio	Road fuel economy		Performance
	SAE urban	SAE-I 55	0–60 mph
8.3	17.4	25.1	18 sec
9.2	18.9	27.9	18.2 sec

[a] Data reported in the literature[9] for vehicle operation with catalytic converters.

lead to the gasoline at essentially no energy loss. The use of lead, however, is not compatible with the maintenance of high catalytic activity, and the engine deposits produced increase hydrocarbon emissions. There is also some concern about the health effects of lead in the atmosphere,[13] although this has been disputed.[14] Methanol, however, can be used to increase octane number and does not deactivate the catalyst, but it presents other problems.

The requirement for high octane fuel by the 9.2 CR engine occurs only during certain driving modes, usually, but not always, during those modes where

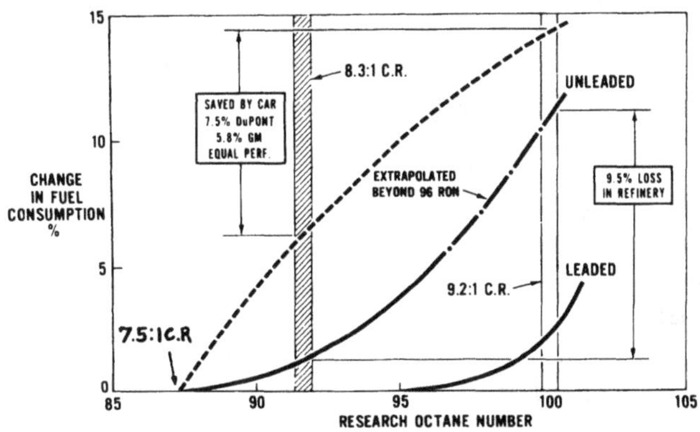

FIGURE 9. Relation between percent decrease in fuel consumption and fuel octane number. 4000-lb vehicle, 350-CID engine, 1977 emission constraints. 7.4, 8.3, and 9.2 compression ratio indicated on graph. Fuel saved by car, – – – –; fuel used in refinery, ——— (from Ref. 9). Reprinted with permission.

the intake manifold pressure is high such as during acceleration. During the other driving modes the engine octane requirements can be met by present ~91 octane fuel. Currie, Grossman, and Gumbleton[7] have described an electronic knock control system which retards the spark for knock control when engine knock is detected (by a detonation sensor) and allows a 9.3 CR engine to run on 91 octane gasoline. For a 4000-lb inertia weight (IW) vehicle with a 350 CID engine and a three-way catalyst system calibrated to give ~0.24 g/mile HC, 2.5 g/mile CO, and 0.63 g/mile NO_x, there was observed a 4.3% gain in EPA composite fuel economy over the value obtained with an 8.4 CR ratio engine in the vehicle (same performance in both cases). Tests on other engine–vehicle systems, however, have not always shown fuel economy improvement with the use of knock sensors.

With regard to a CR increase of ~1 unit, one can say that with an increase in the engine control system complication one can obtain a fuel economy benefit of less than 5% using present fuels. An additional hydrocarbon removal burden will be placed on the aftertreatment device. Although, the broad scale use of higher octane fuel (nonleaded) does not seem practical at this time, it should be borne in mind that the availability of organic antiknock blending agents such as methanol or methyl tertubutyl ether made from coal, and the development of more efficient refinery processes, may change this conclusion.

V. CARBON MONOXIDE EMISSIONS

The emissions of CO, NO, and hydrocarbons from a single-cylinder spark-ignited IC engine operated at optimum spark setting are shown as a function of the air–fuel ratio (A–F) in Fig. 10. At air–fuel ratios fuel rich of stoichiometry, carbon monoxide, and hydrogen are expected as the products of incomplete

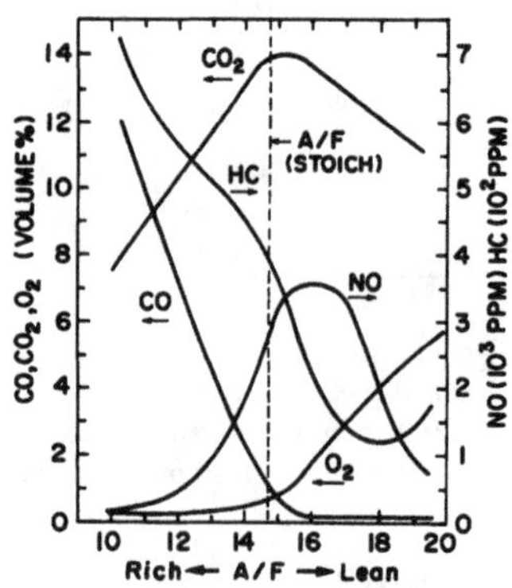

FIGURE 10. The concentration of CO, NO, and hydrocarbons (expressed as hexane) emitted by a single-cylinder spark-ignited IC engine as a function of the air–fuel ratio (pounds of air per pound of fuel).

combustion at the temperatures involved. In practice the ratio of CO to H_2 found in the exhaust is ~3. The chemical reactions which kinetically determine the exhaust CO and H_2 concentrations in the exhaust manifold have been described by Newhall[(15)] and others.[(16)] The exhaust CO concentration is not in equilibrium with the molecular species present (H_2O, CO_2, etc.). At exhaust temperatures equilibrium would favor more H_2 and less CO, but this requires the use of catalytic reactors. Hydrogen is not considered a pollutant.

As can be seen from Fig. 10, at air–fuel ratios greater than ~16, CO emissions are low, and in fact the exhaust CO concentration of an engine operating at steady-state with this lean air–fuel ratio condition can be low enough so that it can approach the *average* vehicle tail pipe CO concentrations required for a 3.4 g/mile emission level (see Table 2). The source of some of the CO in the lean region is thought to be the slow homogeneous partial combustion, during the exhaust event, of some of the hydrocarbons not burned in the main combustion.

In practice, however, there are a number of reasons why we expect higher CO emissions during the FTP even if the engine is calibrated lean of stoichiometry. As pointed out before, very substantial CO emissions can occur during the cold start of the vehicle. Pozniak[(17)] has shown that for a 4000-lb vehicle with a carburetted engine calibrated for a FTP CO emission level, using catalysts, of 3.4 g/mile, the engine-out CO emissions during the first 2 min of operation can account (using the cold-start weighing factor of 0.43) for 80% of the allowable CO emissions during the whole FTP. This number could be smaller if the air–fuel ratio could be better controlled during cold start. Fuel-injected engines have some advantage in this area.[(18)]

Since there is an abrupt increase in CO as one decreases the air–fuel ratio to stoichiometry and into the fuel-rich region, one can expect high CO levels if the engine has sufficiently high cylinder-to-cylinder variation in air–fuel ratio, even though the average is lean of stoichiometry, so that one or more cylinders will run with a fuel-rich ratio. For example, in a paper by Lindsay *et al.*[(19)] in which the CO emission from each cylinder of a four-cylinder commercial engine was measured, it was found that at an average air–fuel ratio of 16.2, the volume percent CO in the exhaust from each cylinder varied from 1.0 to 0.2%. With more uniform mixture distribution to each cylinder the volume percent CO from each cylinder was equal and 0.1% at an air–fuel ratio of 16.5.

During acceleration with a carburetted engine, fuel is injected into the intake manifold by the accelerator pump. This is done to counteract the tendency of the carburetted vehicle to run lean during this transient period because the carburetor system can have poor response to a sudden increase in air flow. All of the injected fuel does not get entrained by the airstream. This leads to a transient period in which the intake charge is fuel rich at the end of the acceleration as the excess fuel on the intake manifold walls evaporates. In many instances this richer than average transient can lead to spikes of high CO emissions.[20] The leaning during acceleration is also true of electronic fuel injection systems but the electronic control systems are better able to compensate for the changes.

If the need for maximum power operation occurs during the FTP (which is infrequent), this will lead to high CO emissions during the period of time that the high power is required. This is because maximum power occurs at A–F 12–13 and present vehicles are calibrated to operate fuel rich near wide-open throttle.

If catalytic NO removal is required, then the engine must be run at stoichiometry or slightly rich of stoichiometry during the time when catalytic NO removal is necessary. This will result in relatively high engine-out CO emissions levels (0.5–1 mole %), which will have to be removed in the catalytic aftertreatment system.

Vehicle fuel economy will be a maximum at an air–fuel ratio in the fuel lean region where the engine-out CO emissions are low as we have seen, although the necessity for NO control may dictate operation in the fuel-rich region. In this respect CO emission control and high fuel economy coincide. However, as we have also seen, it is necessary for the vehicle to have low CO emissions during the cold start, good cylinder-to-cylinder A–F distribution, and low or no CO emission spikes in the exhaust. This objective can be obtained by electronic fuel injection[18] but at a significant economic penalty over the carburetor system.

It is interesting to note that the diesel engine runs lean in all modes and does not have a large cold-start CO emissions problem because of direct cylinder fuel injection. Consequently, vehicles powered with diesel engines meet the federal CO emission standards without the use of aftertreatment devices.

For those vehicles where the engine-out CO emission levels exceed the legal limit, one must employ aftertreatment devices for CO removal, usually catalytic oxidation. Changes in the engine spark calibration have little effect on the CO emissions from the cylinder of a well-designed engine. For homogeneous post combustion of CO, oxygen ($>2\%$ O_2), temperatures of ~ 800 °C, and resident times of 20–50 m/sec are required[21] along with very good mixing.

VI. NITRIC OXIDE EMISSIONS

NO is formed in the cylinder of the IC engine by the higher temperature (~ 2800 °K) reaction between oxygen atoms and nitrogen molecules. The oxygen atoms are formed by thermal dissociation of oxygen containing species in the combustion gases, particularly oxygen.

$$O + N_2 \rightarrow NO + N$$

$$N + O_2 \rightarrow NO + O$$

Predictions of the amount of NO formed during various modes of engine operation have been made using chemical kinetics by Blumberg and Kummer.[23] This work was based on earlier work by Lavoie and Keck.[24] These investigators have shown that over the entire range of air–fuel ratios and engine operating conditions the first elements in the charge to burn contribute most to the overall NO emission level, and that in the area of lean operation, or operation with EGR the NO levels are essentially limited by kinetics of formation. Since the overall activation energy for NO formation is ~130 Kcal/mole, a small decrease in the peak combustion temperature will greatly reduce the rate of NO formation. All methods of NO control in the engine will reduce the NO by either reducing the peak combustion temperatures or the oxygen concentration. NO control by operating in the fuel-rich region in order to reduce the oxygen concentration (and temperature reduction in the highly rich region) will lead to poor fuel economy and should not be considered in light of today's energy problems. Controlling NO by reducing the peak combustion temperature will result in a loss in maximum power capability for an engine of a given displacement, but can result in an increase of the thermal efficiency of the engine provided the correct combustion interval during the engine cycle can be maintained and provided physical characteristics of the working fluid are not degraded. Of course, if the engine combustion temperature is returned to its peak value (rich of stoichiometry: no EGR), when maximum power is required (if this occurs infrequently, as it does, and outside the FTP), then there need be no loss of full power capability. The indicated efficiency of an ideal Otto cycle engine is determined, in theory, only by the compression ratio and the ratio of specific heat, γ. By reducing the peak combustion temperature, the effective γ will tend to increase and give better indicated fuel economy. However, if EGR is used in conventional engines at stoichiometry, this will be offset by the low γ of the recirculated water vapor and carbon dioxide.[22] In addition, for the same engine power output at the lower peak combustion temperature, the intake manifold pressure must be increased. If one can reduce the peak combustion temperature and at the same time increase the intake manifold pressure (increasing this pressure alone will raise the combustion temperature, principally because of less residual gas) then, at equivalent power output, operation at lower temperatures will give an additional increase in fuel economy because of lower pumping losses.

In practice, the combustion temperature can be lowered by a number of different methods. The engine can be operated at very lean A–F, or cooled exhaust gas can be recirculated into the inlet manifold (EGR). The NO reduction obtained is illustrated in Fig. 11. The peak NO production occurs somewhat lean of stoichiometry even though the peak combustion temperature occurs somewhat rich of stoichiometry because of the low oxygen atom concentration rich of stoichiometry. Both very lean operation and EGR are very effective for reducing NO emissions. Since EGR can reduce peak combustion temperatures

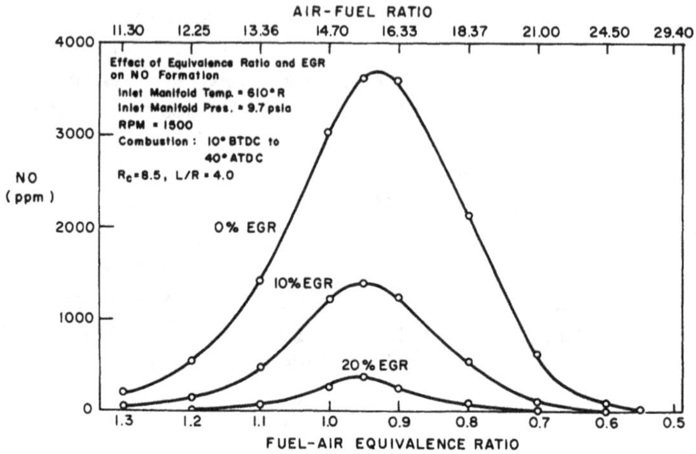

FIGURE 11. Effect of equivalence ratio and EGR on NO formation. Inlet manifold temp. 66 °C, and pressure 9.7 psia. Compression ratio 8.5, rpm 1500, combustion 10° BTDC to 40° ATDC (from Ref. 23).

without at the same time adding oxygen (A–F constant), less EGR is required for a given NO reduction than air, when air is used as a diluent (lean operation). As a consequence, the combustion temperature will be somewhat higher than with air dilution which may increase burning velocity. However, a small penalty in fuel economy occurs when EGR is used in place of lean operation because the exhaust gases contain CO_2 and H_2O which have a lower γ than N_2.

Figure 12 shows that one can maintain fuel economy and control NO by lean operation (or EGR) if the optimum combustion interval is maintained. In practice, however, it is difficult to maintain this correct burning rate because at lower combustion temperatures the burning rate decreases. Special methods such as increased combustion chamber turbulence or dual ignition are required to preserve fuel economy when controlling NO by EGR. In addition to the burning rate problem, the use of EGR will cause the hydrocarbon emissions to increase[(25,26)] and if this increase requires the use of spark retard for some hydrocarbon control then fuel economy will decrease. It is important that the EGR and fresh charge and residual gas and fresh charge be well mixed. Poor mixing will lead to small pockets of exhaust gas containing fuel and oxygen that has diffused into the pocket from the fresh charge. The pocket mixture will be too diluted to sustain a flame and will increase hydrocarbon emissions. This effect is shown in Fig. 13.

By retarding the spark so that combustion is still proceeding during the first part of the expansion process, the peak combustion temperature will be reduced. This will be particularly so for the first elements of the charge to burn which

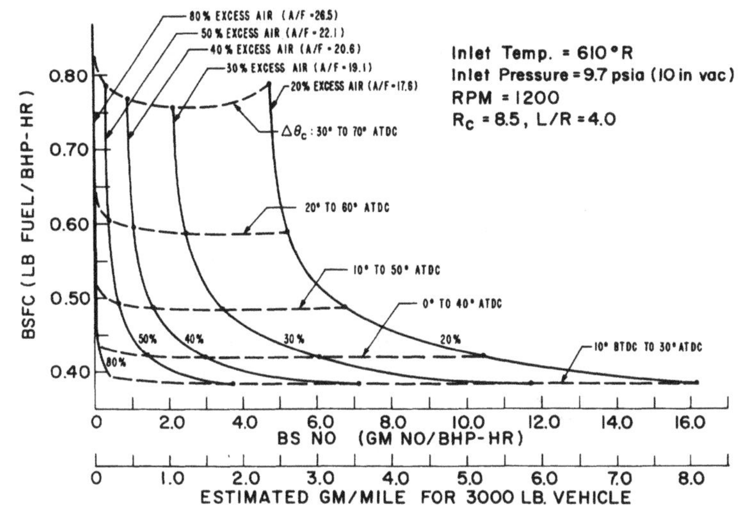

FIGURE 12. Efficiency and NO characteristics in the lean region. Brake-specific fuel consumption as a function of brake-specific NO emissions for various air–fuel ratios in the lean region and at different levels of spark retard (from Ref. 23).

produce the most NO because the recompression and temperature rise of this portion of the charge produced by the subsequent (~constant volume) burning process will be avoided by the expansion process. Figure 14 shows the effect of a progressive delay of the combustion process in the cycle (spark retard) on NO formation. Spark retard is an effective way of reducing NO emissions, but because the burning is proceeding during the expansion stroke (which is necessary in order to produce lower temperatures), all of the burned gases will not be expanded the full compression ratio, and the thermal efficiency of the engine and consequently the fuel economy of the vehicle will be reduced.[27] This is illustrated in Fig. 15. A modest amount of NO control can be obtained with only a small decrease in fuel economy. It is also interesting to note that at high EGR levels spark retard is less effective. From a fuel economy point of view, EGR is a better strategy for NO control. However, it should be noted that spark retard reduces hydrocarbon emissions (see Fig. 3), whereas EGR increases them, and spark retard is easy to implement.

Reducing the temperature of the inlet air to the engine reduces the NO emissions, increases the fuel economy, and lowers the octane requirement of the engine (see Fig. 16). Disadvantages include increased hydrocarbon emissions because of lower exhaust temperatures leading to less combustion of unburned fuel in the exhaust system and decreased fuel volatilization in the intake manifold for fuels such as gasoline with high boiling constitutents. These disadvantages

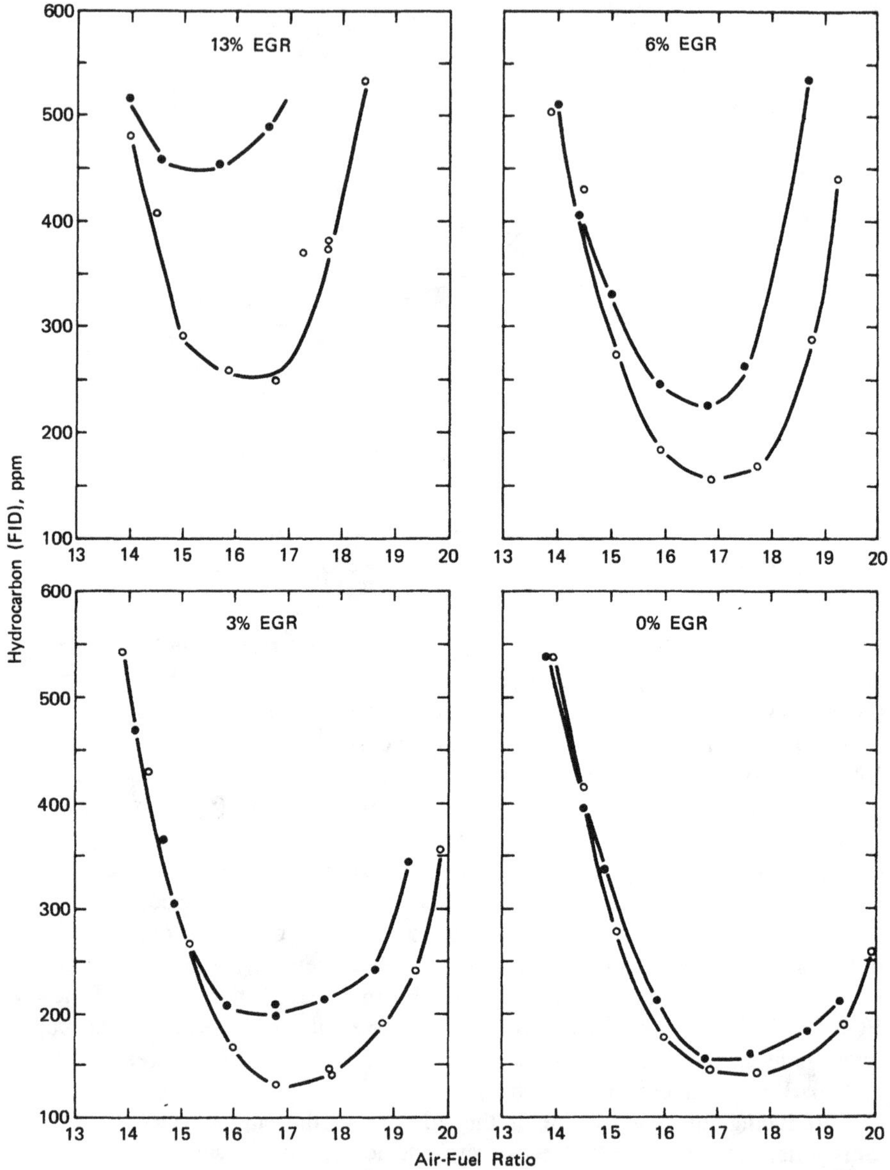

FIGURE 13. The benefit of good mixing in the intake manifold, when EGR is used, on hydrocarbon emissions. Production manifold, ●; turbulent flow manifold (good mixing), ○. Two-liter engine, 9 compression ratio, constant spark timing, 2000 rpm, 10% load (from Ref. 26).

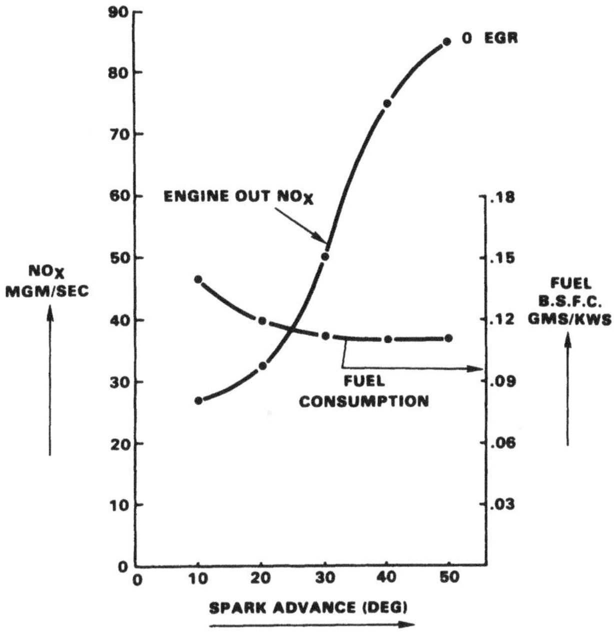

FIGURE 14. NO_x and fuel consumption versus spark advance at stoichiometry; 1600 rpm, 113 N-M torque (from Ref. 27). Reprinted with permission, © 1980, Society of Automotive Engineers, Inc.

disappear when methane is used as a fuel. Because of their lower flame temperatures, methane and methanol produce approximately half the NO emissions as gasoline in normal operation.[23]

As stated before, the first part of the cylinder charge to burn contributes the majority of NO emissions. Blumberg[28] has shown that by producing a stratified charge in the cylinder in which the first element of the charge to burn has a fuel-rich air–fuel ratio and the last element to burn has a fuel-lean air–fuel ratio (the average being stoichiometric) a more effective control of NO can be obtained than with a completely homogeneous charge at stoichiometry. The stratified charge will, however, produce lower fuel economy and increased CO emissions over the homogeneous case.

Increasing the engine compression ratio will increase the maximum combustion temperature and the brake-specific NO (see Fig. 17). This increase will be significantly less at lean operation or with high EGR rates. In Fig. 17 one can see that if the combustion interval is optimum (10° BTDC → 40° ATDC), the NO emissions increase as the compression ratio is increased, but under conditions of spark retard (combustion interval 0° to 50° ATDC) there is very

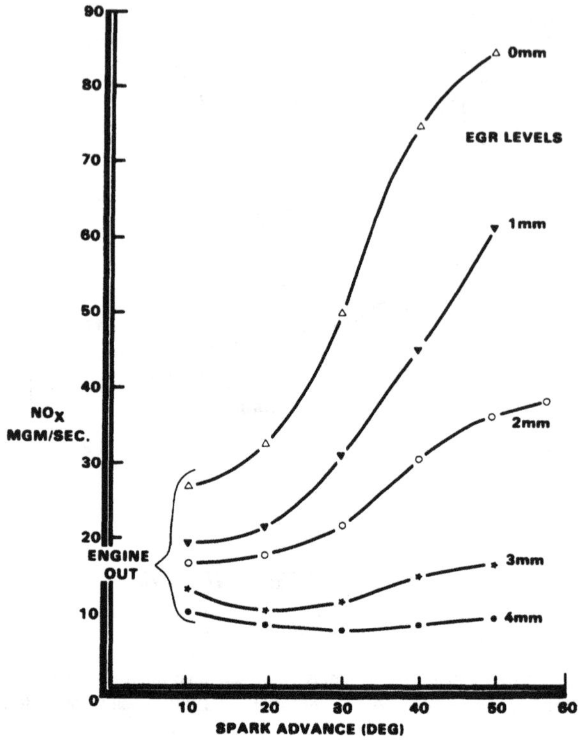

FIGURE 15. NO_x emissions versus spark advance for various EGR levels at stoichiometry; 1600 rpm, 113 N-M torque. EGR valve opening measured in mm. 0 mm represents no EGR, 4 mm represents full EGR (from Ref. 27). Reprinted with permission, © 1980, Society of Automotive Engineers, Inc. % EGR at 4 mm is 11–16%, at 3 mm is 9–12%, at 2 mm is 6–8%, and at 1 mm is 2–3%.

little effect of compression ratio on brake-specific NO emissions. It can also be seen from Fig. 17 that if one increases the compression ratio from 7.5 to 9 and at the same time retards the spark (burning from 0° to 50° ATDC), one can effect considerable NO control (from 3.4 to 1.6 gm NO per BHP-HR) without loss in fuel economy from the 7.5 CR case and without the need for higher octane fuel at the high compression ratio because of the spark retard.

In conclusion, one can say that NO control in the engine will involve reducing peak combustion temperatures. For small to moderate reduction in NO levels one can employ spark retard and incur only a small decrease in fuel economy (see Fig. 2). For larger reductions in NO levels the addition of an EGR strategy to a modest spark retard strategy, together with appropriate methods to increase the burning rate, will be the method of choice in order to preserve fuel

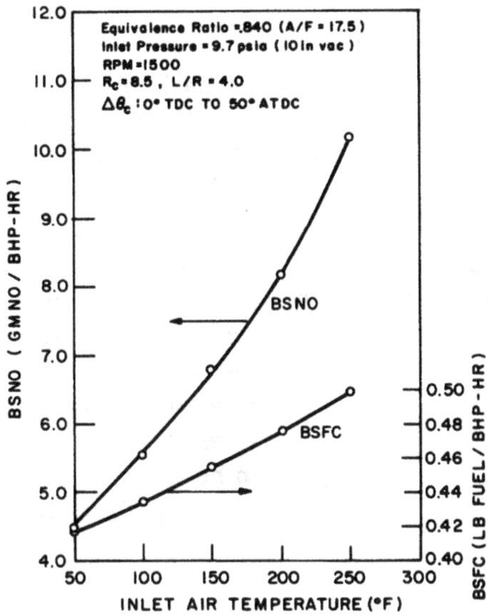

FIGURE 16. Effect of inlet air temperature on brake-specific fuel consumption and brake-specific NO formation (from Ref. 23).

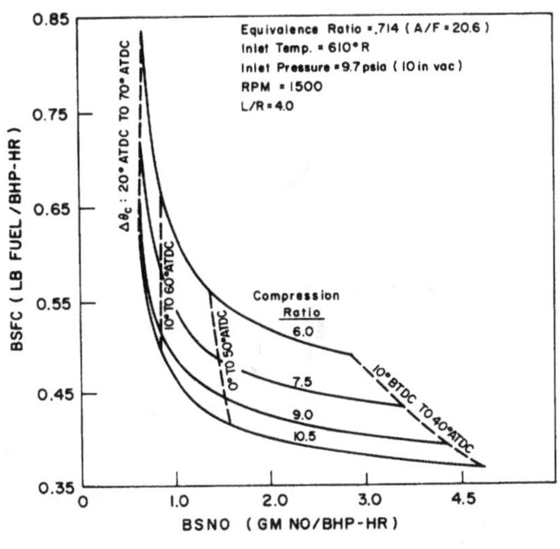

FIGURE 17. Effect of compression ratio on brake-specific fuel consumption and brake-specific NO formation (from Ref. 23).

economy provided that the increase in hydrocarbon emissions does not require loss in fuel economy for its control.

In the course of an FTP emission test it is observed that the NO mass emissions are large during the acceleration modes in the stabilized part of the test. In the cold (or hot) start part of the test the NO mass emissions are more equally divided between cruise and acceleration modes because of the sustained 55 mph cruise in this part. The data in Fig. 18 shows exhaust NO concentrations during the FTP. The peaks correspond to acceleration modes where the exhaust flow is also high. NO concentrations during part of the old seven-mode FTP (now discontinued) are shown in Fig. 19.

The NO emissions are large whenever the intake manifold pressure is high (Fig. 20) principally because at high manifold pressures there is a lower ratio of residual gas, from previously burned charge left in the cylinder, to new charge. This residual gas can be thought of as internal EGR.

The fact that high mass NO emissions occur only when the manifold pressure

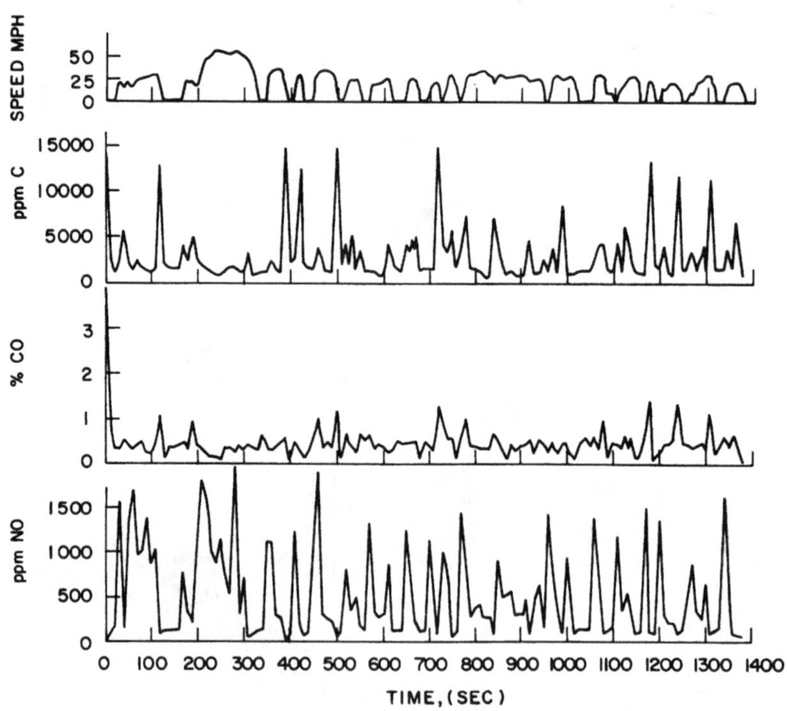

FIGURE 18. Variation of the exhaust concentrations of CO, NO, and hydrocarbons (as hexane) during the federal test procedure.

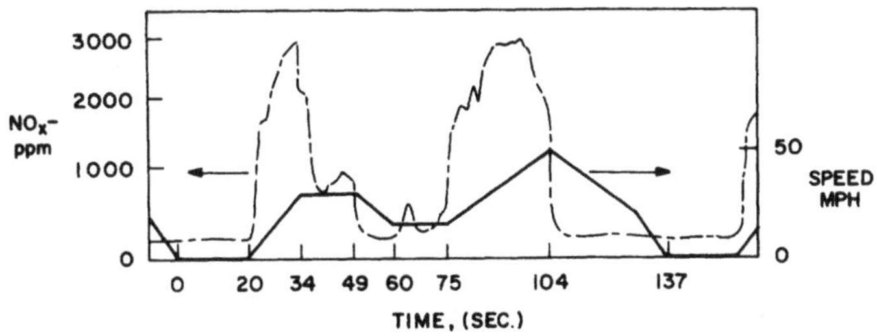

FIGURE 19. Formation of NO during acceleration in a simple repetitive driving cycle from cold start (old LA-7 mode cycle). The depression of NO after the initial acceleration is due to the momentary rich A–F produced at the end of acceleration.

is high and at an air–fuel ratio somewhat lean of stoichiometry suggests that dynamic methods of NO control, where the control methods that were outlined above are introduced as a function of engine speed and load rather than continuously, could be effectively employed. This approach has been used for NO control in production vehicles since the introduction of emission control. EGR and spark retard have been utilized by controlling them as some function of the inlet manifold pressure or of inlet manifold pressure and speed (exhaust back pressure), and attempts are made to avoid the air–fuel ratio for maximum NO during accelerations. With the advent of more sophisticated control systems based

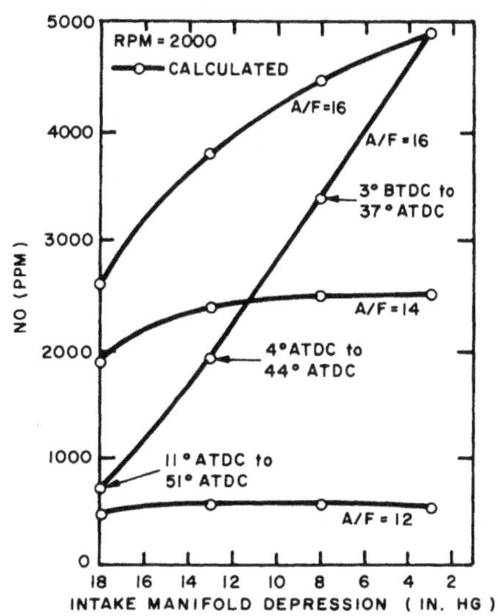

FIGURE 20. Effect of inlet manifold pressure on NO formation (from Ref. 23). Burn from 10° BTDC to 30° ATDC unless indicated.

on microcomputers together with sensors for measuring engine speed and torque, etc., and the ability to rapidly change spark setting, A–F, and EGR, one can control the NO (and other) emissions of a given engine–vehicle system in an optimum way for maximum fuel economy. This will be described in a later chapter.

VII. HYDROCARBON EMISSIONS

Figure 10 shows a typical relationship between the air–fuel ratio and hydrocarbon emissions. There is no air–fuel ratio that will give very low hydrocarbon emissions, and, in fact, for this particular engine and operating mode the hydrocarbon exhaust concentration never reaches the *average* concentration required for emission control to legal limits (Table 2).

The cold-start problem (see Sec. V) leads to high hydrocarbon emissions during the first 100 sec (see Fig. 21) of the FTP. Pozniak[17] estimates that during this time as much as 40% of allowable hydrocarbon emissions during the whole FTP can be emitted from a carburetted engine. This number will vary with the fuel induction system used and will be smaller for electronic fuel injection (EFI)

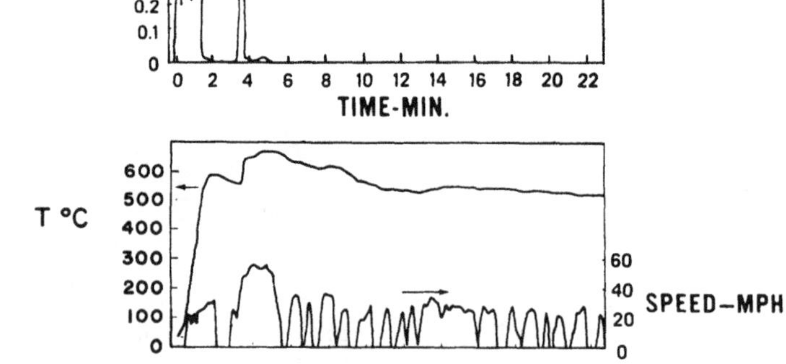

FIGURE 21. The emissions of CO and hydrocarbons from a vehicle equipped with a large underbody pellet catalyst. This catalyst reached operating temperature after the undesirably long time of ~4 min. The importance of rapid catalyst lightoff is evident if a large percentage conversion of engine-out emissions is required.

systems than for carburetted vehicles because of their more accurate fuel control and because their closeness to the intake valve leads to less accumulation of excess liquid fuel in the intake manifold. If the cold start of a vehicle is not well controlled, the hydrocarbon emissions in the first 100 sec can exceed that allowable for the whole test. Control of hydrocarbon emissions from present spark-ignited IC engines is a most difficult problem, and intensive use has been made of catalytic aftertreatment devices for this purpose.[29]

The main combustion event of a well-carburetted spark-ignited engine operated well removed from the misfire limit is thought to leave essentially no hydrocarbons in the residual burned gases. The major source of hydrocarbons found in the exhaust results from that portion of the fuel–air intake charge that is trapped during the compression stroke, and subsequent pressure rise during combustion, inside of crevices in the cylinder into which the main combustion flame cannot enter. In addition to this source it has been thought that a thin layer of unburned air–fuel mixture is produced at the relatively cold walls of the combustion chamber, by thermal quenching of the flame processes as the flame approaches the cold wall. These two sources of unburned hydrocarbons find their way into the burned gases during the later portion of the power stroke and during the exhaust stroke and leave the cylinder during the exhaust stroke. In the exhaust port and exhaust manifold, a portion (10–40%)[30,31] of the residual hydrocarbons are combusted to CO or CO_2. The remainder leave the exhaust manifold to give the hydrocarbon emissions of the engine. Greater than half of these hydrocarbons are unburned fuel.[32,33] Under fuel-rich conditions the crevice volume holds more hydrocarbons, and a smaller fraction of these are combusted before they reach the exhaust manifold outlet due to low oxygen levels. This increases hydrocarbon emissions (see Fig. 10).

The crevices that trap the unburned air–fuel mixture have been identified.[34,35] The principal one is that volume between the piston and the cylinder wall extending from the top compression ring to the top of the piston and on top of and behind the first compression ring, and the volume between the piston and the wall between the two compression rings. In addition there are crevices where the head gasket is used, where the spark plug is inserted, at the valves, and in the porous interior of any engine deposits that may be present.[36]

Recent work[37,38] has questioned the wall quench mechanism as a source of unburned hydrocarbons in the carburetted spark-ignited engine and had led support to an alternate proposal that identifies the unburned hydrocarbons produced on the cylinder walls with the rapid absorption and desorption of hydrocarbons into and out of the oil film on the wall during the compression and expansion stroke.

It is thought, however, that the major sources of unburned hydrocarbons from the carburetted spark-ignited engine are the crevice volumes in the combustion chamber, particularly in the compression ring area,[34] and that control of this hydrocarbon source is the key to reducing hydrocarbon emissions. The hydrocarbon emissions are low from diesel engines where the fuel is not injected until the piston approaches the top of the compression stroke with the crevice volume filled with air, and it is mostly air which enters this volume during the subsequent pressure rise which is small for the diesel cycle.

It is important to have adequate fuel–air mixture preparation[39] so as to avoid the possibility of unevaporated fuel droplets impinging on the cylinder wall and being entrained into the piston crevice volume during the compression stroke, and thus increasing hydrocarbon emissions by producing a very fuel-rich gas in this volume. This can occur during cold start because of a poor air–fuel mixture, but is not thought to be important if the cylinder walls are warm >100 °C and the charge is well carburetted.[40,41]

Wentworth has shown[34] that some reduction of hydrocarbon emissions produced by the ring crevice area can be obtained if the blowby past the compression rings is increased in order to flush some portion of the crevice volume into the crankcase (Fig. 22) or by sealing the ring crevice volume. These approaches have not been used commercially because of associated oil consumption and durability problems and because a large blowby flow would decrease fuel economy. It has been estimated[42] that the blowby present in today's engines reduces fuel economy ~3%. Wentworth has also investigated the problem of exhaust valve leakage as a source of hydrocarbon emissions and has concluded that it is probably small, but that there was little data available for production engines.

There will, of course, be large momentary hydrocarbon emissions if the

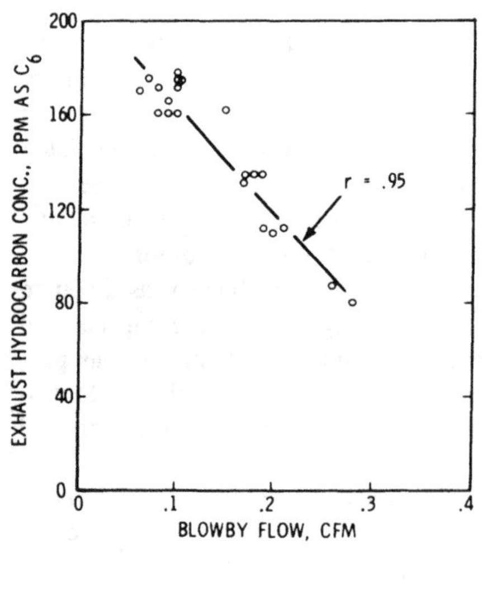

FIGURE 22. The effect of ring blowby flow rate on exhaust hydrocarbon exhaust concentration. 1200 rpm, 12 in. Hg intake depression. Ring gap located 60° from spark plug near exhaust valve. Correlation coefficient of data with line = 0.95 (from Ref. 34). Reprinted with permission, © 1968, Society of Automotive Engineers, Inc.

air–fuel mixture in the cylinder does not ignite during an engine cycle. In general, assuming a well-functioning ignition system, this occurs only if the air–fuel mixture is too lean (>18–20 A–F), too rich (<8–9 A–F), or at too low a cylinder pressure for the plug gap employed (deceleration). The lean limit of engine operation can be extended by increased turbulence in order to increase the burning rate of lean mixtures, and more intense extended ignition sources,[43,59] and ignition during deceleration improved by large plug gaps. The increase in hydrocarbon emission levels for very lean operation, as shown in Fig. 10, is due to intermittent misfiring. Operation at a very lean air–fuel ratio, comfortably smaller than the misfire limit, is attractive for good fuel economy and low NO and CO emissions, but the engine will have low maximum power capability per unit displacement if the air–fuel ratio is invariant, and the hydrocarbon emissions will be too high (without aftertreatment) unless some way can be found to greatly decrease the combustion chamber crevice volumes.

For fuel-injected vehicles it is possible to avoid the hydrocarbons produced by deceleration misfiring[44] by fuel cut off during deceleration. For carburetted vehicles a dashpot is used to prevent rapid throttle closing at high engine speeds during deceleration in order to maintain a satisfactory cylinder pressure,[45] and as a result the vehicle decelerates slowly. The use of a conventional carburetor with a dashpot in order to reduce hydrocarbon emissions will give a small fuel economy loss in urban driving (3–7%) when compared to systems in which complete fuel cut off is possible.[18]

One might expect the quench layer or oil adsorption layer hydrocarbons and the hydrocarbons ejected from the crevice volume during the power stroke to be rapidly consumed in the cylinder at lean air–fuel ratios because they will contact hot oxidizing gas from the main combustion. That this does not occur to completion is obvious since the engine emits hydrocarbons with the hot burned gases. Although the hydrocarbons diffuse rapidly towards the hot oxidizing residual gas, the thermal cooling wave moves at a similar speed.[25] It is estimated that about one-half of the trapped hydrocarbons are burned before they exit the exhaust manifold.

Experimentally, combustion of the hydrocarbons is observed as the exhaust gases traverse the exhaust port[46,47] (probably because mixing on a small scale occurs in the turbulence produced by the exhaust valve orifice) and is a maximum at stoichiometry. The use of port liners in order to reduce heat loss and enlarged ports can reduce hydrocarbon emissions perhaps as much as 10–20%[30,31,46] under some engine conditions, but under the FTP it has been estimated substantially less than this.[30]

For vehicles with cold exhaust such as methanol-fueled vehicles with high (11–13) CR, the exhaust port outlet may be sufficiently low (500–600 °C)[48] so that little oxidation takes place in the port with or without a port liner. In order to oxidize hydrocarbons in the exhaust, manifold temperatures greater than 600°C and residence times of 50 m/sec or greater are required[49] and, if added air is employed, it is essential that it is mixed well with exhaust gas.[30]

Air injection into the exhaust ports of an IC engine (~10% of exhaust flow) has been extensively used to reduce hydrocarbon (and CO) emissions by port oxidation and to some extent manifold oxidation (uninsulated).[50,5] It has been used in the model years 1966–1980. Port air can be effective for rich air–fuel ratios but is considerably less effective for lean air–fuel ratios.[50,51] See Fig. 23 for behavior of insulated manifolds. Spark retard is sometimes used to increase the effectiveness of these systems for idle and low power modes of vehicle operation at the expense of fuel economy. Such systems without extensive spark retard are usually used in conjunction with an oxidation catalyst because the system alone cannot reduce hydrocarbons and CO emissions to the required levels at good fuel economy.

It has been shown (see, for example, Ref. 52) that the mass hydrocarbon emissions from the cylinder of a spark-ignited internal combustion engine during the exhaust stroke will occur in a very nonuniform manner. The preponderance of the mass emissions will occur either at the time the exhaust valve first opens or when the piston reaches the top of the exhaust stroke. This latter mass of hydrocarbons is the greater amount and can reside in the port for the great majority of the cycle time until the exhaust valve opens again. In those cases where EGR is used to control NO emissions, the exhaust gas to be recirculated can be withdrawn from the exhaust port, and, because this gas contains a higher than average hydrocarbon level, some reduction of hydrocarbon emissions as well as NO emissions will occur. This method of hydrocarbon reduction is limited by the fact that EGR is large only during periods of high intake manifold pressure. It is ineffective in controlling hydrocarbon emissions due to partial misfiring or due to the occasional slow burn rate found near the limit of EGR that the engine can tolerate. The exhaust gas in the port can also be recirculated by changing the valve timing so that the exhaust valve closes later in the cycle (increasing valve overlap). This can reduce hydrocarbon and NO emissions[53] in a similar manner as port-extracted EGR. It is thought that cold EGR is preferable to hot EGR.[23]

Hydrocarbon emissions increase as the compression ratio increases because the crevice volume becomes a larger fraction of the final compressed volume, the peak cycle pressure is higher, and the exhaust gas temperature is lower as a result of increased engine efficiency. Because of this lower temperature, a lower fraction of the emitted hydrocarbons are combusted in the port and exhaust manifold. The effect of compression ratio on hydrocarbon emissions is shown in Fig. 24.

As we have seen before, spark retard reduces the thermal efficiency of an IC engine because the charge is still burning during the early portion of the

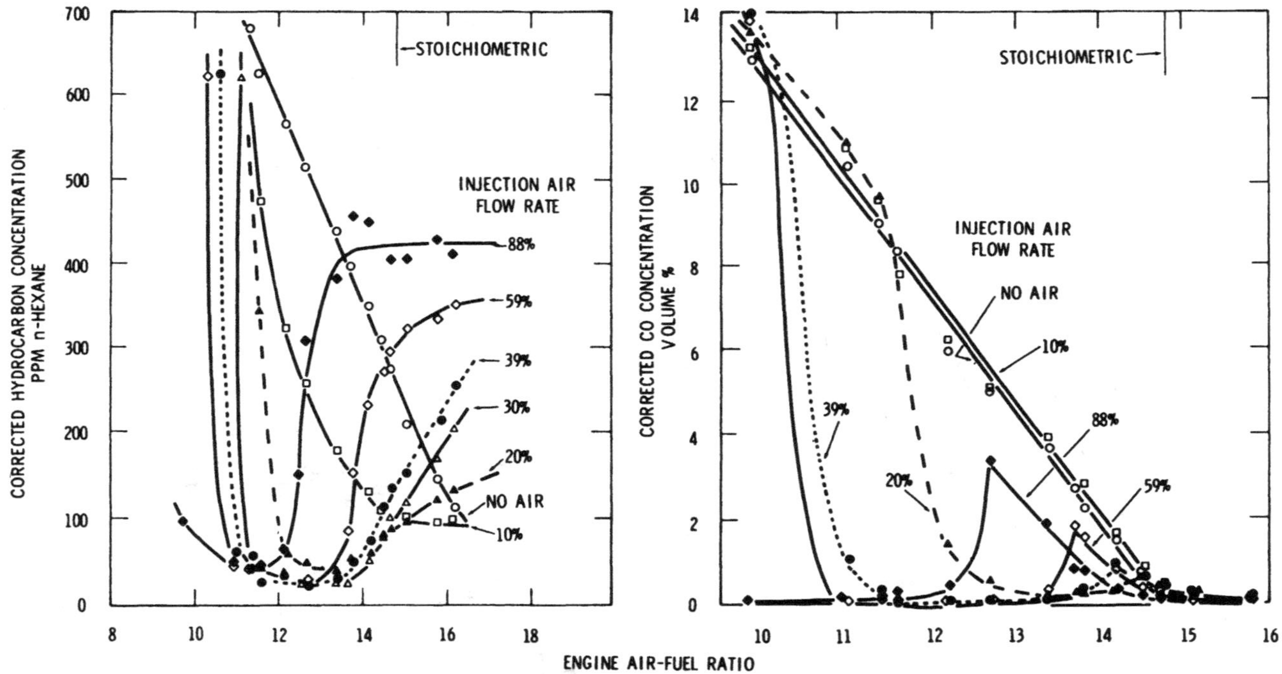

FIGURE 23. The effect of air–fuel ratio and injection air flow rate on exhaust hydrocarbons and CO concentrations. Exhaust port air injection, MBT timing, 9.25 CR, 1200 rpm. Single-cylinder engine (37 in.^{3}), insulated 150 in.^{3} manifold (from Ref. 51). Reprinted with permission, © 1965, Society of Automotive Engineers, Inc.

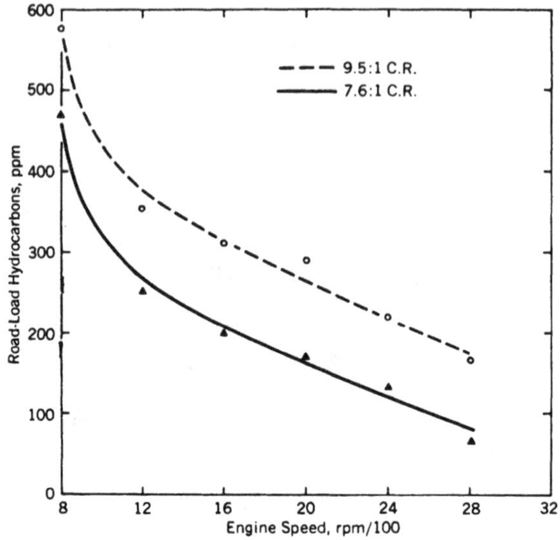

FIGURE 24. The effect of compression ratio and engine speed on hydrocarbon emissions at road load and MBT ignition timing. Felt, A., Krause, S., "Effects of Compression Ratio Changes on Exhaust Emissions," SAE paper 710831. Reprinted with permission, © 1971, Society of Automotive Engineers, Inc.

expansion stroke, rather than near the top of the stroke, and cannot be expanded to the full compression ratio. This produces a hotter exhaust which will produce more combustion of hydrocarbons in the exhaust port and exhaust manifold and hence lower hydrocarbon emissions.[25,47] In addition, the peak cycle pressure will be lower and a smaller amount of hydrocarbons will be forced into the crevice volume by the pressure rise following combustion, and, as a consequence, the hydrocarbon emissions will decrease. This decrease is much more effective with lean air–fuel ratios than with rich ones. Although spark retard is an effective and inexpensive way to reduce hydrocarbon emissions, it does so at the expense of fuel economy (see Figs. 3 and 25). A large retard of the spark for hydrocarbon control is avoided in commercial practice. The EGR rate does not greatly influence the relation between hydrocarbon emissions and spark retard. Although a small amount of spark retard produces a very significant reduction in NO emissions (Fig. 14), it requires larger amounts of spark retard in order to very significantly reduce hydrocarbon emissions (Figs. 3 and 25).

Increasing the back pressure of an IC engine will decrease the flow from the cylinder of the last portion of the exhaust gas which contains high hydrocarbon emissions.[31,47] In addition, the increased back pressure will reduce the expansion

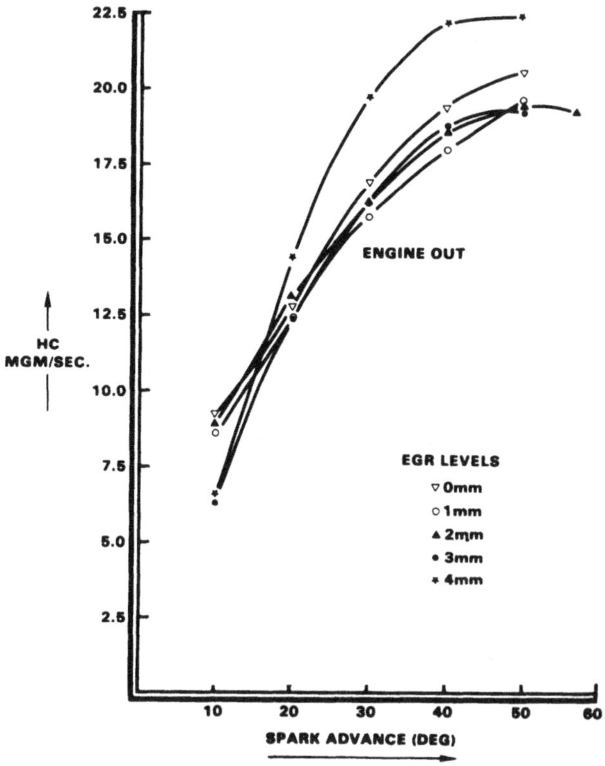

FIGURE 25. Exhaust hydrocarbon emission rate versus spark advance for various EGR rates. EGR valve opening measured in mm. 0 mm represents no EGR, 4 mm represents full EGR (from Ref. 27). Reprinted with permission, © 1980, Society of Automotive Engineers, Inc. See Fig. 15 for % EGR.

ratio of the burned gases which will increase the exhaust temperature and increase the fraction of the exhaust gas combusted in the exhaust port and manifold. Although increasing the exhaust back pressure of an engine is an effective and inexpensive way to reduce hydrocarbon emissions, it does so at the expense of fuel economy and power and, consequently, is not used to any significant extent.

There are, at this time, no known usable changes in engine design or operating conditions that will control the hydrocarbon emissions of present engines (engine out) to the legal limits and at the same time to preserve fuel economy and allow the use of EGR for NO control. However, research in the area of ring crevice and wall quench hydrocarbons (see, for example, Refs. 37 and 38), exhaust port and exhaust manifold burnup (see, for example, Refs. 30

and 46), improved cold start (see, for example, Ref. 17), etc., continue with the expectation that improvements in this area will continue.

In order to meet the hydrocarbon emission standards with good fuel economy, it is necessary to make use of some type of aftertreatment device such as catalysts.

VIII. POSTENGINE EXHAUST TREATMENT FOR EMISSION CONTROL

Since hydrocarbon, carbon monoxide emissions, and in many instances the nitric oxide emissions from present vehicle–engine combinations (using EGR) when run at the best fuel economy levels are not likely to meet the standards, they must be reduced to, or below, these limits by conversion in the exhaust system to carbon dioxide, water, and nitrogen. This reduction can be accomplished by either homogeneous oxidation in an appropriately designed exhaust reactor (thermal reactor) in the case of carbon monoxide and hydrocarbons, or by the use of catalysts in the case of all three pollutants.

A. Thermal Reactors

Initial work with thermal reactors made use of very rich (>3% CO) exhaust gas together with secondary air admission in order to assure complete removal of the pollutants by the high temperatures produced (>900 °C) by the combustion in the thermactor. See, for example, Refs. 54, 51, and Fig. 23. It has been estimated that the fuel economy penalty resulting from the application of this approach is 30–40%.[54,55] In addition, there is an added loss due to the large air pump required, and there was observed an almost insoluble material durability problem for the thermactor parts at the high operating temperatures, particularly in the presence of lead and phosphorus.

Because of these numerous problems, work on the thermal reactor concept in recent years has concentrated on operating thermal reactors in the lean region with the engine operating at air–fuel ratios higher than stoichiometry.[55–57] This can allow vehicle operation at high fuel economy, eliminate the air pump requirement, and increase the durability of the reactors. The problem becomes one of obtaining and maintaining high enough exhaust temperatures so that the reactions can occur in the residence time available. For residence times in the thermal reactor of from 50 to 100 msec, one requires temperatures of about 700 °C in order to reduce the hydrocarbon emissions to approximately 50% of the inlet value, and one can obtain essentially complete conversion at 750–800

°C.[58,55,49] At the lower temperature the product of combustion is largely CO, whereas at the higher temperature it is CO_2. The oxidation of CO is more difficult. Temperatures of ~850 °C are required for high CO conversions. The reactions in the thermal reactor are inhibited by lead in the fuel.[10]

A number of vehicles have been built to demonstrate the usefulness of the lean burn thermal reactor concept.[26,10] As can be seen from Table 5, the thermal reactor concept can be used, on the vehicles tested, to meet 1977–79 standards with only a modest loss from the best fuel economy. The data in Ref. 10 however, shows a gain in vehicle fuel economy (8–14%) in urban driving when catalysts are used at 8:3 CR in place of the thermal reactors used with leaded fuel at 9:2 CR. Attempts to meet 1981 standards with the lean reactor technique would most likely lead to considerably larger fuel economy losses, and possibly poor vehicle driveability. The performance of these thermal reactors is very critically dependent upon the exhaust gas temperature and every effort, such as the use of port liners, insulation, and modest spark retard, must be taken in order to obtain the performance shown. A loss of 50 °C in exhaust temperature can very significantly increase hydrocarbon and CO emissions. Efficient, high compression ratio, unthrottled engines such as stratified charge or diesel engines cannot use this approach because of low exhaust temperatures. Because catalytic devices operate very efficiently and over a much wider temperature range (above 250 °C) and because they can be used to catalytically control NO emissions, they have been used by most automobile manufacturers as the principal component of their emission control systems.[29] These catalysts are very active at normal exhaust temperature (400–800 °C). This allows good fuel economy to be preserved even when controlling to low emission levels in the exhaust because spark retard is either not used or used to only a small extent.

TABLE 5. Lean Thermal Reactors with EGR for NO Control

Ref.	Vehicle wt. (lb)	CR[c]	A–F	Reactor volume	Emissions g/mile HC	CO	NO	CVS fuel economy mpg	Fuel[a] economy loss
10	4500	9.2	16–17	270 in.^{3}	0.85–1	6–8	1.4–1.6	13.5	6–19%
10	4500	8.3	16–17	270 in.^{3}	0.85–1	6–8	1.4–1.6	13.7	—
10	3000	9.2	16–17	200 in.^{3}	0.85–1	6–8	1.4–1.6	20.3–19.1[b]	6–19%
10	3000	8.3	16–17	200 in.^{3}	0.85–1	6–8	1.4–1.6	19.5	—
57	4500	9.3	17	270 in.^{3}	0.9	5	1.7	13.9	2–5%

[a] Loss from best spark with the EGR levels used (5–15% EGR).
[b] The 19.1 value is for leaded fuel which increased hydrocarbon emissions and required spark retard to compensate. At the same emission levels a catalyst-equipped car with 8.3 CR gave 22 mpg and was calibrated for a 3% loss in fuel economy from best spark.
[c] Cars have equal performance at each CR.

B. Catalytic Exhaust Gas Treatment

1. General Comments

Catalytic treatment of the exhaust gas can be used to promote the reactions of CO and hydrocarbons with oxygen and water, and the reactions between NO and reducing agents to convert the NO to nitrogen or other products (NH_3, N_2O). Catalysts have now proven to be very effective in enabling the automotive companies to meet the low emission standards and to preserve good fuel economy for passenger cars. In order to obtain good fuel economy it is necessary to maintain high catalyst activity during 50,000 miles of durability testing since the emission standards must be met at this time. In the course of the AMA durability test the catalyst is exposed to temperatures up to 800–900 °C (higher if engine malfunctions occur) and to catalyst poisons such as compounds of lead, zinc, phosphorous, and sulfur which degrade the catalyst performance.[(60)] During the early periods of catalyst development for automotive use (1955–1964, 1969–1974) it was found that the highly leaded fuel available at the time severely deactivated the catalysts tested both in a physical and chemical sense. This can be particularly severe in some modes of vehicle operation.[(61)] In order to avert the deactivation of catalysts by leaded fuel, the U.S. Environmental Protection Agency specified in 1972 that gasoline with no more than 50 mg/gal of lead be available for use with cars produced in the 1975 model year and beyond. This necessitated a lowering of engine compression ratio ~1–1.5 units to ~8 in order to accommodate the resultant gasoline of lower octane number (91,92 ON). The small amount of lead present in fuels today (1981) of 9–15 mg/gal is due to contamination in the process of transportation from the refinery to the service station, since both leaded and nonleaded fuel are transported over the same system. As the percentage of use of leaded fuel in the United States (~50% in 1980) decreases, it is expected that the lead level in unleaded fuel will reach essentially zero in 5–10 years. The present lead level in unleaded fuel produces little or no deactivation of oxidation catalysts but can produce significant deactivation of catalysts used for NO removal.[(62)]

In addition to chemical deactivation by lead and other compounds in the fuel and motor oil, the catalyst can be deactivated at high temperatures by thermal sintering of the platinum crystallites,[(63)] resulting in a loss of the platinum surface area necessary for catalytic activity. In normal use the *maximum* catalyst temperature expected is 800–900 °C, but the predominant operating temperature is from 400–600 °C. The catalyst deactivation can be substantial during the first 4000–12,000 miles of vehicle use, but it is often small from 12,000–50,000 miles of vehicle use (Fig. 26) unless vehicle malfunctions occur which expose the catalyst to temperatures higher than 1000 °C and which in some cases, can be very harmful (total ignition failure, for example).

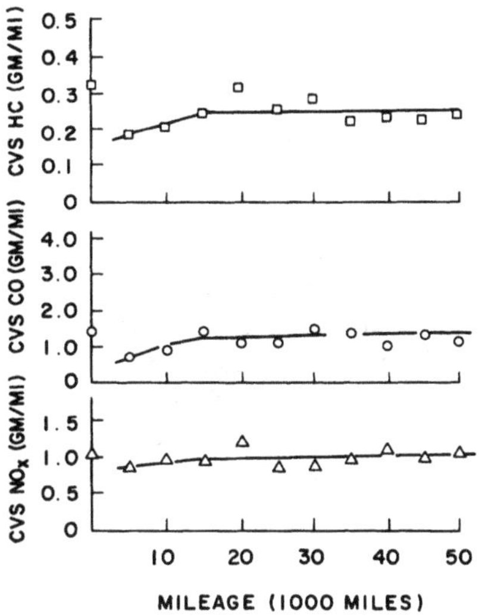

FIGURE 26. FTP emissions from a 5.8-liter 1979 three-way-catalyst equipped vehicle as a function of mileage (from Ref. 75). Reprinted with permission, © 1981, Society of Automotive Engineers, Inc.

As will be shown later, the emission control systems in use today require very high catalyst activity during 50,000 miles use in order to meet emission standards and allow the vehicle to operate at high fuel economy. Because of this, the ability to calibrate a vehicle for high fuel economy is indirectly related to fuel contaminant level, oil consumption (source of Zn and P), catalyst design (more rugged catalysts desirable), and the preservation of modest catalyst operating temperatures. The avoidance of all modes of catalyst deactivation, while possible in theory, entails economic or other constraints. Catalyst activity must be preserved not only for performance at normal operating temperatures but also for the cold-start portion of the FTP. Aged catalysts take longer to light off.

2. *Catalyst Systems*

One can meet the emission standards utilizing catalytic exhaust gas treatment in two different ways. In the first way one can control the NO primarily by employing changes in engine design that will enable one to use EGR effectively[64] (or by very lean operation) and by modest use of spark retard. The hydrocarbons and CO are removed by catalytic oxidation. The second approach makes use of a catalyst or a pair of catalysts together with proper control of the exhaust air–fuel ratio to stoichiometry or rich of stoichiometry in order to remove all three pollutants by the appropriate reactions (see Sec. 4). In the latter approach partial control of NO (and infrequently hydrocarbons) by EGR and spark retard has sometimes been used to assist the catalytic removal of these by reducing the engine-out concentration of these pollutants.

The first approach was used by most production vehicles in the 1975–1980 model years. The latter approach has been used by most production vehicles in the 1981 production year where the legal NO_x emission level dropped from 2 to 1 grams per mile. Some vehicles, however, still use the first approach[(64)] and meet the 1 g NO_x standard with high fuel economy. These vehicles exhibit good fuel economy for a number of reasons (see Table 6), and, because of the fast burn characteristic, they show a lowering in specific fuel consumption as the EGR rate is increased (maximum: 20%).

The active catalytic material (Pt, Pd, Rh, and base metal oxides) used for CO and hydrocarbon oxidation or NO reduction must be held in the vehicle exhaust system in such a manner that the mass-transfer characteristics between the gas phase and the active catalytic surface are sufficient to allow near 100% catalytic conversion assuming high intrinsic activity of the catalytic material itself.[(66)]

Two configurations have been used commercially. One system employs a ceramic honeycomb structure or monolith held in a metal can in the exhaust.[(67)] The other system employs pellets held in a metal can between two retaining screens [~2 in. apart for GM oxidation catalysts[(68)]]. The walls of the honeycomb passageways are coated with gamma alumina containing the catalytic material, and the outer layer of the pellets (~3 mm diameter; gamma of other transition alumina) is coated with the catalytic material.

3. *Oxidation Catalyst Performance*

Oxidation catalysts typically contain platinum and palladium (1–2 g per car) and have been developed to the point where they can, after 50,000 miles

TABLE 6. Breakdown of Improved Fuel Efficiency in the Datsun 510 and 200 SX with NAPS-Z from Ref. 65

Type of improvement	Percentage increase of fuel efficiency
Qualitative improvements of engine proper	10%
Vehicular optimization paralleling increase in engine output (final gear ratio changed from 3.545 to 3.364)	3%
Use of air pump discontinued and EAI adopted	2%
Optimization of ignition timing by adopting catalyst	10%
Shutoff of fuel supply during deceleration	3%
Improvement in fuel consumption during idling	1%

of durability aging, give overall conversions of CO and hydrocarbons during the FTP, including the cold start, of 80–90% and 80–85%, respectively. The higher value requires exceptionally well-controlled cold-start performance not achievable on all cars. This is for well-calibrated hardware and for a durability test where no malfunctions have occurred. A very large fraction of the total amount of CO emitted during the FTP is emitted during the cold start. Catalytic CO removal when the catalyst is hot (>400 °C) can be very high, between 95 and 99%, and is controlled by the mass-transfer characteristics of the catalyst system rather than by the intrinsic activity of the catalytic material itself. The removal of the aromatic and unsaturated hydrocarbons can also be nearly as efficient because of the strong chemisorption of these species on the catalyst surface. However, the saturated hydrocarbons, particularly those of lower molecular weight and most particularly methane, oxidize catalytically more slowly. Their conversions when oxidized at 400–500 °C can range from 10–70% depending upon the degree of abuse that the catalyst has seen during the durability cycle.

Table 7 shows the oxidation catalyst efficiency during that portion of the FTP following the cold start, which is required in order to meet the emission standards for CO and hydrocarbons required for the model year 1981 and beyond. The efficiency is shown for several assumed engine-out and cold-start emission values that cover a range typical of today's vehicles. As can be seen, the required

TABLE 7. Required Efficiency of Oxidation Catalyst after Cold Start

	Emissions g/mile					
	HC[a]	CO	HC	CO	HC	CO
Standards	0.41	3.4	0.41	3.4	0.41	3.4
Engineering objective[b]	0.29	2.4	0.29	2.4	0.29	2.4
Cold-start contribution[c]	0.2	1.5	0.15	1.0	0.10	0.5
Allowable after cold start	0.09	0.9	0.14	1.4	0.19	1.9
Total cycle feedgas	2	10	2	10	2	10
Hot catalyst efficiency required	95%	89.4%	92.5%	84.4%	90.0%	80.0%
Total cycle feedgas	2.5	15	2.5	15	2.5	15
Hot catalyst efficiency required	96%	93.3%	94.1%	90%	92.1%	86.9%
Total cycle feedgas	3	20	3	20	3	20
Hot catalyst efficiency required	96.8%	95.1%	95.1%	92.6%	93.4%	90.3%
Total cycle feedgas	3.5	25	3.5	25	3.5	25
Hot catalyst efficiency required	97.3%	96.2%	95.8%	94.2%	94.4%	92.2%

[a] HC = hydrocarbons.
[b] 70% of standards to account for production and test variability.
[c] A reasonable range of cold-start values.

CO oxidation efficiency over a hot catalyst in the worst case is ~96%. This can be met by catalyst design provided excess oxygen is always present and provided the catalyst has not been severely overtemperatured by engine malfunctions or exposed to excess poisons during the durability test.

There can, however, be considerable difficulty in meeting the high hydrocarbon oxidation efficiencies required after 50,000 miles durability. The allowable percent of the inlet hydrocarbons that can pass through the hot catalyst and not be oxidized varies from 10% for the best case to ~3% for the worst case. The engine-out hydrocarbons from present vehicles can be assumed to be 20% slow-burning hydrocarbons (saturated hydrocarbons) and 80% fast-burning hydrocarbons. For vehicles with colder exhausts (stratified charge or high EGR rate vehicles) one often sees higher hydrocarbon levels but a lower fraction of low molecular weight saturated hydrocarbons. With proper catalyst design with respect to mass transfer (the hydrocarbons diffuse more slowly than CO),[66] high oxidation conversion efficiencies of the *fast-burning* hydrocarbons of >95% can be obtained, but this alone is not sufficient to meet the requirements of allowing only ~3–10% of the inlet hydrocarbons to pass through the catalyst unoxidized. For the best case we must oxidize at least 50% of the slow-burning hydrocarbons. In practice a value of 50% conversion of the slow-burning hydrocarbons during the part of the FTP following the cold start is a reasonable one after 50,000 miles durability. If the hydrocarbon standards cannot be met with the catalytic system employed, then spark retard can be used to assist the hydrocarbon removal with a resultant loss in fuel economy.

The platinum contained in the catalysts used today is the principal component responsible for oxidation of the slow-burning hydrocarbons. Palladium is very much less active, particularly when highly dispersed, although it is highly active for oxidizing CO and fast-burning hydrocarbons. Alloying palladium with platinum reduces the activity of the platinum. During the 50,000-mile durability test, the platinum surface area of the catalyst decreases by one order of magnitude or more[63] which principally increases the lightoff temperature and decreases the rate of oxidation of slow-burning hydrocarbons. At normal operating temperatures the rates of oxidation of CO and fast-burning hydrocarbons are mass transfer limited and show little decrease in normal vehicle use since the catalysts are designed (noble metal content) to function effectively with the reduced surface area. Our ability to obtain high fuel economy with emission standards is related to the catalyst platinum content of today's catalyst formulations and the effectiveness of its utilization.

Catalysts that are located close to the engine exhaust manifold reach lightoff temperature during cold start more quickly than those located further down the exhaust system. This is desirable for the reduction of cold-start emissions. Catalysts located close to the exhaust manifold, however, operate at a higher temperature than those located further downstream. For example, catalysts located

10 in. from the exhaust manifold (close mounted) will run ~100 °C hotter than those mounted 40–50 in. from the exhaust manifold (underbody catalysts). This increased average-operating temperature can cause the catalysts in the close-mounted position to lose cold start and saturated hydrocarbon oxidation activity during durability testing to a greater extent than those mounted underbody. The extent to which this occurs depends upon how well the catalyst is designed to withstand thermal abuse. It is desirable that the increase in lightoff temperature during durability testing be less than 100 °C. Close-mounted catalysts can aid the vehicle calibrator in meeting emission standards with good fuel economy if their design is adequate for that environment. This requirement, however, is not met by all auto exhaust catalysts.

We can conclude from what has been written that in order to meet emission standards and not to decrease fuel economy when using only oxidation catalysts, the emission control system must possess several specific characteristics. First, the engine must be able to accept large amounts of EGR for NO control and in doing so must preserve the correct in-cylinder burning rate for good fuel economy,[(23,64)] and the increase in hydrocarbon emissions due to the EGR should be minimal. Secondly, the cold-start emissions must be kept low by well-performing fuel-metering hardware, a lean air–fuel cold-start strategy, and the use of catalysts with low lightoff temperatures properly located. Finally, the catalyst activity for the oxidation of low molecular weight saturated hydrocarbons must be very good, and large transient spikes of hydrocarbons and CO in the exhaust avoided. Excess oxygen must always be present and the catalyst environment and design and engine reliability be such that catalytic activity is preserved during the durability cycle.

If we are unable to meet emission standards because of lack of performance in any of these areas, then the use of spark retard, increased back pressure, etc., can further reduce the hydrocarbon or NO emissions at the expense of fuel economy, CO emissions during the FTP will be predominantly produced during the cold start.

4. *Catalytic NO Removal*

In order to remove NO catalytically, the engine must be operated at or slightly rich of stoichiometry so that sufficient reducing agents (CO, H_2, hydrocarbons) are present in the exhaust gas for the reduction process to proceed (see Table 8). This is not the optimum air–fuel ratio for best fuel economy. The product of the NO reduction can be either nitrogen or ammonia. At stoichiometry the product is principally nitrogen, but as one operates fuel rich from stoichiometry, more ammonia is made. It is desirable to use catalysts that are selective in the production of nitrogen rather than ammonia. It has been found[(29,69)] that this requires catalysts containing rhodium in addition to platinum. Ruthenium is

TABLE 8. Possible NO Reactions under Reducing Conditions

1. $NO + CO \rightarrow \frac{1}{2}N_2 + CO_2$
2. $2NO + 5CO + 3H_2O \rightarrow 2NH_3 + 5CO_2$
3. $2NO + CO \rightarrow N_2O + CO_2$[a]
4. $NO + H_2 \rightarrow \frac{1}{2}N_2 + H_2O$
5. $2NO + 5H_2 \rightarrow 2NH_3 + 2H_2O$
6. $2NO + H_2 \rightarrow N_2O + H_2O$[a]
7. NO + hydrocarbons $\rightarrow N_2 + H_2O + CO_2 + CO + NH_3$

[a] Occurs at 200 °C, which is below that found in usual auto exhausts.

also effective but can be emitted from the exhaust as RuO_4 and RuO_3 which constitutes a health hazard. Although attempts at stabilization have been made,[70] these have not proven sufficiently effective in practice. The platinum-to-rhodium ratio in commercial use in auto catalysts is often ~12, although the ratio that occurs in the South African mines is ~17–19. A high content of rhodium in the catalyst is desirable in order to suppress ammonia formation but is economically difficult to support.

It has been found that if an engine can be operated at all times with an air–fuel ratio at or near the stoichiometric point (±1% or better), then both NO reduction and CO and hydrocarbon oxidation can be done by one catalyst bed and with good selectivity to nitrogen. In effect the catalyst is used to bring the exhaust gas composition at stoichiometry to equilibrium or very close to equilibrium at the exhaust temperature. This will correspond very nearly to an exhaust composition of only CO_2, H_2O, and N_2.[71] Enough reducing gases will be present to reduce the NO and enough oxygen to oxidize the CO and hydrocarbons. Such a catalyst is called a three-way catalyst (TWC), since it removes all three pollutants simultaneously (see Fig. 27). Such a system usually requires secondary air addition during the cold-start period where the catalyst lightoff is expected. For these systems it is customary to operate with air added to the exhaust during the first ~100 sec or less and then to stop the air addition and switch to stoichiometric A–F control for the remainder of the FTP. No removal of NO occurs during the time with air addition when fortunately the NO production by the engine is not high.

The TWC catalytic system requires very precise air–fuel ratio control at the stoichiometric point or very slightly rich of this point. This is usually done by use of a zirconia (or titania) sensor[72,73] in the exhaust gas to sense the stoichiometric point together with a fuel-metering system (carburetor or injector) controlled by feedback from the sensor in order to maintain stoichiometry. This system requires the maintenance of the TWC activity and the maintenance of the performance of the sensor and fuel-metering hardware with its control system during the 50,000-mile durability test. Under these conditions a single-bed TWC

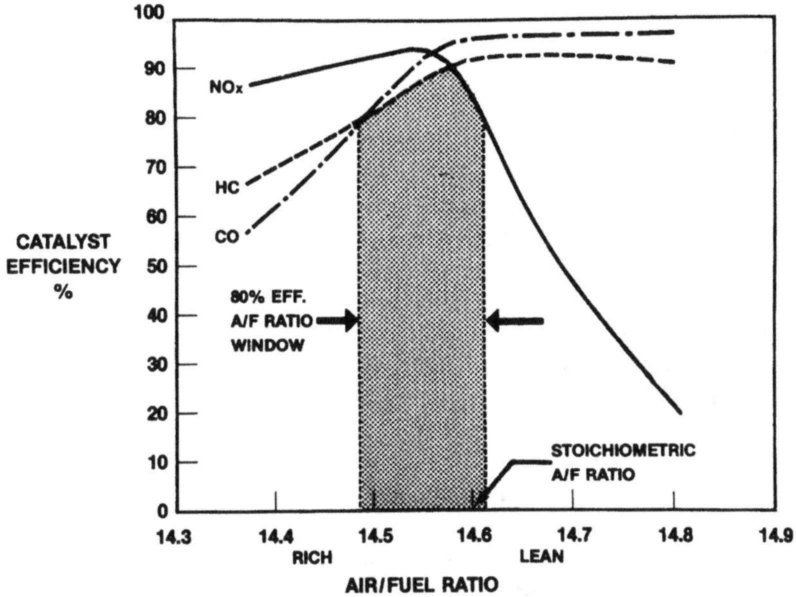

FIGURE 27. Conversion of NO, CO, and hydrocarbons for a three-way catalyst (TWC) as a function of the air–fuel ratio.

catalyst can be employed to meet emission standards with good fuel economy within the air–fuel ratio constraints of this system. Some EGR will be necessary to assist in NO removal since the catalytic NO removal may not be sufficient to reduce NO from an uncontrolled value of as high as 6 g/mile to below 1 g/mile, and the CO removal will be marginal for the 3.4 g/mile standard unless very precise A–F control is used. This system is being used commercially and was first implemented commercially in vehicle use by Volvo.[74]

It often happens, however, during vehicle development that the control to stoichiometry is not precise enough, the engine-out emission levels are sufficiently high at good fuel economy, and the catalyst activity where all three pollutants are removed, sufficiently low after 50,000 miles durability (less than 50 or 60% for one or more pollutant), so that emission standards cannot be met. One solution to this problem is to add to the system a second catalyst bed following the TWC. Secondary air is added before this second bed by means of an air pump or air addition produced by exhaust pulsations (pulsed air system).[6] In this second bed CO and hydrocarbons are removed to meet their emission standards by oxidation. This allows the air–fuel ratio control set point to be set slightly rich to maximize NO conversion at the expense of poorer CO and hydrocarbon control by the TWC since this will be assisted by the following oxidation catalyst. This system has been used by a large number of U.S. vehicles designed to meet the 1 g/mile NO_x standard. Unfortunately, such a system has a serious problem with respect to NO control. As stated before, the first reduction catalyst can reduce a portion of the NO to ammonia rather than nitrogen if one operates fuel rich from the stoichiometric point. The second oxidation bed will very nearly quantitatively oxidize this ammonia back to NO so that the net reduction of NO across such a two-bed system is less than across the TWC alone in proportion to the amount of ammonia made by the TWC. This problem will require more EGR or spark retard for NO control. The retard will cause some loss in fuel economy or the additional EGR may do so also if the engine in-cylinder burn rate decreases significantly. A dual-bed system will insure a substantial degree of CO and hydrocarbon removal for less accurate air–fuel ratio control if the control system allows the air–fuel ratio of the exhaust to spend too much time rich of stoichiometry, but at the expense of poorer NO control. Operation significantly rich of stoichiometry will, of course, result in a fuel economy penalty. With our present catalyst formulations, it is difficult to make up for insufficient air–fuel ratio control by a dual-bed system. Good control is necessary for removal of all three pollutants in a single-bed system and for good control with a dual-bed system. Precise air–fuel ratio control systems, however, add to vehicle cost and are susceptible to misadjustment either accidentally or through changes of component characteristics during aging.

It would be desirable to have available TWCs that do not produce ammonia rich of stoichiometry and oxidation catalysts effective for CO and hydrocarbon oxidation that do not oxidize ammonia to NO. Such developments would reduce the problems encountered in obtaining emission standards at good fuel economy. The use of high concentrations of rhodium (several times greater than the present 0.015 wt %) will help to suppress ammonia formation rich of stoichiometry but is expensive and of limited availability. The use of ruthenium will very greatly suppress ammonia formation,[(69,70)] but its potential volatility mitigates against its use. Research in these areas is continuing.

Low emission levels and good fuel economy in these systems for catalytic NO control are very dependent upon the ability of the fuel-metering system to control the engine exhaust to stoichiometry. Catalysts are available that have high conversions of the three pollutants at the stoichiometric point. Figure 28 shows the performance of a TWC after aging 50,000 miles under conditions where fuel contaminants are low,[(75)] where the maximum catalyst temperature is below 800 °C, and where malfunctions were absent.

The data were obtained with an exhaust composition that did not vary with time (steady-state conversions). Figure 29 shows the conversions of NO, CO, and hydrocarbons observed with a feedgas composition that changes in a periodic way typical of engine operation with feedback control of a carburetor from an

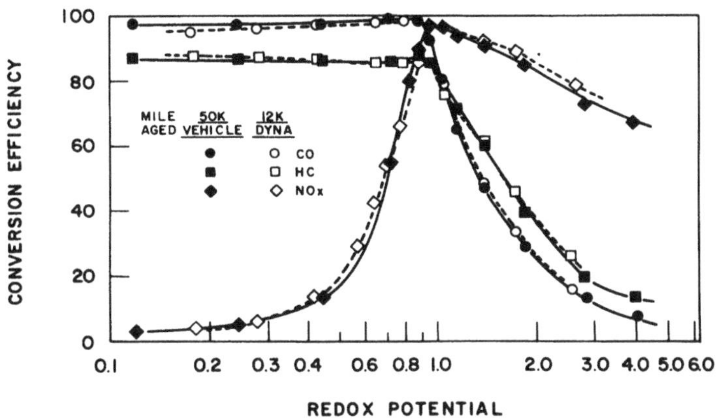

FIGURE 28. Performance of TWCs aged 50,000 miles on vehicle and 12,000 miles on dynamometer. Performance measured at steady state *without* air–fuel ratio modulation. 800 °F, 60,000 1/h space velocity (from Ref. 75). Reprinted with permission, © 1981, Society of Automotive Engineers, Inc.

air–fuel ratio sensor in the exhaust manifold outlet. The sensor response at stoichiometry represents a delta function in that there is an abrupt change in sensor output voltage as the composition of the exhaust gas passes through stoichiometry. The fuel-metering system follows this change in sensor voltage with a corrective action but with a time delay. This results in the exhaust gas composition fluctuating around the set point in a periodic manner, usually with a frequency of 0.5–1 Hz, and with the amplitude of the maximum deviation from the set point of usually from 0.3–1 air–fuel ratios. The maximum TWC performance is obtained when the exhaust gas composition does not deviate from the set point at stoichiometry or very close thereto. High-amplitude deviations and low-cycle frequencies will give poorer TWC performance. The catalyst characteristics can be such that in an exhaust composition produced by combustion fuel rich of stoichiometry it can store on its surface enough reducing capacity to reduce NO to nitrogen for a short time period after the exhaust gas composition moves to lean of stoichiometry. A similar phenomenon (oxygen storage) can be seen as the exhaust gas composition moves from lean to rich of stoichiometry. This characteristic can be intentionally incorporated into the catalyst design and enables the catalyst performance under cyclic conditions to be substantially better than one would expect from a superposition of steady-state values.[76]

As can be seen from Fig. 29, for this particular 50,000-mile aged commercial honeycomb catalyst, one can expect, from a warmed up catalyst, at typical

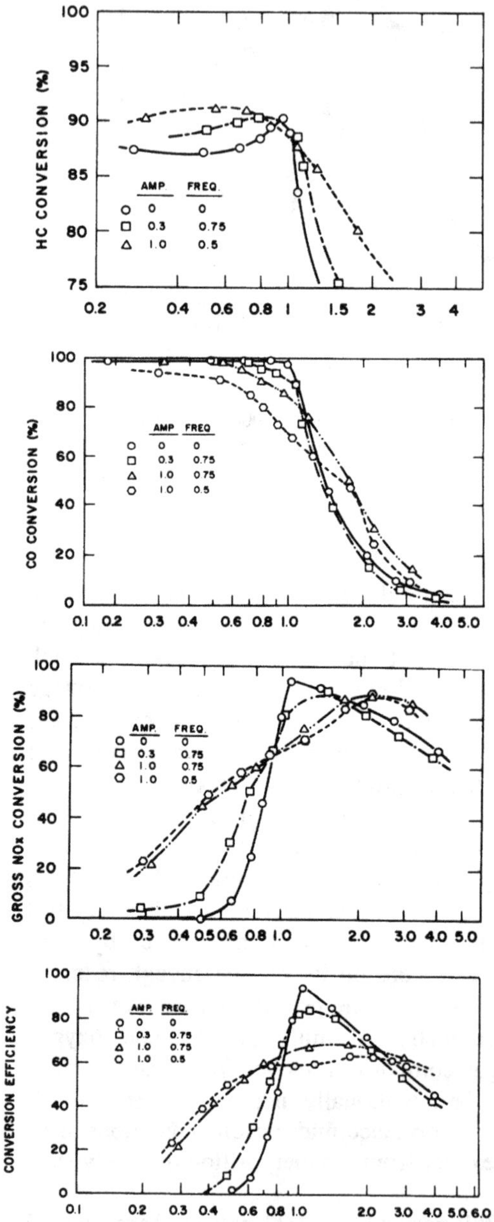

FIGURE 29. Effect of air–fuel ratio modulation on conversion of CO, NO, and hydrocarbons. 1000 °F, 60,000 1/h space velocity. A, hydrocarbon conversion; B, CO conversion; C, gross NO_x conversion (ammonia conversion included); and D, net NO_x conversion (ammonia conversion excluded). (From Ref. 75). Reprinted with permission, © 1981, Society of Automotive Engineers, Inc.

operating temperatures and flow rates and at stoichiometry under cyclic conditions, hydrocarbon conversions of 87–89%, CO conversions of 60–97%, and NO conversions of 65–70% depending upon the amplitude and frequency of the air–fuel control system. Table 9 shows the allowable engine-out emission that one must obtain in order to meet the listed emission standards with the assumed cold-start performance. As can be seen, for this TWC only system and under the simplifying assumptions of using conversions at an average exhaust flow rate and temperature, the allowable engine-out NO_x is ~2 g/mile, and hydrocarbons, ~2.5 g/mile. The allowable CO value is as low as ~10 g/mile for the unfavorable case but very dependent upon the degree of air–fuel ratio control available. For poorer cold-start performance, air–fuel ratio control, and catalyst efficiency the allowable engine-out levels will, of course, decrease and be harder to obtain. For vehicles weighing less than ~3000 lb it is possible with effective fuel-metering systems to obtain the engine-out NO and hydrocarbon emissions given above with little loss of fuel economy from the unemission-controlled case using EGR for NO reduction. For large vehicles (4500–5000 lb) this is difficult if not impossible at this time, but the levels can be met if a loss in fuel economy is acceptable.

IX. SUMMARY: PASSENGER CARS

Considerable time in this chapter has been spent in a very detailed description of the sources of and methods of removal of the three pollutants. This knowledge is necessary if one is to understand, in depth, the relation between emissions

TABLE 9. Allowable Engine-Out Emissions, TWC System Only

Pollutant	Emission standard g/mile	Engineering[a] objective g/mile	Cold-start[b] contribution g/mile	To be removed catalytically g/mile	Warmed-up[c] catalyst efficiency %	Allowable engine-out g/mile
NO	1.0	0.7	0.07	0.63	65–70	2.1–2.4
NO	0.4	0.28	0.07	0.21	65–70	0.9–1.0
CO	7.0	4.9	1.0	3.9	60–95	10–80
CO	3.4	2.4	1.0	1.4	60–95	5–30
Hydrocarbon	0.41	0.29	0.15	0.14	87–89	1.2–1.4

[a] 70% of standards to account for production and test variability.
[b] Typical value 2.3-liter engine, carburetted. This can vary from 0.1–0.2 for HC.
[c] An approximate average value that might be expected for a well-designed catalyst aged without malfunctions at typical temperatures and flow rates (see Figs. 28, 29). First number is air-fuel modulations 0.5 Hz, 1 A–F amplitude. Second number is steady-state data.

and fuel economy. Consider, for example, the case where EGR is used to control NO and an oxidation catalyst is used to control CO and hydrocarbons. The problem areas are the correct use of EGR so as not to lose efficiency (fast burn) and hydrocarbon control. Our carburetted vehicles with catalysts are marginally able to control hydrocarbons to 0.41 grams per mile without a fuel economy loss (depending on vehicle size) but the hydrocarbon control problem relates back, as we have seen, to many other problem areas (see Table 10). For the TWC system more problem areas emerge in addition to the ones in Table 10. Improvements in any of these areas can make the problem of meeting emission standards and maintaining good fuel economy less difficult.

The use of aftertreatment devices has enabled the automobile companies to produce vehicles which meet today's emission standards and that have good fuel economy, albeit at a significant cost. To meet these emission standards by only changes in engine calibrations would entail a severe fuel economy penalty. As

TABLE 10. System Variables that Affect Emissions. Oxidation-Catalyst-Only System. Indirectly Related to Fuel Economy

I. Hydrocarbons from the engine related back to:
   a. Hydrocarbons produced in ring area or by wall quench and oil absorption (ring area design, elimination of crevice volumes, blowby; direct cylinder injection).
   b. Hydrocarbon burnup in port (port design).
   c. The use of port extracted EGR (EGR strategy).
   d. Engine CR and backpressure level (fuel availability, exhaust system design).
   e. Valve overlap, engine deposits (engine design).
   f. Cold-start hydrocarbons, deceleration and acceleration hydrocarbon spikes, cylinder-to-cylinder variation (fuel-metering system).
   g. Hydrocarbon increase with EGR (NO control wanted; mixing of fresh charge and exhaust gas).
   h. Spark calibration.

II. Control of NO emissions from engine:
   a. Maintenance of correct burning rate when EGR added (increased turbulence; possibly dual ignition, A–F)
   b. EGR temperature, fresh charge temperature both of which can affect NO and hydrocarbon emissions as well as fuel economy.
   c. Possible use of stratified charge.
   d. Spark calibration.

III. CO emissions
   a. Cold-start strategy and hardware (fuel-metering system).

IV. Catalyst Efficiency
   a. Noble metal content, mass-transfer characteristics, ability to oxidize slow-burning hydrocarbons (catalyst design).
   b. Resistance to thermal sintering and loss of washcoat, cold-start behavior (catalyst design together with system design).
   c. Catalyst thermal environment in use, and engine malfunctions (system design).
   d. Contaminate level in fuel and oil (fuel under regulations).

has been shown, these aftertreatment devices (exclusively catalysts for the low emission levels) are required and perform well, although they are pressed to the limit of their performance capability.

It is, however, difficult to quantify the extent to which the necessity for meeting the emission standards has caused a decrease in fuel economy from that possible if there were no emission constraints. There is very little data available in which a comparison in fuel economy performance is made between a production vehicle and the same vehicle optimized for maximum fuel economy with no emission constraints and at equal performance and for various vehicle weights (see Ref. 4, for example). The ratio between unconstrained fuel economy and emission-constrained fuel economy, in practice, is very vehicle dependent. It will depend upon desired emission levels, vehicle weight, engine design, type of catalyst system used and catalyst efficiency, the precision of the fuel-metering system and its cold-start performance, overall calibration strategy, transmission characteristics, driving cycle and the driving habits of the driver, etc.

There is, of course, the rather well-defined loss in energy efficiency due to the removal of lead and lowering of the compression ratio (estimated ~3% loss in engine efficiency or ~5% in vehicle fuel economy at equal performance for 1 CR decrease) and the energy loss in the use of air pumps where required (estimated ~2% fuel economy loss). The loss due to dashpots, where used, is less well defined (estimated <3% fuel economy loss). A loss due to the use of EGR is avoidable (in fact, a gain can be achieved but less than with equivalent air dilution). If, however, a particular engine accepts EGR poorly (slow-burning rate), the use of substantial amounts of EGR can lead to a loss in fuel economy. The gain in fuel economy with EGR is only in comparison to the no EGR case because of the lower pumping work required and is, of course, possible to implement in a vehicle with no emission constraints also by using lean operation.

The use of TWCs entail a fuel economy loss[(22)] compared to lean operation. The exhaust CO concentration is controlled between 0.7 and 1.0 mole % with good feedback control for TWC operation. The heating value of 1.0% CO represents ~3% of the heating value of the fuel, and this represents one source of the fuel economy loss. For lean operation the exhaust CO is very low (<0.1%).

With the same type of engine design, calibration, and catalyst position it is more difficult to meet emission requirements and preserve high fuel economy as the vehicle size and weight increases, and engine size (same design) increases to keep the performance constant. The amount of energy (fuel) used to travel 1 mile during the FTP increases with vehicle weight (~0.09 hp hours of energy required per mile, per 1000 lb). Cold-start emissions increase because of the increased thermal inertia of the engine, exhaust manifold, and catalyst system (at constant noble metal loading). For this illustrative system the amount of the three pollutants emitted per gram of fuel consumed is expected to be approximately constant. A car with a FTP fuel economy of 15 mpg will require twice the catalyst efficiency (with respect to material not converted) or more depending upon the cold-start contribution increase as a vehicle with a fuel economy of 30 mpg. Since our catalysts are operated close to their performance limit (after aging), this increase in performance may not be possible within the economic constraints imposed by competition and may require the use of spark retard for partial NO and hydrocarbon control with a resultant fuel economy penalty. For vehicles weighing less than very approximately 3000 lb and with 1981 emission standards, this is probably not the case.

From 1975 to 1981 the weight-normalized fuel economies of vehicles produced for use in California were smaller (in mpg) than those produced for use in the other 49 states. See Fig. 30 from Foster, Murrell and Loos.[77] This would indicate that in these model years there was a fuel economy penalty connected with meeting the California emission standards as compared to meeting the federal (49 state) standards. As can be seen in Table 1, the California standards are lower from 1975–1979 in each of the model years for all three pollutants than the federal standards, and from 1979–1981 California has required lower NO_x standards. However, with the advent of catalytic and EGR techniques which minimize the loss in fuel economy for NO control and the fact that the federal and California standards more closely approach each other, the difference in fuel

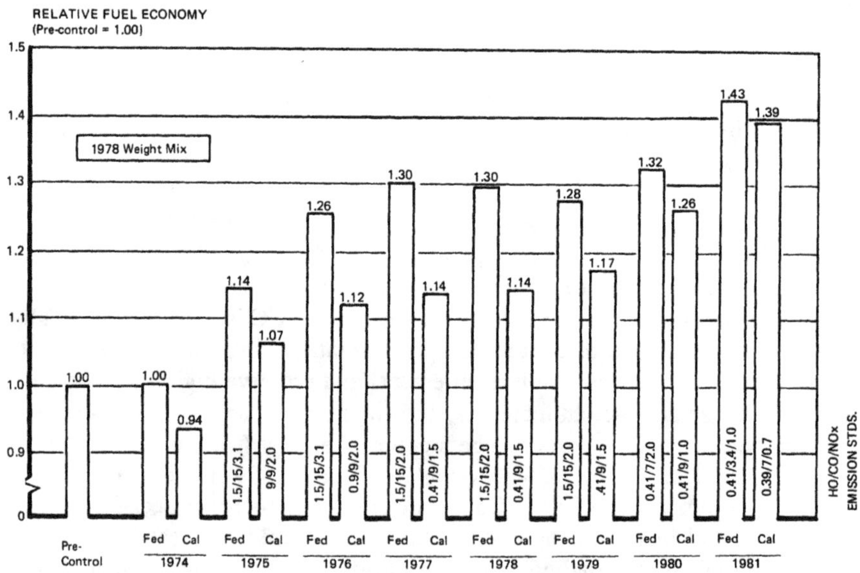

FIGURE 30. Relative fuel economy of the 49-state and California fleets, weight normalized to 1978 weight mix (Ref. 77). Reprinted with permission, © Society of Automotive Engineers, Inc.

economy between vehicles produced for California and the other 49 states has become smaller since 1979 and in 1981 was the smallest for any year shown.

X. TRUCK EMISSIONS AND FUEL ECONOMY—GASOLINE-POWERED VEHICLES

The emission standards for light-duty and heavy-duty trucks are shown in Table 11. The Clean Air Act of 1970 requires a 75% reduction of heavy-duty NO_x emissions from 1969 levels (as compared to 90% for passenger cars). The emission test procedures and durability test for light-duty trucks are similar to those for passenger vehicles. The present heavy-duty truck emission test cycle is a nine-mode engine test cycle run on an engine dynamometer at ~2000 rpm.[78] This nine-mode test has no cold-start portion and the test modes are defined by

TABLE 11. Light- and Heavy-Duty Truck Emission Standards

Light Truck <8500 lb Gross Vehicle Weight			
Federal	Hydrocarbons	CO	NO_x
1979–83	1.7 g/mile	18 g/mile	2.3 g/mile
1984–85	0.8 g/mile	10 g/mile	2.3 g/mile
California	NMHC[a]	CO	NO_x
1982			
<3999 IW	0.39 g/mile	9.0 g/mile	1.0 g/mile
4000–5999 IW	0.50 g/mile	9.0 g/mile	1.5 g/mile
6000–8500 IW	0.60 g/mile	9.0 g/mile	2.0 g/mile
1983			
<3999 IW	0.39 g/mile	9.0 g/mile	1.0 g/mile
4000–5999 IW	0.50 g/mile	9.0 g/mile	1.0 g/mile
6000–8500 IW	0.60 g/mile	9.0 g/mile	1.5 g/mile
Heavy Truck >8500 lb Gross Vehicle Weight[b]			
	Hydrocarbons	CO	NO_x + hydrocarbons
1980–83 Federal	1.5	25	10
	OR no spec HC std	25	5
1982–83 California	1.0	25	6
	OR no spec HC std	25	5
1984–86	To be determined: Standards may be determined that do not require catalysts.		

[a] Nonmethane hydrocarbons.
[b] Standards in grams per Bhp-hr.

specifying the intake manifold pressure during seven modes and by using an idle and throttle cutoff mode. What is thought to be a more realistic transient heavy-duty truck emission cycle has been proposed.[79] This has a cold-start and run portion and a hot-start and run portion with a weighting factor of $^1/_7$for the cold portion and $^6/_7$ for the hot portion. The heavy-duty truck service accumulation durability test is a 1500-hr dynamometer test.[78]

The emission sources from truck gasoline engines are the same as for passenger cars. However, truck engines, particularly heavy-duty truck engines, operate under different speed and load conditions than passenger car engines since the former operate a larger fraction of the time near maximum power or torque. The higher average intake manifold pressures of truck engines will result in higher NO_x emissions, but will also result in higher exhaust temperatures which will make port-injected air somewhat more effective in reducing CO and hydrocarbons. The use of fuel-rich air–fuel ratios for maximum power (particularly for heavy trucks) will cause an increase in CO and hydrocarbon emissions from the cylinders.

The majority of present light-duty gasoline engine trucks employ oxidation catalysts, although TWC systems are in use in California. The loss in fuel economy caused by calibration changes to meet the present (1981) federal emission standards (Table 11), which are higher than passenger cars (Table 1), is thought to be small. There will, of course, be the loss due to unleaded fuel and air pumps. The average passenger vehicle weight is 3100 lb and light truck weight 3870 lb.[77] Light-duty trucks meeting the lower California emission standards of one manufacturer have ~3–5% fuel economy loss due to nonoptimum calibrations from their federal counterparts. An estimate of a loss of 9% on a per ton mile basis has been given.[77] It is estimated that the proposed 1984 federal emission standards will cause a 6% fuel economy loss from 1981 some of which may be recovered by more sophisticated emission control systems.

It is thought that the present heavy-duty truck emission standards have not resulted in a significant loss in fuel economy (<3%), for the gasoline engines employed, due to calibration changes and the use of EGR. There is a small loss due to the use of an air pump, and a loss in power for a given displacement under some operating modes due to EGR. It should be realized, however, that several large-truck manufacturers reduced the compression ratio of their heavy-duty engines to 8 in the early 1970s in order to prepare for the possible use of unleaded fuel (and also to reduce NO and hydrocarbon emissions) and have kept the compression ratio at this value, although they have continued to use leaded fuel. No catalysts have been used, but port-injected air has been employed to produce some reduction of the CO and hydrocarbon levels at the exhaust temperatures involved (up to 800 °C). The EPA adopted stringent HC, CO, and NO standards for heavy-duty gasoline engine trucks produced for the 1984 model year, and later, which have raised technological feasibility issues (1.3 g per Bhp-hr HC, 15.5 g per Bhp-hr CO, and 10.7 g per Bhp-hr NO) because of unproven catalyst durability in the severe operating conditions typical of heavy-duty truck usage and the requirement for the use of unleaded fuel. The EPA recently announced[80] its intent to revise the CO and HC standards so as not to require catalysts. The proposed NO_x standards would also be revised in order to accommodate diesel engines. It appears, however, (1983) that catalysts will be required for heavy-duty trucks with gasoline engines.

REFERENCES

1. U.S. Federal Register, Vol. 37, No. 221, November 15 (1972); Vol. 38, No. 124, June (1973); Vol. 40, part 86, June 30 (1975).
2. Kruse, R., and Huls, T., "Development of the Federal Urban Driving Schedule," SAE paper 730553, May (1973).
3. Austin, T., Hellman, K., and Paulsell, C., "Passenger Car Fuel Economy During Non-Urban Driving," SAE paper 740592, August (1974).
4. Oetting, H., "Impact of Emission Standards on Fuel Economy and Consumer Attributes," SAE paper 790230, February (1979).
5. Brownson, D., Johnson, R., and Candelise, A., "A Progress Report on Man-Air-Ox, Manifold Air Oxidation of Exhaust Gas," SAE paper 486-N, March (1962).
6. Gast, R., "Pulsair—A Method for Exhaust System Induction of Secondary Air for Emission Control," SAE paper 750172, February (1975).
7. Currie, J., Grossman, D., and Gumbleton, J., "Energy Conservation with Increased Compression Ratio and Electronic Knock Control," SAE paper 790173, February (1979).
8. Kirklin, P., and Wise, S., "The Effects of Engine Variables and Exhaust Gas Recirculation on Emissions and Fuel Economy—Part IV," SAE paper 780591 (S.P. 431), June (1978).
9. Gumbleton, J., Niepoth, G., and Currie, J., "Compression Ratio and Fuel Economy with Emission Constraints," *Proc. Amer. Petroleum Inst.*, Refining Dept. 41st Midyear Meeting, API-17-76, pp. 179–188 (1976).
10. Gumbleton, J., Neipoth, G., and Currie, J., "Effect of Energy and Emission Constraints on Compression Ratio," SAE paper 760826, October (1976).
11. Nesmith, T., Tracy, C., Meguerian, G., Keller, B., and Daby, E., "Optimization of Fuel and Vehicle Parameters," SAE paper 780589, June (1978).
12. Cantwell, E., Kinnear, F., and Russell, H., "The Effect of Emission Standards and Gasoline Quality on Fuel Consumption," SAE paper 750671, June (1975).
13. Lin-Fu, Jane S., "Lead Exposure Among Children—A Reassessment," *New England J. of Medicine*, **300**, 731–732 (1979).
14. Assoc. Octel publication, "Leaded Petrol-Energy and Health," December (1979).
15. Newhall, H., "Kinetics of Engine Generated Nitrogen Oxides and Carbon Monoxide," 12th Symposium (International) on Combustion, The Combustion Institute, pp. 603–613 (1968).
16. Delichatsios, M., and Keck, J., "Rate Controlled Constrained Equilibrium Calculations of CO and NO Freezing in Internal Combustion Engines," Amer. Chem. Soc. Div. of Petroleum Chem. Symposium on Chemistry of Combustion in Engines, Philadelphia, PA, April (1975).
17. Pozniak, D., "The Exhaust Emission and Fuel Consumption Characteristics of an Engine During Warmup—A Vehicle Study," SAE paper 800396, February (1980).
18. Glockler, O., Knapp, H., and Manger, H., "Present Status and Future Development of Gasoline Fuel Injection Systems for Passenger Cars," SAE paper 800467, February (1980).

19. Lindsay, R., Thomas, A., Woodworth, J., and Zeschmann, E., "Influence of Homogeneous Charge on the Exhaust Emissions of Hydrocarbons, Carbon Monoxide, and Nitric Oxide from a Multicylinder Engine," SAE paper 710588, June (1971).
20. Hires, S., and Overington, M., "Transient Mixture Strength Excursions—An Investigation of Their Cause and the Development of a Constant Mixture Strength Fueling Strategy," SAE paper 810495, February (1981).
21. Brokaw, R., and Bittker, D., "Carbon Monoxide Oxidation Rates Computed for Automobile Exhaust Manifold Reactor Conditions," NASA Tech. Note D-7024 National Tech. Info. Serv. Springfield, VA 22151.
22. Nicholson, R., "Emissions and Fuel Economy Interactions," SAE paper 780616, June (1978).
23. Blumberg, P., Kummer, J., "Prediction of NO Formation in Spark Ignited Engines—An Analysis of Methods of Control," *Comb. Sci. & Tech.* **4,** 73–95 (1971).
24. Lavoie, G., Heywood, J., and Keck, J., "Experimental and Theoretical Study of Nitric Oxide Formation in Internal Combustion Engines," *Comb. Sci. & Tech.* **1,** 313–326 (1970).
25. Lavoie, G., and Blumberg P., "A Fundamental Model for Predicting Fuel Consumption, NO_x, and HC Emissions of the Conventional Spark-Ignited Engine," *Comb. Sci. & Tech.* **21,** 225–258 (1980).
26. Marsee, F., and Rediker, F., "Emission Control with Lean Mixtures on a Two-Liter High-Performance Engine," Ethyl Corp. Project 12274 (1977).
27. Holl, W., "Air Fuel Control to Reduce Emissions," SAE paper 800051, February (1980).
28. Blumberg, P., "Nitric Oxide Emissions from Stratified Charge Engines," *Comb. Sci. & Tech.* **8,** 5–24 (1973).
29. Kummer, J., "Catalysts for Automobile Emission Control," *Prog. Energy Combust. Sci.* **6,** 177–199 (1980).
30. Novak, J., and Blumberg, P., "Parametric Simulation of Significant Design and Operating Alternatives Affecting Fuel Economy and Emissions of Spark Ignited Engines," SAE paper 780943, November (1978).
31. Eltinge, L., Marsee, F., and Warren, J., "Potentialities of Further Emission Reduction by Engine Modifications," SAE paper 680123, January (1968).
32. Fleming, R., "Effect of Fuel Composition on Exhaust Emissions," Bureau of Mines, Report of Investigation No. 7423 (1970).
33. Daniel, W., "Engine Variable Effects on Exhaust Hydrocarbon Composition," SAE paper 670124, January (1967).
34. Wentworth, J., "Piston and Piston Ring Variables Affect Exhaust Hydrocarbon Emissions," SAE paper 680109, January (1968).
35. Wentworth, J., "More on Origins of Exhaust Hydrocarbons—Effect of Zero Oil Consumption, Deposit Location, and Surface Roughness," SAE paper 720939, October (1972).
36. Gagliardi, J., and Ghannam, F., "Effect of T.E.L. Concentration on Exhaust Emissions in Customer Type Vehicle Operation," SAE paper 690015, January (1969).
37. Kaiser, E., Adamczyk, A., and Lavoie, G., "The Effect of Oil Layers on the Hydrocarbon Emissions Generated During Closed Vessel Combustion," 18th International Symposium on Combustion, paper 188, August (1980).
38. Adamczyk, A., Kaiser, E., Cavolowsky, J., and Lavoie, G., "An Experimental Study of Hydrocarbon Emissions From Closed Vessel Explosions," 18th International Symposium on Combustion, paper 167, August (1980).
39. Freeman, J., Jr., and Stahman, R., "Vehicle Performance and Exhaust Emissions; Carburetor vs. Timed Fuel Injection," SAE paper 650863, November (1965).
40. Wentworth, J., "Effect of Combustion Chamber Surface Temperature on Exhaust Hydrocarbon Concentration," SAE paper 710587, June (1971).

41. Myers, J., and Alkidas, A., "Effects of Combustion Chamber Surface Temperature on the Exhaust Emissions of a Single Cylinder Spark Ignited Engine," SAE paper 780642, June (1978).
42. Blumberg, P., and Lavoie, G., personal communication.
43. Tanuma, T., Sasaki, K., Kaneko, T., and Kawaski, H., "Ignition, Combustion and Exhaust Emissions of Lean Mixtures in Automotive Spark Ignition Engines," SAE paper 710159, January (1971).
44. Hass, G., Bonomassa, F., Newmark, P., and Kayne, N., "The Influence of Vehicle Operating Variables on Exhaust Emissions," *J. Air Poll. Cont. Assoc.*, **17,** 384–387 (1967).
45. Induction System Task Group, "Automotive Exhaust Hydrocarbon Reduction During Deceleration by Induction System Devices," *SAE Transactions*, **66,** 383–396 (1958).
46. Canton, J., and Heywood, J., "Models for Heat Transfer, Mixing and Hydrocarbon Oxidation in an Exhaust Port of a Spark Ignited Engine," SAE paper 800290, February (1980).
47. Huls, T., Myers, P., and Uyehara, O., "Spark Ignition Engine Operation and Design for Minimum Exhaust Emissions," SAE paper 660405, June (1966).
48. Brinkman, N., "Ethanol Fuel—A Single Cylinder Engine Study of Efficiency and Exhaust Emissions," SAE paper 810345, February (1981).
49. Sigworth, H., Jr., Myers, P., and Uyehara, O., "The Disappearance of Ethylene, Propylene, *n*-Butane, and 1-Butene in Spark Ignition Engine Exhaust," SAE paper 700472, May (1970).
50. Steinhagen, W., Niepoth, G., and Mick, S., "Design and Development of the General Motors Air Injection Reactor System," SAE paper 660106, January (1966).
51. Brownson, D., and Stebar, R., "Factors Influencing the Effectiveness of Air Injection in Reducing Exhaust Emissions," SAE paper 650526, May (1965).
52. Tabaczynski, R., Heywood, J., and Keck, J., "Time Resolved Measurements of Hydrocarbon Mass Flow Rate in the Exhaust of a Spark-Ignited Engine," SAE paper 720112, January (1972).
53. Siewert, R., "How Individual Valve Timing Events Affect Exhaust Hydrocarbon Emissions," SAE paper 710609, June (1971).
54. Kuroda, K., Nakajima, Y., Hayashi, Y., and Sugihara, K., "Economical Matching of the Thermal Reactor to Small Engine—Low-Emission Concept Vehicles," SAE paper 720484, May (1972).
55. Herrin, R., "Emission Performance of Lean Thermal Reactors—Effects of Volume Configuration and Heat Loss," SAE paper 780008, February (1978).
56. Adams, W., Marsee, F., Olree, R., and Hamilton, J., "Emissions, Fuel Economy, and Durability of Lean Burn Systems" SAE paper 760227, February (1976).
57. Marsee, F., Olree, R., and Adams, W., "Compression Ratio Effects with Lean Mixtures," SAE paper 770640, June (1977).
58. Sakai, Y., Nakagawa, S., Tange, R., and Maruyama, R., "Fundamental Study of Oxidation in a Lean Reactor," SAE paper 770297, February (1977).
59. Dale, J., Smy, P., and Clements, R., "The Effect of a Coaxial, Spark Igniter on the Performance of and the Emissions from an Internal Combustion Engine," *Combustion and Flame* **31,** 173–185 (1978).
60. Shelef, M., Otto, K., and Otto, N., "Poisoning of Automotive Catalysts," in *Advances in Catalysis*, Vol. 27, pp. 311–365, Academic Press, New York (1978).
61. White, J., Zakrajsek, C., "Implications of Precious Metal Catalysts with Leaded Fuels," SAE paper 810086, February (1981).
62. Williamson, W., Stepien, H., Watkins, W., and Gandhi, H., "Poisoning of Platinum-Rhodium Automotive Three-Way Catalysts by Lead and Phosphorous," *Environmental Sci. and Tech.* **13,** 1109–1113 (1979).
63. DallaBetta, R., McCune, R., and Sprys, J., "Relative Importance of Thermal and Chemical Deactivation of Noble Metal Automotive Oxidation Catalysts," *Ind. Eng. Chem. Prod. Res. & Dev.* **15,** 169–172 (1976).

64. Kuroda, H., Nakajima, Y., Sugihara, K., Takagi, Y., and Muranaka, S., "The Fast Burn with Heavy EGR: New Approach for Low NO_x and Improved Fuel Economy," SAE paper 780006, February (1978).
65. Harada, M., Kadota, T., and Sugiyama, Y., "Nissan NAPS 'Z' Engine Realizes Better Fuel Economy and Low NO_x Emissions," SAE paper 810010, February (1981).
66. Harned, J., and Montgomery, D., "Comparison of Catalyst Substrates for Catalytic Converter Systems" SAE paper 730561, May (1973).
67. Bagley, R., Domain, R., and Duke, D., "Multicellular Ceramics as Catalyst Supports for Controlling Automotive Emissions," SAE paper 730274, January (1973).
68. Zemke, B., and Gumbleton, J., "General Motors Progress Towards the Federal Research Objective Emission Levels," SAE paper 800398, February (1980).
69. Shelef, M., and Gandhi, H., "Ammonia Formation in Catalytic Reduction of Nitric Oxide by Molecular Hydrogen," *Ind. Eng. Chem. Prod. Res. and Dev.* **11,** 393–396 (1972).
70. Shelef, M., and Gandhi, H., "The Reduction of Nitric Oxide in Automobile Emissions, Stabilization of Catalysts Containing Ruthenium," *Platinum, Metals Rev.* **18,** 2–14 (1974).
71. Barnes, G., Klimisch, R., and Krieger, B., "Equilibrium Considerations in Catalytic Emission Control," SAE paper 730200, January (1973).
72. Dueker, H., Friese, K., Haecker, W., "Ceramic Aspects of Bosch Lambda-Sensor," SAE paper 750223, February (1975).
73. Tien, T., Stadler, H., Gibbons, E., and Zacmanidis, P., "TiO_2 as an Air-to-Fuel Ratio Sensor for Automobile Exhausts," *Bull., Amer. Ceramic Soc.*, **54,** 280–282 (1975).
74. Wallman, S., "Development of the Volvo Lambda Sond System," SAE paper 770295, March (1977).
75. Hammerle, R., and Wu, C., "Three-Way Catalyst Performance Characterization," SAE paper 810275, February (1981).
76. Kaneko, Y., Kaneko, H., Kobayashi, H., Komagome, R., Hirako, O., and Nakoyama, O., "Air Fuel Ratio Modulation and Conversion Efficiency of Three-Way Catalysts," SAE paper 780607, June (1978).
77. Foster, J., Murrell, J., and Loos, S., "Light-Duty Automotive Fuel Economy Trends," SAE paper 810386, February (1981).
78. Federal Register Vol. 37, No. 221, November (1972); Federal Register Vol. 41, No. 101, May (1976).
79. Federal Register Vol. 45, No. 14, January (1980).
80. Notice of Intent, Enviromental Protection Agency, April 6,.1981. See also Federal Register Vol. 46, No. 70, April 13, 1981.
81. Wagner, T., Lawrence, D., and Plautz, D., "Diesel: Miles per Barrel," *Hydrocarbon Process.*, 199–204 January (1981).

3

Spark Ignition: Its Physics and Effect on the Internal Combustion Engine

RUDOLF MALY

I. INTRODUCTION

The interest in a better understanding of the processes involved in ignition and self-sustained flame propagation is as old as the internal combustion engine itself. In order to gain a better insight into these processes, the main efforts have been concentrated on experiments in the very engine aiming at a direct and quick success. This, however, also presented its major drawback. The complexity of the mutually dependent influences of the large number of different operation parameters of the real engine prevented any significant breakthrough in the past. Too much room had still to be left by this approach for speculations and interpretations due to serious lacks of accessible information.

Experimental and theoretical work outside the motor on the other hand suffered a long time from inadequate instrumentation and improper simulation of the engine conditions. Thus progress in the understanding of spark ignition was slow and most of the work was concentrated on the accessible and more easily understood generation of the required high-voltage pulses.

General aspects of ignition, combustion, and related topics including reviews of the older literature have been thoroughly treated in the books by Penner and Mullins[1] and by Lewis and von Elbe.[2] A good key to the literature until 1965 relevant to engine-related problems is given in Müller, Rhode, and Klink.[3] An attempt to compile the specific knowledge on spark ignition available from the literature until 1969 into a comprehensive review was made by Conzelmann[4],

RUDOLF MALY • Institut für Physikalische Elektronik, Universität Stuttgart, Böblingerstrasse 70, D-7000 Stuttgart-1, Federal Republic of Germany. Present address: Daimler Benz AG, Research Department, D-7000 Stuttgart-60, Postfach 202, Federal Republic of Germany.

though the author himself felt that major areas in spark ignition were still too little explored to come to any final conclusions.

Therefore, a broad-scale research program* was initiated at the Institut für Physikalische Elektronik at the University of Stuttgart, aiming at the very fundamentals of spark ignition and flame propagation. In the interest of consistency of data all relevant processes were reinvestigated by modern measuring techniques. The goal was to break down the special results of laboratory, engine, and theoretical work into the physics of the involved processes in order to allow for a generalization and utilization of the results for the demands of the near future: sensitive fuel savings by novel combustion concepts combined with effective reductions in exhaust emissions. Major parts of this work have been published in the meantime.[5–11]

It is the purpose of this chapter to provide the reader with a complete and comprehensive survey of the principles and basic properties of spark ignition. As far as possible consistent material from the research program will be used in order to facilitate the reduction to basic principles. Reference to relevant literature is made whenever possible. Additional effects are included which are essential under real engine conditions. As a guideline, we will follow the energy-transfer chain beginning at the storage device for electrical ignition energy. We proceed to the spark plasma, from there to the plasma surface where ignition actually is taking place in a narrow reaction interface, and from there to the self-sustained flame front. We will concentrate on spark ignition. The principal properties, however, and most of the data are also applicable to other ignition processes with little or no adaptation.

II. GENERALIZED PROPERTIES OF IGNITION SPARKS

A. Electrical Characteristics

The electrical properties of the large number of apparently unique ignition systems, commercial as well as experimental,[12–15] may all be described by the simple equivalent circuit presented in Fig. 1. This is due to the fact that changes which eventually have been made (increase of discharge current, increase of discharge time, increase of gap spacing, use of multiple discharges, use of supplementary high current discharges, etc.) will certainly modify the ignition device and its apparent ignition capability but not its basic properties. For a better insight into the basic ignition capability of a system it is sufficient to know its time-dependent voltage and current characteristics which are determined by

* This work has been supported by the Bundesminister für Forschung und Technologie (BMFT), grant no. NTÖ 48, TV 7652 6.

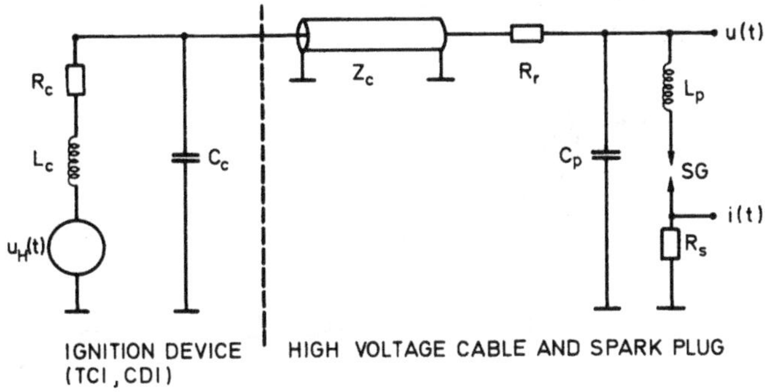

FIGURE 1. Schematic equivalent circuit of technical ignition systems. The high-voltage distribution spark gap is omitted since it does not affect the principal functions of the device. $u_H(t)$, high-voltage signal; L_c, coil inductance; R_c, coil resistance; C_c, coil capacitor; Z_c, high-voltage cable impedance; R_r, radio interference damping resistor; C_p, plug capacitor; L_p, plug inductance; SG: spark gap; R_s, current shunt (incorporated for measuring purposes only); $u(t)$, spark voltage; $i(t)$, spark current.

the components of this discharge circuit. Complete I–V characteristics of two commercial ignition systems (CDI, energy stored in a capacitor on the primary side of a transformer, and TCI, energy stored in the inductance of a coil) are presented in Figs. 2 and 3 together with corresponding power and energy diagrams in Fig. 4.

We recognize a very short (nanoseconds) first phase—the breakdown—during which the spark current rises to a first current maximum of several

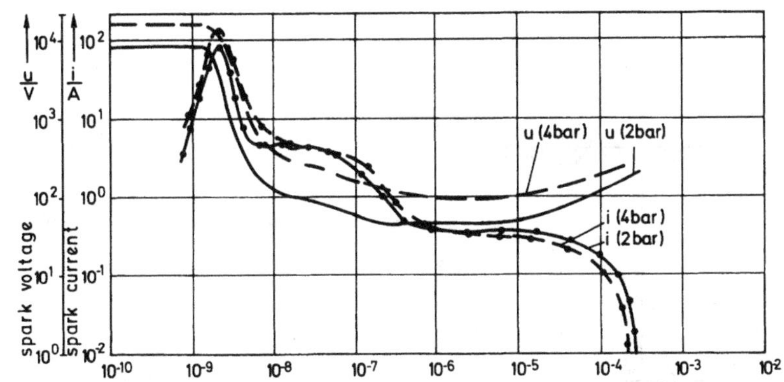

FIGURE 2. I–V time characteristics of a technical CDI (capacitor discharge ignition system) in air at 300 K and 1-mm gap supplying 3 mJ to the gap electrodes.

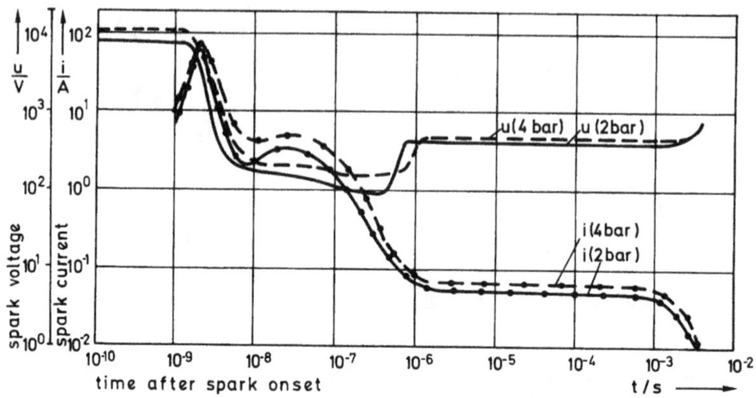

FIGURE 3. *I–V* time characteristics of a technical TCI (transistorized coil ignition system) in air at 300 K and 1-mm gap supplying 30 mJ to the gap electrodes.

hundreds of amperes. Its value is determined by the ignition voltage U_0 and the characteristic impedances Z_p of the near-gap circuit:

$$I_{pb} = U_0/Z_p = 10 \text{ kV}/50\ \Omega = 200 \text{ A}$$

At the same time the gap voltage drops from its initial value to very low values (<100 V). This phase is uniquely determined by the capacitive (5–15 pF) and the inductive components of spark plug and spark (≈5 nH).

This phase is followed by a second one—the arc—existing for roughly 1 μsec. During this time the cable capacity of the high-voltage leads (40–100 pF)

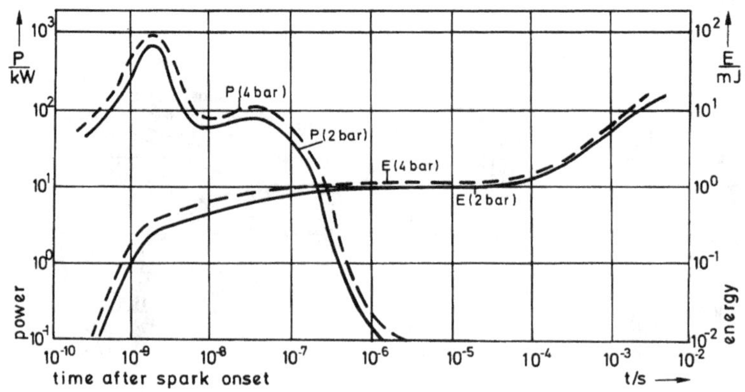

FIGURE 4. E,P: time characteristics of a technical TCI in air at 300 K and 1-mm gap supplying 30 mJ to the gap electrodes.

and the coil capacity ($\approx$50 pF) discharge through the radio interference damping resistor R_r (1–10 kΩ) in series with the characteristic impedance of the high-voltage cable ($Z_c \approx 200\ \Omega$). Typical values for this second peak are

$$I_{pa} = U_0/(R_r + Z_c) = 10\ \text{kV}/2\ \text{k}\Omega = 5\ \text{A}$$

Finally, in the third phase—the glow discharge—the energy storage device proper will dump its energy into the discharge circuit. In case of a coil ignition system this is a glow discharge being maintained for several milliseconds. In CDI systems with high internal impedances this will also be a glow discharge. If the impedance is sufficiently low, however, currents above 200 mA may be sustained and the arc phase will last for up to some hundreds of microseconds. In an intermediate range (100–200 mA) rapid transitions from arc to glow and back again occur. The peak values of the glow and arc currents in this phase are given by:

$$I_{pg} = U_0/Z_c = 10\ \text{kV}/200\ \text{k}\Omega = 50\ \text{mA} \qquad \text{(glow)}$$

$$I_{pa} = U_0/Z_c = 10\ \text{kV}/50\ \text{k}\Omega = 200\ \text{mA} \qquad \text{(arc)}$$

The three phases or discharge modes are also uniquely characterized by their power and energy release functions. The breakdown reaches the highest power level by far (up to megawatts) at rather small energies in commercial systems (0.3–1 mJ). The arc ranges in between, whereas the glow discharge operates at lowest power (tens of watts) and highest energy levels (30–100 mJ). This is primarily due to the extremely long discharge time.

Any ignition system comprises different combinations of these three discharge modes with individually varying energies and discharge periods. If the basic properties of these three discharge modes are known, its ignition capability may be readily deduced from the *I–V* characteristics. Thus it is sufficient to determine the physics of the three discharge modes. The measuring techniques for the long-lasting arc and glow discharges are straightforward. For the breakdown and the intermediate arc phase very special arrangements have to be made in order to avoid erroneous results.[16–18] A detailed description of these techniques (short-time measuring technique for electrical parameters, short-time techniques for spectroscopy, interferometry, and laser methods), however, is beyond the scope of this chapter dealing primarily with the results.

B. Discharge Modes

1. Prebreakdown Phase

Initially the gas volume within the electrode gap represents a perfect insulator. When the spark pulse is applied (rise times for a TCI, $\approx$10 kV/msec, and $\approx$100 kV/msec for a CDI) the randomly existing electrons produced by hard, ionizing radiation from space and earth can gain energy in the rising electric field and are accelerated towards the anode. If the applied electric field has reached sufficiently high values (50–100 kV/cm), the field-accelerated electrons may ionize the gas molecules by collisions generating additional electrons and ions. These charge carriers gain energy from the electrical field as well and further contribute to the rising electron current. The number of electrons and ions increases rapidly very much like in an avalanche.[19] However, as the electrons are collected by the anode additional processes are required to generate enough new starting electrons near or at the cathode in order to make this multiplication process self-sustained. This is accomplished by UV radiation ($\lambda < 200$ nm) emitted from excited ions, as the energy-rich impinging electrons will not only ionize the gas molecules but excite a multitude of electronic levels as well.

In small gaps and at low pressures ($l_g <$ mm, $p < 1$ bar) the radiation may reach the cathode directly and liberate with high efficiency photoelectrons at its surface. At large gap spacings and higher pressures ($l_g >$ mm, $p > 1$ bar) UV radiation is also absorbed by the surrounding gas molecules causing a low percentage of them to become radiation ionized in the cathode region of the primary radiation-emitting volume. Thus the ionization processes initiated by the applied electrical field grow, producing a rapid increase in current flow across the spark gap with an inherent capability to get self-sustained and almost independent from the external electrical field.

As long as the ionizing processes produce less electrons than required for rendering the discharge self-sustained this phase is called the prebreakdown phase. For very slowly rising ignition voltages the prebreakdown phase may last for appreciable intervals of time (minutes and longer) depending on gas composition, pressure, and field configuration.[20] The faster the voltage rise the shorter this phase due to an effective overvoltage which makes the ionizing processes more effective. The gas temperature is still very close to its initial value and the mean electron density less than 10^{15} e/cm^3, though the density in the individual ionizing channels (streamers) may reach 10^{18} e/cm^3. In air and N_2 an intense N_2 molecule radiation is emitted (second positive system) being used in N_2 lasers. Atomic radiation is negligible. The ionization cloud fills the entire space within the gap where sufficiently high field strengths exist.

2. *Breakdown Phase*

As soon as enough feedback electrons are produced, an overexponential increase in discharge current originates assisted effectively by a self-created space charge within the streamers.[21] In practical cases this occurs at currents in excess of ≈ 10 mA. As there is no inherent current-limiting process, the current rises within nanoseconds to values of some hundreds or thousands of amperes until any further increase is limited by the impedances of the discharge and the external circuit close to the gap (i.e., the spark plug). At this stage the voltage across the gap as well as the electrical field drop rapidly to very low values (100 V and 1 kV/cm, respectively, "breakdown"). The minimum energy required to initiate a complete breakdown at 1 bar and a 1-mm gap is 0.3 mJ.

During the breakdown phase a volume segment with a slightly higher electrical conductivity gradually carries more and more of the current flux. The ion density increases rapidly to values of some 10^{19} e/cm^3 so that an energy exchange by Coulomb forces between electrons and ions becomes the dominating process of energy loss for the accelerated electrons.[22] In spite of the unfavorable mass relation electron/ion an effective energy exchange is now already possible in a single collision. Thus electrical energy will be transferred very efficiently from the gap capacitor via the electrical field to the electrons and the ions. Consequently an extremely high degree of ionization (sensible concentrations of the N^{4+} ion have been measured[23]) and electronic excitation occurs with a massive increase in gas temperature (up to 60,000 K). This is illustrated in Fig. 5.

These processes are concentrated to a narrow channel, initially 40 μm in diameter, corresponding to the diameters of the streamers in the prebreakdown phase. All the heavy particles inside this channel are fully dissociated, ionized, and heavily excited so that one heavy particle carries more than 20 eV of internal or potential energy versus only ≈5.5 eV of thermal energy (at 60,000 K). This is shown in Fig. 6. As there are still higher excitation levels available, this readily demonstrates the large amounts of energy a breakdown may store, being much

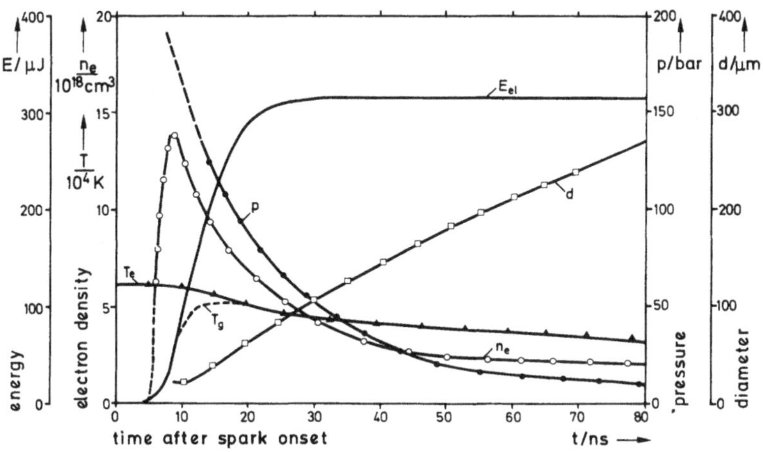

FIGURE 5. Plasma parameters of a breakdown in air at 300 K, 1 bar, and a 1-mm gap. Duration of the breakdown current: 10 nsec. Origin of time scale arbitrary. E_{el}, electrical energy supplied to the gap electrodes; T_e, electron temperature; T_g, gas temperature; n_e, electron density; p, overpressure in the plasma channel; d, diameter of the plasma channel.

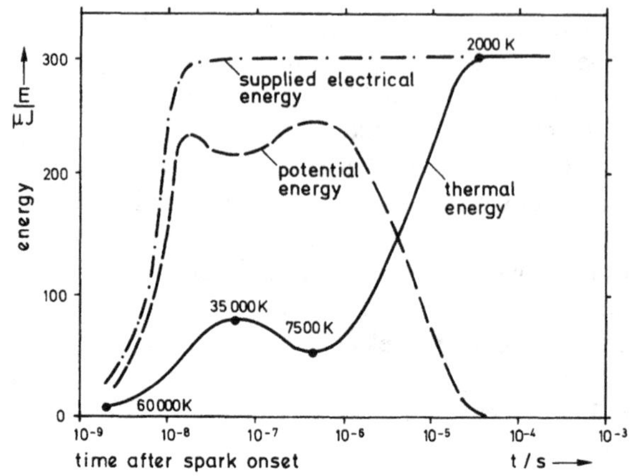

FIGURE 6. Temporal redistribution of supplied electrical energy in a breakdown plasma from potential to thermal energy.

higher than it would be possible by purely thermal means. Due to the extremely rapid transition in gas temperature to 60,000 K the pressure inside the channel rises almost instantaneously to several hundred bars causing the emission of an intense blast (shock wave). The channel expands correspondingly with supersonic speed.

During this expansion the channel cools down and being in thermal equilibrium at each instant, the stored potential energy is converted gradually into thermal energy supporting the expansion process. As recombination is a three-body process overequilibrium concentrations of radicals will survive after the density of atoms has fallen to very low values. The energy fraction ($\simeq$30%) initially carried away by the blast is soon regained since spherical blasts rapidly transfer their energy to the gas molecules within a rather small sphere ($d \simeq 2$ mm) into which the breakdown plasma later expands anyhow.[24]

Cooling the plasma as a function of time is presented in Fig. 7 for two different values of breakdown energies, whereas in Fig. 8 the expansion velocities and channel diameters involved are given. Due to the extremely fast processes there are almost no losses at all. The cathode remains cold and cannot take up appreciable amounts of plasma heat by heat conduction. As the breakdown is an instationary discharge by principle photo, and at later stages field emission are sufficient to supply the necessary cathode mechanisms for carrying its high currents.

However, prolonged duration of a high-current flow leads via a thermionic field to thermionic emission from hot spots. This indicates the end of the breakdown phase and the start of the arc regime. Radiation losses are small as the

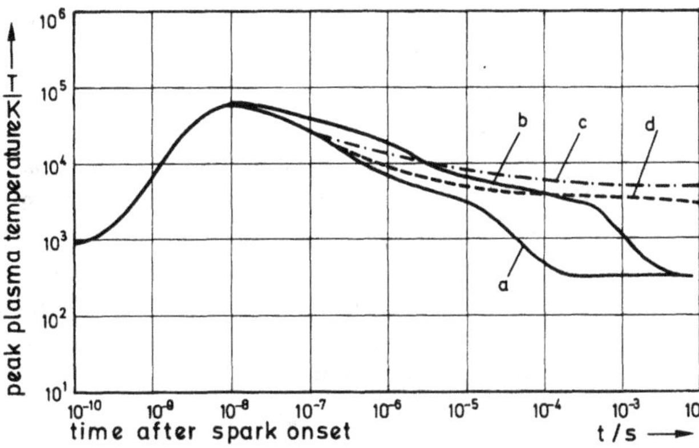

FIGURE 7. Temporal variation of maximum gas temperature in different discharge plasmas. a: CDI, 3 mJ, 100 μsec; b: breakdown, 30 mJ, 60 nsec; c: CDI, 3 mJ, 100 μsec plus superimposed constant current (2A) arc, 30 mJ, 230 μsec, d: CDI, 3 mJ, 100 μsec plus superimposed constant current (60 mA) glow discharge, 30 mJ, 770 μsec.

highly energetic radiation ($\lambda < 200$ nm) is trapped within the channel by reabsorption to ground level states and the time intervals are so short. Thus only minor losses over the surface occur in the low-energy part of the spectrum ($\lambda > 200$ nm). An energy balance is given in Table 1 allowing a direct comparison to arc and glow discharges which will be dealt with below. The end of the breakdown may be taken as the point where the ignition voltage has dropped below 10% of its initial value or more physically as the point from which a hot cathode spot exists and turns the discharge into an arc.

The expansion velocities encountered in breakdown discharges are much higher than those in self-sustained flames. So we must have some idea about the volume filled by the breakdown plasma at that time when the expansion

TABLE 1. Energy Balance for Breakdown, Arc, and Glow Discharge Plasmas under Idealized Conditions (Very Small Electrodes)

	Breakdown	Arc discharge	Glow discharge
Radiation loss	<1%	ca. 5%	<1%
Heat conduction via electrodes	ca. 5%	ca. 45%	70%
Total losses	ca. 6%	ca. 50%	ca. 70%
Total plasma	ca. 94%	ca. 50%	ca. 30%

velocity has dropped to values comparable to those of flames. According to Fig. 8 this is the case for some tens of microseconds. The temperature profile measured across the undisturbed equatorial plane perpendicular to the gap axis is shown in Fig. 9. The corresponding profiles for arc and glow discharge are included for ease of comparison. We see that at this stage the peak plasma temperature has fallen to ≃3500 K where almost all of the potential energy is already converted to thermal energy occupying now a large toroidal volume.

In addition there is a bubble of fresh cold gas in the central region insulating effectively the plasma from the heat-sinking electrodes. This bubble of cold gas is forced to flow into the central region due to the specific properties of spherical blasts. The initial overpressure phase is followed by a suction phase that travels in the same direction as the shock front.[25] This suction phase is caused by the inertia of the gas molecules that are projected outwards with supersonic speed during the early expansion phase of the breakdown channel. Due to friction the expansion velocity near the electrode surfaces is much lower than in the undisturbed central regions so that gas is forced to flow along the electrode surfaces from the exterior into the gap region as soon as the suction phase becomes active. The pressure difference increases with the intensity of the shock. As the viscosity of the gas is rather low, this flux of cold gas will persist even when the suction phase is over.

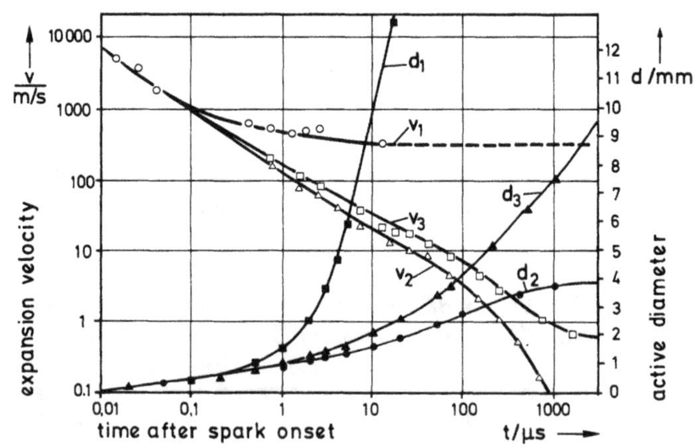

FIGURE 8. Diameters (d) and expansion velocities (v) of volumes activated by a CDI(3 mJ, 100 µsec) in a 1-mm gap. Numbers indicate 1, shock in air at 300 K and 1 bar; 2, plasma channel in air at 300 K and 1 bar; 3, electrical and chemical plasma, respectively, in a stoichiometric CH_4–air mixture at 300 K and 1 bar.

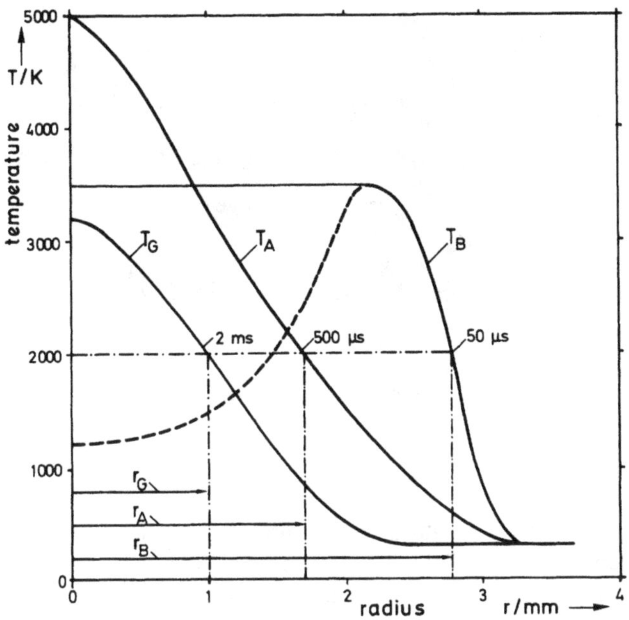

FIGURE 9. Radial temperature profiles T_b, T_a, and T_g for breakdown, arc, and glow discharge plasmas respectively, in air at 300 K, 1 bar, and a 1-mm gap. Due to different power levels final profiles will be reached at different time intervals. Breakdown profile shown at time of ceasing pressure expansion. r_b, r_a, r_g: hypothetical plasma radii for an assumed flame temperature of 2000 K.

3. *Arc Discharge*

An arc must always be preceded by a breakdown in order to create a sufficiently high electrical conductivity to initiate it. It is characterized by currents in excess of 100 mA with the upper limit set only by the external electrical impedances. The burning voltage is very low ($\simeq$50 V at 1 bar in air and 1-mm gap) and splits into $\simeq$15 V for the cathode fall, $\simeq$10 V for the arc plasma, and $\simeq$25 V for the anode fall. The cathode fall is required for sustaining numerous hot cathode spots (T up to 3000 K, order of magnitude: evaporation temperature of the cathode material at actual ambient pressure) of molten material being 10–40 µm in diameter. The electrons emitted from these pools are required to sustain the arc. Figure 10 gives an impression of the surface damage caused by a 10-A arc. Without such pools of molten metal the arc cannot exist so all measures to quench these pools result in rising cathode falls until new ones establish or in extinction of the arc.

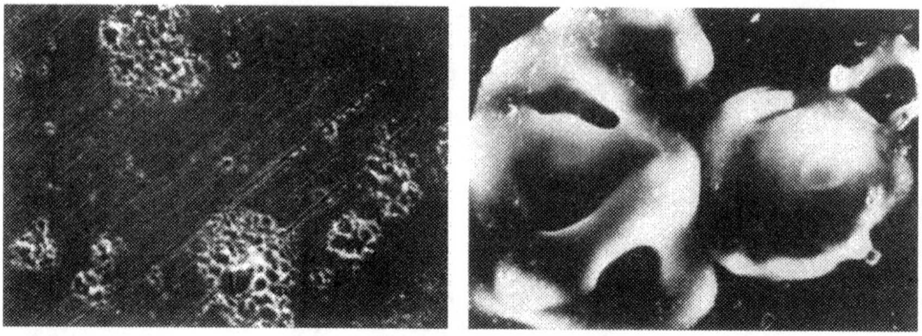

FIGURE 10. Electron microscope prints showing surface damage to a brass electrode by a single 200-V, 225-mJ arc pulse lasting for 100 μsec in air at 300 K, 1 bar, and 1-mm gap. (a) 110 × 90 μm^2 section of the total damaged area of 1 mm ∅. (b) Enlarged detail showing molten structure of individual spots.

Due to the high temperature of the cathode spots heavy evaporation of the cathode material arises. In Table 2 measured erosion rates for all three discharge modes are presented. The erosion rates for arc and glow discharges increase with increasing gap spacing as the higher ignition voltage leads to higher energy fractions stored in the coil and cable capacitors which discharge predominantly in an arc mode. The inverse is true for the breakdown. Larger gaps tend to lengthen the duration of the breakdown phase, thus reducing the energy left for a possibly succeeding arc phase. In optimized designs the undesired contribution of a succeeding arc can be further reduced so that erosion rates may be achieved that are sensibly lower than those of glow discharges. A very promising fact for a real long-life spark plug.

The stationary values of the electron density and the center temperature are 10^{17} e/cm^3 and 5000–6000 K, respectively, and vary only slightly with arc current. The main effect of an arc following a breakdown is therefore not to enhance the breakdown properties but to slow down and limit the electron density and gas temperature to those of the arc. Due to the smaller power inputs and the lower electron densities, only a slow mass-fraction-determined energy exchange between electrons and heavy particles is possible. This renders this discharge an almost purely thermal character with the consequence that energy transfer to the plasma surface is by heat conduction and mass diffusion rather than by overpressure. These processes are comparatively slow (milliseconds) and become the more inefficient the larger the radius grows from the axial energy input to the actual surface. This determines also the "Gaussian" shape of the temperature profiles.

As the temperature is ≃6000 K the degree of dissociation is rather high in spite of low ionization levels. However, this is limited to the very axial region

TABLE 2. Erosion Rates of Brass Cathodes by Breakdown, Arc, and Glow Discharge as a Function of the Electrode Spacing in Air at 300 K and 1 Bar

Device	Gap	Suppl. energy	Erosion per 10^6 discharges	
TCI	1/mm	$\cdot$ E_{el}/mJ	mass/g	mass/g mJ
(glow discharge,				
3. 4 msec, 500 V)	0.5	33	0.89×10^{-4}	2.7×10^{-6}
	1.0	37	1.29×10^{-4}	3.5×10^{-6}
	2.0	41	1.43×10^{-4}	3.5×10^{-6}
CDI				
(low-energy arc,				
100 μsec)	0.5	1.6	0.25×10^{-4}	15.6×10^{-6}
	1.0	1.76	0.32×10^{-4}	18.2×10^{-6}
	2.0	1.17	0.29×10^{-4}	24.8×10^{-6}
CDI + arc				
(arc discharge,				
100 μsec, 100 V)	0.5	185	333×10^{-4}	180×10^{-6}
	1.0	115	242×10^{-4}	210×10^{-6}
	2.0	67	150×10^{-4}	223×10^{-6}
CDI + arc				
(arc discharge,				
100 μsec, 200 V)	0.5	278	467×10^{-4}	168×10^{-6}
	1.0	225	442×10^{-4}	196×10^{-6}
	2.0	236	483×10^{-4}	204×10^{-6}
VFZ				
(breakdown, 5				
nsec, 20 kV)	0.5	40	6.4×10^{-4}	16×10^{-6}
	2.0	40	4.8×10^{-4}	12×10^{-6}
	4.0	40	1.3×10^{-4}	3.3×10^{-6}

due to the rapid falloff with temperature. The temporal dependence of the instationary arc temperature and the final temperature profile have been shown in Figs. 7 and 9, respectively. The plasma expansion is given in Fig. 11. Prolonged and intense contact of the arc plasma with the electrodes leads to serious heat losses via conduction. At the same time appreciable radiation losses also occur from the plasma surfaces as no radiation trapping is possible and sufficient time for emission is available. Detailed data are presented in Table 1.

4. Glow Discharge

The high-pressure glow discharge is very similar to the arc except for a cold cathode. It obtains its feedback electrons from the very inefficient electron

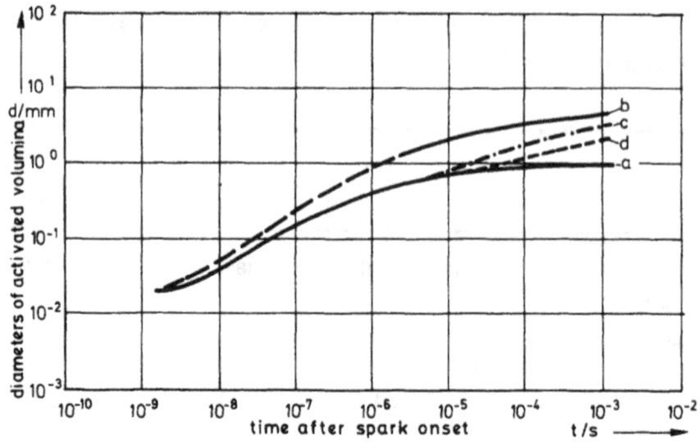

FIGURE 11. Diameters of volumina activated by breakdown, arc, and glow discharge as function of time. (a) Breakdown, 0.3 mJ, 10 nsec; (b) breakdown, 30 mJ, 60 nsec; (c) CDI, 3 mJ, 100 μsec plus superimposed constant current (2A) arc, 30 mJ, 230 μsec; (d) CDI, 3 mJ, 100 μsec plus superimposed constant current (60 mA) glow discharge, 30 mJ, 770 μsec.

liberation by ion impact. Thus only low current densities and currents (<100 mA) can be sustained. At low pressures appreciable parts of the cathode surface are covered by the cathode mechanism. High pressures tend to compress this area and favor therefore a transition to the arc regime. The same occurs if the current is forced to rise above ≃100 mA. Consequently hot cathode spots will develop. Due to the usually long discharge times (up to several milliseconds) it can be regarded as a stationary discharge and therefore most of the literature data on high-pressure discharges may be readily used.[26–29]

In order to emit enough feedback electrons a rather high cathode fall (≃400 V) is necessary. The cathode fall U_c obeys the relation[28]

$$U_c = 3U_i \ln(1 + 1/\Gamma)$$

which gives $U_c = 400$ V for typical values of the ionization voltage U_i (14,534 V for N) and Γ ($\simeq 10^{-4}$). The cathode fall therefore depends not on pressure and temperature but rather sensibly on the gaseous environment (U_i) and on surface layers on the cathode (Γ, oxides, soot, hydrocarbons, etc.).

As the cathode fall region is very close to the cathode surface (<0.1 mm) almost all of the energy input into this layer is conducted into the cathode electrode. Due to the long burning times involved, even appreciable amounts of the energy input to the glow plasma are lost again (up to 50% in quiescent conditions at 1 bar and a 1-mm gap). As the current is low, the field strength in the gap region is in the order of 10^3 V/cm. The anode fall is the same as in the arc. This sums up to ≃500 V for the total burning voltage at 1 bar and a 1-

mm gap. The plasma voltage increases with increasing density of the gas. The proportionality constant is ≃50 V/mm bar determined at 300 K.

The steady-state values for the electron density and the kernel temperature are 2.10^{14} cm^{-3} and 3000 K, respectively. The same arguments as for the arc apply for the energy transport mechanisms from the kernel to the plasma surface. Thus the glow discharge channel may be regarded essentially as a heated wire with a Gaussian temperature profile. The experimental data have been given already in Figs. 7, 9, and 11 and in Table 1.

Erosion rates are low (see Table 2) and are caused mainly by sputtering. The rate is proportional to the product of current and time[30] and independent of the actual values of current and time. Due to the long burning time the glow discharge is very sensitive to flow fields which transport the discharge in the stream from the electrodes. For flow velocities up to 15 m/sec the discharge generates enough new electrons and ions to follow the increase in channel length smoothly and the burning voltage increases proportionally. At higher flow velocities the channel may attain such a length that the gap voltage rises above the value required for generating a new spark directly across the electrodes before the energy in the coil has been consumed. This restriking voltage is low (≃2–3 kV) since sufficient initial electrons and ions are available at the cathode. Thus the old discharge is interrupted and successive sparks, each starting with a "breakdown," are initiated. The breakdown plasmas associated with restriking are less efficient, however, as the lower field strengths and the heavy preionization produce a plasma state corresponding to the transition phase between late breakdown and arc.

In this case the individual channels are separated from each other and contain only fractions of the total electrical energy. The effective discharge length may rise to more than the 20-fold value of the gap spacing. The oldest plasma portion contains then most of the energy whereas the legs near the electrodes contain the least. The increase in burning voltage may conveniently be used to measure flow field properties under real engine conditions.[31,32]

C. Energy-Transfer Efficiency

Due to the physical properties of the discharge modes discussed above, only fractions of the electrical energy supplied to the spark gap may be actually transmitted to the mixture for inflammation. In Fig. 12 data for glow and arc discharges in quiescent air are given. The effects of heat losses to the electrodes and of anode/cathode falls are obvious. Unless discharge times are reduced drastically (<10 μsec) the transfer characteristics are independent of the actual values of current and time.

The breakdown (Fig. 13) allows a very high transfer efficiency ($>80\%$),

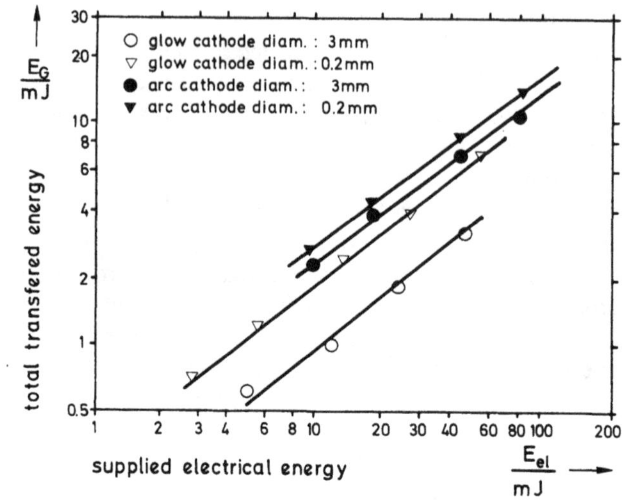

FIGURE 12. Total transferred energy as function of electrical energy supplied to the gap electrodes for arc and glow discharges. Unless discharge times are smaller than ≈10 μsec, no significant dependence of the energy-transfer efficiency on discharge time and/or current are found.

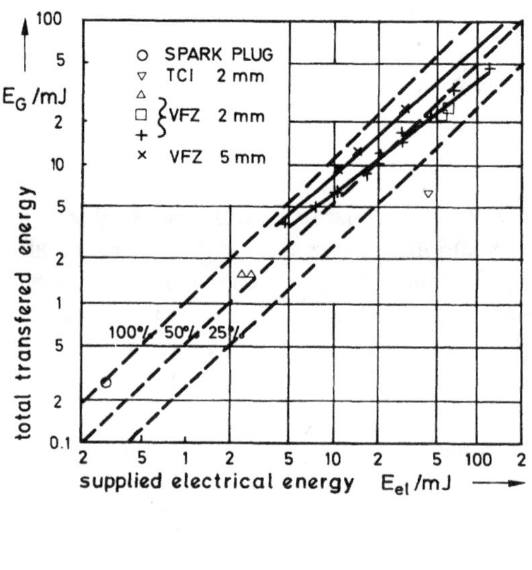

FIGURE 13. Total transferred energy as function of electrical energy supplied to the gap electrodes for breakdown discharges. Efficiency of TCI (30 mJ, 2 msec) shown for comparison.

however, with a falling tendency towards increasing energy inputs. This tendency is caused by an increase of the succeeding arc phase if the design of the breakdown device is not fast enough to supply all the stored energy during the time interval allotted for the breakdown phase. Practical systems will therefore have to be optimized for transferring the energy in extremely short time intervals which is possible by suitable layouts.[33]

In real engines quiescent conditions are rare and flow velocities of the mixture may be quite high especially near top dead center (TDC). In Fig. 14 the effect of a flow field on the energy-transfer efficiency is given. The breakdown is not affected at all as the energy is transferred in the nanosecond range (even at 100 m/sec the channel will not move more than 1 μm in 10 nsec). However, glow and arc modes are favored by flows because the discharge channel is carried away with the flow and lengthened accordingly. This effect is shown in Fig. 15. For flow velocities below 15 m/sec a steady increase in channel length results. Beyond 15 m/sec an increasing number of re-ignitions occur so that the energy input is distributed onto numerous independent discharge channels. Technical arc and glow discharges in flows may therefore be considered as having a "hairpin" geometry almost independent of the actual gap spacing.[34]

By flow lengthening, the ratio of the plasma voltage to anode/cathode falls/ rises considerably and at the same time heat losses due to heat conduction by the electrodes drop. Thus more energy is transmitted to the gas—however, in less concentrated form: the larger the channel length the smaller the radius which affects the ignition criteria, as will be shown in Sec. III.C. The maximum transfer efficiencies under ideal conditions as shown in Table 1 will be approached by

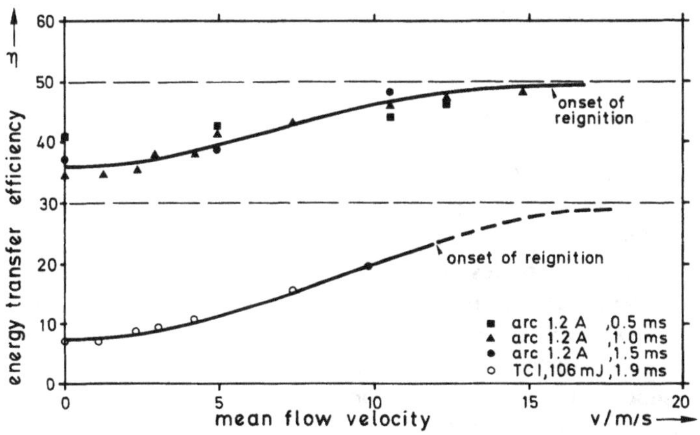

FIGURE 14. Energy-transfer efficiency for arc and glow discharges in flowing air at 300 K, 1 bar, and 1-mm gap.

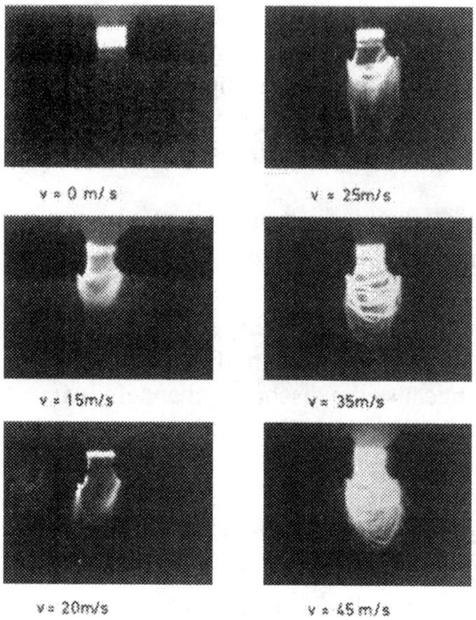

FIGURE 15. Single-shot TCI discharge (glow discharge, 30 mJ, 0.77—1.5 msec) in a flow field in air at 300 K, 2 bars, and a 1.2-mm gap. v_f = mean flow velocity perpendicular to gap axis. Below 15 m/sec almost no multiple discharges, above 15 m/sec only multiple discharges. Glow discharge channels are visible due to excitation by succeeding breakdown phases along the preionized path instead of forming a new channel.

flows around 15 m/sec. At higher velocities the efficiency with respect to individual channels remains near the ideal values of 30% and 50% for glow and arc, respectively, due to re-ignitions as soon as the channel length has grown long enough to assure good efficiency.

It has to be kept in mind, however, that the total transmitted energy is by no means a direct measure for the energy finally available for inflammation at the plasma surface as this is determined additionally by the temperature profile (see Fig. 9) and the plasma geometry (ignition radius and surface area, see Sec. III.).

D. Inflammation by Discharge Plasmas

Chemical reactions (e.g., formation of CN) are observable spectroscopically a few nanoseconds after spark onset (i.e., during the breakdown phase). They are initiated by the extremely high radical density in the breakdown plasma where all the heavy particles N, O, H, C are present as highly excited atoms and ions. However, as the kernel temperatures are still much too high to allow stable molecules to exist these reactions can take place at the low-temperature end of the plasma surface only. The expansion velocity of the plasma at this time is still supersonic so the possible contributions of low-rate chemical reactions to the front velocity are small (see Fig. 8).

All the energy-transfer processes are dominated by the violently expanding plasma, and the small contributions by the chemical reactions during this short time interval are simply added to the total plasma energy. However, irrespective of the interior conditions and the expansion velocity of the physical plasma there will always be a reaction zone at the plasma surface where a sufficient temperature level ($T < 8000$ K) offers ideal conditions for very intense chemical activities. These activities comprise the full scale of all reactions possible between the species present in the temperature interval from 8000 K down to room temperature, though they are still strongly dependent on the energy and particle flux from the plasma.

Sensible contributions of the chemical reactions to the volume expansion will become observable when the plasma velocity has fallen below 100 m/sec and chemical reaction rates for these extreme conditions become comparable to the local expansion rates. At this time the interior of the spark plasma still consists of a completely reacted gas kernel containing most of its energy in the form of potential energy (see Figs. 5 and 6) stored in the radicals. N, O, H, C from the fully dissociated molecules of the mixture. The conditions along the steep temperature profile through the narrow (0.6 mm at 1 bar, 0.15 mm at 4 bars) surface zone—the inflammation zone—may be illustrated by the calculated equilibrium composition of a CH_4–air mixture (given in Table 3) though actually the real concentrations are certainly far from equilibrium.

As the different radicals have different diffusion velocities (the hydrogen radical has a fivefold higher diffusion coefficient than the other species, see Table 4) a layered structure in the reaction zone establishes with the hydrogen radical diffusing farthest into the surrounding mixture. Thus we have at the high-temperature side of the inflammation zone an effective energy input by all kinds of

TABLE 3. Calculated Equilibrium Composition of Stoichiometric CH_4–Air Mixture at Different Temperatures

	Temperature (K)				
Species	1200	2000	3000	4000	5000
CO	—	2.6E-3	5.3E-2	7.2E-2	6.3E-2
CO_2	9.1E-2	8.8E-2	3.6E-2	1.5E-3	7E-5
H	—	6E-5	1.7E-2	1.4E-1	2.3E-2
H_2	—	1.2E-3	1.2E-2	7.3E-3	1.3E-3
H_2O	1.8E-1	1.8E-1	5.8E-2	1.3E-3	1E-5
NO	—	7.5E-4	2.2E-2	2.3E-2	8.9E-3
N_2	7.3E-1	7.2E-1	6.4E-1	5.3E-1	4.8E-1
O	—	3E-5	2.5E-2	1.6E-1	1.9E-1
OH	—	8.3E-4	2.8E-2	1.5E-2	1.8E-3
O_2	—	1.9E-3	4.9E-2	1.1E-2	6.4E-4
N_2H_3	—	7.9E-4	5.7E-2	4.2E-2	4.6E-3

TABLE 4. Diffusion Coefficients for Major Species in Fuel–Air Mixtures[a]

Temperature (K)	Diffusion coefficient $D_{1,2}$ (cm^2/sec at 300 K and 1 bar)		
1000	N_2–N_2	O–N_2	H–N_2
1000	1.55	2.15	8.5
2000	4.93	6.81	26.8
3000	9.58	13.4	53.4
4000	13.2	21.5	84
5000	22.3	31.2	121.5

[a] According to Ref. 24.

radicals carrying many times their thermal energy portion ($E_{th} \simeq 0.7$ eV per particle at 8000 K) in potential form ($E_p \simeq 5$ eV per particle, mostly dissociation energy) which they transfer to the mixture molecules within a few collisions. At the low temperature side overequilibrium concentrations of the start radicals for chain reactions O and H built up. All this is assisted by a high thermal heat flux via gradient processes.

Reactions will therefore run at much higher speeds than usual and a fast liberation of chemical energy results. Due to the high nitrogen content of combustible mixtures with air, reactions with the very energetic N radical are dominating during the inflammation phase where the temperature level is still appreciably higher than in ordinary premixed flames. Molecules such as NO, NH, and CN are readily formed. Being unstable at lower temperatures, they enter the complex reaction schemes of hydrocarbon–air combustion[35,36] and transfer their energy onto the other species as soon as the inflammation zone has cooled down sufficiently during the progressing expansion.

At some 10 μsec after spark onset the plasma temperature has fallen to flame temperatures and the energetic support from the plasma begins to cease. Molecules such as OH, CH, C_2, CO, etc., appear now indicating combustion processes as known from stationary flames. This is the most crucial point in inflammation as the reactions in the inflammation zone—the flame—must be developed by this time so far that they are self-sustained. From now on the chemical reaction rate has to be fast enough to compete successfully with the beginning concurrent heat loss over the front surface by heat diffusion and conduction. Calculated values of heat conductivity of relevant species are given in Table 5.

The processes described above are valid for the initial inflammation processes irrespective of the plasma mode used, be it breakdown, arc, or glow discharge, as sparks always start with a breakdown phase. However, due to the

TABLE 5. Calculated Heat Conductivity of Major Species in Fuel–Air Mixtures[a]

Temperature (K)	Heat conductivity (10^{-6} J/cm sec/K)					
	N_2	O_2	CH_4	CO_2	H_2O	H_2
1000	665	747	1687	672	1159	4279
2000	1141	1262	3453	1182	2630	7520
3000	1511	1728	4827	1413	3904	10.446
4000	1854	2154	5972	1935	4999	13.069
5000	2158	2539	6988	2257	5982	15.496

[a] According to Ref. 49.

technical layout of ignition systems and/or the requirements of the mixture, the amount of energy contained in the breakdown plasma may not be sufficient for rendering the flame self-sustained during the available cooling interval. In this case (which is relevant for arc and glow plasmas) the initial reaction rates may drop prematurely with the onset of thermal heat losses. The reaction zone will then move upwards on the temperature profile (i.e., retracts towards the discharge axis) until more favorable conditions are found again. As arc and glow discharges supply heat mainly by heat conduction from the axis, the possible increase in reaction rates is less effective than in breakdown plasmas where most heat is carried by radicals. If conditions are severe, the inflammation zone may even have to retreat virtually to the discharge axis where temperatures are highest. In the axial region also radicals from the dissociation of oxygen and combustibles will be present though in lower concentrations as in the initial breakdown plasma. In this case the arc will provide more favorable conditions for inflammation than the glow discharge due to the higher axial temperatures and the consequently much higher radical densities.

In stoichiometric mixtures the energy in the preceding breakdown plasma from a conventional spark plug (0.3–1 mJ, depending on ignition voltage and gap capacity) is sufficient for inflammation as shown in Fig. 16. All the measured flame front gradients clearly carry the steep profiles of the early breakdown plasma. Only slight contributions of the arc plasma, and to a lesser extent of the glow plasma, to front temperature and expansion velocity are noticeable. On the other hand increasing the energy input to the breakdown plasma promotes reaction rates considerably.

By leaning of the mixture, the reaction rates may be lowered and the demand on ignition energy raised. In consequence we recognize in Fig. 17 at $\lambda = 1.4$ a distinctly slower smearing out of the temperature valley between arc and breakdown plasma. This indicates that the low energetic breakdown plasma from

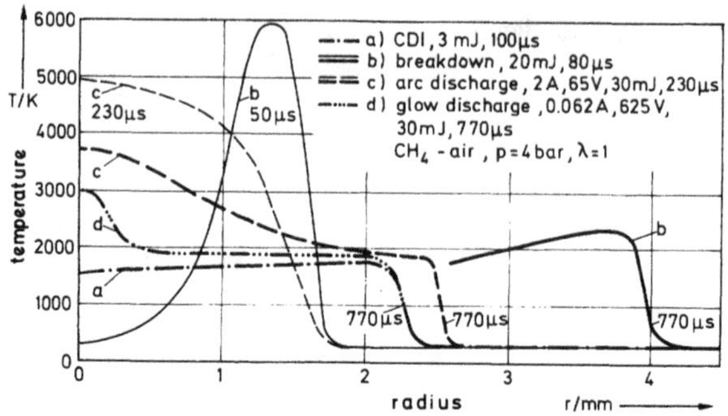

FIGURE 16. Temperature profiles for different discharge modes at different times measured in a stoichiometric CH_4–air mixture at 300 K, 4 bars, and a 1-mm gap. (a) CDI, 3 mJ, 100 μsec; (b) breakdown, 20 mJ, 80 nsec, (c) CDI, 3 mJ, 100 μsec plus superimposed constant current (2 A) arc, 30 mJ, 230 μsec; (d) CDI, 3 mJ, 100 μsec plus superimposed constant current (60 mA) glow discharge, 30 mJ, 770 μsec.

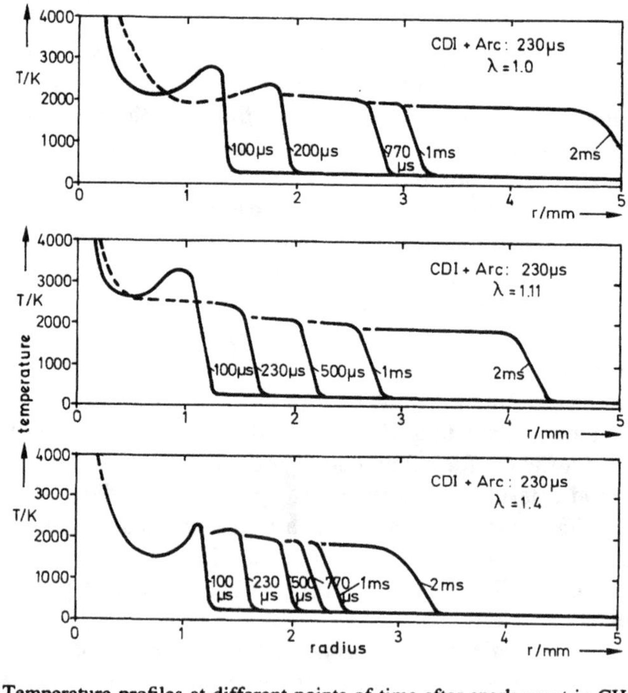

FIGURE 17. Temperature profiles at different points of time after spark onset in CH_4–air mixtures (300 K, 4 bars, 1-mm gap) and different A–F ratios but identical ignition devices: CDI, 3 mJ, 100 μsec plus superimposed constant current (2 A) arc, 30 mJ, 230 μsec.

the plug alone is now insufficient for rendering the flame front self-sustained. Reactions are delayed until sufficient energy from the arc plasma is available for assistance and the resulting temperature profile shows the slower slope of the arc plasma.

Ignition energy is thus transferred from the plasma to the combustible mixture by collisions of energy-rich radicals carrying large amounts of potential energy with the surrounding molecules on the one side and by heat diffusion and conduction on the other side. As the potential energy of the radicals is converted into thermal heat by only a few collisions between heavy particles, and the flame will proceed then as a thermal phenomenon anyhow, there is from a global point of view no need to discern strictly between radical and temperature effects. The goal of inflammation is to start self-sustained reactions which have their own chemistry so both energy forms may be considered equivalent provided they supply the energy density required by the mixture conditions.

As flame propagation is inversely proportional to the thickness of the flame front[(37)], steepest gradients in temperature and thus in particle density will provide highest transport rates and therefore best possible start positions for flame propagation. Excess gradients may act as reserves to overcome adverse effects occurring at later stages of propagation. The slow slopes characteristic for arc and glow plasmas provide thicker fronts with lower velocities and are therefore more sensitive to disturbances.

The high initial flame speeds desirable for efficient engine operation reduce the time available for energy exchange processes in the inflammation zone and the spatial range across which energy may be carried to about 100 μsec and some diffusion lengths, respectively. Ignition energy should therefore be preferably introduced into the narrow inflammation zone alone while the other volume sections can remain at ambient conditions.

E. Discharge Mode and Flame Propagation

1. Quiescent Mixtures

The effect of the discharge mode on inflammation and flame propagation was studied with three specially built ignition systems each supplying the same amount of electrical ignition energy to the gap electrodes in as far as possible only a single discharge mode: breakdown, arc, or glow discharge. A CDI (3 mJ, 100 μsec) was used to initiate the arc and the glow discharge. Some experiments were carried out with devices suitable also for operation under motor conditions: TCI (essentially a glow discharge) and VFZ (essentially a breakdown discharge with the same energy as the TCI).

In Fig. 18 the temporal evolution of a combustion wave in a quiescent CH_4–air mixture is shown by a series of successive interferograms. It serves as

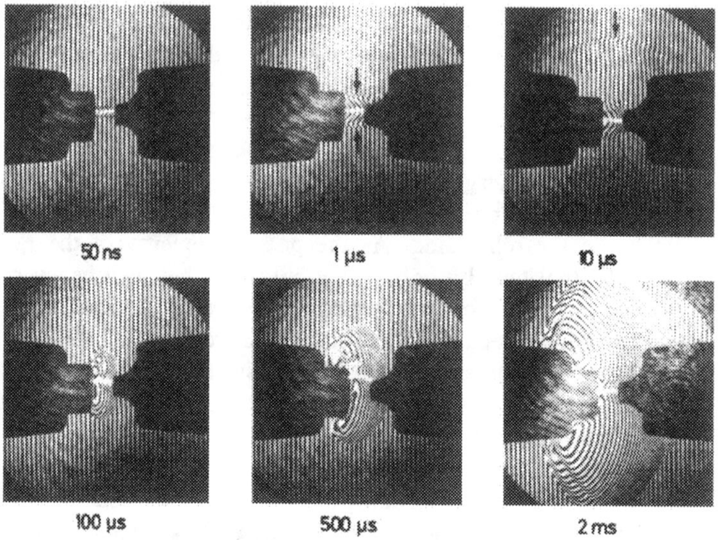

FIGURE 18. Successive interferograms showing the temporal evolution of a combustion in a stoichiometric CH_4–air mixture (300 K, 3 bars, 1-mm gap) ignited by a CDI, 3 mJ, 100 µsec. Arrows indicate blast (shock) fronts.

a quick overview over time and space relations involved during inflammation and early flame propagation. The initially cylindrical discharge channel soon acquires a spherical shape with flat poles at the electrodes due to quenching effects. Arc-and-glow-initiated fronts proceed in spherical form whereas breakdown-initiated fronts go through an intermediate toroidal shape before finally acquiring also a spherical geometry. Diameters and velocities during the initial phases have been given already in Fig. 8. Figure 19 completes this for late-time intervals.

It has been noted already (see Figs. 16 and 17) that the preceding breakdown dominates the volume expansion until some 10 µsec after spark onset because of the overpressure-determined plasma expansion irrespective if the environment is combustible or not. Later expansion in noncombustibles is much slower due to the slow processes of heat conduction and diffusion. Sustained energy inputs beyond 10 µsec as in arc and glow discharges contribute via continuous heating of the central region to somewhat higher expansion rates. However, inflammation has occurred already within less than 10 µsec as has been described above. This is verified in Fig. 19 by the fact that the curves for combustible mixtures differ markedly from that time on from those in air. Again we see that the same energy input in a breakdown creates the largest volume of all discharges at all times. There is no additive initial increase as one might expect but a multiplicative

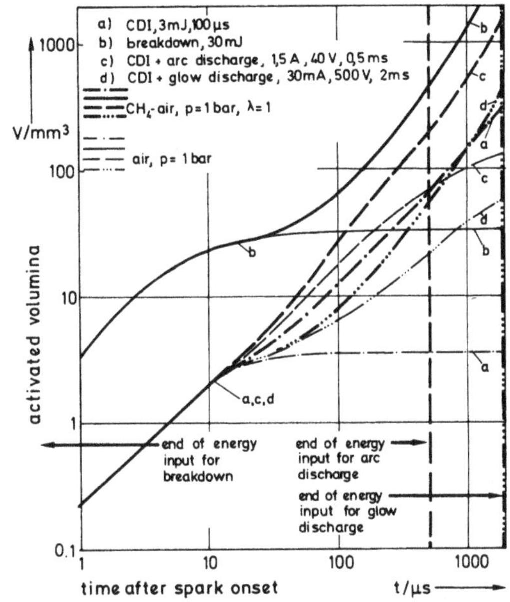

FIGURE 19. Temporal increase in volumes activated by CDI, breakdown, arc and glow discharges in a stoichiometric CH_4–air mixture at 300 K, 1 bar and a 1-mm gap.

persisting until the end of the combustion process. This indicates that there is some effect remembering the initial inflammation process in spite of its short duration.

Figures 16 and 17 explain the different expansion velocities by corresponding temperature profiles. Arc and breakdown achieve almost the same active volumes in air at 50 and 230 μsec, respectively, however, with different slopes of the temperature profiles. In stoichiometric CH_4–air mixtures the flames clearly display the steep gradients of the initial breakdown plasma with only a minor contribution of the arc which heats primarily the central region. The same applies for the glow discharge though with a still further reduced influence. The flame fronts thus travel with velocities determined predominantly by the gradients imprinted by the preceding breakdown phase. This demonstrates clearly that the heat transfer by conduction and diffusion is normally too slow to carry appreciable amounts of energy from the central input region to the active surface zone. Estimates of the time scales involved may be derived from the expansion curves in air.

The higher the energy input in the breakdown phase the faster the flame will travel. This is in contrast to literature on stationary combustion where the burning velocity is considered to be independent from the ignition process and a function of the combustible mixture alone.[2] However, here as well as in engines ignition and flame propagation are instationary processes where initial conditions distinctly affect the entire course of the reaction as will be theoretically shown later. This is illustrated in more detail in Fig. 20 by successive temperature profiles during the course of flame propagation. Arc and glow contribute only to a minor degree to the temperature profile at the surface of the plasma. Here the fronts reflect the much steeper fronts of the preceding breakdown for appreciable time and space intervals.

Expansion velocities twice those of glow and arc may be achieved in addition to a second flame front on the inner surface of the plasma torus from a breakdown plasma. Thus we recognize again the necessity to establish the steepest possible temperature gradients (and thereby also radical gradients) within the narrow inflammation zone in order to inflame most efficiently and to accelerate the reactions as far as required.

An increase of the gap spacing affects flame propagation as shown in Fig. 21. Increasing the spacing from 0.5 mm to 0.7 mm reduces mainly the heat losses to the electrodes. This gain drops off for larger widths (>1 mm). The remaining effect is caused by the increase in ignition voltage with larger gaps

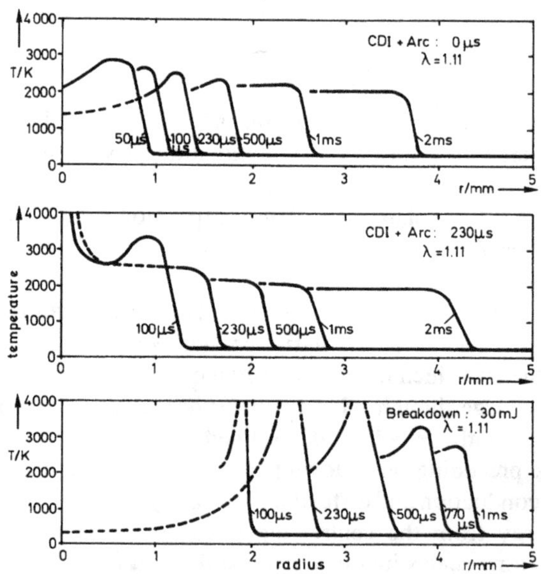

FIGURE 20. Temperature profiles at different points of time after spark onset in a CH_4–air mixture at 300 K, 4 bars, and a 1-mm gap. Ignition sources used were: (a) CDI, 3 mJ, 100 µsec plus superimposed constant current (2 A) arc, 30 mJ, 230 µsec [zero arc time is identical to CDI plus superimposed constant current (60 mA) glow discharge, 30 mJ, 770 µsec]; (b) breakdown, 30 mJ, 60 nsec.

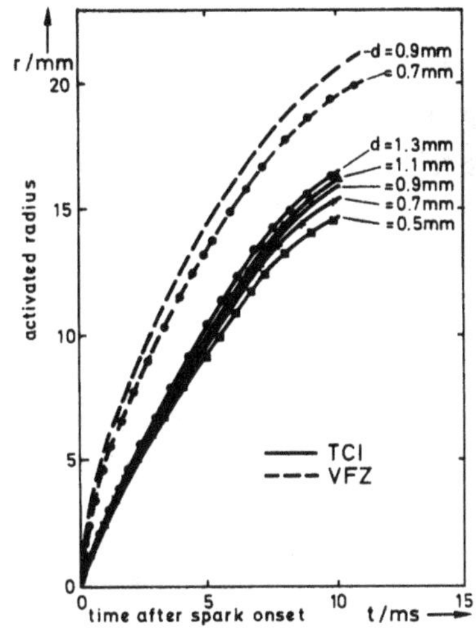

FIGURE 21. Effect of gap spacing and ignition source on radius of activated volumes in a stoichiometric CH_4–air mixture at 300 K and 2 bars. TCI, 30 mJ, 770 μsec (essentially a glow discharge); VFZ, 30 mJ, 60 nsec (essentially a breakdown discharge).

which leads to a redistribution of the ignition energy in favor of the breakdown phase ($E_b = {}^1/_2\ CU_0^2$. The same applies for the VFZ where the increase in gap spacing reduces the energy for the succeeding arc phase also in favor of the energy input to the breakdown phase.

An acceleration of the reaction rates is especially useful in lean mixtures where flame temperatures fall and reaction times rise unduly. In Fig. 22 the effect of the plasma mode on the reaction pressure is shown. Even at $\lambda = 1$, where very small ignition energies are sufficient to start a reaction, distinctly higher rates are obtained. At $\lambda = 1.4$ the relative improvement is much more drastic indicating the need of a substantially increased energy input into the reaction zone in lean mixtures. The lean limit may then be extended, as Fig. 23 illustrates, causing higher thermal efficiencies and lower cyclic pressure variations in engines.

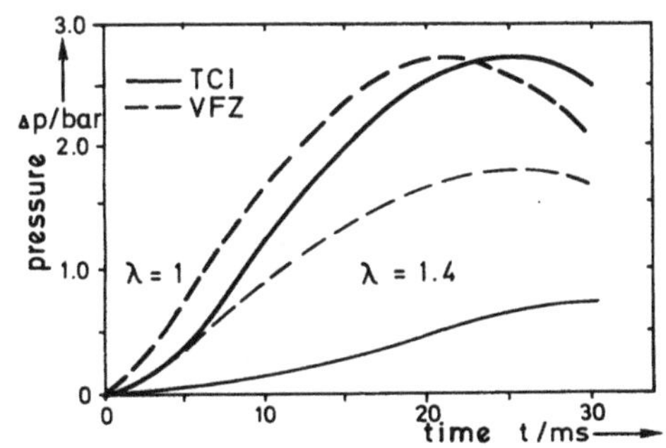

FIGURE 22. Pressure traces from a combustion of CH_4–air mixtures at 300 K, 2 bars, and a 1-mm gap in a disk-shaped bomb. TCI, 30 mJ, 770 μsec, essentially a glow discharge; VFZ, 30 mJ, 60 nsec, essentially a breakdown discharge.

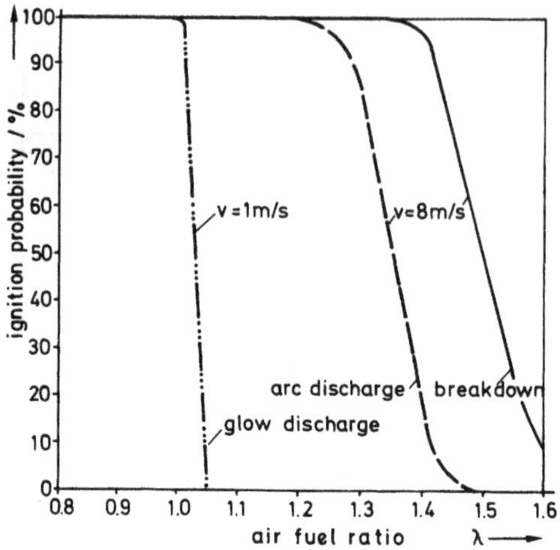

FIGURE 23. Ignition probability for CH_4–air mixtures at 300 K, 4 bars, 1-mm gap, and different mixture velocities *v*. (a) Breakdown, 30 mJ, 60 ns; (b) CDI, 3 mJ, 100 μsec plus superimposed constant current (2 A) arc, 30 mJ, 230 sec; (c) CDI, 3 mJ, 100 μsec plus superimposed constant current (60 mA) glow discharge, 30 mJ, 770 μsec. The glow discharge was extinguished at mixture velocities above 1 m/sec.

If a pure glow discharge is used only a small air–fuel range close to $\lambda = 1$ may be inflamed. This range widens, however, with the energy-transfer capabilities of the discharge mode. The lean limit is, on the other hand, a very sensitive indicator of the amount of energy actually transmitted to the reaction zone:

In Fig. 24 a comparison of the lean limits for TCI and VFZ is shown as a

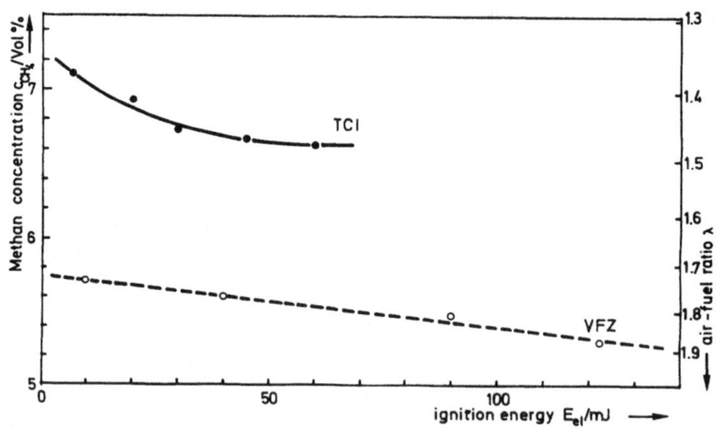

FIGURE 24. Variation of lean limits with ignition energy in quiescent CH_4–air mixtures at 300 K, 1 bar, and 1-mm gap. TCI, 30 mJ, 770 μsec, essentially a glow discharge; VFZ, 30 mJ, 60 nsec, essentially a breakdown discharge.

function of the supplied energy. Whereas the VFZ may virtually push the lean limit to the theoretical one at $\lambda = 1.95$ with only moderate increases in energy inputs, the TCI fails already below $\lambda = 1.5$ due to the insufficient energy transport processes involved. We also see that it is not necessary to resort to unrealistic high energy levels for extending lean engine operation, but that energies already used in modern ignition systems will do.

In practical combustion processes inflammation and flame propagation will always be influenced additionally by the locally inhomogeneous properties of the combustible mixture, be it mixture strength, mixture composition, temperature distribution, or turbulence level. In a statistical model one might attribute different grades of the inhomogeneous property to different spheres of a size corresponding to the statistically mean volume of the inhomogeneity which are distributed at random in space, as shown schematically in Fig. 25.

If ignition diameters of the spark plasma are comparable or smaller than the statistical mean sphere size of the inhomogeneities, a strong dependence on the local properties will result. On the other hand the effect of the inhomogeneous structure vanishes if the plasma volume is sensibly larger than the mean sphere size because then a large enough volume will always be activated, now being representative indeed for the mean and not for the local properties of the mixture. This is fully verified experimentally as shown in Fig. 26. The larger the plasma volume the better all kinds of inhomogeneities will be overcome. It might be useful to recall that it is not necessary to resort to full plasma spheres as shells of sufficient thickness (some flame front thicknesses) will serve the same purpose.

In Fig. 27 finally the effect of wall contacts on flame propagation by corresponding pressure traces is illustrated. Inflammation in the center of the reaction chamber results in faster reaction rates and higher peak pressures. The closer the point of ignition is to the walls the earlier heat losses from the combusted part of the charge arise, which lower thermal efficiencies sensibly. Changes in the pressure slopes clearly indicate the arrival of appreciable parts of the burning

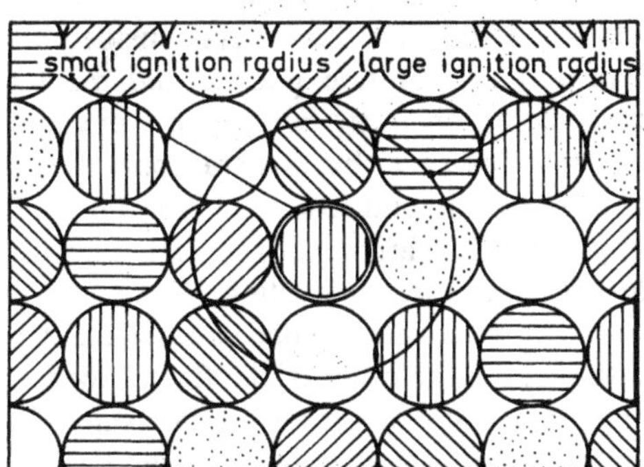

FIGURE 25. Schematic representation of the effect of the size of the ignition radius on inflammation in an inhomogeneous mixture. Different grades of the inhomogeneous property are attributed to different randomly distributed spheres of a constant size equal to the statistically mean size of the inhomogeneity.

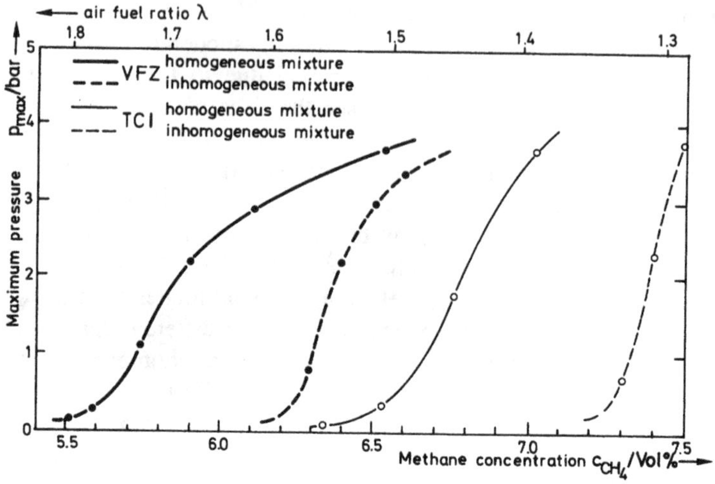

FIGURE 26. Effect of ignition radius on lean limits in homogeneous and inhomogeneous CH_4–air mixtures (300 K, 2 bars, 2-mm gap), respectively. TCI, 30 mJ, 770 μsec, essentially a glow discharge, small ignition radius; VFZ, 30 mJ, 60 nsec, essentially a breakdown discharge, large ignition radius. Reproducible inhomogeneous mixtures were prepared by variable stirring times of the separately but synchronously introduced CH_4 and air fractions.

mixture at the walls. Leaning the mixture lowers the burning velocity and higher heat losses occur. The principal relations remain unchanged however, as shown.

2. *Flowing Mixtures*

Flame velocities as measured in quiescent mixtures are far too low for high-speed engine operation. In order to achieve a faster burning rate turbulent flow fields are used which affect both inflammation and flame propagation.

The mean velocity and the macro-turbulence component (large, slowly rotating eddies with diameters considerably larger than the flame front thickness) transport the ignition plasma and the flame front to other places within the combustion chamber but will not engage themselves into the reaction process *per se*. However, by this transport the reacting volume may come in contact with the walls and lose heat (see Fig. 27). During inflammation this loss will impair the development of the self-sustained flame; at later stages increased heat losses lower the combustion efficiency.

The micro-turbulence, on the other hand (small, fast rotating eddies with diameters in the order of the flame front thickness) intensify the transport of the energy-rich species from the hot side of the flame front to the unreacted side and vice versa so that an increased heat exchange in the front results (turbulent

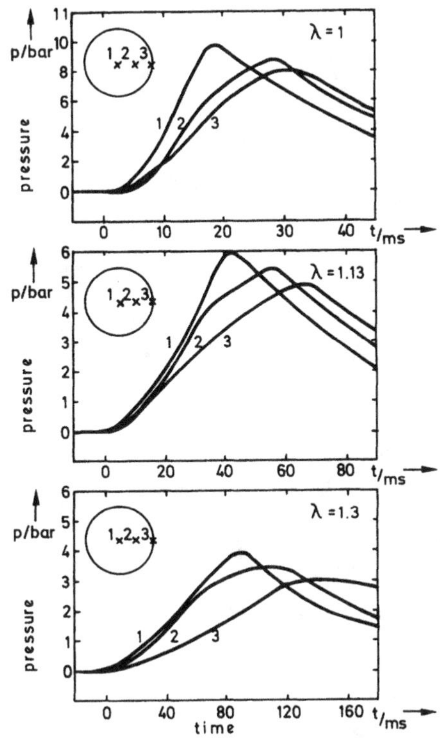

FIGURE 27. Effect of locus of ignition on the combustion pressure in a disk-shaped bomb. Ignition of the quiescent CH_4–air mixture at 300 K, 4 bars, and a 2-mm gap by a TCI (30 mJ, 770 μsec, essentially a glow discharge). The locus of ignition is indicated by crosses.

diffusion and heat conduction) by far exceeding the contributions by conventional diffusion and conduction. Flame reactions will thus be dominated by the turbulent heat exchange and all further propagation will be determined by the turbulence levels. Too high levels during the inflammation phase will unduly lower the temperature in the inflammation zone. In extreme cases the reaction may therefore even cease completely.

If the reaction rates are higher than those of the turbulent heat exchange an efficient increase in flame velocity results. However, as the turbulence cannot be restricted to the moving flame front alone, turbulent heat losses to the walls increase at the same time and lower the thermal efficiency of the combustion process. Turbulent flow fields, therefore, have a strong bivalent effect. Though they do not modify the fundamental chemistry, they speed up the effects described in quiescent mixtures be it for or against the benefit of the overall process. For optimum response the turbulence should always increase at the same rate as the flame gains stability.

Larger flame kernels from large ignition plasmas in which reactions have proceeded faster than in small ones are less sensitive to the disadvantageous

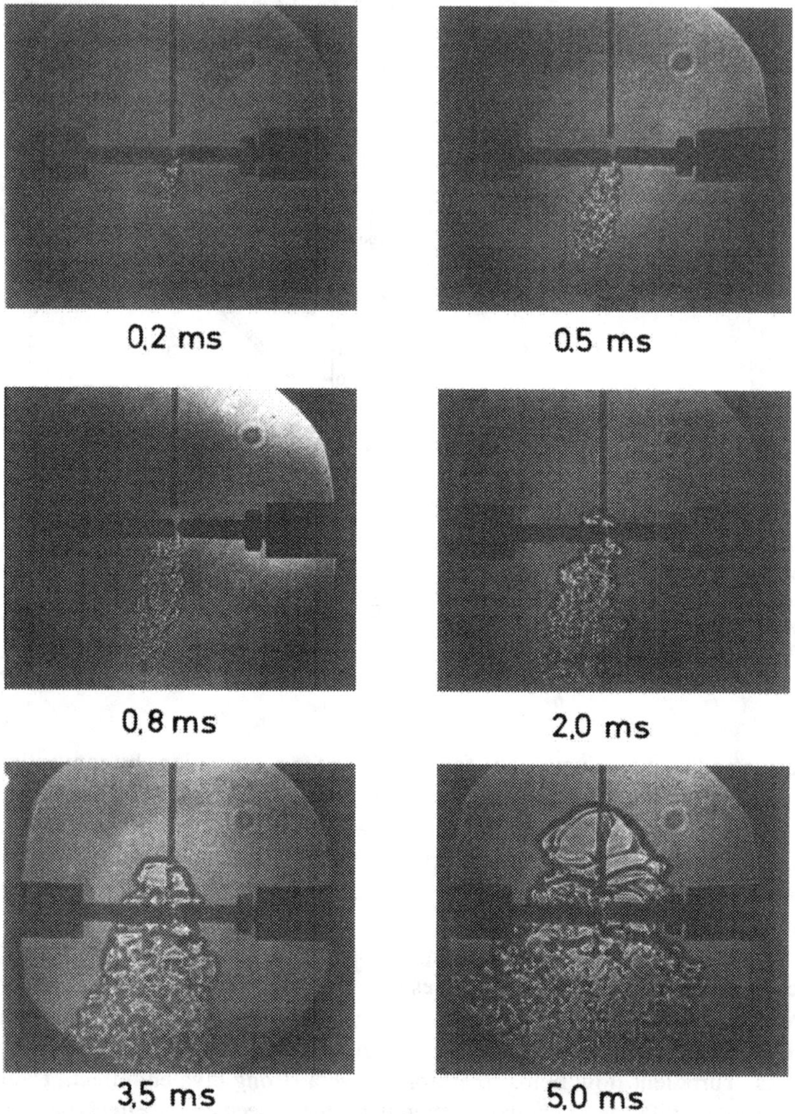

FIGURE 28. Shadowgraphs of the time-dependent combustion in a stoichiometric CH_4–air mixture at 300 K, 2 bars, and a 1-mm gap in a turbulent flow of 50 m/sec, perpendicular to the gap axis. Ignition by TCI, 30 mJ, 770 μsec, essentially a glow discharge.

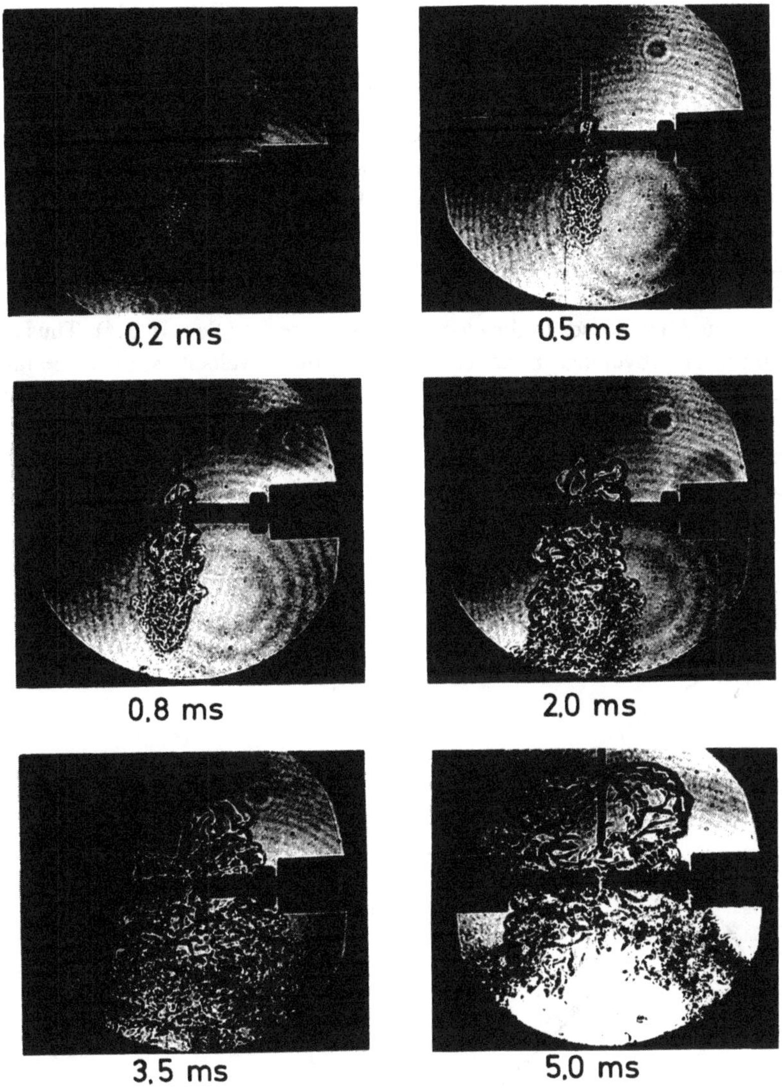

FIGURE 29. Shadowgraphs of the time-dependent combustion in a stoichiometric CH_4–air mixture at 300 K, 2 bars, and a 1-mm gap in a turbulent flow of 50 m/sec perpendicular to the gap axis. Ignition by VFZ, 30 mJ, 60 nsec, essentially a breakdown discharge.

effects of the turbulence. This is illustrated by Figs. 28 and 29. Whereas the TCI-ignited flame kernel cannot burn against the 50 m/sec turbulent gas flow from the nozzle perpendicular to the gap axis, this is easily possible for the VFZ-inflamed reactions which additionally benefit from the turbulence and achieve a faster overall burning rate.

The basic findings from quiescent mixtures are thus valid for turbulent flows as well, though they will be modified as described above. As the micro-turbulence lowers the initial flame temperature in the same way as it is done by leaning the mixture, the lean limit is shifted by the turbulence towards the richer side. This is shown in Fig. 30. The curve for the VFZ displays this effect very clearly because of its constant energy-transfer characteristics. The ignition capability of the TCI, however, benefits from flow velocities up to $\simeq$15 m/sec as the channel lengthens and heat losses to the electrodes decrease (see Sec. II.B.4). Thus there is an initial improvement in the lean limit. At higher velocities, then, the falloff is faster than expected due to the onset of re-ignitions which cut off further energy input to an individual discharge plasma. This effect limits the potential for minimizing turbulent heat losses in TCI-ignited systems due to the degradation of the ignition capability at low flow velocities.

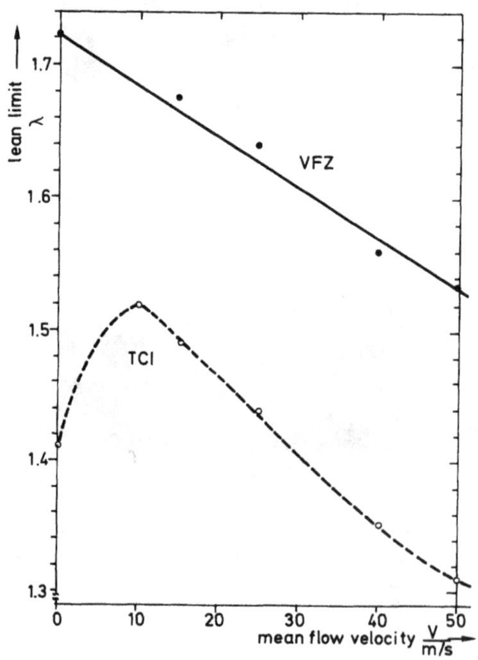

FIGURE 30. Effect of flow velocity on lean limits of a CH_4–air mixture at 300 K, 4 bars, and a 1-mm gap ignited by TCI, 30 mJ, 770 μsec (essentially a glow discharge) and VFZ, 30 mJ, 60 nsec (essentially a breakdown discharge).

III. IGNITION MODELING

A. Inflammation

A number of ignition models are known from the literature[38–42] which have been developed for those special purposes where time for affecting inflammation is not limited. Since we have seen that flame propagation is started at the surface of a violently expanding plasma inflammation, time is limited and the effect of nonzero initial velocity must be accounted for in a model if the instationary processes as spark and flame propagation in engines shall be treated. Though we have seen that the initial inflammation is accomplished primarily by highly energetic radicals one may treat the ignition process nevertheless as a thermal problem in a simple zero-order approximation due to the fact that the processes in the self-sustained flame front are closely linked to the thermal temperature profile within the front. The main purpose of such a model is to compress the experimental findings into principal dependencies which may be generalized and applied to arbitrary conditions rather than to reproduce special experimental results. As criterion for successful ignition, we define that the energy balance within the inflammation zone is identical to that of a self-sustained flame front with expansion velocity v_F. Radiation processes are neglected as radiation from flames is mainly due to the infrared emission from CO_2 and H_2O bands which normally cannot be absorbed by the unreacted mixture as these species are not available ahead of the front.[37]

The contributions to the heat flux across the front by diffusion and heat conduction (ordinary as well as turbulent) are summarized into a single transport coefficient κ_0 with the dimension of a heat conductivity. Assuming spherical geometry and constant pressure (the temperature profile is shown in Fig. 31) we obtain by a linearization of the problem the following expressions for the energy differences in an unreacted volume element ΔV (spherical shell of arbitrary size r_1 of thickness Δr_1) during the time interval Δt. Available energy ΔQ^V in ΔV is calculated as follows:

$$\Delta Q^V = \frac{(xq - \overline{\Delta H_F})}{V_0} 4\pi r_1^2 \, \Delta r_1 = \frac{(xq - \overline{\Delta H_F})p_0}{RT_0} 4\pi r_1^2 \Delta r_1 \tag{1}$$

where xq = molar heat of combustion, $\overline{\Delta H_F}$ = molar enthalpy required to heat the reaction products to flame temperature, and V_0 = molar volume.

Net energy loss over the cold surface (energy lost over the cold surface minus energy gained from the hot side at constant temperature gradients) is calculated as follows:

$$\Delta Q^A = \kappa_0 \frac{(T_F - T_0)}{\Delta r_1} \Delta t \; 8\pi r_1 \Delta r_1 \tag{2}$$

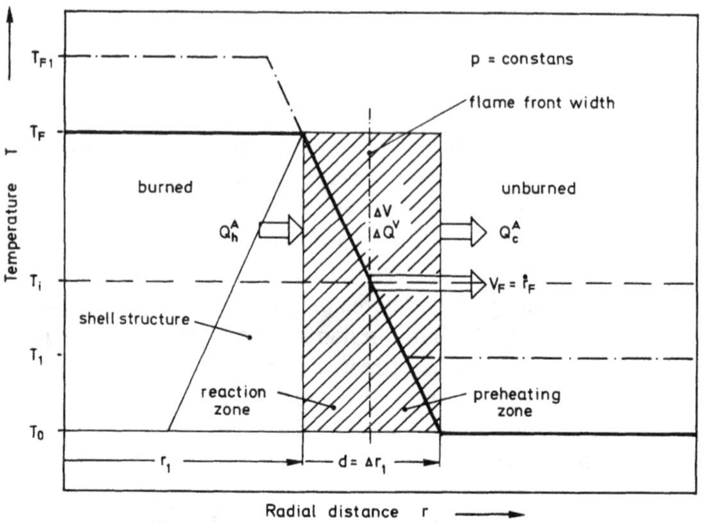

FIGURE 31. Schematic temperature profile and heat balance of a model flame front.

A self-sustained flame propagation occurs only if the net surface loss never exceeds the available energy in the volume element; thus

$$\nu = \Delta Q^V / \Delta Q^A \geqslant 1 \tag{3}$$

or explicitly

$$\nu = \frac{p_0(xq - \overline{\Delta H}_F)r_1\Delta r_1}{2R\kappa_0 T_0(T_F - T_0)\Delta t} = \frac{p_0(xq - \overline{\Delta H}_F)r_1\dot{r}_1}{2R\kappa_0 T_0(T_F - T_0)} \tag{4}$$

This indicates that at constant $\dot{r}_1$ the relative gain in energy available in the inflammation zone increases with the space coordinate r_1. Propagation conditions are the more favorable the larger the distance of the front from the discharge center. Prior to self-sustained propagation $\dot{r}_1 = v_P$ where v_P is the plasma expansion velocity. Therefore one calculates a minimum radius of the plasma required for successful inflammation:

$$r_{\min,s} \geqslant \frac{2R\kappa_0 T_0(T_F - T_0)}{p_0(xq - \overline{\Delta H}_F)v_P} \tag{5}$$

A check of this equation using literature data and the measured lean limits for the discharge profiles in Fig. 9 indicates a good agreement between calculated and measured minimum ignition radii[(10)]:

λ	$r_{\text{min,calc}}$	$r_{\text{min,meas}}$
1.03	0.7 mm	0.9 mm
1.5	2.85 mm	2.8 mm

Whereas all the parameters entering Eq. (5) alter the minimum ignition radius in accordance with experimental experience, special attention must be paid to the dependence on λ. Entering via x into a difference in the denominator it magnifies its influence by virtue of the resulting pole. The ignition radius increases therefore extremely rapidly for only moderate increases in λ in the lean range. As ignition energy is proportional to r^3_{min} a tremendous demand on plasma energy will arise calling for a more economical use of the electrical energy in shell-like structures.

Consequently, for optimum use of limited ignition energy the shortest possible discharge time is mandatory. High pressures (high energy densities) facilitate ignition by lowering the required minimum ignition radius and the inverse is true for heat exchange processes (κ_0). Too high values of κ_0 may cause unrealistic large ignition radii whereas too low values will cause poor propagation velocities.

If the interior of the spark-activated sphere is at $T = T_F$ (full sphere structure, see Fig. 31) the minimum ignition energy will be

$$E_{\text{min},sp} = \frac{4\pi p_0}{3R}\{\overline{\Delta H}_F\, r^3_{\text{min},s}/T_F + [(r_{\text{min},s} + d)^3 - r^3_{\text{min},s}]\,\overline{\Delta H}_F/\overline{T}\} \tag{6}$$

with $\overline{T} = (T_F + T_0)/2$ the approximate average temperature in the inflammation zone. However, as only a thin surface zone is required for inflammation, the kernel may be kept at room temperature in a shell structure and one obtains a real minimum energy:

$$E_{\text{min},sh} = \frac{4\pi p_0 \overline{\Delta H}_F}{R\overline{T}}[(r_{\text{min},s} + d)^3 - (r_{\text{min},s} - d)^3] \tag{7}$$

In case of cylinder geometry (large electrode gaps, arc, or glow plasmas in flows) one obtains

$$r_{\text{min},c} \geqslant (1 + 2r_{\text{min},c}/l)r_{\text{min},s}/2 \tag{8}$$

For short cylinders with $l < 2r_{\min,c}$

$$r_{\min,cs} \approx r_{\min,s} \tag{9}$$

and for long cylinders with $l \gg 2r_{\min,c}$

$$r_{\min,cl} = r_{\min,s}/2 \tag{10}$$

This seems to suggest that long cylinders are superior to spheres. However, this is not the case as minimum energies comparable to those of spheres are obtained only for cylinders with $1 \simeq r_{\min,s}$. This limits in practical ignition systems the possible length of the gap, the maximum tolerable turbulence level, and the discharge time due to the increase in effective length.

For very short cylinders or very large plasma radii, we obtain the quenching distance l_q:

$$l_q = r_{\min,s} = \frac{2RT_0\kappa_0(T_F - T_0)}{p_0(xq - \overline{\Delta H}_F)v_F} \tag{11}$$

Here v_F has been substituted for the plasma velocity. The quenching distance is therefore not a constant but depends strongly on the flame front velocity. This may be useful to reduce adverse effects of gaps and slits on exhaust emissions in engines.

B. Flame Propagation

As inflammation and early flame propagation are not sharply separable, a tendency for the temporal increase in flame radius is useful for interpretations of experimental data. This may be obtained from Eq. (4) by introducing the additional contribution to the volume expansion caused by chemical heat release and solving the differential equation after the time-dependent flame temperature has been removed by use of the excess enthalpy concept of Lewis and von Elbe.[2] Further it is assumed that the release of chemical heat is much faster than heat conduction rates and that the flame velocity can be split into a constant and a variable component. As a general solution one finds for the flame front velocity[43,44]

$$\dot{r} = \frac{\dfrac{p_0\,\kappa_0}{Rh_{F1}^2}[xq - \overline{\Delta H}_i]r_F - \left(2\dfrac{\kappa_0}{h_{F1}}T_0 + \dot{r}_{i0}\right)}{\dfrac{p_0\,C_{pi}}{Rh_{F1}}r_F + 1} \tag{12}$$

$$h_{F1} = \frac{(T_i - T_1)}{(T_i - T_0)} h_{F0} \tag{13}$$

$$\dot{r}_F = \dot{r}_{i0} + \dot{r} \tag{14}$$

which is readily solved on a computer.

For small values of r_F, being encountered in inflammation and early flame propagation, an approximate solution in closed form exists:

$$r_F(t) = \exp\left[\frac{p_0}{Rh_{F0}^2}(xq - \overline{\Delta H}_F)\int \kappa_0\, dt\right] \cdot \left\{C - \left(\frac{2T_0}{h_{F0}} + \dot{r}_{i0}\right)\int \kappa_0 \exp\left[-\frac{p_0}{Rh_{F0}^2}(xq - \overline{\Delta H}_F)\int \kappa_0\, dt\right]dt\right\} \tag{15}$$

By inserting suitable time functions for κ_0 the influence of flow and turbulence may be simulated since κ_0 has been introduced as a factor controlling the heat exchange within the flame front. Using spherical geometry—Fig. 32 will justify this—we now can trace the effects of ignition radius and turbulence on inflammation and flame propagation (see Fig. 33). In quiescent mixtures (turbulence level = 0) the larger the initial plasma radius the faster the flame proceeds. We also see that the initial conditions control the expansion even at later stages to

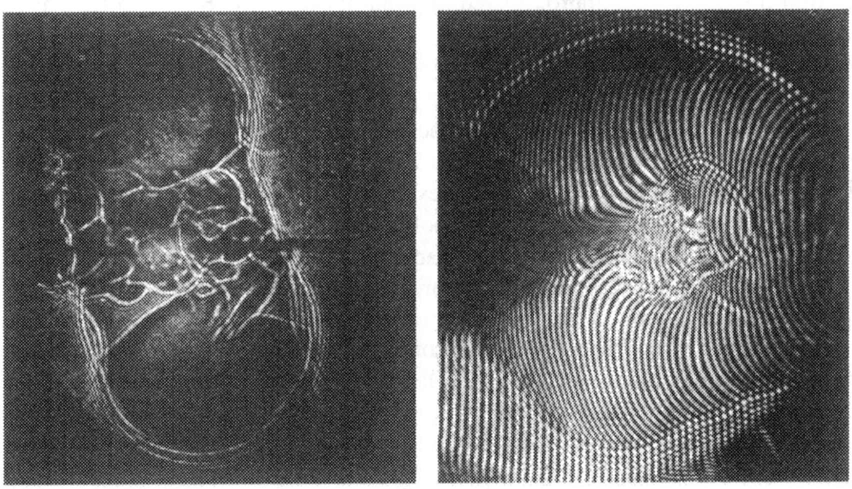

FIGURE 32. Radial and axial view (shadowgraph and interferogram, respectively) of a breakdown-ignited combustion in a stoichiometric, quiescent CH_4–air mixture at 300 K, 4 bars, and a 1-mm gap. Ignition by VFZ, 30 mJ, 60 nsec, essentially a breakdown discharge.

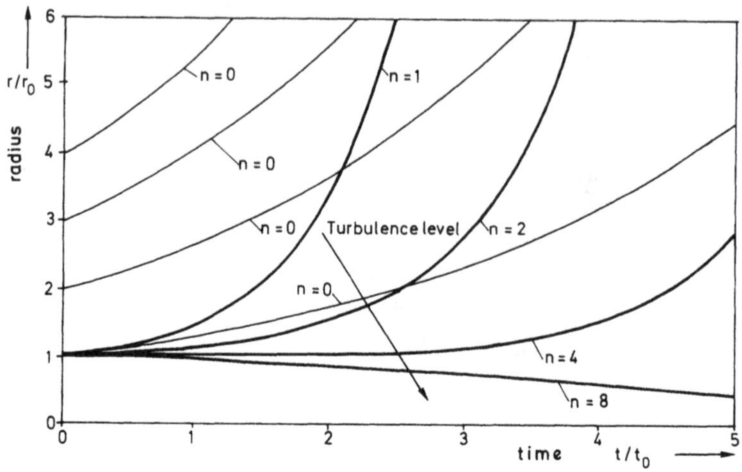

FIGURE 33. Calculated flame front growth for a sphere according to Eq. (15) for different initial ignition radii and turbulence levels at $\lambda = 1$. A linear with time increasing turbulence was simulated by using $\kappa_0(t) = \kappa_0\,(1 + nt/t_0)$.

some extent in agreement with the experimental findings (see Sec. II.E). Too low ignition radii (not shown) will lead to extinction.

The turbulence conditions during the upward motion of the piston near ignition timing may be approximated by a turbulence proportional to time after spark onset. Thus we have $\kappa_0(t) = \kappa_0 + n\kappa_1 t$ where n characterizes the turbulence intensity. The calculation shows (Fig. 33) that the flame front will be promoted by this turbulence function at the given ignition radius if the turbulence intensity is moderate ($n = 1$). With increasing intensity, however, the initial propagation is more and more impaired but is followed by a still more intense rise at late times. Too high intensity levels ($n = 8$) will lead to extinction of the propagation as the flame front is cooled too rapidly to survive.

These calculated tendencies in flame propagation are confirmed experimentally in Fig. 34. With increasing flow velocity the pressure recordings show the predicted behavior. At $v_F = 50$ m/sec no pressure rise occurred though the presence of a plasma kernel was monitored by schlieren records. The effect of the ignition radius (CDI smallest, VFZ largest) on reaction rates is shown in the lower traces. An increase in reaction rates will be achieved by larger plasma volumes similar to the effect of turbulence, however, without any initial degradation (lower part of Fig. 34).

Thus it may be advantageous for ignition systems with limited plasma energy to have an early stage of expansion at low turbulence levels until the flame front has grown so large that it can stand higher levels which might be present for

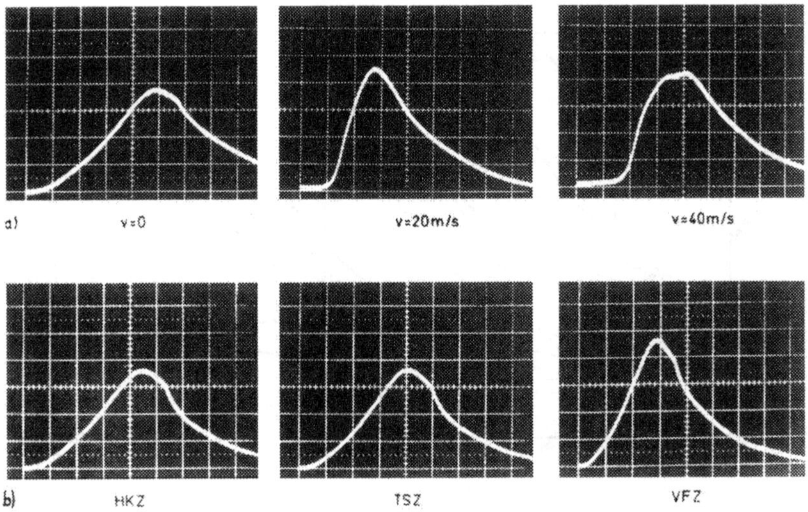

FIGURE 34. Pressure traces from the combustion of a stoichiometric CH_4–air mixture at 300 K, 1 bar, and 1-mm gap. (a) Effect of mean flow velocity of the mixture; (b) effect of ignition device. CDI, 3 mJ, 100 μsec; TCI, 30 mJ, 770 μsec; VFZ, 30 mJ, 60 nsec.

other reasons. This is fully affirmed by shadowgraph studies in an SI engine.[45] Fastest propagation would require an increase in turbulence intensity in the same measure as the flame front expands.

In Fig. 35 the calculated influence of λ is shown. A linear increasing turbulence intensity, as in Fig. 33, has been used additionally. We see that small ignition radii limit the possible leaning of the mixture and large stagnation (induction) times result at high λ values. If λ values are too high, inflammation will take place, but a self-sustained propagation will not be possible. By using larger ignition radii, these restrictions may be removed and extremely lean operation becomes available (compare $\lambda = 1.93$ at $r/r_0 = 1$ and 4).

For large radii Eq. (12) reduces to

$$\dot{r} = \frac{\kappa_0[xq - \overline{\Delta H_i}]\,(T_i - T_0)}{C_{pi}h_{F0}\,(T_i - T_1)}, \qquad \dot{r}_F = \dot{r}_{i0} + \dot{r} \tag{16}$$

This indicates that the flame velocity $v_F = \dot{r}_{i0} + \dot{r}$ attains a constant value at large radii and times as in linear flame expansion. The corresponding maximum flame temperature is then the adiabatic flame temperature. Furthermore we see that at late-time intervals the flame velocity becomes independent of pressure (i.e., engine load) as predicted from the general theory of flame propagation[24] and shown by experimental evidence.[44] Whereas early flame propagation is

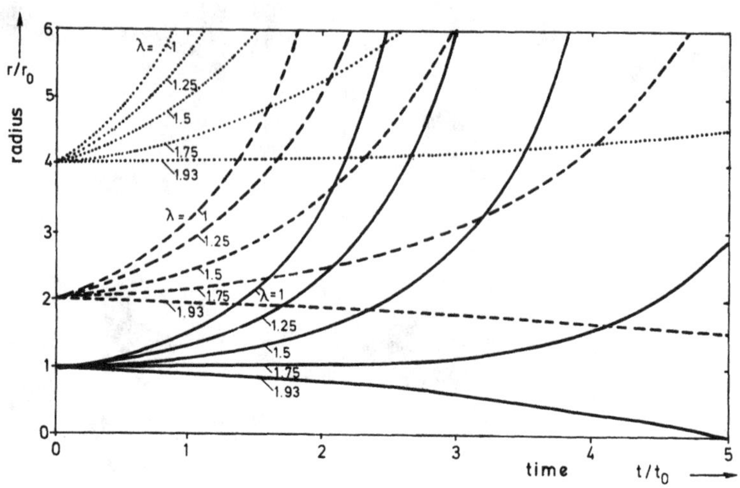

FIGURE 35. Calculated flame front growth for a sphere according to Eq. (15) for different initial ignition radii and λ values at linear with time increasing turbulence. The turbulence was simulated by using $\kappa_0(t) = \kappa_0(1 + nt/t_0)$.

promoted by a rising ambient pressure, it depends now predominantly on the turbulence level (κ_0) and on the heat capacity at flame temperature (C_{pi}), the turbulence being the promoter and the heat capacity the limiter especially in case of onset of appreciable dissociation.

Leaning of the mixture will reduce the flame speed, but since the flame radii are large enough by now, the reduction of flame velocity by an increase in λ can be compensated by higher turbulence levels. Therefore lean operation may be extended by increasing the charge motion in the chamber provided the flame kernel survives the initial stages determined by the ignition process.

Flame propagation must thus be seen as a two-stage process with a first phase sensitive to external conditions especially during inflammation and early flame propagation and a second insensitive one which depends almost exclusively on the time function of the micro-turbulence. In engines the turbulence is mainly generated by the combined action of piston movement and chamber geometry and culminates shortly before and after TDC (10–15° crank angle). This explains why an analysis of burning functions with regard to burn times shows little or no sensitivity to all kinds of parameter changes (ignition, leaning, mixture quality, etc.) in their late phase but a very high response in the first (induction) interval.

The flame front velocity can be increased also by rising T_1, however, without necessarily causing higher heat losses as in rising κ_0. Higher inlet temperatures of the charge are therefore very attractive for speeding up the flame velocity in lean or part load regimes if retention of high thermal efficiencies of the combustion process is desirable. Excessively high temperatures will lead to extremely fast reactions (knock) due to the pole in the denominator because the charge temperature is raised additionally during the combustion process in a sort of feedback process by the adiabatic compression of the end gas by the charge burned: the higher T_1 the faster the flame front, the faster the burning the faster the increase in T_1, etc.

The usual remedy in this case is to increase the charge motion in the chamber (as far as possible ahead of the flame front) and keep T_1 at a safe level by increasing heat losses at the expense of the internal efficiency. However, as the turbulence level cannot be restricted to the end gas alone, flame propagation is supported at the same time so provision of proper flow fields is a little delicate.

Typical results from simulations with the aid of Eqs. (12)–(14) are presented in Figs. 36–38 showing the effect of ignition radius, λ, turbulence level, and initial temperature on flame propagation. Though the relative effect of the ignition radius on flame propagation is largest during the initial phases as has been pointed out already it is still present at very late time intervals as well.

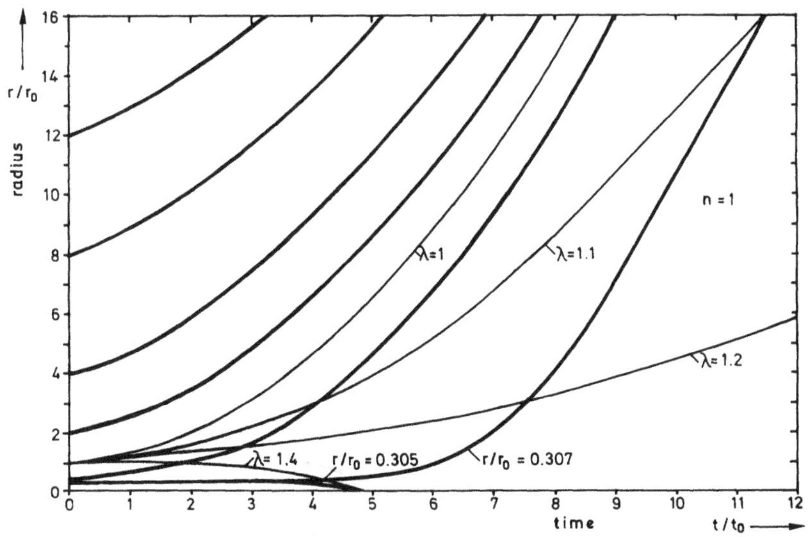

FIGURE 36. Calculated flame front growth for a sphere according to Eqs. (12–14) with $T_l = T_0$ for different initial ignition radii, turbulence levels, and λ values. Thick lines, effect of initial ignition radius; thin lines, effect of λ values for $r/r_0 = 1$. A linear with time increasing turbulence was simulated by using $\kappa_0(t) = \kappa_0(1 + nt/t_0)$.

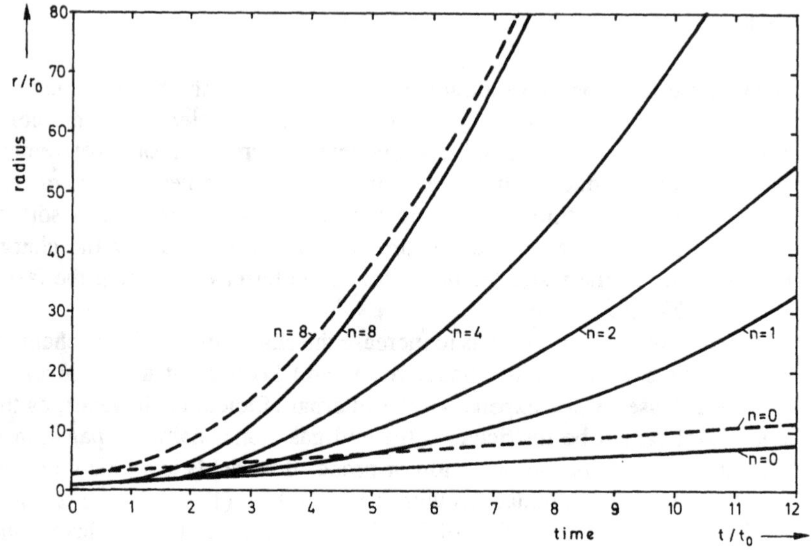

FIGURE 37. Calculated flame front growth for a sphere according to Eqs. (12–14) with $T_1 = T_0$ for different initial ignition radii and turbulence levels. Solid lines, $r/r_0 = 1$; dashed lines, $r/r_0 = 3$. A linear with time increasing turbulence was simulated by using $\kappa_0(t) = \kappa_0(1 + nt/t_0)$.

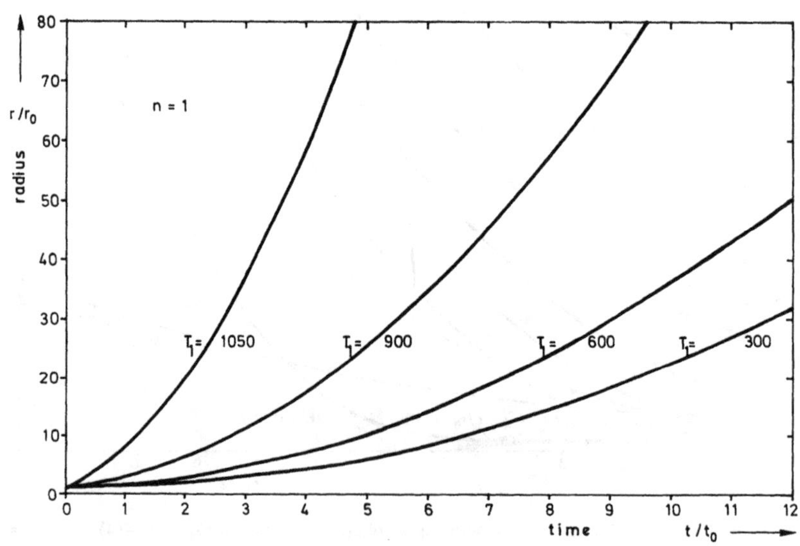

FIGURE 38. Calculated flame front growth for a sphere according to Eqs. (12–14) for different values of initial charge temperature. A linear with time increasing turbulence was simulated by using $\kappa_0(t) = \kappa_0(1 + nt/t_0)$.

In Fig. 39 finally it is illustrated how ignition may help to improve fuel economy also by reducing heat losses. A time-dependent turbulence level according to

$$\kappa_0(t) = \kappa_0[1 + n \exp -(t/t_0 - 5)^2] \tag{17}$$

was used to simulate the turbulence conditions around TDC. Assuming an arbitrary chamber size of $r/r_0 = 63$ we see that too low levels and durations of the turbulence ($n < 100$) will not allow the reaction to be completed close to TDC. A slow burning of an appreciable part of the charge is still necessary, causing a bad thermodynamic efficiency. Turbulence levels that are too high ($n = 360$), on the other hand, will complete the burning too early. Very high turbulence levels are still present causing high heat losses as the loss function is proportional to the mass burned, to the turbulence intensity and the heat-transfer coefficient which rises to more than 14 times the normal value.(46) The gain in fuel economy expected from a specific measure (e.g., increase in compression ratio) may thus be limited by the parallel increasing heat losses.(32,47,48) If a larger ignition radius is used, the optimum turbulence level that was necessary for the smaller ignition radius can be reduced ($n = 80$) and the reaction will

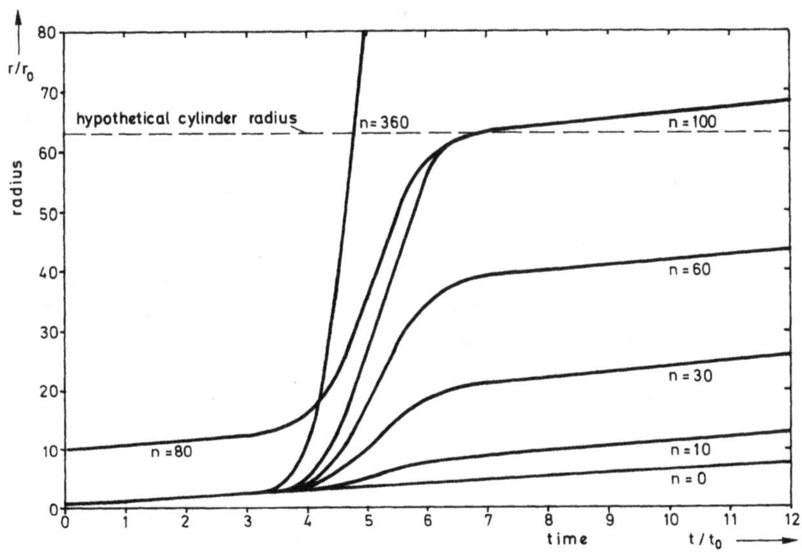

FIGURE 39. Calculated flame front growth for a sphere according to Eqs. (12–14) with $T_1 = T_0$ with a turbulence function simulating conditions around TDC by using $\kappa_0(t) = \kappa_0[1 + n \exp -(t/t_0 - 5)^2]$. Effect of turbulence intensity and initial ignition radius on burn times.

nevertheless be timely finished. Due to the lower turbulence level, heat losses during the reaction will be reduced and better fuel economy is achieved.

In spite of its simplified nature the model is capable of giving the correct tendencies observed in laboratory or engine work. It may thus also be used to simulate quite different effects from those presented in the examples by assuming appropriate time functions for the parameters of interest.

C. Ignition Criteria

From the experimental and theoretical results the following criteria are derived forming the basis for the layout of technical ignition devices with optimum inflammation capabilities.

1. From the total electrical energy supplied to the spark gap only that fraction will be available for inflammation which is contained in the surface sheath of the spark plasma. The thickness of this layer—the inflammation zone—is in the order of the flame front thickness while its energy density is determined by the discharge mode of the spark. Highest energy densities and temperature gradients are achieved in the inflammation zone if the ignition energy is supplied in the shortest time interval possible.
2. Independent of the geometry of the spark plasma (sphere or cylinder) a minimum radius r_{min} of the spark plasma is required which increases rapidly with λ. The minimum ignition radius decreases both with rising ambient pressure and expansion velocity of the plasma, whereas all kinds of effects leading to heat losses from the inflammation zone will increase its value. After inflammation, combustion rates are proportional to the effective surface so plasma geometries providing large r_{min} and large surfaces simultaneously should be used.
3. The time interval available for supplying ignition energy decreases with increasing intended initial flame velocity. Energy supplied after inflammation has been accomplished will not contribute to flame propagation.
4. Larger plasma radii than the minimum ignition radius increase reaction rates, reduce total burn times, and lower any adverse effects.
5. As ignition energy increases with r_{min}^3, unrealistic high values are required for lean operation unless suitable plasma structures are used. An ideal geometrical structure would be a spherical shell containing ignition energy only in the inflammation zone.

These criteria have general validity and may be applied to all kinds of ignition devices from conventional coil systems to torches and plasma jets. A large variety of apparently different conditions for optimum spark discharges can be found in the literature which cannot be discussed here in detail. An example may serve the reader as a guide to the fundamentals of this chapter and how they can be applied to special results.

The optimum ignition conditions most often found in engine work require a certain combination of spark current (i.e., glow current) and discharge time (glow discharge time). Raising the glow current will increase the power input to the glow plasma and thus enlarge the ignition radius. As the discharge voltage drops, however, from ≈500 to ≈50 V if the current exceeds 100–200 mA (transition from glow to arc, see Sec. II.B.4) the effective power input will fall from $P_g = 500 \text{ V} \times 0.1 \text{ mA} \times 0.3 = 15 \text{ W}$ to $P_a = 50 \text{ V} \times 0.1 \text{ mA} \times 0.5 = 2.5 \text{ W}$ and reduce the ignition radius (optimum energy transfer efficiencies of 0.3 and 0.5, respectively, have been assumed). Thus an apparent optimum is found for certain current values. Increasing the current input beyond 1 A, however, will improve the ignition radius above the old optimum as now the arc provides the higher power input. This explains also the effects obtained by high-energy booster discharges added on to conventional systems.

Optimum discharge times in the micro/millisecond range are related to the level of the inhomogeneities present in the charge. Increasing the discharge time will rise the statistical probability to find more favorable start conditions for a plasma with too poor an ignition radius for the conditions given. The possible improvements are limited, however, by the rising turbulence level when TDC is approached, impairing the inflammation and the fall-off in thermal efficiency if ignition is affected too late.

These apparent optima always depend, therefore, on the special conditions of the engine used and differ according to the experimental limitations. If ignition measures are combined with appropriate changes in the engine concept, taking advantage of the mutual dependencies of the operation parameters, much wider margins for optimization can be achieved as has been pointed out throughout this chapter. Thus, essential improvements in fuel economy and exhaust emissions by future burning concepts can be expected.

Some of the results obtained so far along this line have been published in specific papers considering engine aspects in much greater detail.[50–55] Though they are closely related to the physics of sparks and flame propagation they couldn't be discussed properly within the framework of this chapter. The reader is referred to the indicated references.

IV. ENGINE VALIDATION

Engine operation is controlled by a multitude of mutually dependent parameters which are not readily separable under real working conditions. Therefore, samples will be given of how ignition affects essential properties and how these effects can be traced back to the fundamentals presented above. The probing of the motor characteristics was done by special ignition devices providing predominantly a single discharge mode. As reference system, a conventional TCI was used.

In Table 6 a comparison of the influence of experimental breakdown, arc and glow discharge devices on engine characteristics are given. As near stoichiometric mixtures require only small ignition radii which can be supplied by all the systems used, almost no changes show up. A tendency in the NO emissions, however, indicates the still present increase in burning rates by larger plasma volumina. The smallest ignition radius (CDI) causes the lowest NO concentrations, the largest radius (breakdown) causes the highest.

On the lean side the differences caused by the discharge modes become clearly evident in torque and exhaust emissions. Whereas improvements in torque do not correlate with electrical energy input at all, a good agreement is obvious with the properties of the discharge modes as described before. The larger the ignition radius of the device the higher the burning rates, torque, and NO emission and the lower the CH emission.

A more direct insight into the combustion process than by integral data may be accomplished by logarithmic pV diagrams as shown in Fig. 40. The polytropic compression and expansion phases will appear as straight lines. The temperature changes occurring during these time intervals cause only slight curvatures of the effective slope so that the combustion process is clearly separable as the curved part between the limiting slopes. This allows an immediate qualification of the burning function without a need of an explicit calculation requiring additional assumptions about the heat loss mechanisms.

The spread in cycle-to-cycle variation is rather high at this moderately lean point of operation if a glow discharge (TCI) is used for inflammation. Analysis revealed that in glow discharge ignition a very slow cycle is followed by an extremely fast one and vice versa. The reason for this behavior is that a slow reaction with high end temperatures leaves a smaller mass of residual exhaust in the cylinder than a fast one. The following cycle will therefore take in more fresh charge with the consequence of increasing thereby the effective energy density of the resulting mixture. Thus a faster reaction will result with correspondingly lower end temperatures and a higher residual exhaust mass. The succeeding cycle will then take in less fresh charge and a slow reaction results and so on.

The underlying reason for this almost periodic and extremely large variation from cycle-to-cycle is that randomly varying unavoidable inhomogeneities of the cylinder charge (local λ, local temperature, local differences in residual exhaust or mixture quality, local turbulence intensity, etc.) which are comparable in size to the effective ignition radius of the discharge plasma determine the initial expansion velocity of the flame. By a plasma radius which is sufficiently

TABLE 6. Effect of Discharge Mode on Engine Parameters[a]

Ignition	Ignition energy E_{el}/mJ	Peak current i/A	Discharge time $t/\mu sec$	Ignition timing α/°BTDC	Torque M/Nm	NO emission ppm	CH emission ppm
Rich $\lambda = 0.85$							
TCI	30	0.04	1000	−35	67	750	240
CDI + glow	60	0.1	1050	−35	66.5	750	240
CDI + arc	100	375	15	−35	68.5	860	250
Breakdown	20	180	0.02	−35	68	950	270
Lean $\lambda = 1.35$							
TCI	30	0.04	1000	−35	7	45	1350
TCI[b]	30	0.04	1000	−55	10	50	>2000
CDI + glow	60	0.1	1050	−35	7	40	1400
CDI + arc	100	375	15	−35	16	75	1125
Breakdown	20	180	0.02	−35	18	80	500
Breakdown[b]	20	180	0.02	−59	32	230	450

[a] Two-liter four-cylinder engine, partial load, n = 2500 liters/min, gap = 0.6 mm.
[b] Denotes optimum torque timing.

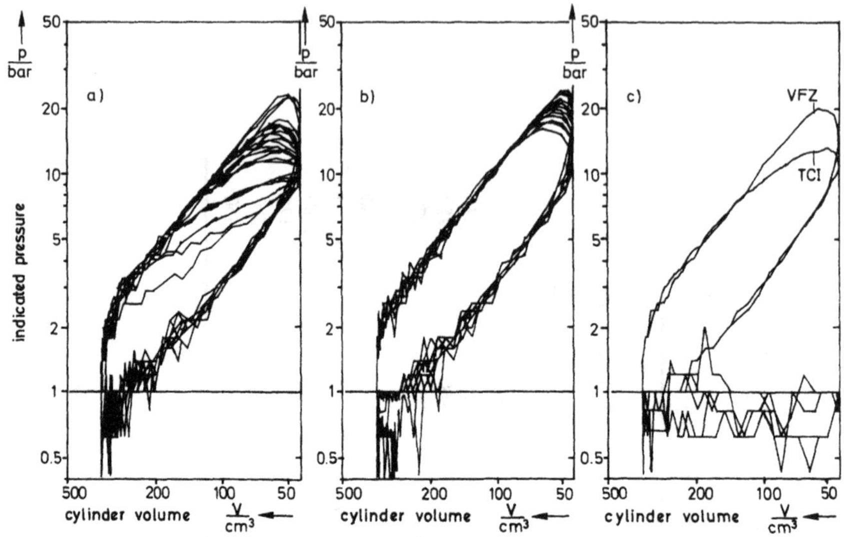

FIGURE 40. Logarithmic pV diagrams of 31 successive cycles of a 1.3-liter four-cylinder engine (BRO = production chamber with squish) taken at constant air mass flux corresponding to half load at $\lambda = 1$, $n = 2000$ 1/min, $\lambda = 1.3$, gap = 0.8 mm, optimum torque timing. (a) TCI, 30 mJ, 770 μsec; (b) VFZ, 30 mJ, 60 nsec; (c) representative (with respect to peak and indicated pressures) diagrams for (a) and (b), respectively.

larger than the average size of the inhomogeneities, this may be drastically reduced (VFZ) leaving the statistically allowed few ones still larger yet. Varying the ignition radius of the discharge plasma allows therefore to probe the size of the inhomogeneity.

These extreme variations put aside, one finds also that average reactions corresponding to the nominal charge conditions are distinctly slower in the case of the TCI when compared to the VFZ. This indicates that the larger plasma volume will raise the reaction rates not only during the inflammation and early propagation phase but in fact also during the full reaction afterwards. So a nanosecond event manifests itself over more than six orders of magnitude.

Improving the motor efficiency by ignition measures will thus always be possible if it is limited by the burn time and the replacing system provides larger ignition radii than the old one. This is especially useful in part load regimes, in lean burn concepts, at high exhaust gas recirculation (EGR), and in combustion chambers with very low or very high turbulence. Reducing cyclic pressure variations (histograms of peak and indicated pressures supplementing the logarithmic pV diagrams are given in Fig. 41) results in better drivability as indicated in Fig. 42. In fact drivability will no longer be the limiting factor in lean operation

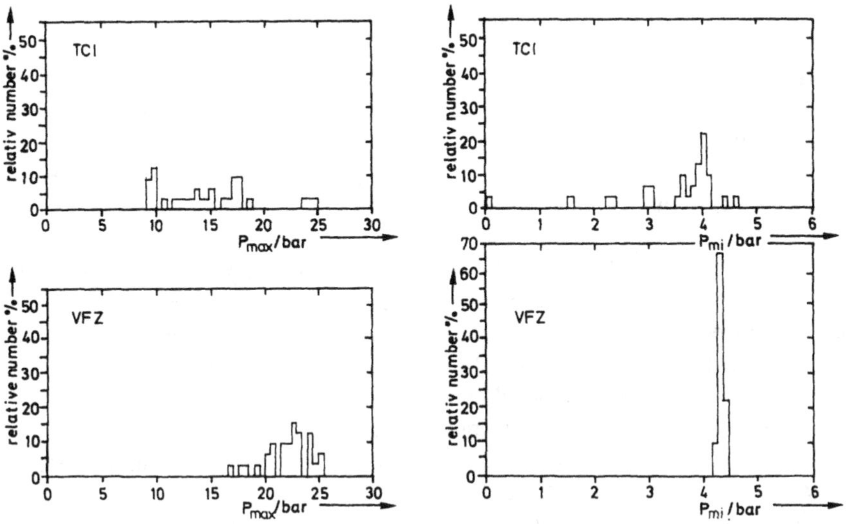

FIGURE 41. Histograms of peak and indicated pressures for pV diagrams shown in Fig. 40.

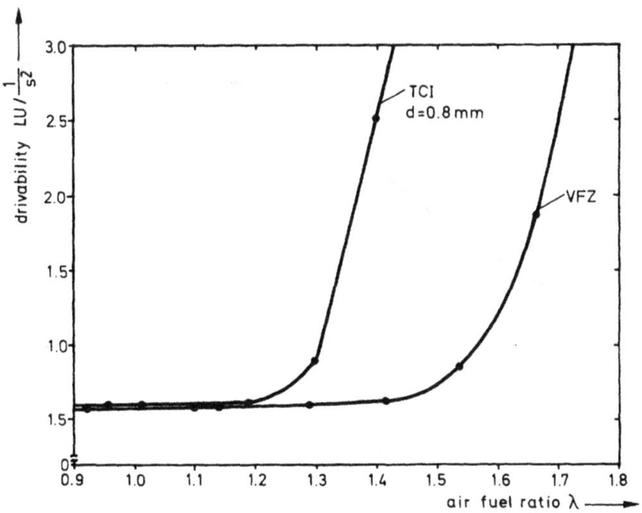

FIGURE 42. Drivability according to Bosch. 1.3-liter four-cylinder engine (BRO = production chamber with squish) measured at constant air mass flux corresponding to half load $\lambda = 1$, $n = 2000$ 1/min, gap = 0.8 mm, optimum torque timing. TCI, 30 mJ, 770 μsec; VFZ, 30 mJ, 60 nsec.

with sufficiently large ignition plasmas as the CH emissions due to flame quenching at the walls (characteristic for the combustion chamber design) start earlier to rise unduly.

The increase in reaction rates related to discharge properties manifests itself consequently also in a corresponding retarding of the optimum spark timing as shown in Fig. 43. The amount of retard is additionally determined by the turbulence level of the combustion chamber as has been pointed out before as flame propagation at late time intervals is determined by the micro-turbulence. A lowering of the turbulence level combined with a rising of the plasma radius will provide an additional retard if the initial propagation was suffering from too high turbulence levels as shown in Fig. 43. The small changes observed in case of the TCI are due to the degrading energy-transfer efficiency of glow discharges if optimum flow levels are lowered (see Fig. 30).

Assuming that the original combustion chamber was certainly optimized for the application of the TCI the flow level of the modified chamber leads to lower ignition radii for the TCI and thus a degradation of its lean limit. According to Fig. 30 the inverse is true for the VFZ due to lower heat losses from the inflammation zone. This effect is readily demonstrated in Fig. 44 showing the influence of combustion chamber turbulence on fuel consumption.

Though no ignition system can save fuel by itself it may very well facilitate to realize burning concepts with inherently higher efficiencies or lower losses

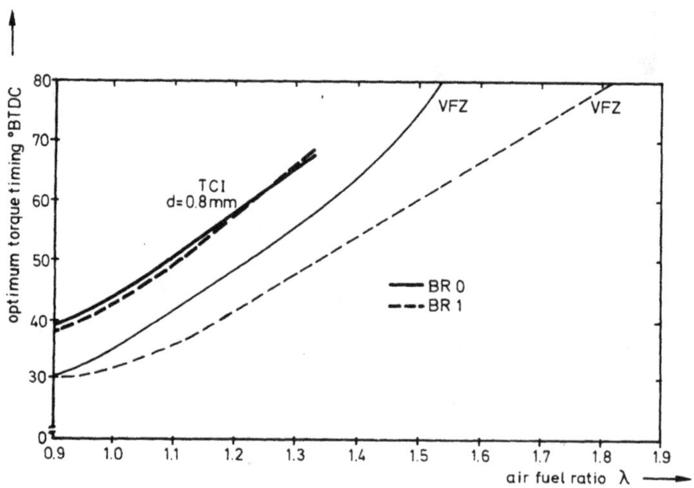

FIGURE 43. Torque optimized timing as function of λ. 1.3-liter four-cylinder engine measured at constant air mass flux corresponding to half load at λ = 1; BRO = production chamber with squish; BR1 = modified, hemispherical chamber without squish; n = 2000 1/min; gap = 1.2 mm; TCI, 30 mJ, 770 μsec; VFZ, 30 mJ, 60 nsec.

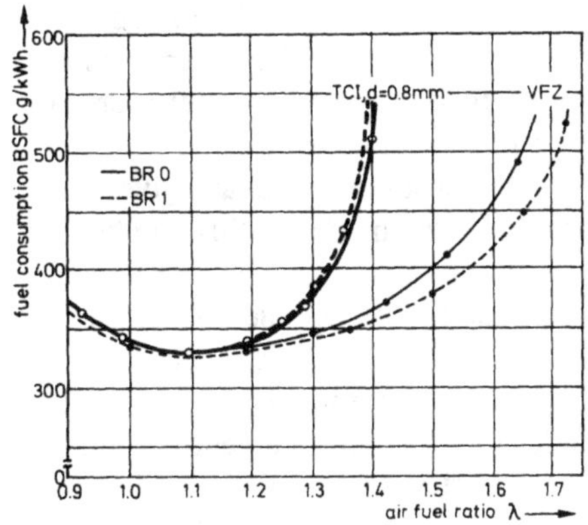

FIGURE 44. BSFC as function of λ for conditions given in Fig. 43.

as it is shown by comparing Figs. 44 and 45. It will thus contribute to fuel savings which otherwise were not possible. Removing the squish areas of the original combustion chamber lowered the compression ratio from 8.2 to 7.2. Using the TCI, optimum BSFC thus dropped by 3% in accordance with theoretical expectations. By the larger plasma volume of the VFZ, however, the flame speed could be increased sufficiently to give a high thermal efficiency at lower general turbulence levels. Thus heat losses were reduced and the same efficiency was obtained as in the old chamber in spite of the lower compression ratio.

This demonstrates once more that effects of inflammation and flame propagation must always be considered in conjunction with all the other relevant

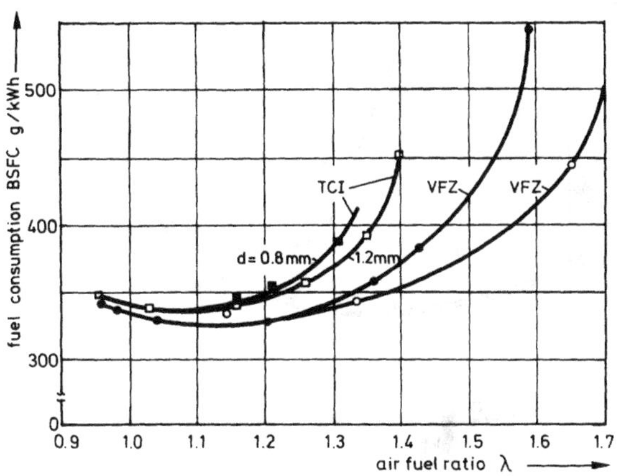

FIGURE 45. BSFC as function of λ for conditions given in Fig. 43 measured with BR1 (modified, hemispherical chamber without squish).

parameters as well as when integral values (such as torque or fuel consumption) are indicators since the parameters are mutually dependent. As has been shown in the fundamentals, the dependencies are rather simple and can be separated by suitable experiments.

Exhaust emissions are given in Fig. 46. CH emissions are not affected by the discharge modes in this combustion chamber. They are due to flame quenching in crevices which cannot be accessed by ignition means. The faster the reactions the higher the NO emissions will be due to the higher flame temperatures. As faster reactions retard the ignition timing, however, the actual increase for optimum torque is less than expected. NO emissions are therefore good indicators for improvements in inflammation and flame propagation. If they are not linked with too high parallel increasing heat losses the relative emissions may even be lower as the efficiency rises at the same time (Fig. 47). By choosing suitable λ values an adaptation to regulations will be possible either for low BSFC or NO emission or both.

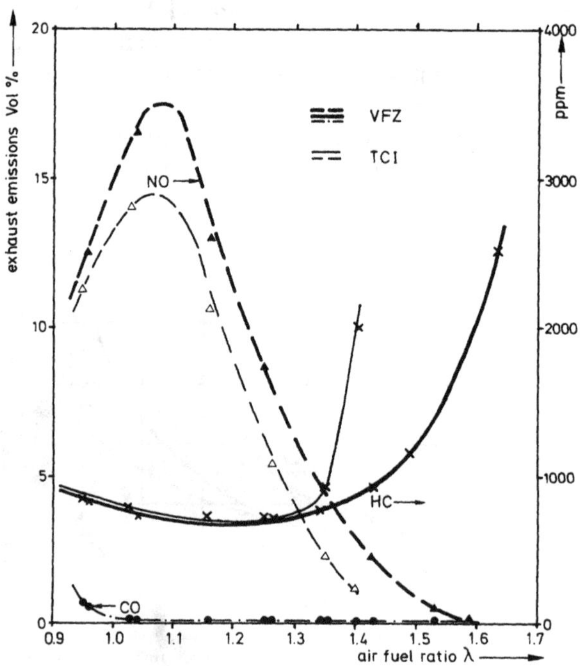

FIGURE 46. Exhaust emissions as function of λ for conditions given in Fig. 45.

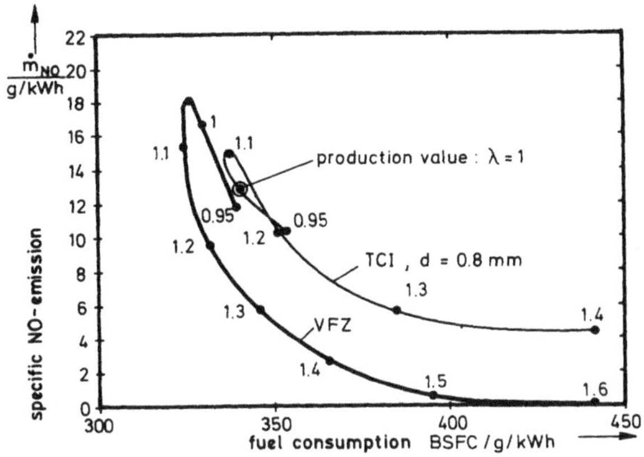

FIGURE 47. Brake-specific NO emissions as function of BSFC for conditions given in Fig. 45.

NOTATION

d	Flame front thickness (cm)
h_{F0}, h_{F1}	Excess enthalpy for $T_F - T_0$ and $T_F - T_1$, respectively, constant and characteristic for the combustible mixture (J/cm^2)
ΔH_F, ΔH_i	Averaged molar enthalpy of reaction products at T_F and T_i, respectively (J/mol)
p_0	Ambient pressure (bar)
q	Lower molar heat of combustion (J/mol)
R	Universal gas constant: 8.3166 (J/mol K)
r_1, r_F	Radius of hot boundary (cm)
$\dot{r}_F$, $\dot{r}_{i0}$	Flame front velocity at T_F and T_i (cm/sec)
T_F, T_i, T_1, T_0	Flame, ignition, ambient, and reference temperature (T_0 = 298 K)
v_F	Flame front velocity (cm/sec)
V_0	Molar volume at T_0 (cm^3/mol)
x	Mol fraction (—)
κ	Effective heat exchange coefficient (J/cm sec K)
λ	A/F ratio normalized to its stoichiometric value (inverse of equivalence ratio ϕ).

REFERENCES

1. Penner, S. S., and Mullins, B. P., "Explosions, Detonations, Flammability and Ignition," Pergamon Press, London (1959).
2. Lewis, B., and von Elbe, G., "Combustion, Flames and Explosions of Gases," 2nd ed., Academic Press, New York (1961).

3. Müller, H., Rhode, S., and Klink, G., "Gemischbildung, Verbrennung und Abgas im Ottomotor," Fachbibliographie mit Referaten bis 1965, Universität Braunschweig, Braunschweig, (1972).
4. Conzelmann, G., "Über die Entflammung des Kraftstoffluftgemisches im Ottomotor," *Bosch Technische Berichte,* **1,** 15–20 (1969).
5. Albrecht, H., Bloss, W. H., Herden, W., Maly, R., Saggau, B., and Wagner, E., "New Aspects on Spark Ignition," SAE paper 770853 (1977).
6. Albrecht, H., Maly, R., Saggau, B., and Wagner, E., "Neue Erkenntnisse über elektrische Zündfunken und ihre Eignung zur Entflammung brennbarer Gemische, 1. Teil," *Automobil Industrie,* **4/77,** 45–50 (1977).
7. Herden, W., Maly, R., Saggau, B., and Wagner, E., "Neue Erkenntnisse über elektrische Zündfunken und ihre Eignung zur Entflammung brennbarer Gemische, 2. Teil," *Automobil Industrie,* **2/78,** 15–21 (1978).
8. Maly, R., Saggau, B., Spaude, H. W., Vogel, M., Wagner E., Greiner, R., and Horch, E. J., "Die Auswirkungen einer verbesserten elektrischen Entflammung auf die Verbrennung im Ottomotor," *Automobil Industrie,* **3/78,** 37–41 (1978).
9. Maly, R., and Vogel, M., "Initiation and Propagation of Flame Fronts in Lean CH_4–Air Mixtures by the Three Modes of the Ignition Spark," *Proceedings of the 17th Symp. (Int.) on Combustion,* The Combustion Institute, **17,** 821–831 (1978).
10. Maly, R., "Ignition Model for Spark Discharges and the Early Phase of Flame Front Growth," *Proceedings of the 18th Symp. (Int.) on Combustion,* The Combustion Institute, **18,** 1747–1754 (1981).
11. Ziegler, G., and Maly, R., "Influence of Ignition on Inflammation and Flame Propagation," *Proceedings of the 1st Specialists Meeting (Int.),* The Combustion Institute, **1,** 89–94 (1981).
12. Müller, H., and Haahtela, O., "Einfluss der elektrischen Zündung auf das Betriebsverhalten und Abgaszusammensetzung im Ottomotor," *MTZ,* **33,** 408–414 (1972).
13. Harrington, J. A., Shishu, R. C., and Asik, J. R., "A Study of Ignition System Effects on Power, Emissions, Lean Misfire Limit, and EGR Tolerance of a Single-Cylinder Engine—Multiple Spark versus Conventional Single-Spark Ignition," SAE paper 740188 (1974).
14. Johnston, R. W., Neuman, J. G., and Agarwal, P. D., "Programmable Energy Ignition System for Engine Optimization," SAE paper 750348 (1975).
15. Schwarz, H., "Ignition Systems for Lean Burn Engines," *Proceedings of the Conference on Fuel Economy and Emissions of Lean Burn Engines,* The Institute of Mechanical Engineers, I Mech E Conference Publications 1979–9, 87–96 (1979).
16. Albrecht, H., Maly, R., Saggau, B., and Wagner, E., "Entladungsvorgänge in Zündkerzen," NTÖ 48, Institut für Physikalische Elektronik, Universität Stuttgart, Germany, May (1974).
17. Albrecht, H., Herden, W., Maly, R., Saggau, B., and Wagner, E., "Entladungsvorgänge in Zündkerzen," NTÖ 48. Institut für Physikalische Elektronik, Universität Stuttgart, Germany, May (1976).
18. Saggau, B., and Ziegler, G., "Time and Space Resolved Temperature Measurements in a Combustion Wave in Methane/Air Mixtures by Spark Excited Rotational Bands," *Proceeding of the 5th International Symposium on Plasma Chemistry,* ISPC-5, Heriot-Watt University, Edinburgh. 558–563 (1981).
19. Raether, H., "Electron Avalanches and Breakdown in Gases," Butterworth, London (1964).
20. Bauer, G. H., "Analyse des Durchbruchs im inhomogenen elektrischen Feld mit Hilfe der Kurzzeitspektroskopie," Thesis, Universität Stuttgart, Germany (1979).
21. Davies, A. J., Davies, C. S., and Evans, C. J., "Computer Simulation of Rapidly Developing Gaseous Discharges," *Proceedings of the IEE,* **118,** 816–823 (1971).
22. Spitzer, L., "Physics of Fully Ionized Gases," Interscience, New York (1956).

23. Albrecht, H., Maly, R., and Wagner, E., "Correlation of Breakdown Mechanisms of Short N_2 Sparks with Typical Line Radiation in the Vacuum–UV Region," *Proceedings of the XIIth International Conference on Phenomena in Ionized Gases,* University of Eindhoven, The Netherlands, p. 153 (1975).
24. Hirschfelder, J. O., Curtiss, C. F., and Bird, R. B., "Molecular Theory of Gases and Liquids," Wiley, New York, pp. 765, 791 (1964).
25. Brode, H. L., "Numerical Solutions of Spherical Blast Waves," *J. Appl. Phys.,* **26,** 766–775 (1955).
26. Gänger, B., "Der elektrische Durchschlag in Gasen," Springer Verlag, Berlin (1953).
27. Finkelnburg, W., and Maeker, M., "Elektrische Bögen und thermische Plasmen," Flügge, S., ed., *Handbuch der Physik,* Vol. 22, pp 254–444 Springer Verlag, Berlin (1956).
28. Cobine, J. D., "Gaseous Conductors," Dover, New York (1958).
29. Brown, S. C., "Basic Data of Plasma Physics," Wiley, New York (1966).
30. Roth, J., Bohdansky, J, and Ottenberger, W., "Data on Low-Energy Light-Ion Sputtering," IPP 9/26, May (1979), Max Planck Institut für Plasmaphysik, Garching bei München, Germany.
31. Maly, R., and Meinel, H., "Determination of Flow Velocity, Turbulence Intensity and Length and Time Scales from Gas Discharge Parameters," *Proceedings of the 5th International Symposium on Plasma Chemistry,* ISPC-5, Heriot-Watt University, Edinburgh, pp. 552–557 (1981).
32. Bloss, W. H., Maly, R., Saggau, B., and Vogel, M., "Experimentelle und theoretische Analyse der Verdichtungserhöhung bei Ottomotoren mit dem Audi-Brennverfahren," 2. Teil: "Strömungs und Zündungseinflüsse," *Automobil-Industrie,* **2/82,** 165–171 (1982).
33. Bloss, W. H., patent application DE-P 28 10 159.
34. Hattari, T., Goto, K., and Ohigashi, S., "Study of Spark Ignition in Flowing Lean Mixtures," *Proceedings of the Conference on Fuel Economy and Emissions of Lean Burn Engines,* The Institute of Mechanical Engineers, I Mech E Conference Publication 1979–9, 153–164 (1979).
35. Bowman, C. T., "Non Equilibrium Radical Concentrations in Shock-Initiated Methane Oxidation," *Proceedings of the 15th Symp. (Int.) on Combustion,* The Combustion Institute, **15,** 869–882 (1974).
36. Westbrook, C. K., and Dryer, F. L., "Prediction of Laminar Flame Properties of Methanol–Air Mixtures," *Combustion and Flame,* **37,** 171–192 (1980).
37. Gaydon A. G., and Wolfhardt, H. G., "Flames, Their Structure, Radiation and Temperature," Chapman & Hall, Ltd., London (1970).
38. Swett, C. C., "Spark Ignition of Flowing Gases," NACA-TR 1287, 1956, Lewis Flight Propulsion Laboratory, Cleveland, Ohio.
39. Spalding, D. B., and Janin, V. K., "The Theory of Steady Laminar Spherical Flame Propagation," *Combustion and Flame,* **5,** 11–25 (1961).
40. Yang, C. H., "Theory of Ignition and Autoignition," *Combustion and Flame,* **6,** 215–225 (1962).
41. Tiggelen, P. J. v., "On the Minimal Initial Size of an Explosive Reaction Center," *Combustion Sci. and Techn.,* **1,** 225–232 (1969).
42. Lefebvre, A. H., "Ignition Theory and its Application to the High Altitude Relighting Performance of Gas Turbine Combustors," *Cranfield International Symposium Series,* The Cranfield Institute, Vol. 11, 105–119 (1971).
43. Maly, R., and Ziegler, G., "Theoretical and Experimental Investigation of the Knocking Process," Symposium: Klopfen von Verbrennungsmotoren, Volkswagenwerk AG, Wolfsburg (1981).
44. Maly, R., Bloss, W. H., Meinel, H., Saggau, B., Wagner, E., and Ziegler, G., "Möglichkeiten zur Beeinflussung des Brennprozesses alternativer Kraftstoffe," BMFT/DGMK-Gemeinschaftstagung, November 11–12, 1981, Deutsche Gesellschaft für Mineralölwissenschaft und Kohlechemie, Hamburg, Germany.

45. Witze, P. O., and Vilchis, F. R., "Stroboscopic Laser Shadowgraph Study of the Effect of Swirl on the Homogeneous Combustion in a Spark Ignited Engine," SAE paper 810226 (1981).
46. Woschni, G., and Fieger, J., "Experimentelle Bestimmung der örtlich gemittelten Wärmeübergangskoeffizienten in Ottomotor, *MTZ,* **42,** 229–234 (1981).
47. van Basshuysen, R, Krömer, G., and Bluhm, K., "Experimentelle und theoretische Analyse der Verdichtungserhöhung bei Ottomotoren mit dem Audi-Brennverfahren," 1. Teil: "Versuchsergebnisse," *Automobil Industrie,* **1/82,** 65–70 (1982).
48. Henning, H., and Vogt, R., "Experimentelle und theoretische Analyse der Verdichtungserhöhung bei Ottomotoren," 3. Teil: "Thermodynamische Analyse," *Automobil Industrie,* **3/82,** 285–288 (1982).
49. Svehla, R. A., "Thermodynamic and Transport Properties of the Hydrogen and Oxygen System," NASA SP-3011, 1964, Lewis Research Center, Cleveland, Ohio.
50. Maly, R., and Ziegler, G., "Thermal Combustion Modeling—Theoretical and Experimental Investigation of the Knocking Process," SAE paper 820759 (1982).
51. Maly, R., Wagner, E., and Ziegler, G., "Modeling of Flame Propagation and Energy Conversion Rates in SI-Engines," SAE paper 830331 (1983).
52. Maly, R., Saggau, B., Wagner, E., and Ziegler, G., "Prospects of Ignition Enhancement," SAE paper 830478 (1983).
53. Maly, R., Meinel, H., and Wagner, E., "Novel Method for Determining General Flow Parameters from Conventional Spark Discharges," Int. Conference on Combustion in Engineering, Oxford, April 1983, I Mech E Publication C268.
54. Ziegler, G., Maly, R., and Wagner, E., "Effect of Ignition System Design on Inflammation Requirements in Ultra-Lean Turbulent Mixtures," Int. Conference on Combustion in Engineering, Oxford, April 1983, I Mech E Publication C268.
55. Maly, R., "General Prospects of Improving Combustion Efficiency of I.C. Engines," invited special paper SP14 for the 11th World Petroleum Congress, August 28–September 2, London (1983).

4

Effects of Valve Events on Engine Operation

T. W. ASMUS

I. INTRODUCTION

The modern spark ignition automotive engine, though fundamentally similar to those having diverse nonautomotive applications, is distinct with respect to many details. The distinct features of automotive engines reflect characteristics of their duty cycle which impose a unique set of design constraints. It is within the realm of these design constraints that this treatment of the subject of valve events is based. Throughout this discussion the term "valve event" denotes the creation of flow area in a phased relationship with piston motion through which the gas exchange process is conducted. This is a general term which includes the timing, lift and size of engine valves.

To satisfy these unique requirements the automotive engine designer must place high priority on all of the following items:

1. High power-to-weight ratio
2. High torque at low engine speeds
3. Low fuel consumption at idle and part load
4. Good idle quality
5. Adaptability to emissions control
6. Compatibility with available road fuels
7. Mechanical simplicity
8. Acceptable durability
9. Ease of service
10. Low cost

T. W. ASMUS • Advanced Power Plant Engineering, Chrysler Corporation, Highland Park, Michigan 48288.

As many of these priorities conflict, it is essential that comprehensive strategies be developed to achieve system optimization. Among the conflicts implicit in the above list are those involving low-speed torque, power output, and idle fuel consumption. Valve events can influence the relationship between these priorities; and as these items can influence vehicle fuel economy, this subject merits careful attention as a part of any comprehensive effort to increase fuel efficiency. Engine performance is a factor which may influence the selection of drive ratios and engine sizes based on vehicle performance and fuel economy priorities. Therefore, specific power output must be regarded as a fuel economy-related factor. Similarly, idle fuel consumption is a significant determinant of vehicle fuel economy particularly over various urban-type drive cycles.

Thus the simultaneous attainment of high specific power output and low idle fuel consumption can be facilitated by developing a comprehensive strategy to optimize valve events. A scheme is discussed that is based upon a framework of understanding of the specific roles of the various valve events. This can provide a basis for selecting camshaft designs which yield minimum valve overlap to increase combustion stability at idle while maintaining or even increasing specific output at wide-open throttle (WOT), i.e., the long-standing notion that valve overlap is a key to high performance is put into perspective.

II. BACKGROUND

Early in the automotive era high priority was placed on developing engines that would operate at higher speed than those typically employed in stationary or other nonautomotive applications. This gave way to early recognition that valve events were important to making high-speed operation feasible. Provision for both intake and exhaust valve lift duration significantly longer than their respective piston events was crucial to achieving this goal. In terms of the phased relationship between valves and pistons, this feature enabled continued cylinder filling following the intake stroke but prior to intake valve closing (IVC). At high engine speeds this is beneficial to volumetric efficiency as cylinder pressure is often below that of the intake system at bottom dead center (BDC) following the intake stroke. Likewise it was found that high-speed operation was additionally facilitated by opening the exhaust valve well before BDC (BBDC) between the expansion and exhaust strokes. This feature aided the scavenging process by exploiting blowdown to accelerate the column of out-flowing exhaust and thus to reduce the pumping work associated with scavenging at high engine speeds. Though these valve-timing features significantly increase performance at higher engine speeds, they incur torque penalties at low speeds. Late IVC allows some fresh charge to escape back into the intake system at the beginning of the compression stroke and thereby reduces low-speed volumetric efficiency.

Early exhaust valve opening (EVO) lowers the expansion ratio and thus lowers cycle efficiency. Though by current standards the Model T Ford engine can hardly be construed as one having high performance, it did incorporate these valve-timing principles, i.e., IVC was 50° after BDC (ABDC) and EVO was 38° BBDC. The timing of the intake valve opening (IVO) and exhaust valve closing (EVC) were 15° after top dead center (ATDC) and 2° BTDC, respectively, which combine to yield 17° of negative overlap or underlap! This is in sharp contrast to a trend which began in the 1930s and accelerated in the high performance era where camshaft designs included significant valve overlap duration, i.e., subtended crank angle between IVO and EVC.

As ever-increasing pressures to improve fuel economy has forced the demise of the automotive V-8 and the reappearance of the four-cylinder engine, increased attention is being directed toward the implementation of more fuel efficient idle control strategies, i.e., less rich idle fuel-air mixtures, more spark advance, and reduced speeds. If idle quality is to be maintained in the presence of these transitions, consideration must be given to valve events or more specifically valve overlap. The feasibility of employing such calibrations is enhanced if valve overlap is minimized.

The subject of valve timing is treated briefly in many textbooks on internal combustion (IC) engines under the general heading of volumetric efficiency. Review of several of these classical treatments[1–6] yields the following areas of general agreement:

1. The principal determinants of high-speed volumetric efficiency are the timing of IVC and valve overlap.*
2. The timing of IVC in conjunction with the diameter and length of the inlet pipe in relation to swept volume establishes the volumetric efficiency–speed relationship of an engine. This characteristic is based on the inertia of the flowing inlet gas column.
3. The timing of EVC should occur sufficiently far beyond TDC that cylinder pressure does not rise near the end of the exhaust stroke.
4. The timing of IVO should occur sufficiently far before TDC that cylinder pressure does not dip early in the intake stroke.
5. Valve overlap is recognized as a factor which can influence the burned gas fraction (BGF) particularly under light load conditions.
6. In supercharged engines operated under boost, valve overlap provides a leakage pathway from the inlet to the exhaust system thereby allowing some fresh charge to escape unburned. Where turbochargers are employed this effect may be exploited to reduce turbine inlet temperatures.

* Valve overlap is the consequence of early IVO and late EVC, and it is mainly the latter of these features which influences high-speed volumetric efficiency.

Other literature on this subject tends to fall into the following three general categories:

1. Experimental studies of valve timing as it relates to emissions control and fuel consumption.[7–11]
2. Modified valve events for load control.[12–13]
3. Analytical studies of flow as related to performance, charge composition, and cycle efficiency.[14–22]

The means by which valve timing can affect emissions and fuel consumption is through the regulation of combustion dilution or BGF, pumping losses, the fluid mechanical state of the charge, and the effective compression and expansion ratios. The valve events mainly responsible for regulation of the BGF are IVO and EVC or more precisely the combination of these as (1) valve overlap and (2) EVC independent of valve overlap. Under conditions of low to moderate engine speeds and loads, sufficient time and intake manifold vacuum are available to cause significant quantities of exhaust to flow into the intake system during the overlap period. This exhaust quantity is subsequently induced along with fresh charge and thus contributes to the BGF. Independent of overlap the timing of EVC can influence the BGF, i.e., assume the IVO to be sufficiently late that variations in EVC do not affect overlap. If EVC is early, some additional exhaust may be trapped in the cylinder as pressure rises late in the exhaust stroke, and if EVC is late exhaust may be induced through the exhaust valve early in the intake stroke. At either extreme the BGF will be high passing through a minimum at some intermediate timing. Precisely where this minimum occurs is a function of engine speed and certain design specifics. The combustion process responds to the total BGF in a manner somewhat independent of the means by which this total accurred, i.e., it makeș little difference whether the total BGF was established solely by internal means or by a combination of internal means plus some form of external exhaust gas recirculation (EGR) system. Several qualifications to this assertion are as follows:

1. Uniform distribution of recycled exhaust, both cylinder-to-cylinder and cycle-to-cycle.
2. Equal resultant charge temperature.
3. No major fluid mechanical factors resulting from the different flow pathways.

Pumping losses occur at all speeds and loads, though their effects are most prominent at light loads where this parasitic loss is caused mainly by intake throttling. Though all valve events can influence pumping losses, those capable of affecting them at light loads are those which establish the level of throttling required to maintain a particular power level. The timing of IVC helps to establish the mass of charge retained in the cylinder while those which regulate the BGF (overlap and EVC) affect internal recirculation of exhaust. Part-load pumping losses may be minimized by delaying IVC and by increasing the BGF. Though each of these means produces the same results in terms of pumping losses, their respective effects on charge composition and combustion are different. At high speeds and WOT the pumping work may be of greater magnitude than at idle, though it is less significant as a fuel economy factor as it is a smaller fraction of the indicated work.

The fluid mechanical state of the charge may be affected by the nature of the relationship between the intake valve flow area and piston motion during the intake stroke. This may affect in-cylinder mixing of the charge components and the burn rate if turbulence levels during combustion are appreciably altered.

The expansion ratio and thus the cycle thermal efficiency may be influenced by the timing of EVO. Also the timing of the EVO may exert some effect on the time–temperature history of the exhaust products which may influence the degree to which HC and CO are burned off.

The results of several experimental studies[(7–11)] have been reported which deal specifically with the effects of valve timing on emissions and fuel economy. Siewert[(7)] describes an extensive battery of experiments in which each of the four valve-timing points was independently varied for a single, low-speed, part-load engine operating condition at stoichiometric F/A with spark timing at MBT. The emissions results mainly reflect the effects of valve timing on the BGF. Studies of the effects of valve timing on emissions and fuel economy at the vehicle level is the subject of Ref. 8. A broad range of part-load engine operating conditions were evaluated, and it was concluded that NO_x can be significantly reduced by increasing valve overlap, although changes to valve timing alone are insufficient to control emissions to any reasonable level. References 9–11 are oriented toward the execution of variable valve-timing hardware for the control of NO_x emissions.

The control of engine load by employing alternate valve event schemes is the subject of Refs. 12–13. Intake valve throttling by means of lift modulation, as a means of load control, is described in Ref. 12. An experimental program based on a fuel-injected engine fitted with a variable fulcrum-type of valve train demonstrated that at light loads the small-scale turbulence generated by high intake flow velocities associated with low valve lift increased the burn rate and extended the lean limit. Load control via late IVC is described in Ref. 13. It was demonstrated experimentally that throttling losses and brake specific fuel consumption (BSFC), at part load were reduced. Additionally, it was shown that this approach had to be combined with some intake throttling to achieve the necessary range of load control for the automotive duty cycle. Early IVC is another means of load control which has been considered.[(14)] By this scheme the mass of trapped charge is established before the end of the intake stroke, then an expansion and recompression occurs for which there is no throttling loss.

The reduced, overall throttling losses associated with this approach should yield a part-load fuel economy improvement.

References 14–21 describe various approaches to the analysis of flows in engines. Details of the gas exchange processes as they relate to part-load emissions and fuel consumption are the subject of Refs. 14–16. Analysis of various aspects of engine breathing where only the induction and exhaust systems are limiting volumetric efficiency is presented in Refs. 17–21. Reference 22 includes an analytical treatment of unsteady flows in exhaust systems of multicylinder engines.

III. PERSPECTIVES ON VALVE EVENTS AND ENGINE OPERATION

Discussion will be confined to mechanically actuated poppet valves as there appears to be no reasonable alternate valve schemes for four-stroke cycle, spark-ignited engines suitable for automotive applications. With these constraints in mind application of any alternatives representing major departures from current practice, i.e., fixed valve timing and lift, must necessarily include consideration of the following questions:

1. Does the alternate scheme satisfy all duty cycle requirements?
2. Does the alternate scheme require a mechanism to vary the valve timing and/or lift?
3. If a variable timing and/or lift mechanism is required, is it dynamically stable over the entire engine speed range?
4. Are the accrued benefits of a variable timing and/or lift mechanism sufficient to justify any added friction and weight.
5. Does the alternate scheme require valve return springs having substantially different loads? (This may be a significant friction factor.)
6. What is the net effect of the alternate scheme on idle fuel consumption? (Many valve train types show an inverse speed–friction relationship owing to boundary-layer lubrication of the sliding surfaces; therefore, friction assessment at idle should be included.)

The alternate valve event strategies discussed in the cited literature are summarized as follows:

1. High valve overlap for the regulation of BGF and NO_x emissions control.[7–11]
2. Load control
   a. Late or early IVC for reduced pumping losses without increasing BGF to increase fuel efficiency[13,14]
   b. Intake valve throttling as a fast-burn feature[12]

3. Volumetric efficiency maximization of WOT for all engine speeds for improved overall performance.[15]

None of these strategies could be implemented with fixed valve timing and valve lift in such a manner that all other duty cycle requirements could be satisfied. Therefore, variable valve-timing and/or variable valve lift mechanisms would be required; and thus all of the above issues must be considered. There are well-known engine features which accomplish approximately the same objectives as all of the aforementioned valve event strategies, and justification of these alternate schemes must be additionally qualified vis-à-vis established techniques.

Examples are as follows:

1. Valve overlap for NO_x control versus EGR.
2. Late or early IVC for part-load fuel efficiency versus reduced displacement with supercharging.
3. Intake valve throttling for fast-burn versus a fast-burn combustion chamber feature.
4. Variable valve timing for performance versus increased displacement or supercharging.

Mass movement within the automotive industry toward any of the previously described alternate valve event features is seen as rather unlikely. The probable reason for this is the failure of these systems to show sufficient net benefits compared to other approaches. Although these alternate schemes may never reach the mass market, many of the principles established during their investigation have contributed to the general understanding of valve events. This understanding is key to the optimization of valve events for engines having more conventional valve gear.

IV. VALVE EVENT FUNDAMENTALS

A. General Discussion

The treatment which follows is intended to provide a framework within which to understand the consequences of moderate changes in valve events but does not include rigorous analysis of flows. Flow analysis requires detailed knowledge of instantaneous flow coefficients and gas density gradients which are not generally available over the broad range of operating conditions typical of automotive engines. Rather, this discussion will center on the relationship between instantaneous changes in cylinder volume, due to piston motion and the valve flow area available for gas exchange. This will yield a sense of direction in the establishment of optimum valve events for modern, high-speed automotive

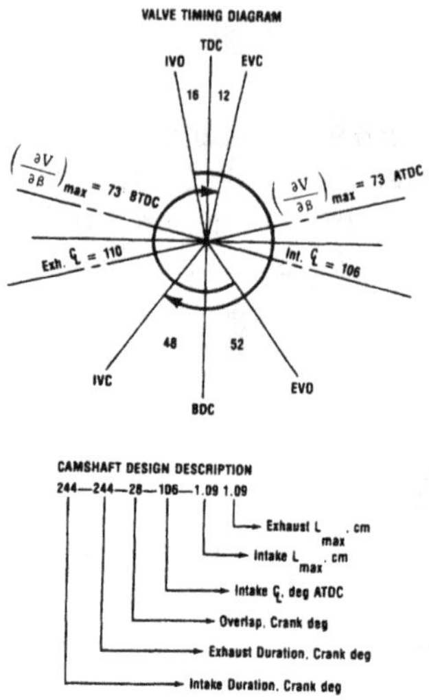

FIGURE 1. Typical valve-timing diagram and camshaft design description for a modern, high-speed, four-cylinder engine (See Table 1).

engines. Figure 1 is a valve-timing diagram illustrating the timing points for a typical, modern automotive engine.*

The maximum piston velocity (designated as $[\partial V(\beta)/\partial\beta]_{max}$ does not, and within the framework of practical engine design constraints cannot, coincide with maximum valve lift (designated as intake and exhaust centerlines, respectively). On the basis of this picture two *a priori* deductions are made as follows:

1. The scavenging process is fundamentally inhibited because maximum piston velocity occurs as the exhaust valve is closing, and this will lead to an incremental trapping of exhaust at high engine speeds.
2. The charging process is facilitated, or at least not adversely affected, by the occurrence of maximum valve lift following maximum piston velocity as there is a natural tendency toward recovery of any flow loss

* It is noteworthy that there is a need to define some valve-timing criterion because of the nature of valve lift profiles. To satisfy high-speed valve gear dynamic requirements, valve lift profiles include long, constant velocity ramps attached to both the opening and closing sides of a symmetrical or slightly skewed, exponentially described lift curve. As the ramps provide very small amounts of lift over large subtended crank angles, timing points based on minimum discernable lift would have little value in terms of the gas exchange process.

The timing points shown are based on the arbitrary criterion of 4 camshaft degrees off of the ramps or into the exponentially described part of the lift curve, i.e., inside of the ramps. There is no universally accepted criterion for valve-timing points. Some are based upon angular criteria along the lift curve, like the one described above, while others are based upon some specific lift criterion.

TABLE 1. Engine Specifications—Chrysler 2.2 Liter

Bore (b) = 8.75 cm
Stroke ($2r$) = 9.2 cm
Connecting rod length (R) = 15.1 cm
Swept volume/cylinder (V_s) = 553.2 cm^3
Clearance volume @ r_e = 8.6 (V_c) = 72.8 cm^3

incurred early in the intake stroke where the intake valve flow area may be low in relation to piston motion.

To quantify this premise the instantaneous valve flow areas and the rate of change of cylinder volume may be established. Valve flow areas must be based upon known valve lift data and some geometric details of the specific valves and seats. Figure 2 shows three separate stages of flow area in the valve lift process. In the first stage of valve lift the minimum flow area corresponds to a frustum of a right circular cone where the conical face between the valve and the seat defines the flow area. As valve lift proceeds, the minimum area is no longer between the valve face and the seat but rather between the seat and the nearest edge of the valve and may be represented as the frustum of a right circular cone having an ever-increasing base angle, i.e., approaching a cylinder. This is denoted as stage II. In the third stage of valve lift the minimum flow area is that of the throat minus the sectional area of the valve stem.

Figure 3(a)–(c) shows a typical valve lift curve, the effective flow area of an intake valve having a throat diameter of 3.5 cm, and the effective flow area of an exhaust valve having a throat diameter of 3.1 cm, respectively. The extended flat sections on the flow area curves reflect the magnitude of the ratio of maximum valve lift L_{max} to throat diameter D. It is apparent from these curves that increasing valve lift beyond a point where $L_{max}/D = 0.3$ is likely to yield relatively little in terms of flow area and incremental volumetric efficiency. Flow rig data often suggest that higher lifts yield significantly higher flows up to L_{max}/D values approaching 0.45. This typically results not from incremental flow area

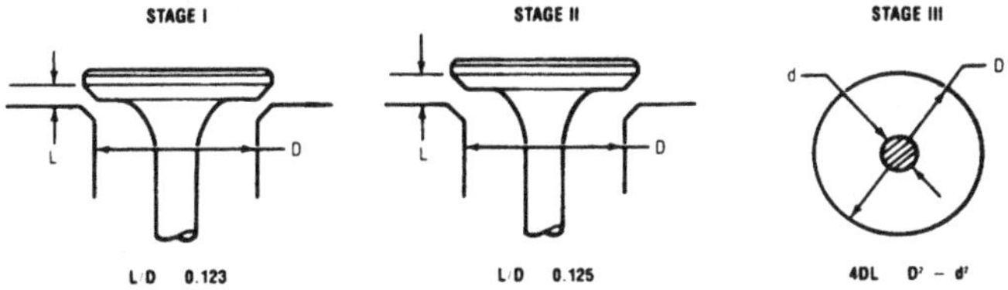

FIGURE 2. Schematic showing three stages of valve lift.

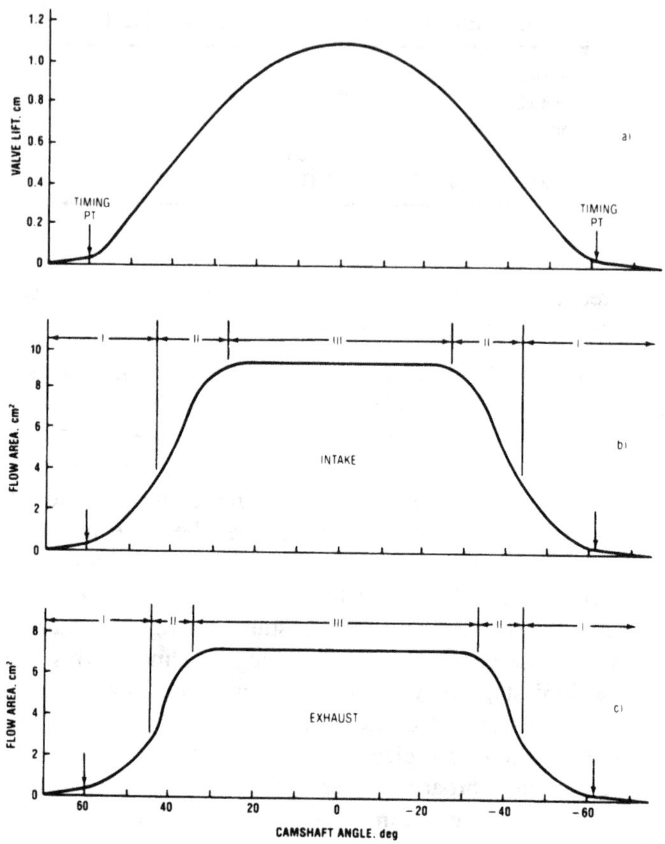

FIGURE 3. (a) Typical 244° duration valve lift curve, (b) and (c) corresponding intake and exhaust valve flow areas for $D = 3.6$ cm and 3.1 cm, respectively.

but rather increased flow coefficients and/or decreased shrouding by combustion chamber walls. As will be discussed later, it often happens that increased valve lift, though providing some apparent flow benefit, does not yield corresponding engine performance benefits. This may result from the fact that the flow benefit does not occur at an appropriate time during the engine cycles; this is particularly true of the exhaust process.

The instantaneous piston velocity in an engine is dependent upon engine speed N, stroke $2r$, the connecting rod length R and the angular crankshaft displacement from TDC β. Neglecting piston pin offset this may be developed as follows:

$$l(\beta) = R + r - r \cdot \cos\beta - (R^2 - r^2 \cdot \sin^2\beta)^{1/2} \tag{1}$$

where $l(\beta)$ = instantaneous linear piston displacement. Differentiation of Eq. (1) with respect to β gives a pseudovelocity

$$\left[\frac{\partial l(\beta)}{\partial\beta}\right] = \left[r \cdot \sin\beta + \frac{r^2 \cdot \sin\beta\cos\beta}{(R^2 - r^2 \cdot \sin^2\beta)^{1/2}}\right] \cdot \frac{\pi}{180°} \tag{2}$$

which, when multiplied by engine speed, gives the real instantaneous piston velocity v

$$v = \left[\frac{\partial l(\beta)}{\partial\beta}\right] \cdot \frac{360°}{\text{rev}} \cdot N \tag{3}$$

The pseudoinstantaneous rate of change of volume due to piston motion is

$$\left[\frac{\partial V(\beta)}{\partial\beta}\right] = \frac{\pi b^2}{4} \cdot \left[\frac{\partial l(\beta)}{\partial\beta}\right] \tag{4}$$

where b = bore. The pseudoinstantaneous rate of acceleration is given as

$$\left[\frac{\partial^2 l(\beta)}{\partial\beta^2}\right] = \left[r \cdot \cos\beta + \frac{r^2}{(R^2 - r^2 \cdot \sin^2\beta)^{1/2}}\cos^2\beta - \sin^2\beta + \frac{r^2 \cdot \sin^2\beta\cos^2\beta}{R^2 - r^2 \cdot \sin^2\beta}\right]\left(\frac{\pi}{180°}\right)^2 \tag{5}$$

Plots of Eqs. (1) and (4), for a typical, modern high-speed engine are shown in Fig. 4. Note that maximum piston velocity occurs well before the midpoint in the stroke and at a crank position of less than 90° from TDC.

As a means of conceptualizing the instantaneous flow conditions at the valves, the rate of change of cylinder volume may be divided by the instantaneous valve flow area to yield pseudoflow velocities across the valves.

This pseudoflow velocity may be expressed as

$$\left[\frac{\partial V(\beta)}{\partial\beta} \cdot \frac{1}{A}\right] = \frac{\pi b^2}{4A}\left[\frac{\partial l(\beta)}{\partial\beta}\right] \tag{6}$$

Figure 5(a) shows the rate of change of cylinder volume and valve flow area over most of the open-valve portion of the cycle, while Fig. 5(b) shows the result of dividing the former by the latter, i.e., Eq. (6). Note the appearance of

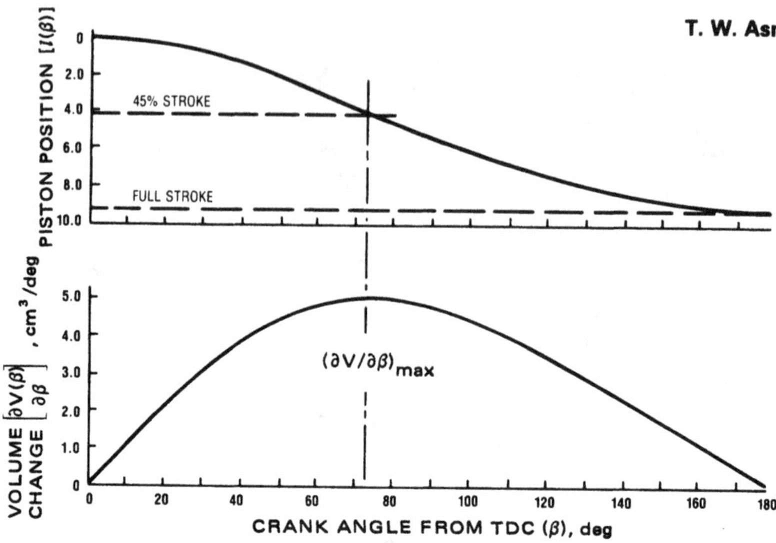

FIGURE 4. Instantaneous piston position and volume change for the Chrysler 2.2-liter engine. (See Table 1 for specifications.)

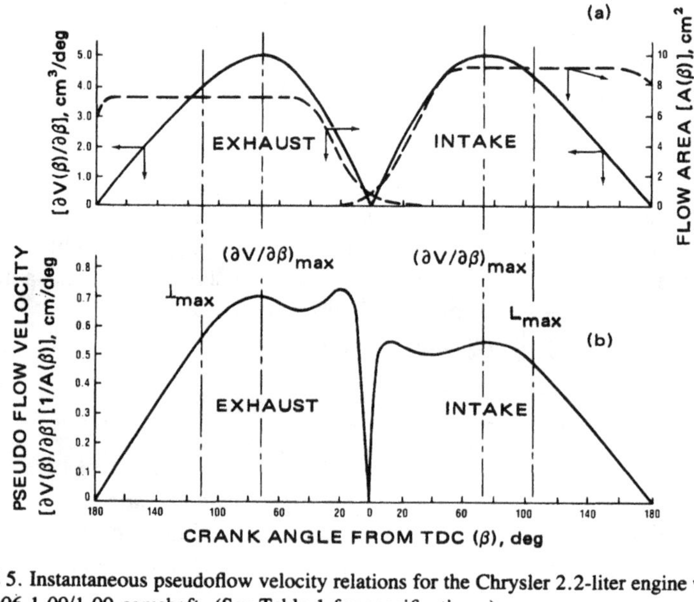

FIGURE 5. Instantaneous pseudoflow velocity relations for the Chrysler 2.2-liter engine with a 244-244-28-106-1.09/1.09 camshaft. (See Table 1 for specifications.)

two peaks in pseudoflow velocity for both the exhaust and intake strokes occurring at maximum piston velocity and within 20° of TDC. The broad peaks occurring at maximum piston velocity reflect the constancy of the flow area at this crank position. The peaks occurring near TDC reflect the reality that the exhaust and intake valves are moving toward closure and opening, respectively, at exponential rates while piston motion is nearly linear at these points. (Ideally there would be constant pseudovelocity where these peaks occur.) Reference is made to the first of the two *a priori* deductions discussed earlier in light of these pseudoflow velocity relationships. That is, the two peaks in pseudoflow velocity occurring during the exhaust stroke suggest that a pressure gradient in accordance with this relationship will exist, and its magnitude will increase with engine speed. The occurrence of pressure gradients across the exhaust valve will yield an irrevocable trapping of exhaust. The consequence of this reduces volumetric efficiency as trapped exhaust displaces fresh charge, and any added heat content to the fresh charge is a factor which may contribute to high-octane requirement. Indicator or pressure–volume (P–V) diagrams focused on the open-valve portion of the cycle, to be shown later, further illustrate this effect. The available options for minimizing exhaust valve throttling near the end of the exhaust stroke is to further delay EVC, employ larger exhaust valves, or employ a skewed exhaust lift curve with maximum lift biased toward the EVC side.

Intake valve throttling early in the intake stroke is seen as inconsequential provided that sufficient time and flow capacity are available prior to IVC. It is to be noted that the most significant determinant of volumetric efficiency is the mass of trapped charge at IVC—not any instantaneous flow loss. It is conceivable that any intake valve throttling which occurs early in the intake stroke will yield turbulence which may contribute to in-cylinder mixing of charge components as

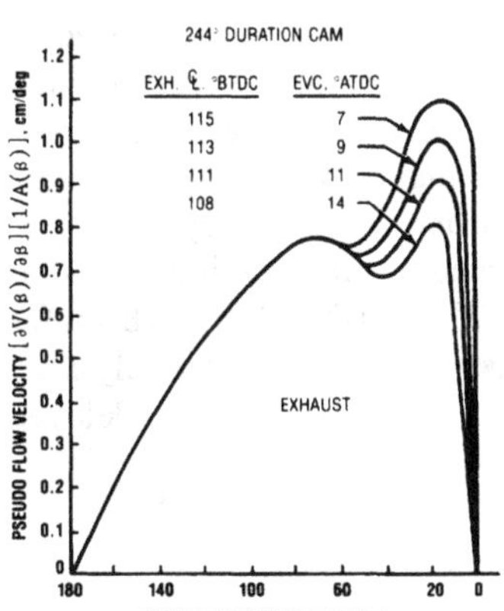

FIGURE 6. Instantaneous pseudoflow velocity as influenced by exhaust cam phasing for the Chrysler 2.2-liter engine with a 244-244-28-106-1.09/1.09 camshaft. (See Table 1 for specifications.)

well as to increase the turbulent flame speed. Therefore, under certain circumstances this unavoidable intake valve throttling may be used to advantage.

Figure 6 illustrates the effect of relatively small changes in the timing of EVC on the pseudoflow velocity curve. In a similar fashion comparative levels of intake valve throttling owing to changes in the timing of IVO or intake valve lift may be assessed.

B. Discussion of Valve Timing

Each of the four valve-timing points will be discussed in turn.

1. *Intake Valve Opening (IVO)*

This timing point must be seen not only as the beginning of intake process but also as the beginning of the valve overlap period in those instances where valve overlap is present. When IVO occurs substantially before TDC in naturally aspirated engines, exhaust flows from the combustion chamber into the inlet system driven by a pressure gradient. The exhaust-diluted fresh charge begins to flow into the cylinder only after the cylinder pressure is reduced to some level below that of the intake manifold by virtue of piston motion. The point at which this occurs depends upon the timing of EVC, the pressures in the intake and exhaust manifolds, and the compression ratio. Typical conditions most conducive to extensive exhaust back flow are low engine speeds, low intake manifold pressures, early IVO, late EVC, and low compression ratios. The engine operating condition generally associated with the consequences of extensive exhaust back flow is idle, although back flow may occur to a smaller degree at all speeds and loads and may even have adverse effects on WOT performance.

Delaying the IVO tends to isolate the intake manifold from the cylinder until the cylinder and the intake manifold pressures are more nearly equal. Further delays in the IVO may actually cause the cylinder pressure to dip below the intake manifold pressure momentarily. Aside from any incremental pumping work which may result from this condition, it is not necessarily seen as detrimental to volumetric efficiency as there may be adequate time later in intake stroke for recovery of any flow loss. (Recall that maximum intake valve lift occurs after maximum piston velocity.) IVO typically occurs in the range of 10°–25° BTDC. It has been found experimentally* and analytically[15] that engine performance is relatively insensitive to this timing point and that it may, under certain conditions, be delayed significantly without incurring performance penalties.

* Unpublished results from Chrysler Corporation.

2. *Intake Valve Closing (IVC)*

This timing point always occurs beyond BDC following the intake stroke providing more time for cylinder filling under conditions where cylinder pressure is below the intake manifold pressure at BDC, i.e., at high engine speeds. Delaying IVC causes the volumetric efficiency to increase at higher engine speeds while penalizing it at lower speeds. At lower engine speeds where the intake manifold and cylinder pressure are equal at BDC, some fresh charge is pushed back into the intake manifold prior to IVC. To assess the relative fractional losses in low-speed volumetric efficiency attributable to this loss mechanism, consider the following relationship:

$$\% \text{ Loss} = \left[\frac{\partial V(\beta)}{\partial \beta}\right] \cdot \frac{100}{V_s} \tag{7}$$

where V_s = swept volume/cylinder.

Equation (7) is plotted for a particular engine in Fig. 7. Note the typical range for the timing of IVC (40°–60° ABDC) and how the specific loss due to late IVC falls in the range of 0.42–0.65%/deg. This argument is based on the assumption that the engine speed is sufficiently slow that there is no valve throttling and that the valve-timing point coincides with the cessation of flow. While these assumptions limit the absolute accuracy of the results, they are useful and interesting on a comparative basis and can aid the establishment of effective valve-timing strategies.

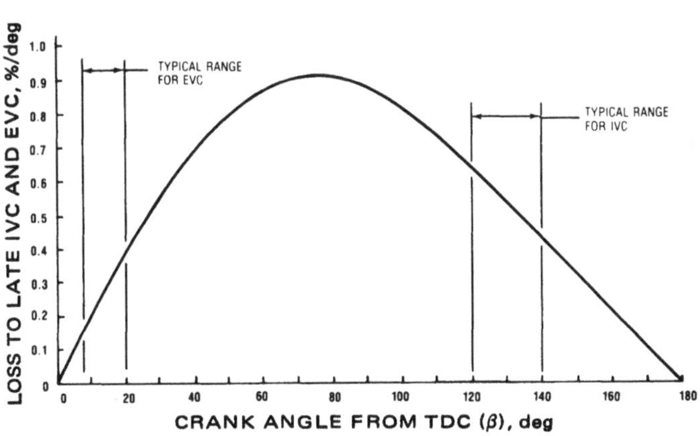

FIGURE 7. Low-speed volumetric efficiency loss to late EVC and late IVC for the Chrysler 2.2-liter engine. (See Table 1 for specifications.)

3. *Exhaust Valve Closing (EVC)*

This timing point must be seen not only as the end of the exhaust stroke but also as the end of the valve overlap period in those instances where valve overlap is present. EVC typically falls in the range of 8°–20° ATDC and therefore may significantly affect BGF both at idle and at WOT. At idle it regulates the quantity of exhaust allowed to flow back into the combustion chamber through the exhaust valve under the influence of intake manifold vacuum. At WOT and high engine speeds it regulates how much exhaust is allowed to escape. Late EVC favors power at the expense of low-speed torque and the idle condition. Referring once again to Eq. (7) and Fig. 7 note that the specific loss due to late EVC is in the range of 0.15–0.35%/deg., i.e., the rate at which EVC affects low-speed volumetric efficiency is about one-half that of IVC. Note also that late EVC reduces low-speed volumetric efficiency by increasing the BGF while late IVC regulates the mass of trapped charge. This timing point should be established such that at the desired power peak the cylinder pressure is minimally above the exhaust back pressure at IVO. This will be discussed later in terms of P–V diagrams of pumping loops.

4. *Exhaust Valve Opening (EVO)*

This timing point is fixed well before the end of expansion stroke such that blowdown can assist the scavenging process. The major performance criterion for selecting the timing of EVO is to ensure that the cylinder pressure is reduced to near the exhaust manifold pressure at or as soon as possible after BDC following the expansion stroke over the entire engine speed range. The pumping loss is minimized and the scavenging efficiency maximized if the cylinder pressure during the exhaust stroke does not rise appreciably above the exhaust manifold pressure. The timing point of EVO affects the cycle thermal efficiency as it regulates the expansion ratio. Consider the simple expression for Otto cycle thermal efficiency.

$$\eta(\beta) = 1 - \left[\frac{1}{r_e(\beta)}\right]^{k-1} \tag{8}$$

and

$$[r_e(\beta)] = 1 + \frac{\pi b^2}{4V_c}[R + r - r \cdot \cos\beta - (R^2 - r^2 \cdot \sin^2\beta)^{1/2}] \tag{9}$$

where V_c = clearance volume at TDC and $r_e(\beta)$ = instantaneous expansion ratio.

Differentiation of Eq. (8) with appropriate substitution of Eq. (9) gives

$$\left[\frac{\partial\eta(\beta)}{\partial\beta}\right] = (k - 1)\left\{1 + \left(\frac{\pi b^2}{4V_c}\right)[R + r - r \cdot \cos\beta - (R^2 - r^2 \cdot \sin^2\beta)^{1/2}]\right)^{-k}$$

$$\times\left[r \cdot \sin\beta + \frac{r^2 \cdot \sin\beta\cos\beta}{(R^2 - r^2 \cdot \sin^2\beta)^{1/2}}\right]\left(\frac{\pi}{180°}\right)\left(\frac{\pi b^2}{4V_c}\right) \tag{10}$$

Plots of Eqs. (8)–(10) for a particular engine are shown in Fig. 8. Note in Fig. 8(a) and (b) how all three dependent variables are relatively invariant between 120° and 180° ATDC and how, for the normal range of EVO, the rate at which efficiency is lost due to early EVO is in the range of 0.07–0.12%/deg. The assumptions involved are instantaneous heat release at TDC and cessation of expansion at the timing point of EVO. These loss rates are directly comparable with those derived previously in reference to IVC and EVC, i.e., both volumetric and thermal efficiency losses have a direct and comparable bearing on indicated torque or IMEP. It is clear that, within the normal range of EVO, this timing point is less significant in terms of lost low-speed performance than either IVC

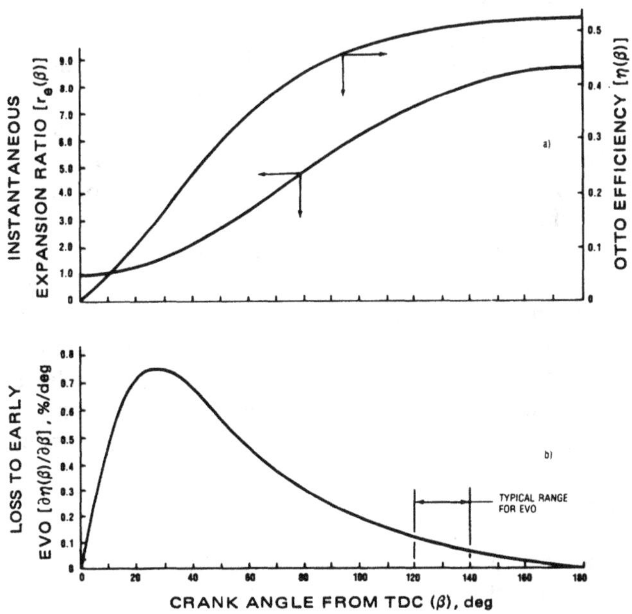

FIGURE 8. Instantaneous Otto cycle loss to early EVO for the Chrysler 2.2-liter engine. (See Table 1 for specifications.)

or EVC. Therefore, the timing of EVO should be sufficiently early to satisfy high-speed performance criteria, though no earlier than what can be justified on this basis because thermal efficiency at all engine operating conditions will be adversely affected though to a small degree. This will be discussed later in terms of P–V diagrams of pumping loops.

C. Discussion of Valve Lift and Valve Sizes

Valve timing, valve lift, and valve sizes combine to yield what, for the purpose of this discussion, are termed valve events. These valve event parameters are interrelated making it impossible to establish, *a priori*, an optimum set of design criteria.

1. Valve Lift

It should be recognized that the flow area created by valve lift may be limited by the minimum flow area in either the valve throat or in some port section. Though at the maximum lift point any "excess" lift may yield only a small incremental flow capacity, it does ensure that the maximum effective flow area is available longer, i.e., over a broader range of piston travel. Consider, for example, the lift and flow area curves shown in Fig. 3. A reduction in lift would have the effect of shortening the plateau on the flow area curves. Any observed changes in engine performance resulting from lift modifications may be a result of changes in flow capacity on the respective closing sides of the flow area curves. Reference is made to Sec. IV.A and the discussion of the basic mismatch between valve flow areas and piston motion. This mismatch, at high speeds, tends to yield a constriction during the final portion of the exhaust stroke which may be affected by closing-side lifts. Similarly, during the intake stroke the occurrence of high volumetric efficiency at high speeds depends upon the intake valve flow capacity as the piston approaches BDC and as the cylinder pressure rises from its depressed state early in the intake stroke. It is never desirable to provide more valve lift than what can be justified on the basis of engine performance. To do so invariably raises valve gear acceleration rates and surface stresses and imposes the need for high-load valve springs to prevent valve floating. This may yield unnecessarily high valve gear friction.

2. Valve Sizes

Within the constraints imposed by cylinder head construction, engines are generally designed with the maximum combined valve sizes, and the ratio of intake-to-exhaust valve sizes is typically based on past experience. Stroke–bore ratio is one of the principal factors determining the total valve area in relation to swept volume. Though there is no universally accepted valve size ratio, the range in modern high-speed, four-cylinder engines is approximately 1.1 to 1.3 intake/exhaust. It is interesting to bear in mind that the volume of exhaust which must flow out of the cylinder through the exhaust valve is approximately 2.75 times the volume of fresh charge that flows into the cylinder by way of the intake valve. Some insight can be gained as to the appropriateness of a given valve size ratio by analysis of P–V diagrams of pumping loops. This subject will be discussed later.

V. VALVE EVENT STRATEGIES

A. Overview

The discussion which follows concentrates on the optimization of valve events for automotive-type engines employing well-known types of valve gear. It is recognized that there are many engine-specific items which preclude *a priori* specification of optimum valve event parameters. Therefore, emphasis will be placed on the establishment of general direction.

It is widely recognized that valve events affect engine operation at idle and at WOT. Valve overlap affects combustion at idle while all aspects of valve events may influence volumetric efficiency at WOT and therefore, are determinants of engine performance. In the intermediate, part-throttle range valve overlap affects combustion as it does at idle but to a smaller degree. It is fundamental that combustion at idle is marginal because of excessive exhaust dilution, i.e., the BGF is too high. Valve overlap increases the BGF and is detrimental to the idle condition. At many part-throttle conditions the inherent BGF is sufficiently low that additional quantities of exhaust can be added to control NO_x without adversely affecting combustion. It makes little difference whether additional exhaust is recycled via valve overlap or externally via an EGR system providing both methods yield equivalent cylinder-to-cylinder distributing of EGR. Valve overlap tends to provide its maximum dosage of EGR when it can least be tolerated and provides for only small dosages when the engine can best assimilate it. Therefore, valve overlap is not seen as a practical means of controlling NO_x at part throttle since EGR systems perform this function very well. Volumetric efficiency as determined by valve events is clearly not important at part throttle as inlet throttling is a deliberate means of reducing volumetric efficiency to control power.

Therefore, idle and WOT performance are the rightful concerns of valve event optimization, and any part-throttle effects must be seen as lower priority. Optimization of valve events must then be seen as a task involving the minimization of valve overlap while maximizing WOT performance.

B. The Idle Condition

Engine combustion at idle is inherently slow owing to excessive exhaust dilution, low pressure, and a quiescent fluid mechanical state resulting from the low engine speed. Intake throttling reduces the intake manifold pressures to about one-half (or less) that of the exhaust manifold with the result that the retained exhaust must be expanded by this pressure ratio prior to any net induction of fresh charge. The presence of any valve overlap provides a leakage pathway by which additional quantities of exhaust can back flow from the exhaust manifold through the combustion chamber and into the intake manifold where it can mix with fresh charge and be induced during the intake stroke. This latter source of residual can be regulated by controlling the amount of valve overlap. As additional diluent can least be tolerated at idle, minimization of overlap is often beneficial. Improvements in the idle condition will result in any or all of the following:

1. Reduced incidence of misfire or slow-burn cycles;
2. Tolerance for less rich fuel–air mixture;
3. Tolerance for lower speeds;
4. Tolerance for optimum spark advance; and
5. Greater stability upon low-speed lug during clutch engagement.

From the greater margin of combustion stability afforded by reduced valve overlap, improvements in idle quality can be realized and/or more fuel efficient idle control strategies may be employed. Figure 9 shows the effect of spark timing on idle fuel consumption, intake vacuum, and emissions for two different camshafts having 28 and 14 mm-deg of valve overlap, respectively.* Note that the reduced overlap camshaft under the simulated idle-drive conditions, gave increased intake vacuum and NO_x while reducing HC. These trends are similar to what would be expected if the EGR level was reduced at some part-load conditions. Though idle fuel consumption was not directly affected in a significant way by the reduced overlap feature, there were obvious signs of improved idle quality—most prominent among these was speed stability. Despite the absence of a direct change in idle fuel consumption attending the effects of reduced overlap, the indicated thermal efficiency must have increased fractionally by the incremental indicated load associated with the change in intake manifold pres-

* Previous reference was made to valve overlap as the subtended crank anagle between IVO and EVC. Subsequent discussions of the consequences of this factor will include valve overlap expressed as a more meaningful physical area, i.e., mm of actual valve lift over the actual subtended crank angle or mm-deg. The relationship between the two expressions for overlap is rather nonlinear particularly in the low overlap range as an increasing fraction of the overlap arises from ramps on the ends of the lift curves.

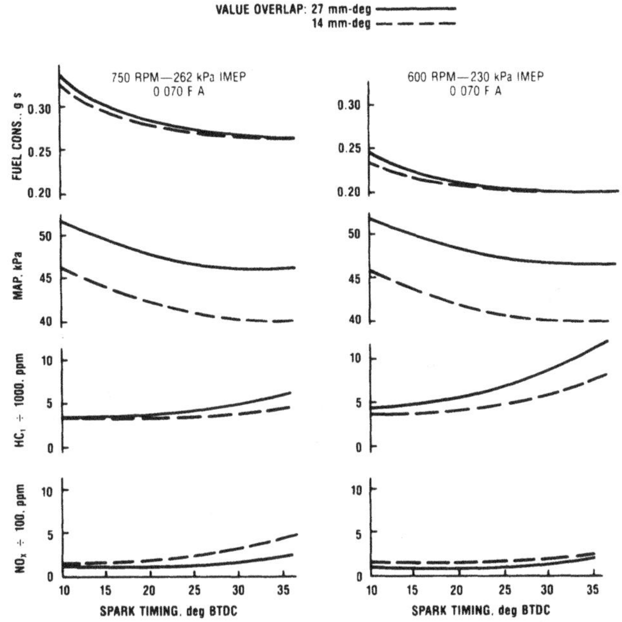

FIGURE 9. Spark loops at simulated idle-drive loads for the Chrysler 2.2-liter engine. Camshafts: 244-244-28-106-1.09/1.09 and 232-244-16-110-1.09/1.09 giving 27 and 14 mm-deg of overlap, respectively (Chrysler Corporation).

sure. One of the benefits of the reduced overlap feature is that less rich idle fuel–air mixtures can be used without adversely affecting idle quality. Figure 10 illustrates the effect of F–A on idle fuel consumption at optimum spark timing.

Figure 11 shows schematics of P–V diagrams to aid the interpretation of those to be presented later. The first is like those showing various idle configurations while the second has the pressure axis expanded to facilitate interpre-

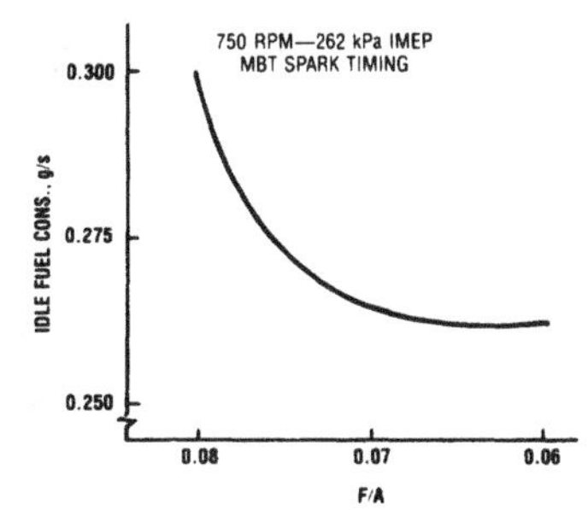

FIGURE 10. Idle fuel consumption versus F/A for the Chrysler 2.2-liter engine (Chrysler Corporation).

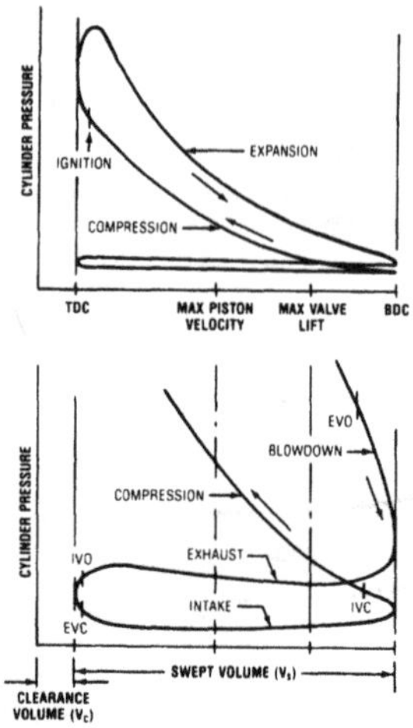

FIGURE 11. Schematic of P–V diagrams showing the entire engine cycle and focused on the pumping loop portion.

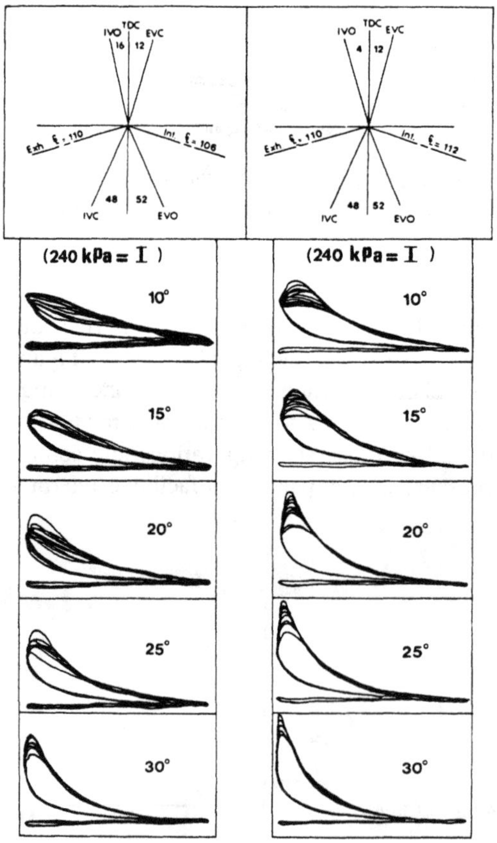

FIGURE 12. Idle P–V diagrams over spark loops from 10° to 30° BTDC for 244-244-28-106-1.09/1.09 and 232-244-16-112-1.09/1.09 camshafts showing the effect of valve overlap in a Chrysler 2.2-liter engine (Chrysler Corporation).

tation of WOT pumping loop details. Figure 12 shows P–V diagrams at simulated idle-drive loads. Note how the reduced overlap camshaft promotes faster combustion as suggested by the higher rates of pressure rise following ignition at a given spark timing. Figure 13 illustrates the effect of changes in phasing of a given camshaft having very low overlap, i.e., 0° per the convention stated in footnote on page 156 or 7 mm-deg. The apparent differences in burn rate, independent of overlap, show the effect of the timing of EVC on the BGF. As discussed in Sec. IV.B.3 the slower burn rates associated with late EVC, i.e., 15° ATDC versus 6° BTDC reflect an increase in BGF as a result of reintroduction of exhaust through the exhaust valve early in the intake stroke.

With the goal of reducing the BGF at idle by modifying valve timing, two basic options are available. These are (1) reduce overlap and/or (2) modify the EVC. Modification of EVC in the direction of reduced BGF will be detrimental

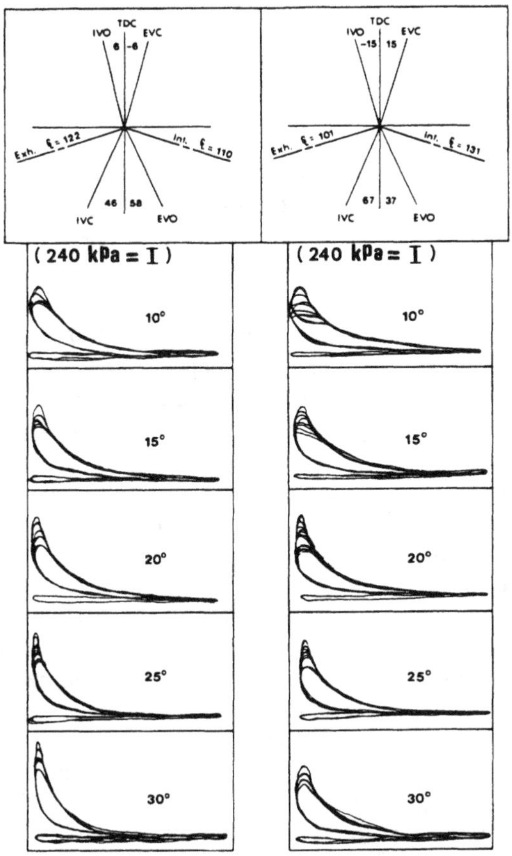

FIGURE 13. Idle P–V diagrams over spark loops from 10° to 30° BTDC for a 232-232-0-( )-1.09/1.09 camshaft phased at 110° and 131° intake centerlines showing the effects of changes in EVC on the idle condition for a Chrysler 2.2-liter engine (Chrysler Corporation).

to WOT performance at high speeds as was discussed in Sec. IV.B.3. This will be discussed further in Sec. V.C. Minimization of valve overlap to reduce the BGF at idle need not affect WOT performance if it is done by delaying IVO while fixing EVC at the earliest possible timing compatible with high-speed, WOT performance criteria.

C. WOT Performance

The discussion that follows deals with the establishment of a framework of understanding engine breathing processes as affected by valve events. A technique will be described wherein the inadvertent throttling caused by valves and ports may be minimized to possibly increase engine performance in such a way that the idle condition is not adversely affected. Key to this technique is the expansion of P–V diagrams to focus on the open-valve portion of the engine cycle. This involves installing an appropriate pressure transducer directly in the combustion chamber and mounting a crankshaft encoder on the subject engine in such a manner that a precise record of the phased relationship between pressure and volume is attainable. (Though it is not essential that volume be used as one of the coordinates it does provide a convenient means for interpreting results.) Figure 14 shows P–V diagrams of the open-valve portion of the cycle over a WOT power run from 1200 to 5600 rpm in 400 rpm increments. (See Fig. 11 as a guide to the interpretation of these diagrams.) Note the ever-increasing "bulge" during the exhaust stroke which becomes particularly apparent at about 4000 rpm leading to a maximum pumping loop pressure drop of about 120 kPa at 5600 rpm. For the particular example cited this number is to be contrasted with the nominal measured pressure drop between the intake and exhaust ports of 52 kPa, i.e., the intake depression was 7 kPa, and the exhaust back pressure measured at the port was 45 kPa. The 68 kPa difference is due to valve and/or port throttling. As the cylinder pressure over the intake stroke is relatively unchanged over the engine speed range, and as there is no significant "bulge" occurring near the maximum piston velocity (73° ATDC or 45% swept volume) it is concluded that little of this pressure gradient occurs across the intake valve. On the other hand the appearance of an ever-increasing pressure "bulge" near maximum piston velocity during the exhaust stroke implies that most of the valve throttling occurs at the exhaust valve. The implication of these observations is that a performance improvement may be realized if the exhaust valve size were increased. Furthermore, if the intake valve size was reduced by the same amount, no adverse effect would be expected to occur. It should be realized that the exhaust pressure "bulge" occurs near maximum piston velocity as the exhaust valve is closing and is related to peaks in the pseudovelocity relations for the exhaust process referred to in Sec. IV.A and Fig. 5. Evidence of excessive flow restriction at the intake valve would yield a downward "bulge" near maximum

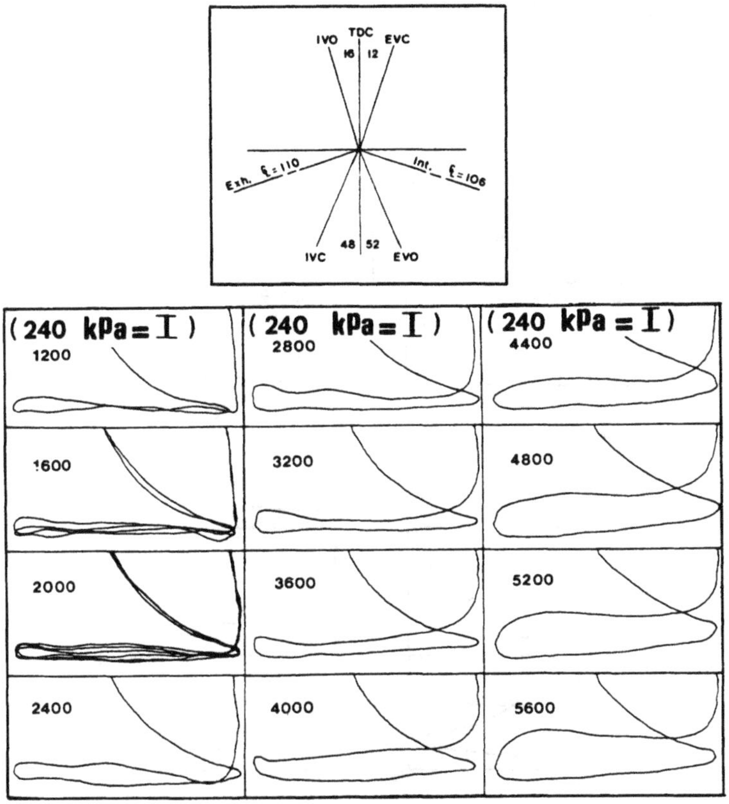

FIGURE 14. WOT, pumping loop, P–V diagrams over the speed range; net power run from 1200 to 5600 rpm at MBT with maximum power F/A. Chrysler 2.2-liter engine with a 244-244-28-106-1.09/1.09 camshaft configuration (Chrysler Corporation).

piston velocity during the intake stroke, becoming increasingly pronounced at high engine speeds. Increasing the flow capacity of an exhaust port is expected to yield diminished performance benefits as most of the exhaust constriction occurs at the valve itself as it is closing. On the intake side, however, improvements in port flow capacity may significantly increase performance as the inlet flow follows the prevailing pressure gradient even at maximum valve lift as the system tries to recover from flow losses incurred early in the intake stroke. (Recall that maximum piston velocity occurs prior to maximum valve lift.)

Figures 15 and 16 illustrate the effect of a 4° (crankshaft) change in camshaft phasing for two different camshafts having high and low overlap, respectively. In particular these show how the timing of EVC can affect cylinder pressure near the end of the exhaust stroke. Only pumping loop P–V diagrams for the

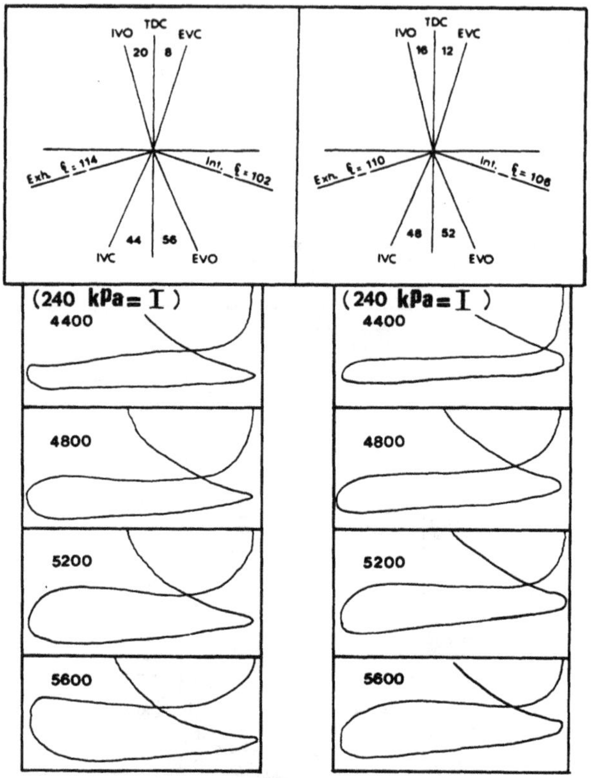

FIGURE 15. WOT, pumping loop, P–V diagrams over partial, net power runs from 4400 to 5600 RPM at MBT with maximum power F/A showing mainly the effects of EVC on the pressure histories for a Chrysler 2.2-liter engine with a 244-244-28-( )-1.09/1.09 camshaft phased at 102° and 106° intake centerlines, respectively (Chrysler Corporation).

higher speeds are shown as these are most illustrative. Note the critical nature of EVC in terms of cylinder pressure at the end of the exhaust stroke. High cylinder pressures at this point as indicated by an apex is detrimental to high-speed performance because it leads to increases in trapped exhaust which will displace fresh charge and also increase the heat content of the same with the result that volumetric efficiency is reduced. Added heat content of the fresh charge may also increase octane requirement. Note that the low overlap camshaft produced much sharper apices than did the high overlap one; this reflects a more abrupt transition from exhaust to intake flow passages. These pumping loop P–V diagrams provide a basis for monitoring the status of the EVC event.

As discussed in Sec. IV.B.4, the timing of EVO should be established on the basis of WOT performance such that the exhaust valve opens just early

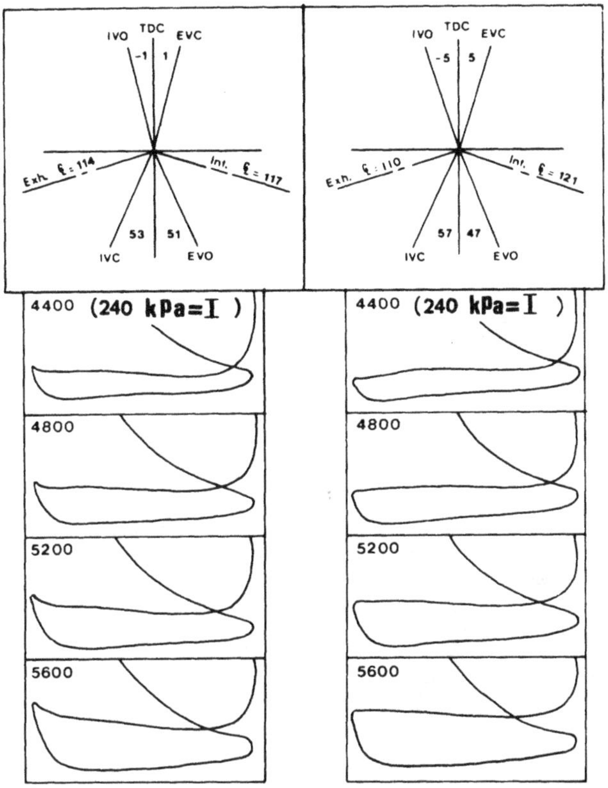

FIGURE 16. WOT, pumping loop, P–V diagrams over partial, net power runs from 4400 to 5600 rpm at MBT with maximum power F/A showing mainly the effects of EVC on the pressure histories for a Chrysler 2.2-liter engine with a 232-232-0-( )-1.09/1.09 camshaft phased at 117° and 121° intake centerlines, respectively (Chrysler Corporation).

enough to insure that blowdown is complete as soon as possible after BDC. Pumping loop P–V diagrams over power runs provide a basis for precisely monitoring the status of the timing of EVO. Figure 17 illustrates the consequence of delaying EVO too much to insure efficient blowdown, i.e., 37° BBDC. Note the slow rate of blowdown shown in Fig. 17 and how this contributes to pumping work in comparison with the higher rates of blowdown illustrated in Figs. 14–16. Despite an abnormally late IVO, i.e., – 15° BTDC (15° ATDC), there are only slight indications that a "bulge" is beginning to appear during the intake stroke. This observation tends to confirm the basic insensitivity of the breathing process to the timing of IVO as discussed in Sec. IV.B.1. Fig. 18 shows the result of analysis of pumping loop P–V diagrams as the timing of EVO was systematically varied over a 24° range at 5200 RPM WOT. The attendant pumping loss was

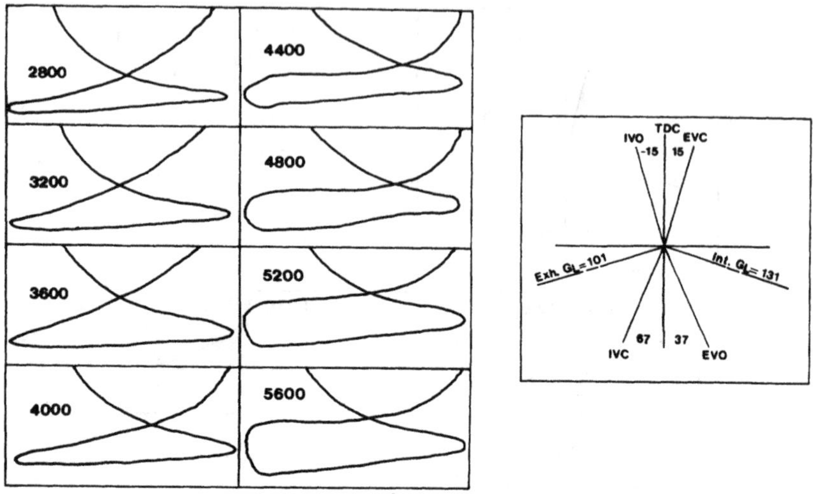

FIGURE 17. WOT, pumping loop, P–V diagrams over a partial, net power run from 2800 to 5600 rpm at MBT with maximum power F/A showing mainly the effect of EVO on pressure histories for a Chrysler 2.2-liter engine with a 232-232-0-131-1.09/1.09 camshaft configuration (Chrysler Corporation).

measured, and the thermal efficiency was calculated using an appropriate form of Eq. (10). As the pumping and thermal efficiency losses are countervailing, their sum is relatively insensitive to EVO and yields a minimum at about 45° BBDC, i.e., the power-optimized EVO is about 45° BBDC for the subject engine. The timing of EVO may be further delayed to promote fuel economy while incurring only small performance losses.

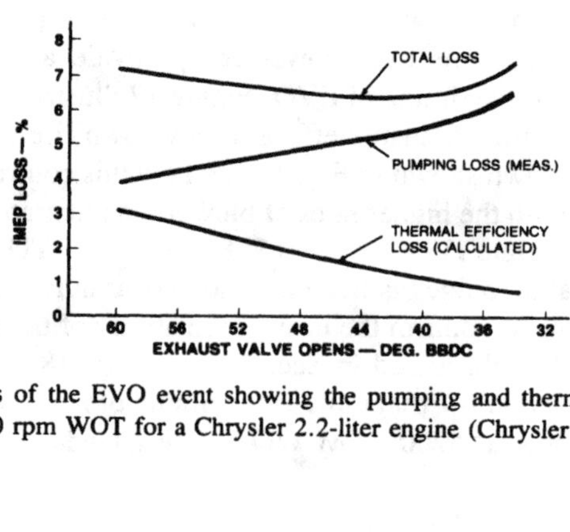

FIGURE 18. Analysis of the EVO event showing the pumping and thermal efficiency effects as EVO is varied at 5200 rpm WOT for a Chrysler 2.2-liter engine (Chrysler Corporation).

D. Synopsis

Observation of the differences in pressure history over the heat release portion of P–V diagrams such as those presented in Figs. 12 and 13 enable qualitative assessment of cycle-to-cycle variability in burn rate. (The slow-sweep method for assessing cycle-to-cycle variability, i.e., peak cylinder pressure versus time over many cycles, is sensitive to this variable only near optimum spark timing.)

P–V diagrams focused in the open-valve portion of the engine cycle can provide valuable diagnostic information for identifying flow losses which are limiting engine performance. Losses associated with valve timing as well as those relating to port and valve sizes may be pinpointed using this technique. The following specific examples are cited:

(i) The appearance of a top-side apex in the pumping loop P–V diagram near the end of the exhaust stroke signals an EVC that is too early. To avoid unnecessary power penalities, EVC should be delayed just enough to eliminate this type of apex at the maximum useful engine speed.

(ii) The timing of EVO should be just early enough to promote complete blowdown within 10–20% swept volume after BDC at all engine speeds. Compliance with this guideline may be determined either at the maximum useful speed or near peak volumetric efficiency.

(iii) Experimental evidence has been generated which supports the basic premise that IVO may be delayed further than what might be regarded as standard practice. The degree to which this option may be exercised is dependent upon many engine specifics, principal among which are intake valve size and port flow capacity. The appearance of a downward pressure "bulge" representing an abnormal depression early in the intake stroke signals a limit beyond which significant flow losses may be encountered. This is key to reducing overlap without incurring performance penalties.

(iv) The timing of IVC is the single most significant determinant of the balance between low-speed and high-speed volumetric efficiency and, therefore, must be established on the basis of desired engine performance characteristics. It is common practice to change the phasing between the camshaft and crankshaft to alter the torque-speed relationship at WOT. If this is done over too broad a range, the result may be an excessive power loss when phasing for low-speed torque and conversely an excessive loss in low-speed torque when phasing for power.

(v) Valve size optimization can be guided by analysis of pumping loop P–V diagrams. The appearance of pressure "bulges" predominating either late in the exhaust stroke or early in the intake stroke which increase with engine speed may signal a valve size ratio which is not appropriate for maximum performance.

VI. VEHICLE CONSEQUENCES OF VALVE EVENT OPTIMIZATION

As discussed earlier, the fundamental consequences of developing an engine to a state approaching optimum with respect to valve events is most likely to yield fuel economy benefits resulting from (1) improvements in the idle condition such that a more fuel efficient idle control strategy may be adopted and (2) increases in WOT performance which may affect either the selection of drive ratios or engine sizes in the direction of improved vehicle fuel economy. Absolute levels of incremental fuel economy cannot be given as these depend upon baseline conditions and many items unique to specific engines, drive trains, and vehicles. However, some general "rules of thumb" based on FTP urban and highway drive cycles are given as follows:

(i) One-fourth to one-sixth of the fractional reduction in closed-throttle fuel consumption (both idle and deceleration) is a reasonable expectation for the urban cycle fuel economy improvement, e.g., a 10% reduction in closed-throttle fuel consumption may yield a 1.7 to 2.5% increase in urban cycle fuel economy. The expected highway cycle fuel economy improvement may be about 1.0%.

(ii) If the WOT torque curve for an engine is uniformly elevated across the speed range and if the drive ratios are adjusted such that vehicle performance is held constant, then one-fourth to one-half of the fractional engine performance increase represents a reasonable expectation for the combined urban and highway fuel economy improvement. If, on the other hand, an increase in engine performance leads to the selection of a smaller engine for a given vehicle application and performance level, then the fractional increase in specific output may yield a vehicle fuel economy improvement approaching an equal fraction. For example, if the WOT torque curve for an engine is uniformly elevated by 10%, then a 2.5 to 5.0% combined fuel economy increase may be expected on the basis of drive ratio modification while if the engine performance increase leads to the selection of a smaller engine, then a vehicle fuel economy improvement approaching 10% may be expected.

A. Idle Quality

Idle quality is that property of an engine–vehicle system which arbitrarily and subjectively reflects noise and vibration from the engine while idling. In the era of six- and eight-cylinder engines idle quality was seldom a problem, although it was during this period that customer expectations were established. As the automotive industry has responded to the fuel economy challenge with new four-cylinder engines, idle quality has become a significant issue. Inasmuch as two- and three-cylinder engines appear to have a place in the future so does the idle quality challenge.

Idle quality considerations oftentimes influence idle control strategies and therefore idle fuel consumption, i.e., increased speeds, richer mixtures, and retarded spark timing favor idle quality and adversely affect idle fuel consumption.

As discussed in Sec V.B the high levels of combustion dilution (BGF) have a dominant effect on combustion at idle with the result that there is typically a high degree of cycle-to-cycle variability in burn rate; this is reflected in the idle P–V diagrams shown in Fig. 12 as a broad range of cylinder pressure histories over the combustion and expansion events. This gives rise to variability in the torque impulses associated with the engine firing events causing the engine structure to vibrate accordingly. Figure 19 illustrates vibrations from an idling, in-line four-cylinder engine as determined by an accelerometer mounted normal to a plane defined by the bore centers. Note the difference in vibrational amplitude for this automatic transmission vehicle between idle neutral and idle drive and also the degree of vibrational regularity associated with the two different camshafts having 27 and 14 mm-deg of overlap, respectively. Note the lower value of the standard deviation (σ) and the coefficient of variation (COV) for the case involving the low overlap camshaft. The reduced overlap camshaft lowers the BGF such that there is greater tolerance for those factors which yield variability in burn rate, e.g., mixture maldistribution, inadequate mixture preparation, random fluid mechanical characteristics, etc.

For four-cylinder engines the fundamental vibrational frequency at idle is second order, i.e., twice engine speed. Cylinder-to-cylinder variations in vibrational amplitude yield a one-half order component, i.e., one-half engine speed. These low frequency, one-half order disturbances may be significant contributors to objectionable idle quality.

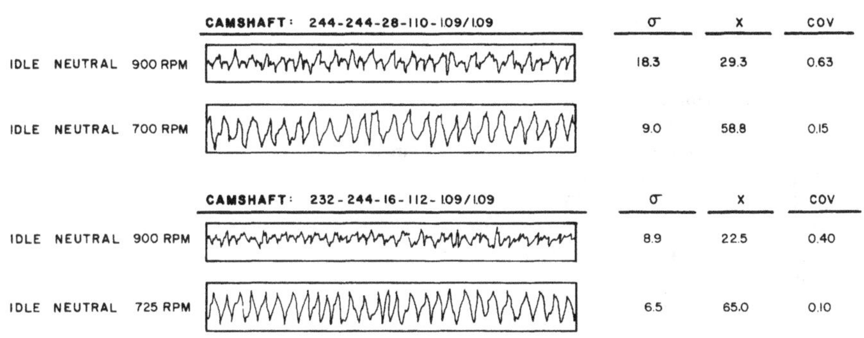

FIGURE 19. Accelerometer traces showing idle vibrations in both neutral and drive at 0.070 F/A and 10° BTDC spark timing for a Chrysler 2.2-liter engine having camshafts with 27 and 14 mm-deg of overlap, respectively (Chrysler Corporation).

The fundamental second-order idle vibrations are mainly influenced by the indicated load on the engine at idle, i.e., the IMEP at which the engine must operate at idle. Therefore, assessory and torque convertor loads as well as engine friction are factors which must be considered in this regard.

B. Idle Fuel Consumption

Adaptation of camshaft designs having low overlap may enable the implementation of fuel efficient idle control strategies without adverse effects on idle quality. That is, less rich (or even net lean) fuel–air mixtures, lower speeds, and more nearly optimum spark timing may be employed while maintaining acceptable idle quality.

Minimization of valve spring loads must be seen as a factor which is capable of affecting both idle fuel consumption and idle quality. As discussed in Sec. IV, optimized valve events must include the minimum valve lift which is compatible with WOT performance criteria. Minimized valve lift may be exploited to yield minimum valve gear acceleration rates which, in turn, can be exploited to minimize valve spring loads within the constraint imposed by high-speed valve gear dynamics. As most valve gear types have substantial sliding friction, minimized valve spring loads may yield reductions in valve gear friction. Due to boundary layer lubrication of these sliding surfaces, valve gear frictional torque has an inverse speed relationship, i.e., valve gear frictional torque is highest at minimum speed and may represent a significant fraction of the total engine friction at idle.

NOTATION

$A(\beta)$	Instantaneous valve flow area
b	Bore
β	Subtended crank angle from TDC
D	Valve throat diameter
d	Valve steam diameter at the minimum valve throat diameter
η	Otto cycle efficiency
k	Polytropic exponent ($\simeq 1.35$)
L	Valve lift
$l(\beta)$	Instantaneous linear piston displacement
R	Connecting rod length
r	Crank throw (one half stroke)
r_e	Expansion ratio
σ	Standard deviation of cycle-to-cycle idle vibrational impulses (arbitrary units)
$V(\beta)$	Instantaneous cylinder volume [$V(0) = 0$]
V_c	Clearance volume at TDC
V_s	Swept volume/cylinder
$\overline{X}$	Mean amplitude of cycle-to-cycle idle vibrational impulses (arbitrary units)

REFERENCES

1. Fraas, A. P., *Combustion Engines,* 1st ed., pp. 135–154, McGraw-Hill, New York (1948).
2. Polson, J. A., *Internal Combustion Engines,* 1st ed., pp. 380–413, Wiley, New York (1933).
3. Taylor, C. F., and Taylor, E. S., *The Internal Combustion Engine in Theory and Practice,* 2nd ed., Vol. I, pp. 147–205 MIT Press, Cambridge, MA (1966).
4. Ricardo, H. R. and Glyde, H. S., *The High-Speed Internal-Combustion Engine,* pp. 226–241, Interscience, New York (1941).
5. Heldt, P. M., *High-Speed Combustion Engines,* pp. 358–361, P. M. Heldt, Nyack, NY (1944).
6. Obert, E. F. *Internal Combustion Engine and Air Pollution,* 3rd ed., pp. 469–478, Intext Educational Publishers, New York (1973).
7. Siewert, R. M., "How Individual Valve Timing Events Affect Exhaust Emissions," SAE paper 710609 (1971).
8. Bechtold, R. L., and Marshall, W. F., "Valve Timing—Its Effect on Emissions and Fuel Economy," BERC/RI-77/9, Energy Research and Development Administration, Bartlesville, Oklahoma, August (1977).
9. Freeman, M. A., and Nicholson, R. C., "Valve Timing for Control of Oxides of Nitrogen (NOx)," SAE paper 720121 (1972).
10. Meacham, G. B. K., "Variable Cam Timing as an Emission Control Tool," SAE paper 700673 (1970).
11. Schiele, C. A., DeNagel, S. F., and Bennethum, J. E., "Design and Development of a Variable Valve Timing Camshaft," SAE paper 740102 (1974).
12. Stivender, D. L., "Intake Valve Throttling—A Sonic Throttling Intake Valve Engine," SAE paper 680399 (1968).
13. Tuttle, J. H., "Controlling Engine Load by Means of Late Intake-Valve Closing," SAE paper 800794 (1980).
14. Sherman, R. H., and Blumberg, P. N., "The Influence of Induction and Exhaust Processes on Emissions and Fuel Consumption in Spark Ignited Engine," SAE paper 770880 (1977).
15. Tasaka, H., and Matsuoka, S., "Predictions of Combustion, Fuel Economy and Emission Characteristics Influences by the Gas Exchange Process of S. I. Engines," SAE paper 810821 (1981).
16. Tasaka, H., and Matsuoka, S., "Analysis of Gas Exchange Process of S.I. Engines Under City-Driving Conditions," SAE paper 800535 (1980).
17. Wolgemuth, C. H., and Olson, D. R., "A Study of Engine Breathing Characteristics," SAE paper 650448 (1965).
18. Huber, P., and Brown, J. R., "Computation of Instantaneous Air Flow and Volumetric Efficiency," SAE paper 812B (1964).
19. Brandstetter, W. R., "Simularity Laws for Four-Stroke Engines and Numerical Results for the Intake Process Calculated with the Method of Characteristics" SAE paper 690466 (1969).
20. Kastner, L. J., Williams, T. J., and White, J. B., "Poppet Inlet Valve Characteristics and Their Influence on the Induction Process," *Proc. Instn. Mech. Engrs.,* **178,** Pt. 1, No. 36, 955–975, July (1963).
21. Benson, R. S., and Whitehouse, N. D., "Internal Combustion Engines," Vol. II, pp. 205–269, Pergamon Press, Oxford (1979).
22. Blair, G. P., and Goulburn, J. R., "An Unsteady Flow Analysis of Exhaust Systems for the Multicylinder Automobile Engine," SAE paper 690469 (1969).

5

Flame Propagation and Heat-Transfer Effects in Spark Ignition Engines

CLAUS BORGNAKKE

I. INTRODUCTION

The essential characteristics of the internal combustion engine are ultimately determined by the processes that take place in the combustion chamber. It is these processes that generate the power output, the heat losses, and the formation of pollutants and therefore determine the trade-off between efficiency and emissions. The need thus arises to identify and qualify the significant physical processes that take place in the combustion chamber. Only after this accomplishment is it possible to establish a control of the major parameters that influence the combustion process so desired design goals can be achieved. In the past an acceptable level of understanding of the phenomenon involved has been reached through an analysis of experimental evidence and the pursuit of theoretical investigations. The processes inside the combustion chamber of an internal combustion engine involve a broad range of different subjects. The characterization includes chemistry, thermodynamics, fluid mechanics, and heat transfer, to mention a few of the most important fields as outlined in Fig. 1. Furthermore, the combustion process takes place in such a highly complex environment that an accurate and detailed description on a fundamental level cannot be done with present-day knowledge. Through numerous experiments a large amount of information has been acquired allowing the key processes to be identified and investigated in more detail. Guided by these experiments and the theoretical analysis, semiempirical methods have been used to develop the internal combustion engine, and such two-sided investigations are necessary for the successful development of the internal combustion engine. Though today's engine is highly

---

CLAUS BORGNAKKE • Automotive Laboratory, University of Michigan, Ann Arbor, Michigan 48109.

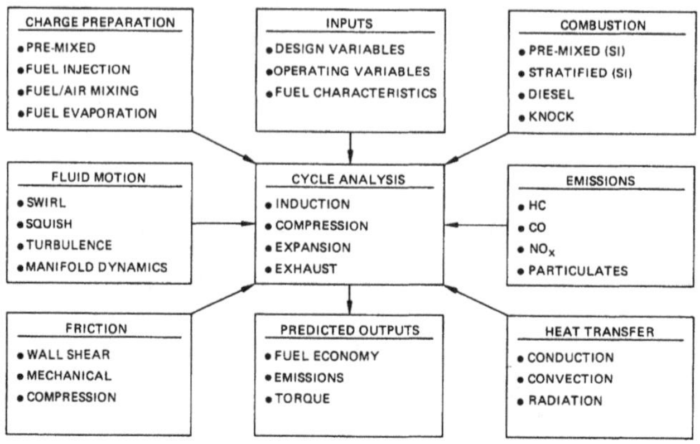

FIGURE 1. Subjects in engine analysis.

sophisticated, continued research efforts produce an increased understanding of the different processes leading to an improvement of the overall engine performance.

In this chapter we will focus on the aspects of fluid mechanics and heat transfer that influence the combustion process. Examination of the different phenomena listed in Fig. 1 will reveal that nearly all of them are strongly dependent upon the behavior of the gases explaining the importance of the fluid mechanical processes. Together with the description of these processes some theoretical models will be shown so the reader may gain an impression of the subject and an entry to the current literature available.

II. MODES OF COMBUSTION

During the induction stroke a fresh charge is taken into the combustion chamber and starts mixing with the residuals that are present. The mixing process continues in the compression stroke where the combustion process begins with the discharge of energy in the spark plug gap. After the ignition of the gases has created a volume of plasma, the combustion process proceeds to establish a flame kernel that is able to propagate into the unburned charge. The ignition process is, in general, controlled by chemical effects and the fluid properties, which is also the case for the initial combustion that creates the flame kernel. However, the latter process is more sensitive to the flow field in and around the burning gases. Once the flame kernel has grown large enough, it is gradually transformed into a fully developed propagating flame. In the traveling flame fluid mechanics usually dominates, but chemistry may become important, depending on the properties of the flow field and the charge composition. Eventually the flame has engulfed most of the gases and during the final stage of the combustion process, it is slowed down near the walls where it is quenched by the wall heat transfer. After the quenching of the flame the remaining unburned gases will burn in a process that is controlled by diffusion.

The whole combustion process is transient, but from the short description given above it may be classified according to the nature of the development of the combustion zone as

1. Ignition
2. Flame development
3. Flame propagation
4. Flame termination

This classification applies to a normal burning situation and does not include effects such as misfire, partial burning, or knocking. The latter processes are failures in the combustion process to proceed in a desirable way, and they represent the limits of operation for a given set of conditions in the engine. Since different effects are dominant in the four combustion phases, they will be discussed separately in the following sections.

III. IGNITION AND FLAME DEVELOPMENT

The ignition process and the influence of engine-operating parameters, as related to the chemistry and fluid mechanics, will not be covered here. A short description of some of the fluid mechanical and chemical aspects that are important for the flame propagation mechanism also will be given below.

The combustion process is initiated by the discharge of energy in the spark plug gap. The energy released to the gas mixture generates a volume of plasma in which a rapid breakdown of fuel occurs. The chemical conversion of the fuel and the oxidizer shifts energy from chemical binding energy to internal (thermal) energy of the gases and creates a range of different species. Some of the products in the reactions act as active species that initiate the chain-breaking process of the unburned fuel. These active species, mostly light radicals, are very important for the propagation of the combustion process since they lower the apparent overall activation energy. The energy and the active species created in the plasma diffuse to the surroundings as shown in Fig. 2, with the surroundings as the unburned gas mixture or the solid surfaces. Propagation of the combustion process then occurs if the unburned gas next to the plasma can be brought to a condition with a high reaction rate. This ignition of the surrounding gas mixture takes place when sufficient fluxes of energy, q, and active species, J, reach the unburned gas. The fluxes q and J are driven by the temperature and the concentration gradients between the burned gas and the unburned gas and their local magnitude depends on the overall geometry of the interface.

Since the fluxes out of the combustion volume will lower the temperature

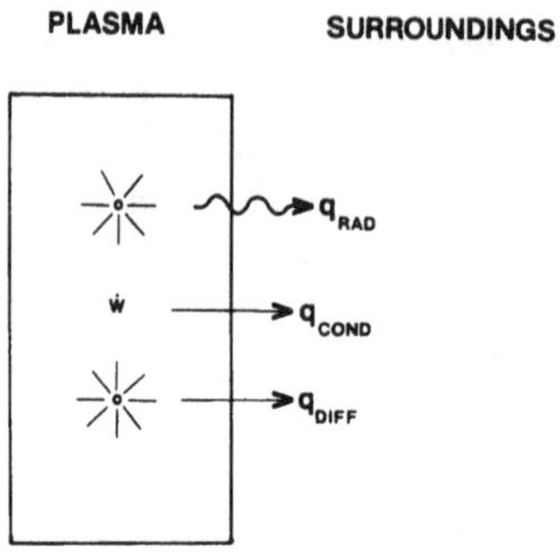

FIGURE 2. Ignition plasma.

and the concentration of active species, an equilibrium situation exists if the volume production equals the loss by the fluxes. Initially the plasma is at a very high temperature compared with the flame temperature for the equilibrium situation. Due to the finite chemical reaction rates the species flux is being built up in a time that corresponds to the chemical induction time. During this time the plasma loses energy by heat transfer and radiation so the temperature will drop, as tentatively shown in Fig. 3, until the necessary species flux is established, curve *a*. If this is not the case the temperature will continue to drop and no ignition will occur, curve *b*.

The successful development of the ignition plasma into a flame kernel depends on a large number of parameters such as ignition energy, plasma volume and location, chemical reactions, mean flow field, and turbulence around the spark plug location. The effect of turbulence on ignition has been investigated by Ballal and Lefebvre,[1,2,3] De Soete,[4,5] and Karpov and Sokolik,[6,7] among others. The studies show that the ignition limit in terms of air–fuel ratio is very strongly influenced by the presence of turbulence in the flow. One example of the ignition limits of a propane–air mixture is shown in Fig. 4, from the studies of Karpov and Sokolik.[6] Both the lean and the rich limit are affected, making

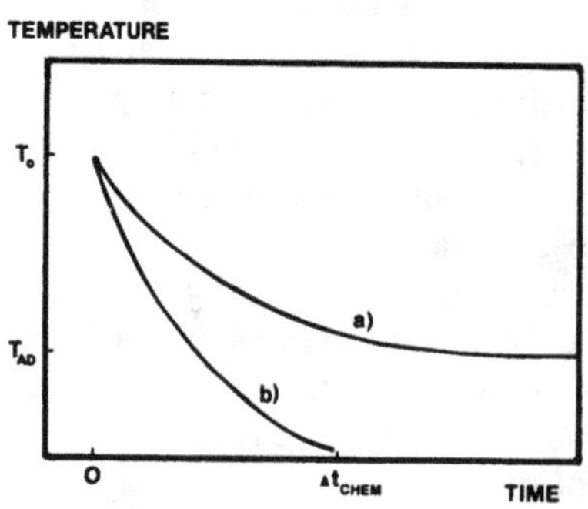

FIGURE 3. Kernel temperature versus time; (a) ignition, (b) no ignition.

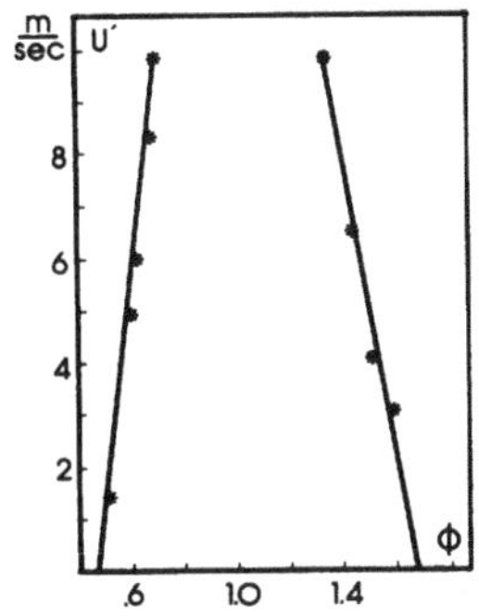

FIGURE 4. Ignition limits with turbulence intensity.

the window in ignitable air–fuel ratios smaller the higher the turbulence intensity u'. In these experiments the limits for nearly stoichiometric mixtures were not obtained due to the range of turbulence intensities that could be obtained in the experiment. However, for stoichiometric mixtures, they obtained ignition limits in terms of the turbulence intensity when diluting with argon and helium. The ignition limits are greatly reduced with the dilution and, for the same dilution and therefore same flame temperature, the limiting value of u' is much lower for helium than it is for argon. This difference may be attributed to the change in the thermal conductivity and diffusivity resulting in different Lewis numbers for the two species.

Ballal and Lefebvre[1,2,3] measured minimum ignition energy as a function of turbulence intensity, turbulence length scale, and equivalence ratio. The minimum ignition energy, as shown in Fig. 5, increases sharply as the air–fuel ratio is changed from stoichiometry and the increase becomes more pronounced with a higher turbulence intensity. Around stoichiometry, where the ignition energy is minimum, there is an increase in minimum ignition energy by a factor of 8 in going from 1 to 22% in the turbulence intensity. The influence of the turbulence length scale indicates the complex nature of turbulent combustion. From the measurements of the minimum ignition energy and previous measurements of flame propagation,[8] Ballal and Lefebvre find a transition region when the turbulent intensity u' is around two times the laminar flame speed S_L. For very small ratios of u'/S_L, which correspond to very weak turbulence, the minimum ignition energy decreases with turbulence length scale and the turbulent flame speed increases with scale. For levels of turbulence intensity of one to four times the laminar flame speed, the minimum ignition energy increases with scale, and for higher levels of u' the behavior reverses and the minimum ignition energy decreases with scale. Part of this behavior may be caused by the sensitivity of the flame to disturbances of different wavelengths, as will be explained later. Ballal and Lefebvre hypothesize that the transition is associated with the ratio

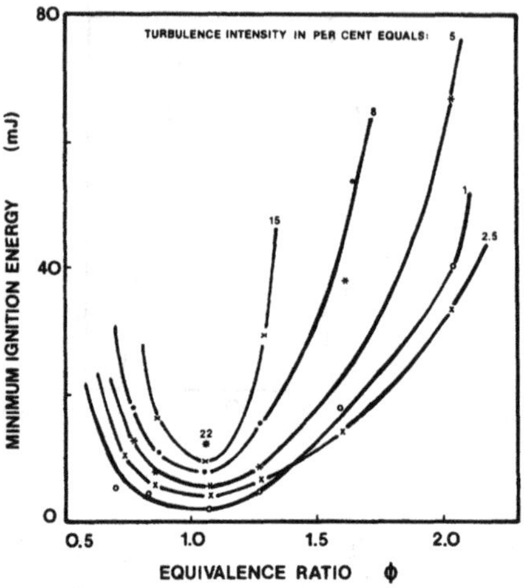

FIGURE 5. Minimum ignition energy as a function of equivalence ratio and turbulence intensity.

of the Kolmogorov length scale of turbulence to the laminar flame thickness. For large values of the Kolmogorov scale (weak turbulence), the flame is still laminar but is wrinkled by the disturbances and the flame area is therefore increased. For smaller values of the Kolmogorov scale (strong turbulence), the flame is stretched and distorted so much that it can no longer be identified with a laminar flame.

The flame development can be characterized as the transition of the established flame kernel into a fully developed propagating flame. During this period the flame size increases to be of the order of the chamber dimensions, and the influence of the flow field during this transition will change due to the change in dimensions of the flame relative to the flow structures. The effect of the chemistry will decrease as the flame develops, and the flame propagation will be more influenced by the fluid mechanics. Three regions in the ratio of turbulence length scale, λ_T, to kernel size or flame thickness, δ_{FL}, may be recognized:

$$\lambda_T/\delta_{FL} = \sim 0,\ \sim 1,\ \text{and} \gg 1,$$

which corresponds to diffusion, distortion, and convection, respectively. When the flame is fully developed, the flame thickness is of the order of the turbulence integral length scale, implying that the structures are highly distorted and the propagating velocity is in the order of the turbulence convection velocity.

The flame development is very important for the subsequent combustion of the rest of the gas mixture. A large flame area in a favorable location produces a rapid combustion, whereas a smaller area near a solid surface or separate smaller areas creates slower combustion or even extinction of the flame. Such variations in the growth and position of the flame are one of the major contributions of cycle-to-cycle variations. Since the initial flame kernel is small compared with the larger "eddies" in the turbulent flow field, the kernel is convected around in the cylinder by the mean flow and by the larger eddies. This motion by the turbulence is stochastic in nature, and it can be viewed as a random walk, which has been shown by schlieren photographs taken during the early combustion stage in investigations by Iinuma and Iba[9,10] and Namazian *et al.*[11] Quantitative correlations between velocity and peak pressure were obtained by Cole and Swords[12] in an investigation of pressure fluctuations. From these studies it is clear that there is a strong correlation between the convection of the flame kernel and the later combustion process. In cycles where the flame kernel is convected closer to the combustion chamber walls, a lower peak pressure is observed, corresponding to a slower combustion process. And it is indicated from the data that the fastest burn occurs for a flame kernel which moves towards the center of the combustion chamber, thus decreasing the flame travel distances. However, a number of different phenomena take place at the same time, so the above observation is somewhat simplistic and does not explain all of the experimental observations.

In the initial development of the flame kernel the growth is dominated by the laminar flame speed and the kernel surface area. This means that the influence of some of the engine-operating conditions, such as compression ratio, exhaust gas recirculation (EGR) level and air–fuel ratio, can be seen from the effects on the laminar flame speed. For the hydrocarbon fuels the laminar flame speed has been correlated as

$$S_L = C\left(\frac{P}{P_0}\right)^{\alpha}\left[Y_F^a\, Y_{02}^b \exp\left(\frac{-E}{RT_M}\right)\right]^{1/2} \tag{1}$$

by Van Tiggelen and Deckers, [13] Tabaczynski *et al.*[14], and Lavoie.[15] In this expression the mean flame temperature T_M is

$$T_M = 0.74(T_b^0 - T_u) + T_{u'} \tag{2}$$

with the unburned gas temperature T_u and the adiabatic flame temperature T_b^0. The powers a and b are orders of the reaction with respect to fuel (concentration Y_F) and oxygen (concentration Y_{02}), respectively. From the functional form of the correlation in Eq. (1) it can be seen that the greatest influence on the laminar flame speed is exerted by parameters which change the mean flame temperature. For instance, variations in equivalence ratio and dilution will alter the concentrations, but the major influence is caused by the associated change in the adiabatic flame temperature. These effects and others can easily be evaluated through

Eqs. (1) and (2) once the chemical reaction scheme is known so the adiabatic flame temperature can be calculated.

Besides the convection of the flame kernel by mean velocity or large-scale turbulence, the kernel becomes highly deformed and stretched by the fluctuating shear stress regions in the flow. The perturbations and deformations of the developing flame changes the local fluxes of energy and species from the flame front to the unburned gases. A curvature of the flame front will also change the velocity field due to the expansion of the gases as they burn, resulting in a nonuniform pressure field around the flame front. These two phenomena may lead to instability of the flame propagation. The diffusional–thermal instability is caused by the fluxes of energy and species for nonstoichiometric mixtures, where the diffusivity of the limiting reactant is high. This condition is normally not present for the fuel–air mixtures used in internal combustion engines, so this instability should not occur. The second instability is a hydrodynamic instability caused by the pressure field. This instability has been studied by Markstein,[(16)] Lewis and Von Elbe,[(17)] and others, see Williams.[(18)] Flames were found to be unstable for all wavelengths, or a limited range of wavelengths, dependent upon the sensitivity of the flame speed to the curvature. Since a large spectrum of wavelengths are present in the turbulent flow field, a possible instability will always be triggered. The flame propagation itself, however, acts as a stabilizing effect on the hydrodynamic instability, such that if the flame is not extinguished, a change in the overall flame speed is the net result. This effect is very important for the explanation of the fully developed turbulent flame propagation speed, as done by Ballal and Lefebvre,[(8)] Chomiak,[(19)] Williams,[(20)] and Boyer *et al.*[(21)]

During the growth of the flame kernel, the effect of turbulence changes due to the size of the kernel relative to the length scales in the flow field. The part of the turbulence that has a small scale ($\lambda_T < \delta_{FL}$) appears as an increase in effective viscosity and diffusivity, such that the transport properties are greatly increased. For mixtures that are not near a combustion limit, the increased transport properties enhance the propagation of the combustion process, and the kernel grows more rapidly. This is clearly evident from the measurements of flame propagation speeds by Lancaster *et al.*,[(22)] as shown in Fig. 6. However, near a combustion limit it is possible that the turbulence mixes burned and unburned gas so quickly that the flame kernel is cooled down and the reaction rates decrease. If the reactions are terminated, we have a bulk quenching that corresponds to a partial burn limit, which is not strictly the same as an ignition limit. Experimental verification of the difference in the limits for an engine is very difficult to obtain due to the small amount of mass burned. All the limits associated with the failure of the initial burning process to progress are therefore normally classified as ignition limit or misfire limit.

In a series of experimental verification of the lean combustion limits Quader[(23)] and Peters and Quader [(24)] were able to identify two limits. By the use of several

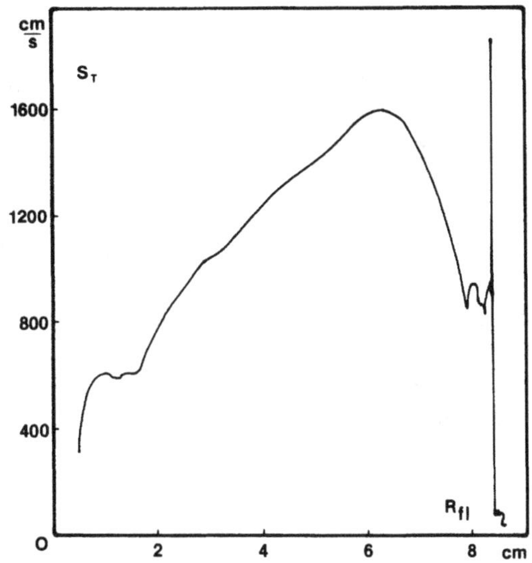

FIGURE 6. Flame speed versus flame radius.

ion gap probes placed in the combustion chamber the partial burn limits and the ignition limits could be detected. The ignition limit is found for advanced spark timing, and the partial burn limit is reached for retarded spark timing. The window between the two spark timings gives the possible range of timings for a given equivalence ratio, and establishes together with the MBT timings the limits for a given engine. This range in spark timings narrows down to zero as the mixture is made leaner, and shows the leanest possible mixture that will ignite and burn completely.

IV. MATHEMATICAL MODELS FOR THE INITIAL FLAME

Several investigators have proposed mathematical models for the ignition and the early flame development. The models for minimum ignition energy will not be covered here. From the experimental evidence and considerations about the initial kernel development, it is clear that the early combustion takes place under nearly laminar conditions. The turbulence may convect the gases around the spark plug such that the initial volume in which the spark energy is released is increased, and as the kernel grows the influence of the turbulence increases with time. These effects have been included in the discussion of the flame establishment and propagation by De Soete,[4,5] Chomiak,[25] Maly,[26] and Adelman.[27] In the latter investigation the kernel size is found as a function of time to be

$$r_k = (A + Bt)^{1/2} \tag{3}$$

where A and B are functions of the energy input and the thermodynamic state. By including the effect of turbulent diffusion, Maly[26] developed a differential equation for the kernel radius as

$$\frac{dr_k}{dt} + \left(1 + \frac{\nu_t}{\nu}\right)(B - Ar_k) = 0 \tag{4}$$

in which ν and ν_t are the laminar and turbulent viscosities, respectively. If ν_t is taken as an increasing function of time, Eq. (4) gives a kernel radius that is enhanced by the turbulence or extinguished if ν_t is too high. The ratio is equal to a turbulent Reynolds number Re_T:

$$\mathrm{Re}_T = \nu_t/\nu = u'l/\nu = K^{1/2}\, l/\nu \tag{5}$$

where $u' = K^{1/2}$, K is the turbulent kinetic energy, and l is the integral length scale of turbulence. For larger times Eq. (4) is not realistic with respect to the assumptions for the flame thickness and the behavior used by Maly for the turbulent viscosity as a function of time.

Phenomenological models have been very successful in the description of the combustion process by assumptions of the general physical phenomena. Neglecting the detailed description of the transient behavior of the initial flame establishment, Blizard and Keck[28] write the entrainment and burn up as

$$\frac{dm_e}{dt} = \rho_u A_{FL} u_e \tag{6}$$

and

$$\frac{dm_b}{dt} = (m_e - m_b)/\tau_b \tag{7}$$

where m_e is the mass entrained by the flame front and m_b the mass burned. The entrainment of fresh charge with density ρ_u takes place over the flame surface, A_{FL}, with the entrainment velocity u_e, followed by the burning of the mass in a characteristic time τ_b. In a refinement of the model Tabaczynski *et al.*[14,29] and Hires *et al.*[30] wrote the combustion as Eqs. (6) and (7) for a single "eddy" of size l with

$$u_e = u' + S_L \tag{8}$$

and

$$\tau_c = \lambda/S_L \tag{9}$$

The Taylor microscale λ can be calculated[31] from the integral length scale and the turbulence Reynolds number under isotropic conditions as

$$\lambda = l \cdot \left(\frac{15}{A}\right)^{1/2} \mathrm{Re}_T^{-1/2} \tag{10}$$

where A is a constant. The eddies are assumed to have internal turbulence structures of size λ over which the burning propagates as a laminar flame with speed S_L. By performing calculations over a range of length scales λ, l, turbulence intensities u', and equivalence ratios ϕ, a correlation was developed for the characteristic time τ_b.

$$\tau_b = 1.8\beta^{2/3}\tau_e \tag{11}$$

where β is the ratio of the laminar burntime τ_c to eddy half time τ_e:

$$\beta = \tau_c/\tau_e, \qquad \tau_e = l/u' \tag{12}$$

The ignition time can be estimated from the values of l and u'. A reasonable assumption for these variables is to associate the integral length scale with the chamber height h, and to relate the turbulence intensity to the mean piston speed V_p. The resulting expression for the ignition time can be written as

$$\Delta\theta_{\mathrm{ig}} = C_{\mathrm{ig}}(V_p\nu)^{1/3}\,(h/S_L)^{2/3} \tag{13}$$

This model shows good agreement with some data for the ignition delay when exhaust gas recirculation (EGR), air–fuel ratio, and compression ratio are varied.[30] For spark-timing variations the model predicts the trend for ignition delay, but not the main combustion duration, as compared with experiments by McCuiston *et al.*[32] The comparison is shown in Fig. 7, where this model is labeled Model 2. A better agreement with the data is obtained if the model includes the decay of turbulence with time, as done by Borgnakke *et al.*[33] and shown in Fig. 7 as Model 1.

The flame speed as a function of flame radius has been measured indirectly by Lancaster *et al.*[22] from experimental pressure–time traces. One example of the variation in the flame speed is shown in Fig. 6, and the ratio of the turbulent flame speed to the laminar flame speed is shown in Fig. 8 for two different levels

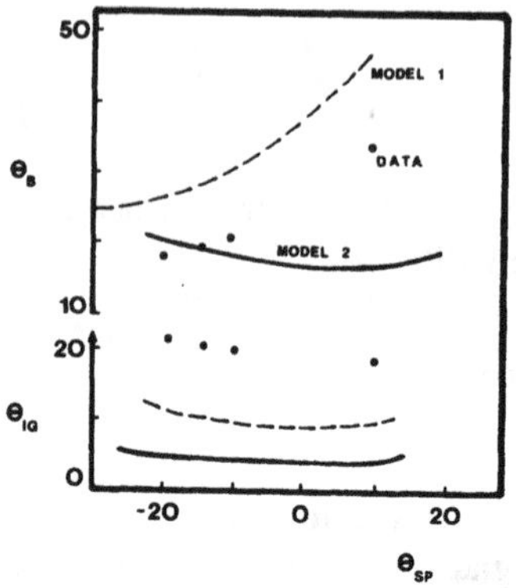

FIGURE 7. Ignition delay and burn duration versus spark timing.

of turbulence. Based on their data they proposed a model for the ratio of the turbulent flame speed to the laminar flame speed as

$$R_{FS} = S_T/S_L = R_{FSM}\,(r_{FL}/r_{FLM})^{(m-1)/2} \tag{14}$$

where R_{FSM} and r_{FLM} refer to the fully developed flame. In the development of Eq. (14) it is assumed that the turbulent flame speed can be written as a function of u':

$$S_T = C_1 u' + S_L \tag{15}$$

and then $u'(r)$ is found from the energy distribution. For a particular engine geometry r_{FLM} is fitted to the data (~3 cm) as approximately twice the clearance height and m was found to be 2 for best agreement with the data. This model

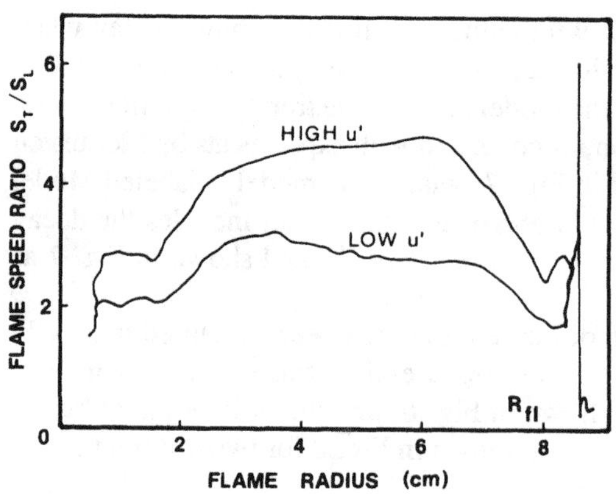

FIGURE 8. Flame speed ratio versus radius for two levels of turbulence.

is one of the few examples of a flame propagation model that depends on the size of the flame relative to the size of the flow structures, which is an important effect to include.

In conclusion it may be said that only a few measurements and models are available for the initial flame propagation. The lack of experiments under engine conditions is due to the difficulty in obtaining enough information about the statistics for the flow field and the combustion process. Extraction of the mass burned fraction x_b from a pressure–time trace has large errors for the initial combustion, and the assumption of a negligible flame thickness to calculate the flame speed is not valid.[22,34] More recent experiments with measurements of the velocities, temperatures, and visualization, as reported by Dyer,[35] show the feasibility of methods that may eventually be applied simultaneously.

V. FULLY DEVELOPED FLAME PROPAGATION

A fully developed flame is established once the flame has grown to a size comparable to the flow dimensions. This stage of the flame propagation is mainly controlled by the geometry and the flow field. The flame becomes thicker and there is a nonhomogeneous distribution of properties through the flame. The flame speed may become 5 to 10 times the laminar flame speed dependent upon the level of turbulence, as shown in Fig. 8 from Lancaster *et al.*[22]

In measurements of the combustion in an engine it is common to reduce the data by an engine simulation model. In the simulation everything is modeled except the rate of combustion, which then is extracted with the measured pressure–time trace as input (see Refs. 14, 29, 30, 32, 34, 36, 37). From such calculations the mass fraction burned x_b versus time can be obtained, from which the ignition delay is defined as the

$$x_b = m_b/m_{\text{tot}}, \qquad 0 \leqslant x_b \leqslant 1$$

time θ_{ig} (in crank angles) from the spark timing to the time where x_b has reached 2 or 10%. Similarly the main burn duration θ_b is given by the 2–98% or 10–90% time.

$$\begin{aligned} \theta_{\text{ig}} &= \Delta\theta \quad \text{for} \quad 0 \leqslant x_b \leqslant 0.02 \quad \text{or} \quad 0.1 \\ \theta_b &= \Delta\theta \quad \text{for} \quad \begin{cases} 0.02 \leqslant x_b \leqslant 0.98 \\ 0.10 \leqslant x_b \leqslant 0.90 \end{cases} \quad \text{or} \end{aligned}$$

An example of curves for the mass fraction entrained and the mass fraction burned is shown in Fig. 9 taken from Ref. 32, where McCuiston *et al.* validated an engine simulation model with measurements. From these data it can be seen

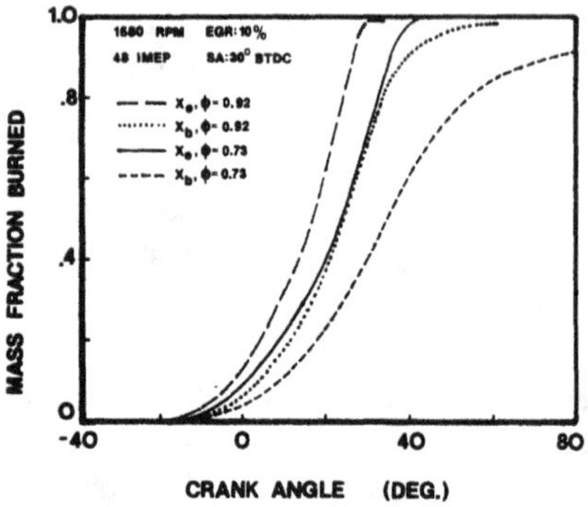

FIGURE 9. Mass fraction burned versus crank angle.

that the fully developed flame propagation occurs approximately when the burned mass fraction, x_b, changes from 10 to 60%. At a burned mass fraction of 50–70% the flame front has entrained nearly 98% of the fresh charge and the front is in the close vicinity of the walls. Since the velocity fluctuations are damped near any solid surfaces the flame front is slowed down before the walls are reached, which is clearly evident from Figs. 6 and 8.

Measurements of θ_b indicate that the burn duration is nearly constant with rpm, as reported by Lancaster *et al.*[(22)] and McCuiston *et al.*,[(32)] or it may increase slightly as found for a Combustion Fuel Research (CFR) engine by Hires *et al.*[(30)] In these studies and others[(37,38)] the influence of the equivalance ratio ϕ was found to be very small for near stoichiometric mixtures as $0.8 < \phi < 1.2$. As the example from McCuiston *et al.*[(32)] in Fig. 9 indicates, the burn duration (here 2 to 98%) increases with larger deviations from stoichiometry due to the drop in the laminar flame speed (or equivalently the reaction rate). The increase in burn duration for $\phi > 1$ or $\phi < 1$ continues until the extinction or misfire limit is reached. The burn duration varies significantly less over the same range of air–fuel ratios if the spark timing is varied to MBT, such that the combustion takes place closer to top dead center (TDC). This indicates that the increase is mainly caused by the initial combustion process where the laminar flame speed is important. The burn duration is also increased when the fresh charge is diluted by EGR as illustrated in Fig. 10, obtained from measurements[(30)] performed in a Ford PROCO engine. The increase in the combustion duration is determined from the influence of the chemical kinetics and is more pronounced for low levels of turbulence, as found in the single-cylinder CFR engine. Investigation of the influence of other parameters such as compression ratio, spark timing, and chamber shape show that the flow field is the dominant factor for the flame propagation process. Factors that influence the chemical kinetics, and thereby the laminar flame speed, are of secondary importance, as long as the engine conditions do not approach any of the combustion limits.

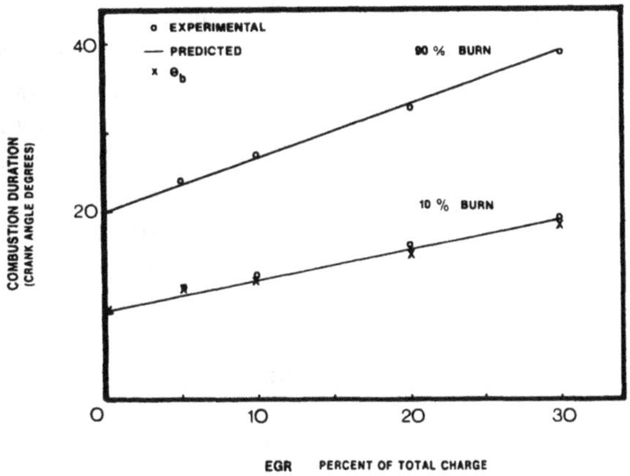

FIGURE 10. Combustion duration versus EGR.

The flow field influence can be divided in small-scale effects and large-scale effects. The small-scale effects are local in nature, such as part of the turbulence, air–fuel ratio fluctuations, and heat-transfer rates. These smaller-scale effects are highly influenced by the larger-scale effects, such as mean flow structures, geometry, and overall air–fuel ratio. The turbulence, for instance, enhances the local flame propagation, but the mean flow structures affect the total flame front area and the flame position within the combustion chamber. The total combustion duration is then determined by the combined effect of propagation speed and necessary flame travel distances to burn up the charge. The best performance is obtained when both the small-scale structure and the large-scale structure are favorable for the combustion process.

The influence of turbulence on the flame propagation has been studied extensively. Measurements in engines[34,36,39–45] and combustion bombs[35] have been performed, as well as for open or ducted flames.[46–53] Though some of the data are contradictory, the major effect of turbulence for high turbulence levels is through the turbulence intensity. The turbulent flame speed varies approximately linearly with u' and is independent upon scale. However, it should be realized that in most flows, there is a range of scales determined by u' and flow geometry, so scale and intensity are not independent. When $u' > S_L$ the turbulent fluctuations convect the flame and entrain fresh mixture at a rate determined by u'. The flamelets that are stretched and protruded into the unburned mixture must be large enough to avoid extinction from loss of active species and energy. Therefore, the part of the turbulence length scale spectrum that is below a critical size for flamelet extinction does not contribute to the flame convection, but it does increase the mixing and the overall preheat zone. The larger scales of turbulence draw parts of the flame into the unburned mixture and unburned mixture into the flame. The process generates a nonhomogeneous field with regions of burned, burning, and unburned gases through the flame. Measurements with ionization probes[7,19,34,51] clearly shows an intermittent signal corresponding to this inhomogeneous structure shown in Fig. 11. These measurements and others,[46–48] obtained for a ducted flame, support the view that the combustion process takes place in thin interface structures and not as a more homogeneous volume process. The local rate of combustion then depends on the rate at which interfaces between burned and unburned gases are generated by the flow field. These interfaces are created by the interaction of different flow structures in a cascading process. The continued diminishing of flow structure sizes ultimately leads to the destruction of the structures by viscous forces in a rate given by the turbulent time scale τ. In an equilibrium situation the rate of dissipation of the turbulent kinetic energy, ε, is determined by the rate of energy that the larger scales feeds into the cascading process, and τ^{-1} can therefore be used as an estimate for the interface generation rate. The models related to this interpretation will be given in the following section.

The large-scale fluid motion such as swirl and squish affects the overall location and spread of the combustion zone. Swirl is the motion in the tangential direction, whereas squish is the radial motion induced by the combustion chamber shape. Besides squish, there is another radial motion called slosh, which is the motion caused by the expansion of the gases as they burn. These mean motions generate and/or sustain the turbulence level well into the compression and may alter the turbulence present around TDC, such that the turbulence is close to isotropic conditions and the repeatability is increased. The reduced cyclic dispersion of the time of ignition improves the usable lean limit of operation as well as the smoothness of the engine. Another benefit with a high swirl level comes from the tendency of the burned gases to center around the swirl axis. This is due to the gravitational force and the difference between the densities of

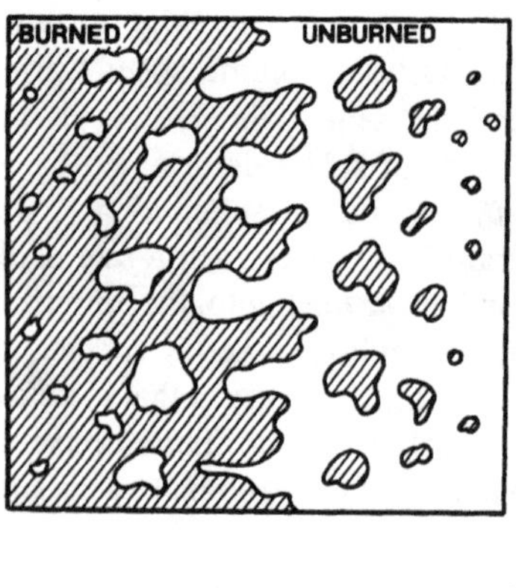

FIGURE 11. Flame structure.

the burned and unburned gases. The burned gases are then moved away from the walls reducing the heat losses to the walls and increasing the contact area between the burned and unburned gas regions. Some of these effects of the mean motions have been shown by Witze and Vilchis,[54] Witze,[55] and Inoue *et al.*[56] with laser shadowgraphs of the combustion zone location. These studies also show the importance of the spark plug location and the differences in the burn durations for cases where the flame is attached or not attached to the wall. When the flame is attached to the wall by the spark plug, it is highly deformed and stretched by the swirling motion generating a large flame area. This behavior offsets the increase in burn duration from long flame travel distances with a wall plug; for high enough swirl levels the burn duration becomes less than for a centrally located spark plug. If the flame is not attached to the wall or a protruding spark plug, it is generally better to have a central ignition position with the corresponding shorter flame travel distances.

The flow field in a motored engine is fairly well understood from a number of measurements[12,24–36,40–45] and some theoretical investigations[33,57–60] using simple models. There is, however, a conceptual difficulty in the interpretation of the data from velocity measurements whether done by hot-wire or Laser Doppler Anemometer. Besides the corrections needed for the hot-wire due to change in temperature and pressure, a filter function must be used to split the velocities into a mean flow and a fluctuating flow. An ensemble average over a large number of cycles have been used to define the mean flow, and the difference between the instantaneous velocity and the ensemble-averaged value is then the fluctuating component. This procedure will lead to reasonable values for the mean flow and the turbulence in the motored engine, where the cycle-to-cycle variation is small. In the fired engine randomness in the combustion process introduces a cycle-to-cycle variation in the mean flow driven by the expansion, and the averaging over a large number of cycles interpretates this as a contribution to turbulence. Under these circumstances a careful analysis is necessary to extract the proper mean and turbulence fields from measurements of the flame propagation process.

The concepts of using both the small- and large-scale flow to promote the combustion have already been implemented. Recently, a number of fast-burn engines[60] have been developed with an improvement in the fuel consumption and a control of the emissions. Yet many of the detailed interactions between the mean flow, the turbulence and the combustion is not completely clarified.[12,18,34,37,60] The measurements show the importance of the small recess volumes near the spark plug and between piston and cylinder, where the compression introduces a squish velocity that may be quite nonuniform over space. Such irregularities in the geometry around the spark plug is important for the motion of the initial flame kernel and all the crevices are important for the emission of unburned hydrocarbons; however they may be difficult to predict in the design analysis. To aid in the analysis of design changes in the combustion chamber shape and the inlet conditions, models for the detailed flow field calculations have been developed. These models are able to calculate a two-dimensional flow field, and investigations are in progress[61–64] to evaluate some of these effects. The models are still not developed to a stage where they can be used in the actual design process, but with an increased accuracy and resolution they may eventually show some of the interaction that occurs in the flow field.

VI. MODELS FOR THE TURBULENT FLAME

A large number of theories for the turbulent flame propagation have been developed[16–18,34,37,39] from theoretical considerations and experimental evidence. No model seems to be clearly better than any other, and they are only valid within certain limits of the turbulence and chemical parameters. The models can be divided into two classes; one class of models is based on the concept of a disturbed laminar flame, so the laminar speed becomes an important parameter. The other class of models corresponds to an infinite reaction rate, where the turbulent mixing is dominant and the flame propagation is determined by the turbulence, also referred to as mixing controlled models. A few models exist which are in between these two classifications with a finite chemical reaction rate, but without the assumption of a laminar flame.

If we look at a simple plane flame propagation in a one-dimensional flow, a balance of convection and diffusion for the energy can be written down, see Williams,[18] for the preheat zone as

$$\text{Convection} \sim \text{diffusion: } S_T \sim \Gamma/\delta_{FL}$$

Here S_T is the flame propagation speed, Γ is an effective transport coefficient, and the flame thickness is δ_{FL}. The average reaction rate $\overline{W}$ can be expressed from the total flow

$$\overline{W}\delta_{FL} \sim S_T$$

so the propagation speed becomes

$$S_T \sim (\Gamma\overline{W})^{1/2} \tag{16}$$

and the flame thickness is

$$\delta_{FL} \sim (\Gamma/\overline{W})^{1/2} \tag{17}$$

In general we may write the flame propagation speed as

$$S_T = (\Gamma \overline{W}/\gamma)^{1/2} \tag{18}$$

where γ is the eigenvalue for the propagation problem. In the limit of large activation energy and an Arrhenius single-step reaction rate, the eigenvalue can be found for the laminar flame by a perturbation method.[18,65–67] Similar solutions have been tried for the turbulent flame,[20,67,68] but the results inherently contain some uncertain assumptions for the statistical correlation between the velocity and the concentrations fluctuations.

For the turbulent flow the transport coefficient Γ can be expressed by the intensity and the length scale of the flow as

$$\Gamma_{\text{eff}} = \Gamma_{\text{lam}} + K^{1/2}\, l/\sigma_T \tag{19}$$

where σ_T is a generalized Prandtl–Schmidt number of order 1. The dissipation of turbulent kinetic energy ε is

$$\varepsilon = K/\tau \sim K^{3/2}/l \tag{20}$$

and the time scale of turbulence or eddy half time is

$$\tau \sim l/K^{1/2} \tag{21}$$

Following the experimental findings, as mentioned previously, the average reaction rate may be estimated by the mixing time as

$$\overline{W} \sim \tau^{-1} \sim K^{1/2}/l \tag{22}$$

for infinitely fast reactions. The estimates for S_T and δ_{FL} are thus

$$S_T \sim (\Gamma_{\text{eff}} \overline{W})^{1/2} \sim K^{1/2} \tag{23}$$

$$\delta_{FL} \sim (\Gamma_{\text{eff}}/\overline{W})^{1/2} \sim l \tag{24}$$

and completely determined by the turbulence field. Equation (23) gives a linear relation between the turbulence intensity and the turbulent flame velocity, which have been verified for engine operating conditions by Lancaster *et al.*[22,36] This relation is similar to Blizard and Keck[28] entrainment and burn-up model given by Eqs. (6) and (7) for a steady flame propagation with the entrainment velocity as $K^{1/2}$. In the limit of a laminar flame we have from Eq. (16)

$$S_L \sim (\Gamma_{\text{lam}}\overline{W})^{1/2}$$

so that for a small turbulence intensity, where the reaction rate presumably retains its laminar value, we would find

$$S_T \sim S_L\,(1 + \mathrm{Re}_T\sigma_T/\sigma)^{1/2} \tag{25}$$

A similar equation was proposed by Damköhler[69] from the consideration of a wrinkled laminar flame stated as

$$S_T \sim aS_L + bu' \tag{26}$$

which is a typical form of the correlations obtained by experiments. All the mechanical theories emphasize the existence of a laminar flame and therefore predict a turbulent flame speed as a close correlation to the laminar flame speed

$$S_T \sim S_L \cdot f(S_L, u', U) \tag{27}$$

For most of these models the turbulent flame speed is considered to be independent of the scale of turbulence. A discussion of these correlations as well as other models is given by Andrews *et al.*,[50] Shet *et al.*,[52] and Abdel-Gayed *et al.*[53] Examples of the utilization of the correlations like Eqs. (25)–(27) for engine simulation models is the Blizard and Keck model[28,14,32,33] mentioned previously, and a model by Heikal *et al.*[70] using an equation of the form of Eq. (25). The models show good agreement with data for engines if the turbulence variation with time is taken into account.[33,34]

The influence of the turbulence scale is included in some of the phenomenological models that consider the flow structures. Using the concept of large- and small-scale eddies, Sokolik *et al.*[6,7] developed a model based on the mixing of burned and unburned gases as a pulsating continuous ignition. Similar ideas have been used by Chomiak[19,25,51] to develop a model that gives an intermittent flame structure of the size of Taylor's microscale λ. From the consideration of the structures it is clear that several parameters relating to the energy and characteristic size of the structures can be used to correlate the flame speed as

$$S_T \sim f(S_L, u', \lambda, l) \tag{28}$$

Chomiak gave the form as

$$S_T \sim u'^{3/4}\left(\frac{\nu}{l}\right)^{1/4}\frac{\delta_{\text{LAM}}}{\eta} \tag{29}$$

where η is the Kolmogorov length scale. Abdel-Gayed *et al.*[53] examined some data in view of a two-scale eddy model, and they found the following form:

$$S_T \sim S_L \cdot (u'/S_L)(\nu_T/\nu\tau_f)^{1/2} \tag{30}$$

with τ_f as a dimensionless time that expresses the burning in the large and small eddies. The previous correlations are difficult to verify due to the lack of experimental information about the turbulent properties and the statistics for the structures. At best the correlations may be calibrated with turbulence data for the cold approach flow in a continuous combustion process.

All the previous models for the turbulent combustion process were developed for the overall propagation process. For models that describe the flow in one or more dimensions by differential equations the local reaction rate per unit mass or per unit volume is needed. For a laminar combustion process the chemical kinetics can be written down to yield a set of coupled rate equations. For simple systems it may reduce to an Arrhenius type of reaction rate of the form

$$R = c_i^{\alpha_i} B \exp(-E/RT) \tag{31}$$

where B is the frequency factor, R the gas constant, and E is an overall (or apparent) activation energy. The powers of the concentrations c_i indicate the order of the reaction schemes. For the turbulent combustion process the equations are averaged over an ensemble, so the average of Eq. (31) will give the local average reaction rate. With the intermittent nature of the turbulence and the magnitude of the flow scales relative to the chemical scales (length or time scales), the averaging becomes much more difficult than developing the correlations for the overall flame speed. It should be emphasized that the average (indicated by the angle brackets $\langle\ \rangle$) of Eq. (31) as

$$\langle R \rangle = \langle c_i^{\alpha_i} B \exp(-E/RT) \rangle$$

cannot just be written in terms of the averaged variables, so an expression with the mean concentrations and the mean temperature is not valid.

$$\langle R \rangle \neq \langle c_i \rangle^{\alpha_i} B \exp(-E/R\langle T \rangle)$$

Some attempts have been made to expand the highly nonlinear terms, see Bray and Moss[71] and Libby *et al.*,[72,73] but this method has not proven to be generally applicable. In the limit of infinite chemical reaction rates the reaction becomes mixing controlled and the reaction rate may be approximated by the turbulence time scale

$$R \sim \tau^{-1} \sim K^{1/2}/l \tag{32}$$

as mentioned previously. Bray and Moss[71] examined the transport equation for the concentration and concluded that a suitable model for the reaction rate could be expressed as

$$R = \rho K^{1/2}\langle c'^2 \rangle/l \tag{33}$$

through the mean square of the concentration fluctuations. With a normalized concentration the maximum value for the autocorrelation of the fluctuating concentration is given by the mean concentration C

$$\langle c'^2 \rangle_{\max} = C(1 - C) \tag{34}$$

so it is zero in the unburned and fully burned gases. Equation (33) is equivalent to the dissipation rate of the concentration fluctuations and the autocorrelation for c' is expected to come close to its maximum value for a high-turbulence level. In that case Eq. (34) can be used as an estimate for c' and substituted into Eq. (33); the mean reaction rate may therefore be expressed in terms of the mean concentration without the need to solve any equation for c'. Similar models have previously been proposed, see Spalding,[74] based on the arguments leading to Eq. (32). From a more physical point of view Spalding[75] and Howe and Shipman[47] developed and Eddy-Break-Up model of the turbulent combustion in parcels of fluid with different "ages" with respect to the entrainment into the flame. The resulting form of Spalding's model resembles the forms in Eqs. (33) and (34), though it contains the rate of shear instead of the turbulent kinetic energy and the length scale.

The mathematical description of the propagation problem can be written from the conservation equations for energy and species, see Williams.[18] This formulation contains the flame speed as an eigenvalue for the problem. With the reaction rate expression from Eq. (31) a cold boundary difficulty arises[18] due to the exponential form of the temperature dependence. The reaction rate is not zero in the cold mixture so a cut-off temperature must be specified, not unlike some kind of ignition temperature. On the other hand with Eq. (34) the form of the temperature (or concentration) dependence leads to an infinite flame speed. This indicates that the expression for the reaction rate may be a good approximation for the most part of the combustion zone, but it cannot be valid for the flame front. This is also confirmed by the work of Howe and Shipman,[47] where they compared their model with experiments.

Instead of modeling the reaction rate term by mean quantities, statistical approaches have been developed. Dopazo[76] and Pope[77,78] describe the reaction rate from the ensemble average using the joint probability function for velocity and concentration. A transport equation for this probability function can be solved by a Monte Carlo simulation[78] using the ensembles to describe the distribution. This method has successfully been used for continuous combustion problems by Pratt,[79] Pope,[78] Flagan and Appleton,[80] and Radhakrishnan and Heywood.[81] In the latter investigations the spatial distribution was retained, but the method has also been used without any spatial resolution as a coalescence and dispersion model by Corrsin[82] and applied for an engine by Mansouri *et al.*[83] This type of model, although costly in computational effort, may be useful for inhomogeneous or stratified charges where the resulting combustion temperature will vary stochastically in space and time. The more proper characterization of the field is important for the calculation of the processes that are very temperature sensitive, such as the reactions rates, the NO_x formation rates, and soot or particulate formation.

The modeling of the reaction rate or flame propagation speed requires the turbulence as input, so a model for the turbulence is needed. Since the flow field and the turbulence in the combustion period around TDC is established from the intake flow, we may examine the creation and development of the flow from the induction process. The measurements show that during the intake period the flow field is highly anisotropic and nonhomogeneous, but during the compression it becomes more isotropic and homogeneous. It is then possible to characterize the turbulence from parameters related to the intake process and account for the relaxation phenomena. Several models of turbulence have been proposed with a difference in the choice of parameters and the number of parameters used. Most models use the turbulent kinetic energy K (or equivalent u'), and with more than one parameter the second is usually taken as the length scale l or the dissipation ε. Here only the models that have been used in engine simulations will be covered.

In the development of an engine simulation model Blizard and Keck,[28] Tabaczynski *et al.*,[14,29] and Hires *et al.*[30] used the turbulence intensity and the length scale to describe the influence of turbulence. The turbulence present in the combustion chamber after the induction is estimated from an average of the unsteady jetlike intake flow past the valve. Scaling the turbulence to the mean flow, the turbulence intensity is proportional to the mean inlet velocity

$$u'_0 = C_u V_p B^2 / L_v D_v \tag{35}$$

and the length scale is proportional to the flow dimension

$$l_0 = C_l L \tag{36}$$

where V_p is mean piston speed, B is bore, L_v, D_v are valve lift and valve diameter, and L may be taken as the valve lift or the clearance height h. In the subsequent compression of the gases, u' and l are assumed to follow an isotropic rapid distortion. This implies that the compression is so fast that the angular momentum of the "eddies" is conserved and the length scale follows the size of the "eddy." Under this assumption the variation of the parameters will be given by the density

$$u' = u'_0(\rho/\rho_0)^{1/3} \tag{37}$$

$$l = l_0(\rho_0/\rho)^{1/3} \tag{38}$$

change from the end of induction. The rapid distortion variation given in Eqs. (37) and (38) also keeps the turbulent viscosity ν_T constant, and the turbulent Reynolds number varies only due to the laminar viscosity. This model predicts an increase in the turbulence intensity and a decrease in the length scale with the compression, which is only partly confirmed by experiment.[22,34] A high level of turbulence with a small scale is generated by most intake systems, so the turbulence intensity will decay very rapidly early in the compression and the length scale will increase. This behavior can be included in the model if the dissipation of turbulence is included.

One way of describing the dynamic behavior of the turbulence is to use a set of rate equations for the turbulent parameters in the model. The turbulence may be generated, diffused, and dissipated by the flow field, so a rate equation for the turbulent kinetic energy can be written as a balance of these terms[31,33,62]

$$\rho \frac{dK}{dt} = P_k - D_k - \rho\varepsilon \tag{39}$$

The production P_k is given by the work of the turbulent shear stresses and the diffusion D_k can be modeled as a gradient diffusion with an effective viscosity as

$$P_k = \sigma_{ij}^T \frac{\partial U_i}{\partial x_j} = \tau_{ij}^T \frac{\partial U_i}{\partial x_j} + \frac{2}{3} K \frac{d\rho}{dt}$$

$$D_k = \frac{\partial}{\partial x_j}\left(\rho \frac{\nu_T}{\sigma_k} \frac{\partial K}{\partial x_j}\right)$$

In a K model with only one rate equation, the dissipation ε is given from the length scale, as in Eq. (20)

$$\varepsilon = C_D K^{3/2}/l, \qquad C_D = 0.09 \tag{40}$$

and an algebraic expression is used for the length scale

$$l = C_l L, \qquad C_l \cong 0.22 \tag{41}$$

where the flow dimension L in most cases has been taken as the clearance height. This expression for the length scale is taken from the correlations developed for fully developed flow and it does therefore require that the flow be relaxed to a stage where it is in equilibrium and the influence of the boundaries is felt. Inspection of Eq. (39) now shows that the level of K may increase if the gain by production is larger than the loss by dissipation and diffusion. Even with a positive production, K may decrease due to a high dissipation, corresponding to a small-length scale, which is the case for the inlet flow. It may further be noted that in the absence of a shear production, diffusion, and dissipation, only the dilatation effect remains, and Eq. (39) degenerates to the solution given in Eq. (37). The one equation model in a lumped form can be formulated from Eq. (39) by an integration over the total volume. In this form (see also Refs. (33, 59), the diffusion term becomes a boundary flux and both the production and the dissipation terms are then taken as the volume average. This is a reasonable model if the turbulence is nearly uniform in the combustion chamber, otherwise the spatial distribution must be taken into consideration due to the nonlinearities in the model.

If the flow is not near equilibrium the length scale should not necessarily follow the flow dimension, as implied by Eq. (41). An improvement of the model description of the turbulence with respect to this deficiency can be done by adding another rate equation for the second turbulence parameter instead of the algebraic expression for the length scale. The most common choice in the past has been to use the dissipation as the second parameter for which a rate equation is modeled. In the K–ε model, the rate equation for ε is very similar to the equation for K as

$$\rho \frac{d\varepsilon}{dt} = P_\varepsilon + D_\varepsilon - C_\varepsilon \rho \varepsilon^2/K \tag{42}$$

again with production, diffusion, and dissipation terms. The length scale is now calculated from K and ε as

$$l = C_D K^{3/2}/\varepsilon \tag{43}$$

and the variation in the length scale may be different than the equilibrium value in Eq. (41). If the shear production and the diffusion is neglected in Eqs. (39) and (42) and the dissipation is small relative to the dilatation, it can be shown that the model will predict a variation of K and ε as

$$K \sim \rho^{2/3}, \qquad \varepsilon \sim \rho^{4/3} \tag{44}$$

which is the same as stated in Eqs. (37) and (38) for the rapid distortion. The K–ε model have been used in the lumped form for engine simulations by Borgnakke *et al.*[33,59] and Davis and Borgnakke[84] and in the local form for all of the two-dimensional flow calculations shown by Witze,[61] Gosman and Harvey,[63] and Diwakar.[64] Though some uncertainties still remain in the modeling of the terms, particularly in the ε equation, it is fairly accurate and the most versatile model to date.

VII. HEAT TRANSFER

The presence of solid walls in the combustion chamber influences both the momentum and the energy fields. Wall stresses and wall heat transfer profoundly alter the levels and the distributions in space of the velocities and the temperature or energy of the gases. At the walls all tangential velocities are zero and the normal velocity is equal to the wall velocity, which also implies that all the fluctuating velocity components, i.e. turbulence, vanishes at the walls. The temperature in the gases continuously approaches the wall temperature as the position comes close to the wall, tentatively shown in Fig. 12. This change in momentum and energy from the prevailing levels of the bulk gases to the wall levels take place in a thin gas layer next to the walls, and these boundary layers are important for the analysis of the heat transfer to the walls. During the combustion period there is a distribution of boundary layers in the burned gas zone and the unburned gas zone as indicated in Fig. 13. Since the bulk gas conditions are changing throughout the cycle, the boundary layers will change due to both the fluxes to the walls and the changes in the bulk gases. The heat transfer at any given wall location is then a function of the history of the nearby gases, the instantaneous bulk gas temperature, wall temperature, and the transport properties of the thermal boundary layer. This dependence of the actual levels

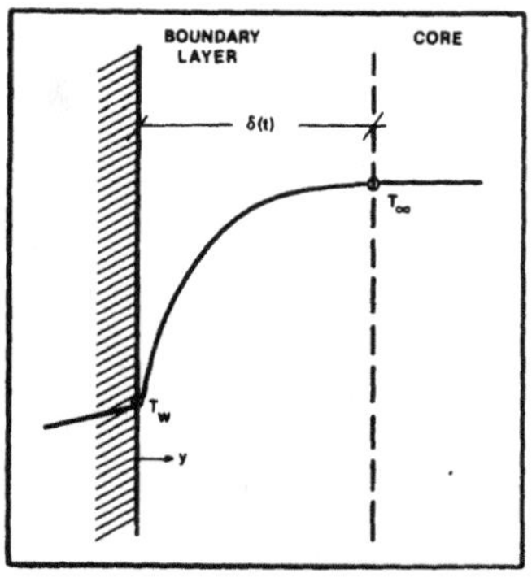

FIGURE 12. Thermal boundary layer.

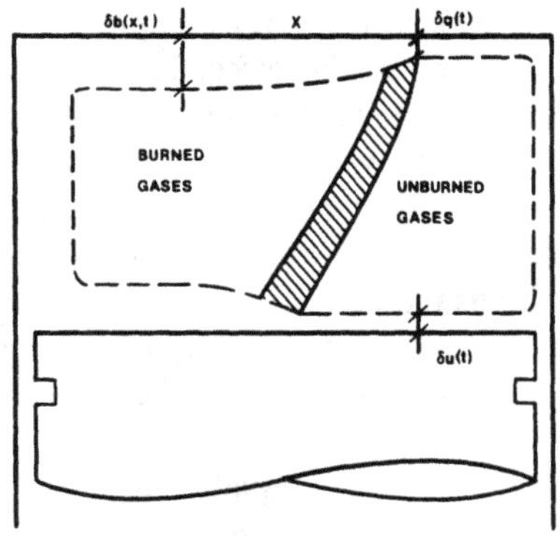

FIGURE 13. Thermal boundary layers in the combustion chamber.

and the transport properties causes the behavior of the thermal boundary layer to be coupled to the momentum boundary layer. From these considerations it is therefore clear that an analysis of the heat transfer cannot be done without some analysis of the momentum boundary layer.

To illustrate some of the features of these boundary layers we may write the one-dimensional energy equation for the gases assuming ideal gas behavior as

$$\rho C_p \frac{\partial T}{\partial t} + \rho C_p U \frac{\partial T}{\partial y} = \frac{dP}{dt} - \frac{\partial q}{\partial y} \tag{45}$$

The boundary condition for this equation is

$$T = T_W \quad \text{for} \quad y = 0, \qquad T = T_\infty \quad \text{for} \quad y \to \infty$$

and the heat transfer to the wall can be found from the temperature gradient at the wall

$$q_w = -k \frac{\partial T}{\partial y} \quad \text{for} \quad y = 0 \tag{46}$$

Before the energy equation and the evaluation of the heat transfer can be done, it is necessary to know how the effective transport coefficient varies as a function of y. If the flow field is purely laminar, the effective transport coefficient is equal to the laminar thermal conductivity and can be calculated as any other property of the gases.

The laminar unsteady heat-transfer calculations have been compared with measurements by Nikanjam and Greif[(85)] and by Greif *et al.*[(86)] for a piston compression of air and argon. They found that the thermal boundary layer starts growing in the early compression part due to the heat flux, but at a later time the compression effect becomes stronger and forces the boundary layer to diminish. Since they had no dynamic intake process as in an engine the flow conditions were purely laminar and quiescent except for the piston-induced motion. In these circumstances very good agreement was found between the measurements and the theoretical calculations. However, under actual engine conditions the flow will be expectd to be highly turbulent with possible strong mean motions like swirl and squish. The problem formulations then have to be modified to include the effect of the turbulent convection of energy, as we will describe next.

The energy equation (45) can be integrated over the thermal boundary layer thickness δ. Assuming the boundary layer can be regarded as existing in a pseudo-steady state, the heat-transfer flux q is constant cross the layer. With this condition the integral form of the energy equation becomes

$$\frac{d}{dt}\int_0^\delta \rho C_p T\, dy - C_p T_\infty \frac{d}{dt}\int_0^\delta \rho\, dy = \delta \frac{dP}{dt} - q_w \tag{47}$$

as an equation for the thermal boundary layer thickness δ, and the heat transfer rate is written as

$$q = k_{\text{eff}} \frac{\partial T}{\partial y} \tag{48}$$

However, the effective conductivity in a turbulent flow field is not a gas property, but may be modeled as a sum of a laminar and a turbulent contribution:

$$k_{\text{eff}} = \rho C_p \nu_{\text{eff}}/\sigma = k_L + k_T \tag{49}$$

where the turbulent contribution will dominate away from the wall. A model for the turbulent conductivity must be used and this model can then be examined for the behavior close to the wall. With a constant q and an effective conductivity that is changing across the layer, we may integrate Eq. (48) to express the heat transfer as

$$q = k_{\text{avg}}\,(\mathrm{T}_\infty - \mathrm{T}_w)/\delta \tag{50}$$

from the average conductivity

$$k_{\text{avg}} = \delta \Big/ \int_0^\delta \frac{\sigma}{\rho C_p \nu_{\text{eff}}}\, dy \tag{51}$$

This form for the heat-transfer rate is the lumped form and the ratio k_{avg}/δ corresponds to a heat-transfer coefficient commonly found in the literature. It can now be seen that it is necessary to evaluate the momentum boundary layer for the spatial variation of the effective viscosity before the integral in Eq. (51) can be calculated. If we recall the form for the effective transport coefficient, Eq. (19), and write it for the viscosity it gives

$$\nu_{\text{eff}} = \nu + K^{1/2} l \tag{52}$$

and the variation of K and l will provide the desired information. Following the work of Wolfsthein[87] and Spalding,[88] a balance between diffusion and dissipation of K exists in the boundary layer. A similarity solution for K in terms of a power function can be found if the length scale is assumed to vary linearly across the boundary layer. This variation allows the integral in Eq. (51) to be of the form

$$k_{\text{avg}} = k_L \Big/ \int_0^1 [1 + C_h Z^{3/2}]^{-1}\, dZ \tag{53}$$

with Z as the dimensionless distance

$$Z = y/\delta$$

and C_h is a function of the boundary layer thickness and the bulk gas level of K and l as

$$C_h = \delta^{3/2}(K/l)^{1/2} \,\text{Pr}\, C_l/\nu$$

It is interesting to note the two contributions to the average conductivity, one coming from the laminar conductivity and the other from the turbulent conductivity. The turbulent contribution dominates over most of the boundary layer except in the limit of $Z \to 0$, where the turbulent contribution is zero. When the thermal boundary layer is sufficiently thick, the laminar contribution may be neglected over most of the boundary layer and the integral may be approximated to yield a form for the heat-transfer rate as

$$q_w = k_L \frac{T_\infty - T_w}{l} (\text{Re}_T)^{2/3} f(\delta/l) \tag{54}$$

The form of Eq. (54) resembles the standard Nusselt number correlations with the Reynolds number, though it is the turbulent Reynolds number that appears in the formula. In most cases the turbulent intensity is proportional to the mean velocity and the length scale proportional to the chamber dimensions so the difference between the two Reynolds numbers is small.

The change in the heat transfer with time can be solved from Eqs. (47) and (53). For a very small thermal boundary layer thickness we have $C_h < 1$ and the thermal boundary layer is within the laminar region of the momentum boundary layer. Under these circumstances Eq. (53) shows that the average conductivity is equal to the laminar conductivity, and Eq. (47) gives a thermal boundary layer growth as the square root of time. As the boundary layer grows out through the momentum boundary layer it becomes more affected by the turbulence and the rate of growth changes to be as $t^{0.8}$. For the Nusselt number the variation is as $t^{-1/2}$ in the laminar region, and once the thermal boundary layer is in the fully turbulent region it changes to follow $t^{-0.2}$. These variations correspond to the limiting cases for the solution to Eqs. (47) and (53).

This transition in the behavior of the unsteady thermal boundary layer in part forms the measurements by Summers[89] taken from Annand[90] and shown in Fig. 14. The measurements were performed in a rapid compression machine where the piston was locked after arriving at TDC, and the machine was operated with and without combustion. During the rapid compression the bulk gas temperature increases following an isentropic process, and a very thin boundary layer is formed. At TDC where the piston motion is stopped, the boundary layer grows due to the heat transfer to the walls with only small convection velocities since there is no expansion of the gases. If an expansion had taken place, the growth of the thermal boundary layer would have been enhanced by the expansion-induced velocities. As seen from Fig. 14, the Nusselt number starts falling off as $t^{-1/2}$, corresponding to a laminar thermal boundary layer. This variation is followed by a steeper decrease with time which for large times levels out and approaches a variation as $t^{-0.25}$. The variation of the Nusselt number in this case may be explained, if the decay of the turbulence and the mean flow in the chamber is taken into consideration. As the turbulence level decreases with time,

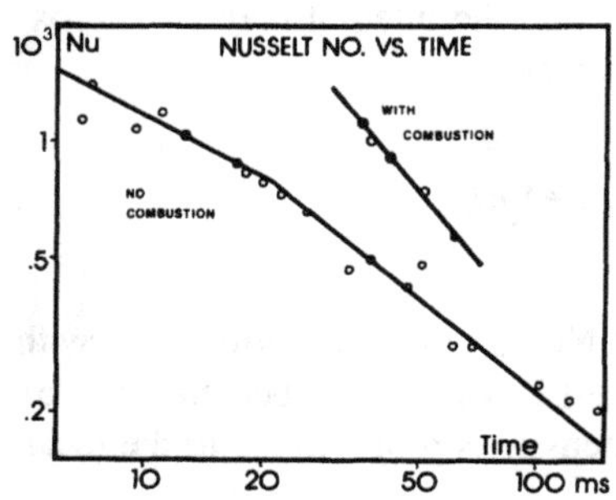

FIGURE 14. Heat transfer versus time in rapid compression machine.

C_h will decrease and the average conductivity, Eq. (53), will be lower. This dynamic behavior of both the flow field and the heat transfer may be the reason for the sharper drop-off of the Nusselt number for the intermediate times. The behavior seems to be the same for both the cases with and without combustion, though the Nusselt number is higher with combustion, as shown in Fig. 14. It should be recalled that with the combustion there is an additional compression of the gases and a quenching of the flame at the wall so the initial thermal boundary layer may be thinner than the corresponding boundary layer without combustion. This would explain the increase in the Nusselt number, but it is also possible that the flow field is slightly different for the two cases.

The previous example showed part of the dynamics of the thermal boundary layer behavior in the pseudo-steady state. In the engine rapid compressions and expansions occur which may render the assumption of a constant heat transfer q across the boundary layer invalid. The boundary layers can have a local minimum or maximum for the temperature, so the wall heat flux is out of phase with the overall temperature difference between the wall and the bulk gases. At some crank angle during the compression of the fresh charge, which initially is below the wall temperature, the heat transfer changes sign before the bulk gas temperature is above the wall temperature. And in the expansion of the hot combustion products, the heat-transfer rate will drop faster than the temperature difference. This compressibility effect will be less pronounced, if present at all, for high transport properties. Since the tranport properties are highly affected by the flow field, the variation of the turbulence and the mean flow, such as swirl, will have a dominating effect on the heat-transfer rate to the walls and it will also vary with position along the combustion chamber walls.

Experimental verification of the heat-transfer rates in a motored engine have been done by Hassan,[(91)] Dao *et al.*,[(92)] Dent and Suliaman,[(93)] and Annand and Pinfold[(94)] among others. The main conclusion from these studies is that the pseudo-steady state Nusselt–Reynolds number correlation can be applied relating the heat-transfer flux to the momentum flux, as done for boundary layer flows over flat plates.

$$\mathrm{Nu} \sim \mathrm{Re}^n \tag{55}$$

The wall heat flux can then be written as

$$q = C \cdot \frac{k_L}{B} \left(\frac{V_p B}{\nu} \right)^n \cdot (T - T_w) \tag{56}$$

where the exponent n is in the range of 0.7 to 0.8. The effect of turbulence is incorporated in the value of the constant C and the exponent, so the turbulence should be in equilibrium with the mean flow characterized by the free-stream velocity V_p as a prerequisite for Eqs. (55) and (56) to be valid. In the circumstances of a strong swirling flow the Reynolds number may be based on the swirling velocity, which under the assumption of a solid body rotation becomes

$$V_{sw} = \omega r \tag{57}$$

$$\mathrm{Re} = \omega r^2/\nu \tag{58}$$

As shown in Eq. (58) the variation across the radius of the chamber can be included, as done by Dent and Suliaman,[93] or it may be taken with the radius equal to the bore for a bulk average value.

From the measurements in the fired engine, as done by Woschni,[95] Annand and Ma,[96] and Dent and Suliaman,[93] it was found that the form of the heat-transfer correlation in Eqs. (55) and (56) required a slight modification due to the combustion. Woschni included the combustion-induced motion into an effective Reynolds number as

$$\mathrm{Re} = (V_p + V_e)B/\nu \tag{59}$$

$$V_e = C\left(\frac{VT_0}{V_0}\right)\left(\frac{P - P_m}{P_0}\right) \tag{60}$$

where the expansion velocity V_e is determined from the difference between the actual pressure and the corresponding motored pressure P_m. The inclusion of the expansion velocity enhances the sensitivity of the heat-transfer correlation to the speed of the combustion process. Indirectly it therefore also accounts for changes in the flow field that will affect the propagation of the combustion process, as shown in the previous section. An explicit account for the influence of the swirling motion has been done by Le Feuvre *et al.*,[97] Dao *et al.*,[92] and Dent and Suliaman[93] by substitution of the swirling velocity instead of the mean piston speed and then omitting the expansion velocity. The phase lag between the heat-transfer rate and the temperature difference were included in the investigations by Annand and Ma[95] and Annand and Pinfold.[93] They used the basic Nusselt–Reynolds correlation, but instead of using $T - T_w$ as the driving force, it was modified with an unsteady term as[93]

$$q = C\frac{k_L}{B}\mathrm{Re}^n\left(T - T_w + a\frac{B}{V_p}\frac{dT}{dt}\right) \tag{61}$$

or with a different factor for the unsteadiness.[95] These relations gave an improved agreement with the data with respect to the rate of increase and decrease of the heat-transfer rate around TDC.

For all of the correlations based on the Nusselt–Reynolds number relation, as given above, the analysis was done from a global point of view. They represent the average heat transfer in the combustion chamber, and it is assumed that they can be expressed in terms of the bulk gas average temperature. This temperature can be calculated from the pressure trace, the thermodynamic state, and the composition, which makes it reasonably simple in mathematical terms and readily available from the experimental data. Another assumption involved in the use of the correlations is the choice of the conditions at which the gas properties are taken. Before the combustion starts, the nearly uniform temperature is equal to the bulk gas average temperature and the composition is given by that of the unburned charge. After the spark the bulk gas temperatures have been used with the composition given by the combustion products for most of the models. In a few studies[(97)] a film temperature was used instead of the bulk gas temperature for the driving force and in the calculation of the gas properties. The film temperature is found as the average between the wall temperature and the bulk gas temperature, and it has been used quite often for the heat-transfer calculation in hot-wire calibrations. It does not appear, however, that the models with the film temperature are superior to any of the models that use the bulk temperature. In the comparison of all of these models with measurements, discrepancies are found due to fairly large variation of the heat-transfer rates with the location in the combustion chamber. During the combustion period the cylinder charge is divided into two more or less distinct zones, one with unburned gases and one with burned gases. The two zones are separated by the propagating flame front so their location within the combustion chamber is determined by the spread of the flame as influenced by the mean flow field and the geometry. At any given location the wall will then be exposed to either the hot combustion products or the still unburned charge with a large difference in the heat-transfer rate. The position of the flame front at a given crank angle will vary due to cycle-to-cycle variations in the combustion and the turbulence, so the measurement of the heat transfer over many cycles gives the statistical average. Though the bulk average gas temperature includes an average over the whole charge, it may not account for the averaging over the areas in this nonlinear process. It is therefore difficult to assess the validity of the correlations from single-point measurements and the comparison should be made with averages over several locations within the combustion chamber.

The wall area may be split into two areas wetted by each of the two zones if the zones can be located in space as shown in Fig. 13.

$$A_{\text{tot}} = A_b + A_u$$

Assuming that each of the two regions can be described by their bulk average temperature and composition, we would get the total heat transfer as

$$Q = q_b A_b + q_u A_u$$

where the heat-transfer rates in the burned and unburned zones are

$$q_u = C \cdot \frac{k_{L,u}}{B} \cdot \mathrm{Re}_u^n (T_u - T_w)$$

$$q_b = C \cdot \frac{k_{L,b}}{B} \cdot \mathrm{Re}_b^n (T_b - T_w)$$

The usefulness of such a split, of course, depends on how well the two zones resemble the actual field. Without going to the two-dimensional calculations(61–64) a geometrical shape must be assigned to the flame front and for this a spherical(14,28,32,37) or cylindrical(30,37) form has been used in two-zone thermodynamic models. The effect of the shape and the areas becomes important for the heat-transfer calculations when the geometry is changed. A study of the effect of bore–stroke ratio by Siewert(98) showed a significant increase in the overall heat losses as the bore–stroke ratio is increased. The total heat losses increased from 15 to 25% of the fuel energy, when the bore–stroke ratio varied from 1.1 to 3.3 as shown in Fig. 15. With a two-zone thermodynamic model and assumed spherical flame propagation, Borgnakke *et al.*(33) simulated the bore–stroke ratio study and compared different heat-transfer correlations. The heat transfer predicted by Woschni's correlation, Eqs. (59) and (60), and Annand's correlation, Eq. (56), shows the improvements in the heat-transfer calculation when the expansion velocity is included.

A model based on the unsteady heat transfer from Eqs. (47)–(51) together with an averaged K–ε model of turbulence was developed by Borgnakke *et al.*(33) and used in some engine simulations. The thermal boundary layer thickness was calculated from Eq. (47) for several locations in the combustion chamber, and the local heat transfer at those locations was used to estimate the total wall heat transfer. The heat transfer is shown in Fig. 16 for some fixed locations and for

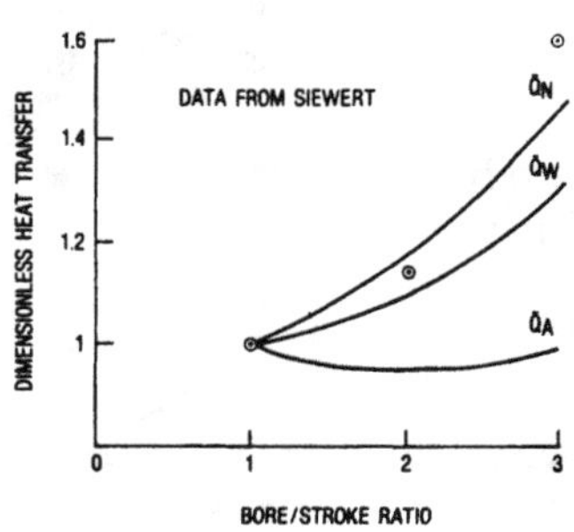

FIGURE 15. Dimensionless heat transfer versus bore–stroke ratio.

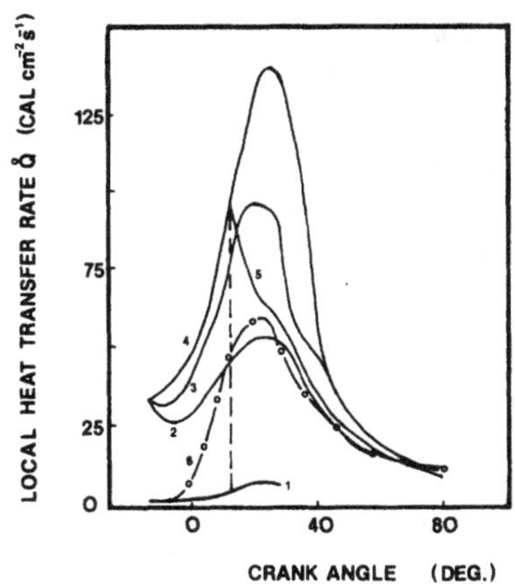

FIGURE 16. Heat transfer versus crank angle at the following locations: (1) unburned region, (2) spark location, (3) trailing flame edge, (4) flame front, (5) at r = 0.6 m from spark, and (6) data at loc. 5.

some locations relative to the moving flame. At the instantaneous flame location the thermal boundary layer thickness is assumed to be equal to the quench distance and the heat transfer is at the maximum value. With the growth of the thermal boundary layer the heat-transfer rate drops off behind the flame, but it may increase due to the continued compression as evident for the spark plug location shown as curve 2 in Fig. 16. The local instantaneous heat transfer calculated from the model agrees well with data obtained for the same location displayed as curve 6 corresponding to the model in curve 1 and 5. Since the data was collected as an average over 256 cycles, the finite probe size and the cycle-to-cycle variations gave a smearing effect, so some average of the heat transfer in the burned and unburned gases (curve 1 and 5) will be expected in the neighborhood of the flame front. The total heat transfer from this model is also shown in Fig. 15, and it can be seen that this model performs slightly better than the other pseudo-steady state correlations.

The heat transfer is generally enhanced by turbulence and strong mean motions as swirl and squish. Some of these effects are predicted reasonably well by the model of Woschni, Eq. (59), and to a lesser extent by the simple correlation in Eq. (56). A higher swirl level will increase the turbulence in the boundary layers resulting in an enhanced transport of momentum and energy. This effect is included in the heat-transfer correlations by the use of the swirl velocity instead of the mean piston speed, as already shown in Eq. (58). Presently no model includes the combination of swirl, piston speed, or expansion velocity, which would be required if an optimization should be done. Since a higher flame propagation speed will increase the efficiency and a higher heat transfer will decrease the efficiency there is an optimum level of swirl, squish, and turbulence. Some measurements of the indicated specific fuel consumption (ISFC) by Lee *et al.*,[99] Nagayam *et al.*,[100] and Mayo[101] show the effects of swirl and squish.

The interactions between these phenomena was investigated by changing the geometry of the combustion chamber, and the results indicate a very complicated influence from the combination of these effects. Increasing swirl and squish improves the efficiency up to a point where the heat transfer becomes so large that the efficiency starts to drop. In a study of the influence from the flow field with a model by Davis and Borgnakke[84] several of the experimental observations were predicted qualitatively. The model predicts a minimum in ISFC with variations in percent EGR and spark plug location that is in accordance with the experimental data. The optimum squish level could be predicted as a 30% squish area whereas the measurements indicate an optimum at 40 to 45% squish area. For all of these cases it was concluded that the limits for the efficiency are determined by the flame travel distances and the heat-transfer rate. An improvement in the capability of such models towards a quantitatively correct prediction may be possible with a better heat-transfer model and an understanding of the influence of the mean motion on the combustion process.

NOTATION

A_{tot}	Total wall area
A_B	Wall area in contact with burned gases
A_u	Wall area in contact with unburned gases
A_{FL}	Flame front area
B	Bore
B	Constant for reaction rate
C	Mean concentration
c'	Concentration fluctuation rms
c_i	Concentration of species i
C_D, C_l	Constants for turbulent model
C_p	Heat capacity
C_h	Constant for heat flux
D_v	Valve diameter
D_k, D_ε	Diffusion of K and ε
E	Activation energy
h	Clearance height
K	Turbulent kinetic energy $\langle \rho u_i' u_i' \rangle / 2 \langle \rho \rangle$
k_L, k_T	Thermal conductivity laminar, turbulent
k_{avg}	Average boundary layer conductivity
L	Flow or chamber dimension
L_v	Valve lift
l	Turbulent length scale
m	Exponent
m_e	Mass entrained by flame
m_b	Mass burned
m_{tot}	Total mass in charge
n	Exponent

Nu	Nusselt number $= qB/k\Delta T$
P	Pressure
P_k, P_ε	Production of K and ε
Pr	Prandtl number
q	Heat-transfer rate per unit area
Q	Total heat-transfer rate
r	Radius
r_k	Flame kernel radius
R	Gas constant
R	Reaction rate per unit mass
R_{FS}	Flame speed ratio S_T/S_L
Re	Reynolds number VB/ν
Re_T	Turbulent Reynolds number $u'l/\nu$
S_L	Laminar flame speed
S_T	Turbulent flame speed
T	Temperature
t	Time
T_B, T_u	Temperature in burned/unburned gases
T_w, T_∞	Temperature at wall, in core gases
U	Mean velocity
u'	Fluctuating velocity rms
u_e	Entrainment velocity
V	Volume
V_p	Mean piston speed
W	Average reaction rate
x_b	Mass fraction burned m_b/m_{tot}
x_e	Mass fraction entrained m_e/m_{tot}
Y	Concentration
y	Distance from wall
Z	Dimensionless distance y/δ
β	Dimensionless time ratio
δ	Thermal boundary layer thickness
δ_{FL}	Flame thickness
ε	Dissipation of K
Γ	Transport coefficient
γ	Eigenvalue for flame speed
λ	Taylor microscale
η	Kolmogorov length scale
ρ	Density
τ	Turbulent time scale
τ_e	Eddy half time
τ_b	Burning time
τ_{ij}^t	Turbulent shear stresses
ϕ	Equivalence ratio
θ	Crank angle
θ_{ig}	Ignition delay
θ_b	Combustion duration
σ_t	Prandtl–Schmidt number
σ_{ij}^t	Turbulent stresses
ν, ν_t	Viscosity laminar, turbulent

REFERENCES

1. Ballal, D. R., and Lefebvre, A. H., "The Influence of Flow Parameters on Minimum Ignition Energy and Quenching Distance," *15th Symp. on Combustion*, p. 1473, The Combustion Institute (1975).
2. Ballal, D. R., and Lefebvre, A. H., "Ignition and Flame Quenching in Flowing Gaseous Mixtures," *Proc. Roy. Soc. London, A* **357,** 163 (1977).
3. Ballal, Dr R., and Lefebvre, A. H., "A General Model of Spark Ignition for Gaseous and Liquid Fuel-Air Mixtures," *18th Int. Symp. on Combustion*, p. 1737, The Combustion Institute (1981).
4. DeSoete, G. G., "The Influence of Isotropic Turbulence on the Critical Ignition Energy," *13th Int. Symp. on Combustion*, p. 735, The Combustion Institute (1973).
5. DeSoete, G. G., "Initial Burning Velocity of Spark Ignited Flames Measured by a Laser Tomography Technique," paper presented at Spring Technical Meeting of The Combustion Institute, Warren, Mich., March (1981).
6. Karpov, V. P., and Sokolik, A. S., "Ignition Limits in Turbulent Gas Mixtures," Institute of Chemical Physics, Academy of Sciences, USSR. Translated from *Dokl. Akad. Nauk. SSSR*, **141**(2), 393–396 (1961).
7. Sokolik, A. S., Karpov, V. P., and Semenov, E. S., "Turbulent Combustion of Gases," *Fizika Goreniya Vzryva*, **3**(1), 61–76 (1967).
8. Ballal, D. R., and Lefebvre, A. H., "The Structure and Propagation of Turbulent Flames," *Proc. Roy. Soc. London, A* **344,** 217 (1975).
9. Iinuma, K., and Iba, Y., "Studies of Flame Propagation Process," *JARI Tech. Memo* **10** (1972).
10. Iinuma, K., and Iba, Y., "Studies of Flame Propagation," *JARI Tech. Memo* **15** (1973).
11. Namazian, M., Hansen, S., Lyford-Pike, E., Sanchez-Barsse, J., Heywood, J., and Rife, J., "Schlieren Visualization of the Flow and Density Fields in the Cylinder of a Spark-Ignition Engine," SAE paper 800044 (1980).
12. Cole, J. B., and Swords, M. D., "On the Correlation Between Gas Velocity and Combustion Pressure Fluctuations in a Spark-Ignition Engine," 18th Int. Symp. on Combustion, p. 1837, The Combustion Institute (1981).
13. Van Tiggelen, A., and Deckers, J., "Chain Branching and Flame Propagation," *6th Int. Symp. on Combustion* p. 61, The Combustion Institute (1957).
14. Tabaczynski, R. J., Ferguson, C. R., and Radhakrishnan, K., "A Turbulent Entrainment Model for Spark Ignition Engine Combustion," SAE paper 770647 (1977).
15. Lavoie, G. A., "Correlations of Combustion Data for S. I. Engine Calculations—Laminar Flame Speed, Quench Distance and Global Reaction Rates," SAE paper 780229 (1978).
16. Markstein, G. H., *"Non-Steady Flame Propagation,"* Pergamon Press, New York (1964).
17. Lewis, B., and Von Elbe, G., *"Combustion, Flames and Explosions of Gases,"* Academic, New York (1961).
18. Williams, F. A., *Combustion Theory*, Addison-Wesley, Reading, Massachusetts (1965).
19. Chomiak, J., "Dissipation Fluctuation and the Structure and Propagation of Turbulent Flames in Premixed Gases at High Reynolds Numbers," *16th Int. Symp. on Combustion* p. 1665, The Combustion Institute (1977).
20. Williams, F. A., "A Review of Some Theoretical Considerations of Turbulent Flame Structure," *AGARD Conf. Proc.*, CP-164, III-1 (1975).
21. Boyer, L., Clavin, P., and Sabathier, F., "Dynamic Behavior of a Premixed Turbulent Flame Front," *18th Int. Symp. on Combustion*, p. 1041, The Combustion Institute (1981).
22. Lancaster, D. R., Krieger, R. B., Sorenson, S. C., and Hull, W. L., "Effects of Turbulence on Spark-Ignition Engine Combustion," SAE paper 760160 (1976).

23. Quader, A. A., "What Limits Lean Operation in Spark Ignition Engines—Flame Initiation or Propagation?" SAE paper 760760 (1976).
24. Peters, B. D., and Quader, A. A., "Wetting the Appetite of Spark Ignition Engines for Lean Combustion," SAE paper 780234 (1978).
25. Chomiak, J., "Flame Development from an Ignition Kernel in Laminar and Turbulent Homogeneous Mixtures," *17th Int. Symp. on Combustion*, The Combustion Institute (1979).
26. Maly, R., "Ignition Model for Spark Discharges and the Early Phase of Flame Front Growth," *18th Int. Symp. on Combustion*, p. 1747, The Combustion Institute (1981).
27. Adelman, H. E., "A Time Dependent Theory of Spark Institute," *18th Int. Symp. on Combustion*, The Combustion Institute, p. 1333 (1981).
28. Blizard, N. C., and Keck, J. C., "Experimental and Theoretical Investigation of Turbulent Burning Model for Internal Combustion Engines," SAE paper 740191 (1974).
29. Tabaczynski, R. J., Trinker, F. H., and Shannon, B. A. S., "Further Refinement and Validation of a Turbulent Flame Propagation Model for Spark-Ignition Engines," *Combust. Flame*, **39,** 111 (1980).
30. Hires, S. D., Tabaczynski, R. J., and Novak, J. M., "The Prediction of Ignition Delay and Combustion Intervals for a Homogeneous Charge, Spark Ignition Engine," SAE paper 780232 (1978).
31. Hinze, J. O., *Turbulence*, McGraw-Hill, New York (1979).
32. McCuiston, F. D., Lavoie, G. A., and Kauffman, C. W., "Validation of a Turbulent Flame Propagation Model for a Spark Ignition Engine," SAE paper 770045 (1977).
33. Borgnakke, C., Arpaci, V. S., and Tabaczynski, R. J., "A Model for the Instantaneous Heat Transfer and Turbulent in a Spark Ignition Engine," SAE paper 800287 (1980).
34. Tabaczynski, R. J., "Turbulence and Turbulent Combustion in Spark-Ignition Engines," *Prog. Energy Combust. Sci.* **2,** 143 (1976).
35. Dyer, T. M., "Characterization of One and Two-Dimensional Homogeneous Combustion Phenomena in a Constant Volume Bomb," SAE paper 790353 (1979).
36. Lancaster, D. R., "Effects of Engine Variables on Turbulence in a Spark-Ignition Engine," SAE paper 760159 (1976).
37. Blumberg, P. N., Lavoie, G. A., and Tabaczynski, R. J., "Phenomenological Models for Reciprocating Internal Combustion Engines," *Prog. Energy Combust. Sci.* **5,** 123 (1979).
38. Taylor, C. F., *The Internal Combustion Engine in Theory and Practice*, Vol. 2, MIT Press, Cambridge, MA (1979).
39. Groff, E. G., and Matekunas, F. A., "The Nature of Turbulent Flame Propagation in a Homogeneous Spark-Ignited Engine," SAE paper 800133 (1980).
40. Semenov, E. S., "Studies of Turbulent Gas Flow in Piston Engines," *NASA Tech. Trans.* **F97** (1963).
41. Dent, J. C., and Salama, N. S., "The Measurement of the Turbulent Characteristics in an Internal Combustion Engine Cylinder," SAE paper 750886 (1975).
42. Witze, P. O., "Measurements of the Spatial Distribution and Engine Speed Dependence of Turbulent Air Motion in an I.C. Engine," SAE paper 770220 (1977).
43. Johnston, S. C., Robinson, C. W., Rorke, W. S., Smith, J. R., and Witze, P. O., "Application of Laser Diagnostics to an Injected Engine," SAE paper 790092 (1979).
44. Rask, R. B., "Laser Doppler Anemometer Measurements in an Internal Combustion Engine," SAE paper 790094 (1979).
45. Morse, A. P., Whitelaw, J. H., Yianneskis, M., "Turbulent Flow Measurements by Laser-Doppler Anemometry in Motored Piston-Cylinder Assemblies," *J. Fluids Eng.* **101,** 208 (1979).
46. Howe, N. M., Shipman, C. W., and Vranos, A., "Turbulent Mass Transfer and Rates of Combustion in Confined Turbulent Flames," *9th Int. Symp. on Combustion*, p. 36, The Combustion Institute (1963).

47. Howe, N. M., and Shipman, C. W., "A Tentative Model for Rates of Combustion in Confined Turbulent Flames," *10th Int. Symp. on Combustion* p. 1139, The Combustion Institute (1965).
48. Cushing, B. S., Faucher, J. E., Gandbhir, S., and Shipman, C. W., "Turbulent Mass Transfer and Rates of Combustion in Confined, Turbulent Flames II," *11th Int. Symp. on Combustion* p. 817, The Combustion Institute (1967).
49. Batt, R. G., "Turbulent Mixing of Passive and Chemically Reacting Species in a Low-Speed Shear Layer," *Jour. of Fluid Mechanics* **82,** part 1, 53 (1977).
50. Andrews, G. E., Bradley, D., and Lwakabamba, S. B., "Turbulence and Turbulent Flame Propagation—A Critical Appraisal," *Combust. Flame* **24,** 285 (1975).
51. Chomiak, J., "Basic Considerations in the Turbulent Flame Propagation in Premixed Gases," *Prog. Energy Combust. Sci.* **5** (3), 207 (1979).
52. Shet, U. S. P., Sriramulu, V., and Gupta, M. C., "A New Approach to the Correlation of Turbulent Burning Velocity Data," *18th Int. Symp. on Combustion,* p. 1073, The Combustion Institute (1979).
53. Abdel-Gayed, R. G., Bradley, D., and McMahon, M., "Turbulent Flame Propagation in Premixed Gases: Theory and Experiment," *17th Int. Symp. on Combustion,* p. 245, The Combustion Institute.
54. Witze, P. O., and Vilchis, F. R., "Stroboscopic Laser Shadowgraph Study of the Effect of Swirl on Homogeneous Combustion in a Spark-Ignition Engine," SAE paper 810226 (1981).
55. Witze, P. O., "The Effect of Spark Location on Combustion in a Variable Swirl Engine," SAE paper 820044 (1982).
56. Inoue, I., Nakanishi, K., Noguchi, H., and Iguchi, S., "The Role of Swirl and Squish in Combustion of the SI Engine," *VDI-Berichte* **370,** 181 (1980).
57. Dent, J. C., and Derham, J. A., "Air Motion in a Four-Stroke Direct Injection Diesel Engine," *Proc. Instn. Mech. Engrs.* **188** (21/74), 269 (1974).
58. Kido, H., Wakuri, Y., Ono, S., and Murase, E., "Prediction of In-Cylinder Gas Motion in Engines by an Energy Method," SAE paper 800985 (1980).
59. Borgnakke, C., Davis, G. C., and Tabaczynski, R. J., "Predictions of In-Cylinder Swirl Velocity and Turbulence Intensity for an Open Chamber Cup in Piston Engine," p. 964, SAE paper 810224 (1982).
60. Mattavi, J. N., et al., "Combustion Modeling in Reciprocating Engines," (Mattavi, J. N., and Amann, C. A., eds.), Plenum, New York (1980).
61. Witze, P. O., "Comparison Between Measurements and Analysis of Fluid Motion in Internal Combustion Engines," Sandia Rep. SAND81-8242 (1981).
62. Reynolds, W. C., "Modeling of Fluid Motions in Engines—An Introductory Overview," in "Combustion Modeling in Reciprocating Engines," (Mattavi, J. N., and Amann, C. A., eds.), Plenum, New York (1980).
63. Gosman, A. D., and Harvey, P. S., "Computer Analysis of Fuel-Air Mixing and Combustion in an Axisymmetric D.I. Diesel," SAE paper 820036 (1982).
64. Diwakar, R., "Direct-Injection Stratified-Charge Engine Computations with Improved Submodels for Turbulence and Wall Heat Transfer," SAE paper 820039 (1982).
65. Clarke, J. F., "Parameter Perturbations in Flame Theory," *Prog. Aerospace Sci.* **16,** 3 (1975).
66. Bush, W. B., and Fendell, F. E., "Asymptotic Analysis of Laminar Flame Propagation for General Lewis Numbers," *Combust. Sci. Technol.* **1,** 421 (1970).
67. Clavin, P., and Williams, F. A., "Theory of Premixed Flame Propagation in Large-Scale Turbulence," *Jour. Fluid Mechanics* **90,** 589 (1979).
68. Peters, N., Hocks, W., and Mohiuddin, G., "Turbulent Mean Reaction Rates in the Limit of Large Activation Energies," *Jour. Fluid Mechanics* **110,** 411 (1981).
69. Damkohler, G., *Zeit. fur Electrochem.* **46,** 601 (1940); also NACA TM 1112 (1947).
70. Heikal, M. R., Benson, R. S., and Annand, W. J. D., "A Model for Turbulent Burning Speed in Spark Ignition Engines," *Proc. Instn. Mech. Engrs.* **C115/79,** 195, (1979).

71. Bray, K. N. C., and Moss, J. B., "A Unified Statistical Model of the Premixed Turbulent Flame," *Acta Astronautics* **4,** 291 (1977).
72. Libby, P.A., and Bray, K. N. C., "Implications of the Laminar Flamelet Model in Premixed Turbulent Combustion," *Combust. Flame* **39,** 33 (1980).
73. Libby, P. A., Bray, K. N. C., and Moss, J. B., "Effects of Finite Reaction Rate and Molecular Transport in Premixed Turbulent Combustion," *Combust. Flame* **34,** 285 (1979).
74. Spalding, D. B., "Development of the Eddy-Break-up Model of Turbulent Combustion," *16th Int. Symp. on Combustion,* p. 1657, The Combustion Institute (1976).
75. Spalding, D. B., "A General Theory of Turbulent Combustion the LaGrangian Aspects," AIAA paper 77-141 (1977).
76. Dopazo, C., and O'Brien, E. E., "Statistical Treatment of Non-Isothermal Chemical Reactions in Turbulence," *Combust. Sci. Technol.* **13,** 99 (1976).
77. Pope, S. B., "The Probability Approach to the Modelling of Turbulent Reacting Flows," *Combust. Flame* **27,** 299 (1976).
78. Pope, S. B., "A Monte Carlo Method for the PDF Equations of Turbulent Reactive Flow," *Combust. Sci. Technol.* **25,** 159 (1981).
79. Pratt, D. T., "Mixing and Chemical Reaction in Continuous Combustion," *Prog. Energy Combust. Sci.* **1,** 73 (1976).
80. Flagan, R. C., and Appleton, J. P., "A Stochastic Model of Turbulent Mixing with Chemical Reaction: Nitric Oxide Formation in a Plug-Flow Burner," *Combust. Flame* **23,** 249 (1974).
81. Radhakrishnan, K., and Heywood, J. B., "Effects of Combustor Inlet Conditions on Flame Stability," *Combust. Sci. Technol.* **24,** 165 (1981).
82. Corrsin, S. C., "Statistical Behavior of a Reacting Mixture in Isotropic Turbulence," *Phys. Fluids* **1**(1) (1958).
83. Mansouri, S. H., Heywood, J. B., and Radhakrishnan, K., "Divided-Chamber Diesel Engine, Part 1: A Cycle-Simulation Which Predicts Performance and Emissions," SAE paper 820273 (1982).
84. Davis, G. C., and Borgnakke, C., "The Effect of In-Cylinder Flow Processes (Swirl, Squish and Turbulence Intensity) on Engine Efficiency—Model Predictions," SAE paper 820045 (1982).
85. Nikanjam, M., and Greif, R., "Heat Transfer During Piston Compression," *J. Heat Transfer* **100,** 527 (1978).
86. Greif, R., Namba, T., and Nikanjam, M., "Heat Transfer during Piston Compression Including Side Wall and Convection Effects," *Int. Jour. Heat Mass Transfer* **22,** 901 (1979).
87. Wolfshtein, M., "The Velocity and Temperature Distribution in One-Dimensional Flow with Turbulence Augmentation and Pressure Gradient," *Int. J. Heat Mass Transfer* **12,** 301 (1969).
88 Spalding, D. B., "Heat Transfer from Turbulent Separated Flows," *J. Fluid Mech.* **27,** 97 (1967).
89. Summers, I. G. S., "Convective Heat Transfer in a Rapid Compression Machine Simulating a Spark-Ignition Engine," M.Sc. Thesis, University of Manchester (1970).
90. Annand, W. J. D., "Heat Transfer from Flames in Internal Combustion Engines," in "Heat Transfer from Flames," (N. H. Afgan and J. M. Beer, eds.), p. 377, Wiley,New York (1974).
91. Hassan, H., "Unsteady Heat Transfer in a Motored I.C. Engine Cylinder," *Proc. Instn. Mech. Eng.* **185,** 1139 (1971).
92. Dao, K., Uyehara, O. A., and Myers, P. S., "Heat Transfer Rates at Gas-Wall Interfaces in Motored Piston Engine," SAE paper 730632 (1973).
93. Dent, J. C., and Suliaman, S. J., "Convection and Radiative Heat Transfer in a High Swirl Direct Injection Diesel Engine," SAE paper 770407 (1977).
94. Annand, W. J. D., and Pinfold, D., "Heat Transfer in the Cylinder of a Motored Reciprocating Engine," SAE paper 800457 (1980).

95. Woschni, G., "A Universally Applicable Equation for the Instantaneous Heat Transfer Coefficient in the Internal Combustion Engine," SAE paper 670931 (1967).
96. Annand, W. J. D., and Ma, T. H., "Instantaneous Heat Transfer Rates to the Cylinder Head Surface of a Small Compression Ignition Engine," *Proc. Instn. Mech. Eng.* **185,** 976 (1971).
97. LeFeuvre, T., Myers, P. S., and Uyehara, O. A., "Experimental Instantaneous Heat Fluxes in a Diesel Engine and Their Correlation," *SAE Transactions,* **78,** paper 690464 (1969).
98. Siewert, R. M., "Engine Combustion at Large Bore-to-Stroke Ratios," SAE paper no. 780968 (1978).
99. Lee, W., Schafer, H. J., and Schapertons, H., "Investigation of High Compression Ratio S.I. Engine by a Two-Dimensional Model," *5th Intl. Automotive Propulsion System Symp.* (1980).
100. Nagayama, I., Araki, Y., and Lioka, Y., "Effects of Swirl and Squish on S.I. Engine Combustion and Emission," SAE paper 770217 (1977).
101. Mayo, J., "The Effect of Engine Design Parameters on Combustion Rate in Spark-Ignited Engines," SAE paper 750355 (1975).

6

Abnormal Combustion Effects on Economy

R. W. WHEELER

I. INTRODUCTION AND HISTORICAL BACKGROUND

A. Introduction

The fact that the efficiency of the spark ignition (SI) engine depends on the usable value of compression ratio is well known. The thermodynamic reasons for this are explained in various textbooks,[1,2,3] while the practical consequences of raising compression ratio have been exhaustively researched, particularly by Caris and Nelson.[4]

Typical gains to be obtained in high-compression engines are shown in Fig. 1 for representative cool-running modern units of overhead-valve design. Gains in power output are also obtained and with low-cost fuel this was a potent incentive for raising compression ratio, but with the energy crisis this has changed.

Caris and Nelson could not define the optimum compression ratio closely but did show that efficiency was actually falling at 17:1, which is considerably lower than the maximum possible commercial compression ratio which could be anything up to a limit of approximately 24:1. This physical limit is set by production engineering considerations and in fact is only used in small high-speed diesel engines of the swirl and prechamber varieties where it is needed for ease of starting and for good combustion.

An ultrahigh compression ratio in SI engines brings with it the problems of excessive peak cylinder pressures which necessitate a heavy engine structure and a consequent high frictional loss, both of which must be paid for in fuel consumption in the automobile.

R. W. Wheeler • Atlantic Research Associates, 59 Goods Station Road, Tunbridge Wells, Kent TN1 2DH England

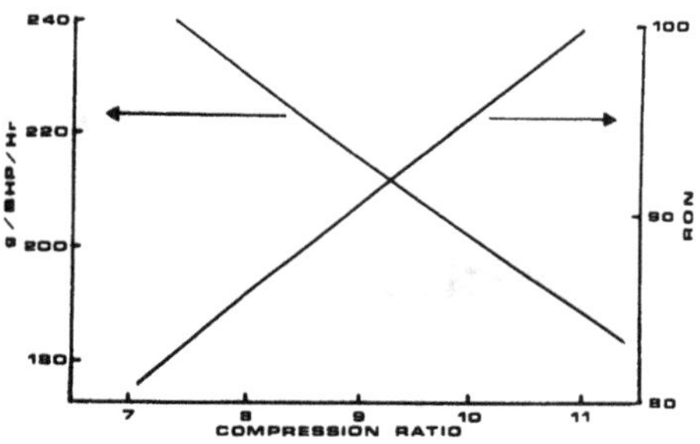

FIGURE 1. Effect of compression ratio on brake-specific fuel consumption and octane requirement.

Present-day engineering technology points toward an optimum usable compression ratio of from 11:1 to 14:1 for SI engines used in light passenger cars. However, in most cases the fuels available have inadequate resistance to the high cylinder pressures and concomitant high temperatures and burn in an abnormal fashion. Thus, in practice, excepting high-speed racing engines, compression ratios rarely exceed 10:1 and are more usually 9:1. Despite the law of diminishing returns, there is a considerable economy gain in moving up from 9:1 to 14:1,[(5)] and therefore a study of the factors involved in abnormal combustion which disallow this gain is clearly worthwhile.

The two major elements of abnormal combustion are knock and preignition, which have a history as long as the SI engine itself, and although Dugald Clerk reports that he first observed knock in 1882[(6)] much of it is ill documented. However, a brief historical study is a valuable introduction to the subject.

B. Relationship between Preignition and Knock

The first engines with primitive carburetion, sometimes of the surface evaporation or wick design, needed a light gasoline of a low (100 °C) final boiling point. This was a straight run of petroleum crude, and the "cut" between it and kerosene was often flared away. We now know this fuel to be the worst possible for abnormal combustion, and in fact, many of these engines would lose power and finally stop, although combustion was still occurring. Other engines developed very noisy combustion and continued running.

A researcher in England, Professor Bertram Hopkinson of Cambridge University, showed that two quite distinct phenomena were involved: preignition, which stopped the engine, and detonation, which caused the noise.

By inserting a red-hot rod into a running engine he caused the ignition to occur before the spark. He could observe this preignition on his crude indicators and see how the negative work on the piston quietly stopped the engine. When he caused the engine to "detonate," however, the indicators showed very high pressure peaks on otherwise normal diagrams which finally shattered his pressure pickups. He considered that a phenomenon akin to detonation in long tubes was present.

Undoubtedly other workers in other areas reached the same conclusions at the same time (1904), but certainly it was Hopkinson who inspired his pupil H. R. Ricardo to build a research engine in which he demonstrated how detonation raised the temperature of the cylinder head causing the onset of preignition—a most important sequence of events. He also observed the highly "anti-detonant" character of pure benzene. His next engine, built in 1913, gave a variable compression facility by raising variable supercharge.[(7)] This allowed him to rate fuels for resistance to "detonation." He found that toluene was the best available fuel, and that high-gravity gasolines were better than low. These tests made him reach the firm conclusion that this fundamental limit on engine compression ratio was the greatest barrier to engine efficiency.

After World War I, during which engine design and construction materials had advanced greatly in the field of aircraft and tanks,[(8)] the new automotive engines demanded a great research effort on the combustion side, a challenge taken up with enormous success in the United States particularly at the General Motors Corporation research laboratories under C. F. Kettering. There Thomas Midgley discovered the antiknock properties of tetraethyllead (TEL) and pushed back the compression ratio barrier at least for petroleum-based fuels.[(9)]

Also the herculean task of the knock rating of all available pure hydrocarbons was commenced by the Cooperative Fuel Research (CFR), committee of the U.S. petroleum and automotive industry, culminating in the octane–heptane scale before the end of the twenties.[(10)] Contemporaneously, in the United Kingdom the Shell Company had promoted work by H. R. Ricardo, H. T. Tizard, and D. R. Pye using probably the first true variable compression engine designed by Ricardo—the E35 single-cylinder, 2-liter, 4-valve o.h.c. engine.[(11)] They discussed the theoretical efficiency of the engine in relation to the fuels, the compression ratios, and the thermodynamic properties of the working fluids. Tests on the E35 showed good agreement between prediction and observation, thus putting engine thermodynamics on a sound footing.

Toluene again showed the best antidetonant properties and a scale was proposed using a particular aromatic-free gasoline as zero and toluene as 100. This scale predated, but was clearly inferior to, the CFR committee scale in the United States.

Ethanol also showed up well as a fuel, causing Ricardo to become a lifelong enthusiast, saying at the beginning of World War II, "By the use of a fuel (ethyl alcohol) derived from vegetation, mankind is adapting the sun's heat to the development of motive power, as it becomes available from day to day by using mineral fuels he is consuming a legacy—and a limited legacy at that."[(12)]

By the year 1927, also chosen by W. G. Agnew as the commencement of a great period in GM Research,[(13)] it had become clear that preignition could practically be eliminated by engine design, but that detonation, which was now more often called "knock" to specify it as a unique engine phenomenon, was a fundamental problem which would have only a limited solution.

1. Internal Combustion—Turbulence and Knock

The importance of turbulence in the engine cylinder was realized by the pioneers because it explained the apparent paradox of a high-speed engine running faster than the flame speed of the mixture appeared to allow. Hopkinson showed how a variable-speed fan in a bomb experiment could raise the flame speed of the mixture at will, and his contemporary Sir Dugald Clerk demonstrated the opposite effect in an engine by allowing the induction turbulence to stagnate over several cycles. When it was finally ignited the charge would only partially burn.

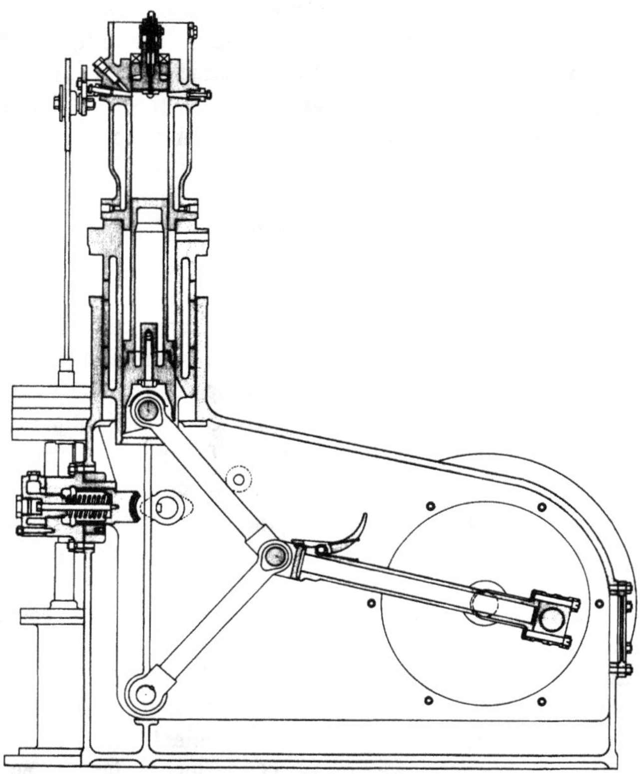

FIGURE 2. Rapid-compression machine (Sphinx).

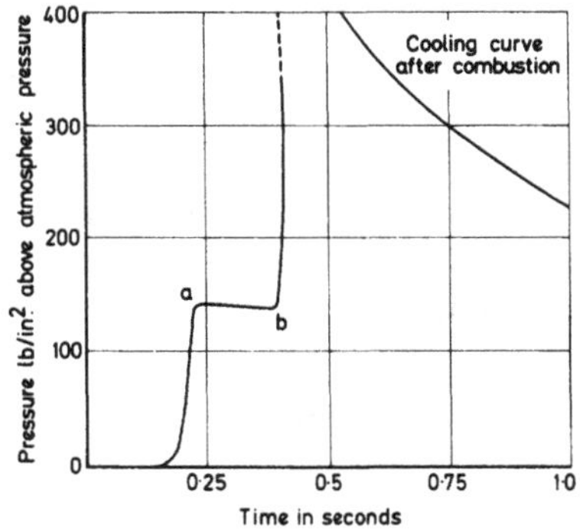

FIGURE 3. Indicator diagram of ignition of *n*-heptane in air in a quiescent chamber.

During this early period it was inevitable that theories of knock should proliferate. Tizard[14] considered that it was caused by high ratios of flame acceleration leading to very high flame temperatures. To examine how such a high flame speed came about, he used a rapid-compression machine, built on engine principles, shown in Fig. 2. The piston latched at top dead center (TDC) and disconnected from the drive crank by virtue of the telescopic connecting rod. A weight-driven fan supplied variable turbulence in the combustion chamber. Compression autoignition, which he considered to be the cause of the high flame speeds, occurred and was recorded on indicator diagrams.

The effect of the fan-induced turbulence on the ignition delay of a heptane–air mixture is beautifully shown by comparing Figs. 3 and 4. The two-stage nature of autoignition and its dependence on fuel molecular structure was observed, but the role of turbulence was complex as, on the one hand, the higher heat losses to cold cylinder walls reduced charge temperatures, while on the other hand, increased turbulence promoted the diffusion of autoignition-prone areas throughout the charge.

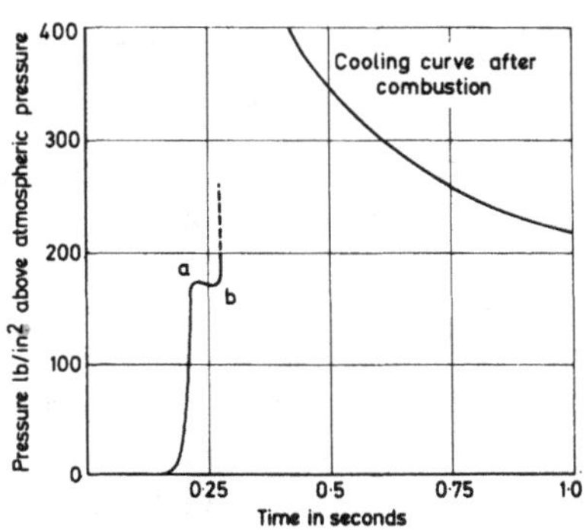

FIGURE 4. Indicator diagram of ignition of *n*-heptane in air in a turbulent chamber.

These experiments along with the discovery of the dramatic effect of TEL caused Prof. Callendar to propose that during the delay period, active species, which he called "nuclear drops," formed and promoted knock.[15] E. W. S. Mardles showed that in all probability the active species were organic peroxides, and at about the same time, A. Egerton promoted the theory that antiknocks are molecularly dispersed multivalency metal oxides which by a simple redox reaction destroyed the active oxygenated molecules.[16]

2. *Combustion Chamber Design*

Concurrent with the emergence of these fundamental notions of the knocking process, Ricardo designed the first logically argued and researched combustion chamber design for the high-efficiency operation of mass-produced cars—the turbulent head side-valve or L-head. This engine configuration was indicated by production engineering constraints. The whole development is well documented[17] and placed in context with its time.[8]

Salient points that emerged were:

1. the use of precisely controlled turbulence to increase the flame speed such that the delay period for autoignition was overtaken;
2. the maximum distance of flame travel was limited by the use of a flame-quenching piston-head clearance of 0.14 in.;
3. the use of the rate of pressure rise to define turbulence and engine roughness. (This chamber had an optimum rate of pressure rate of 35 psi/crank degree giving a flame speed rising to 300 ft/sec and a total combustion time of 25 crank degrees.);
4. the spark plug was sited in the hottest region of the chamber, over the exhaust valve, for least knock. Knock itself occurred in the last part of the charge to burn, which was situated over the piston crown in a relatively cool area, and displayed itself by metal erosion and pitting of the piston within the knock zone;
5. cylinder heads fitted with six equally spaced windows across the major axis were used to measure the flame speed stroboscopically and relate it to the pressure diagram.[18]

During these tests it was found under knocking conditions that the flame traversed windows 1–5 normally, but when it reached 6 a bright white flash was seen which lit all six windows simultaneously. Thus piston-pitting observations of knock were fully confirmed.

The charge motion in these engines was very confused due to the asymmetric design, but in a smooth cylindrical chamber a free-wheeling swirling motion could be set up with low energy loss. Alcock[19] used a single-sleeve valve engine in which baffles in the circumferential inlet ports could vary the swirl from 0 to

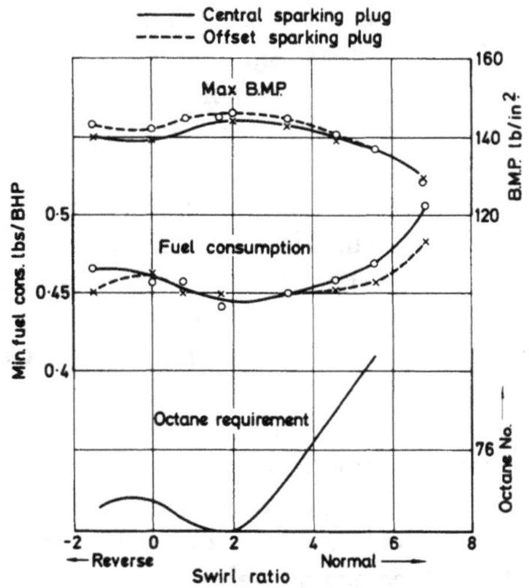

FIGURE 5. The effect of swirl on the performance of a single-sleeve valve engine.

7 revolutions per crank revolution. He showed (Fig. 5) a clear optimum at a swirl ratio of 2 to 1 for power economy and knock.

Offsetting the spark plug gave a benefit at high swirl because a central flame kernel was reluctant to travel centrifugally due to its low density. Alcock named this effect thermo-centrifugal convection,[20] and it has been used in setting up combustion stratification. Towards the end of the twenties there was enough information on knock and preignition to satisfy the immediate needs of engine designers. TEL had been discovered as an antiknock, a start had been made on determining the antiknock qualities of hydrocarbons, and a knock scale had been invented to help refiners and designers; however, scientists had scarcely scratched the surface of the problem.

II. DEFINITION AND MEASUREMENT OF ABNORMAL COMBUSTION

A. Definition of Terms

1. Firstly, normal combustion must be defined as that which stems only from the spark at its appointed timing, progressing in the form of a flame front across the chamber and dying out smoothly when the mixture is completely consumed. There may, however, be noise from the engine structure being driven by high rates of pressure rise.

2. Abnormal combustion is either that which is initiated by hot surfaces at timings before or after the spark, hence yielding incorrect pressure diagrams, or

that which occurs ahead of the flame front at extremely high rates. All forms of surface or hot-spot ignition or knock are included.

3. Knock is the noise filtering through the engine structure caused by a spontaneous ignition (autoignition) of a part of the cylinder charge trapped ahead of a flame front in what is called an end-gas zone. The flame front may be initiated by the spark or by a surface ignition. The former is sensitive to spark timing, the latter is less sensitive.

4. Hot-spot or surface ignition originates from high-temperature areas of chamber walls, exhaust valve heads, spark plug electrodes or insulators, or incandescent deposits. It propagates at normal flame speed from its initiation which may be prior to the spark, in which case it is called preignition, or after the spark, in which case it is sometimes called postignition. In either case it disturbs the normal engine pressure diagram.

5. Running-on or after-running is typically an irregular firing of an engine for a limited time after the ignition has been switched off with the throttle at idle and the engine warm.

Depending on its build, its operation, and the quantity and quality of deposits an engine may demonstrate various mixes of these abnormalities in its combustion and various names have been given to these over the years. The descriptive title wild ping was given to irregular sharp knocks observed on high-compression engines with unstable deposit formations. It is a surface ignition-induced knock.[(21)]

Rumble is another term for a fairly stable surface ignition of a nonrunaway variety. High rates of pressure rise too early in the cycle may be responsible.[(22)]

High-speed knock is the latest in the line. It is something of a misnomer because it happens at high engine speed and load, often without the driver being aware of it, and results finally in eroded or even melted pistons.

Operation of the engine on fuels of insufficient motor octane number seems to be the cause, sometimes with a runaway preignition the denouement.

B. Detection and Measurement of Knock and Preignition

First and foremost the human ear is the natural knock detector and intensity estimator[(25)] and is often used as a final arbiter even in modern research. It is difficult to best a well-trained ear even with sophisticated electronic processing, but even so knockmeters have many advantages in eliminating human prejudice and fatigue.

Direct knock detection by the observation of piston pitting can be extremely useful to delineate the end-gas area but is time consuming. The author has needed a 10-h run on a robust test engine at 1500 rpm to make slight surface erosion when the knock intensity positively scared lay observers.

Observation of heat transfer in the end-gas zone gives a very definite indication of knock by a sudden increase in rate over and above the normal increase caused by the factor causing knock (e.g., compression ratio increase or spark advance).[26]

Soot emissions are commonly seen when engines knock at a high intensity but attempts to monitor knock by soot detection have not led anywhere.

If a window is available in the cylinder, knocking can be detected by observation of the high-intensity flash with knock, but unfortunately windows and practical high-performance cylinders are incompatible. Coupled with the flash a sharp increase in ionization is observed when knock occurs. This can be detected by an ionization probe, which is usually negatively polarized from 100–300 V. The ionization current is detected using an electrometer amplifier and oscilloscope. This engine spark plug can also be used as an ionization gap with isolation circuits to avoid the high voltage.[27,28] Flame speed, mapping, and quenching can be analyzed by this technique[29–31] which can readily be applied to a test engine because only a small hole (~3 mm) need be drilled into the chamber.

Characteristic knock vibrations can be seen on the back slope of ionization diagrams. However, the technique is little used in knock research. For preignition detection the use of strategically placed ion probes, with their diagrams shown on a crank angle time base, have been much used.

The use of rapid-response quartz piezoelectric pressure transducers has advanced knock research greatly. Their small size, thermal stability, and robust nature make their use possible even in unmodified commercial engines in which they are used attached to a spark plug body using a narrow tube,[33] taking care to avoid pipe resonances. Some authorities use knock overpressure to define its intensity.[34] Others prefer to relate knock intensity to the rate of pressure rise [35].

Preignition can very conveniently be studied using piezogauges to indicate the advancing pressure diagram due to its onset.[36]

Accelerometers. The pressure shock waves within the cylinder following the knock reaction excite the engine structure to transmit, or emit, vibrations of from 3–10 kHz depending loosely on the combustion chamber size and geometry. The first practical knockmeter used on the early CFR octane rating engines was a "bouncing-pin" device which resonated at the vibration frequency. This caused contact points to close passing a current through a crude integrating circuit which gave a measure of the knock intensity.[37,38]

Present-day accelerometers are often clamped to, or under, a critical part of the structure, such as a head-fixing bolt, the best position discovered by cut-and-try because each engine design will vary in its spectrum and location of extraneous noise caused by valve gear, chains, gears, etc.

Accelerometer design has improved greatly because of their commercial use in engines.[39]

C. Rating Fuels

1. Octane Rating

The use of the primary reference scale, that is *n*-heptane = 0 and isooctane (2-2-4 trimethylpentane) = 100,[10] has worldwide acceptance. It depends entirely on the use of a closely specified engine, the CFR, which is arranged to run at two levels of thermal severity. The knock on the test fuel is matched with primary reference blends, the proportion of isooctane, which in the low-severity case is called the *R*esearch *O*ctane *N*umber (RON), and in the high-severity case, the *M*otor *O*ctane *N*umber (MON).[40]

Generally speaking a cool-running engine (racing and aircraft) will conform to RON, whereas most automobile engines conform closer to MON. The sensitivity of a fuel (to thermal severity) is given by RON–MON. So-called road octane numbers can be obtained by accelerating a car in top gear with a range of fixed ignition advances and observing the engine speed for borderline knock. A set of curves of engine rpm versus spark advance for this light level of knock is developed using primary reference blends. The test fuel can then be compared by simple bracketing.[41]

If CFR engines are not available, then any variable compression engine may be used comparing the fuels on an audibility basis. This is the highest useful compression ratio (HUCR) method.[25]

Of course, only the CFR gives true universal octane numbers. It is possible to obtain a close approximation to RON and MON using hydrocarbon analysis by nuclear magnetic resonance (NMR), coupled with lead and sulfur analyses of the test fuel.[42]

2. High-Speed Knock

At present there is no fully agreed engine-rating procedure for the high-speed knocking propensity of motor fuels because the phenomenon is difficult to observe and even more difficult to relate to engine damage.[43,44]

The indications are that MON may be more important than RON and that in its very severe stages an element of preignition is involved. The problem is most severe in Europe where the major researches are in progress. The Coordinating European Council (CEC) is actively encouraging the development of methods for assessing the relationship of knock and engine damage in commercial engines. CFC Project Group CF-23 has developed a "draft" test method that involves stabilizing the engine knock, the selection of engine speed, throttle setting and knock intensity, with an endurance test at these fixed conditions and knock intensity.[45]

The knock intensity setting method that is preferred is by a variable band-pass accelerometer device by Assorini (Italy) based on a model by the Snam-

Progetti Company.[46] A tuned microphone system has also proved useful in this context.[46,47]

Important advances in relating engine damage to knock intensities, which have been calculated in various ways, have been made in Italy.[23,46] The whole concept of knock rating has been questioned on the basis that primary reference fuels are too sensitive to changes of engine conditions and do not behave in the motor method as do premium gasolines which contain aromatics and olefins.[48]

There is a case for using PRFs, which have chemical kinetics in the high-speed knock regime closer to those of commercial gasolines.

3. Preignition Rating

Because of the highly sensitive surface catalytic effects and the inevitable presence of surface deposits of varying nature in crucial areas of the chamber, it is exceedingly difficult to rate fuels for preignition tendency. Downs used isooctane as having preignition rating 100 and either cyclohexane or cumene rated as zero.[49] He induced preignition in a Napier Dagger aircraft engine cylinder using an air-cooled hot thimble and in a laboratory research engine (Ricardo E6) using an electrically heated coil incorporated a thermocouple. The energy required, as heating or cooling, to bring the preigniter to its standard preignition temperature from its normal running temperature was used to assess preignition tendency. In the case of the hot thimble it normally ran at a temperature well above that required for preignition, therefore the volume of cooling air became an inverse function of the energy requirement. In the case of the heated coil it was relatively cool running and well below preignition temperature, therefore a wattmeter in the AC heating circuit became a direct measure of the energy input. Arrigoni[50] used a similar technique but controlled the AC heated coil to a fixed preset temperature and statistically analyzed the cylinder pressure diagrams for those originating from true preignition cycles. Where it is desired to research the effect of hot spots in the engine structure, or in deposit formations, recourse can be had to cutting the ignition while running at high load and counting the preignitions (using an ionization probe) before the hot spots have time to cool.[28] The spark plug itself can be used as the ionization probe in this case so that no engine modifications are required.

Engine "rumble," a variety of deposit-induced preignition, was shown at GM to be induced by benzene and reduced by isooctane fuel, thus allowing an engine severity scale to be constructed.[22]

D. Engine Parameters and Knock

A spark ignition engine, particularly if it is designed to run at the knock limit, has a large number of interdependent factors making its performance almost impossible to predict except in the light of educated experience. The more important parameters can be listed:

1. Compression ratio } charge density
2. Throttle position } charge density
3. Volumetric efficiency } charge density
4. Spark timing and quality
5. Engine speed
6. Flame travel distance
7. Mixture quality and strength
8. Engine temperatures
9. Residual gases

Textbooks must be consulted for a full account of their intricacies.[1–3] Only a brief discussion of some major points can be given here.

1. Charge Density

It is usual to discuss knock in terms of compression ratio and load, but of course the governing parameters are charge density, end-gas temperature, and time. Charge density is a function of the inlet pressure (throttle opening), the compression ratio, and volumetric efficiency. The temperature is dependent on the adiabatic compression, the heat losses of combustion, the intake charge temperatures, the end-gas wall temperature, as well as any preflame oxidation in the end gases. The time for the development of knock depends on the engine speed, flame speed, spark advance, and turbulence, which are highly interdependent factors. Researchers at NACA explored the use of knock-limited charge density and temperature in rating aircraft engine fuels.[51,52] They demonstrated that for a given fuel there is a limiting end-gas density for any given end-gas temperature, and that the knock-limited charge density fell with compression ratio increase at a fixed spark timing.

Heron modified the concept of knock-limited end-gas density index using a volumetric efficiency derived from the IMEP. This avoided the measurement of mass air flow.[53] Where a number of chamber designs are compared for knock-limited performance these methods prove very useful.

2. Spark Timing

The spark timing is the most readily adaptable control variable for knock. Retarding the spark reduces end-gas temperature and pressure, so delaying the onset of knock. Spark plug heat factors have an effect on knock because a hot-running plug will ignite the flame kernel earlier, simulating spark advance. A cool-running plug may tend to quench the kernel, delaying its transition to a flame front and thus retarding combustion. Similarly a used spark plug that has deposits in its cavity and eroded points behaves as a cool-running plug with a lower knock level.[36]

Many manufacturers set their engines ~5° retarded from the minimum advance for best torque (MBT) to give some leeway against knock.[54] At high engine speed the ignition or prepropagation delay can be a very high proportion of the total flame travel time,[55] and its cyclic irregularity an important contributor to overall irregularity. At low-speed rpm (~1000) it is less important.

The statistics of cyclic irregularity have received much attention[56,57,29] and the lean inflammability limits of a range of hydrocarbons and gasoline blends under enginelike conditions and spark ignition have been shown to extend to ~33 : 1 air–fuel ratio.[58]

3. *Engine Start*

The most dramatic effect of increasing engine speed is an almost parallel increase in flame speed. At very high engine speeds the effect of turbulence on flame initiation and the falling volumetric efficiency cause the apparent flame speed to drop below proportionality with an increasing spark advance requirement. Even so Honda has operated small four-valve cylinders up to 25,000 rpm with only ~50° spark advance.[55]

They calculate a combustion efficiency which is constant up to maximum engine speed if the apparent flame speed is corrected for the ignition delay to give true flame speed. In these free-flowing chambers turbulence is a function of chamber shape and piston speed, with squish adding an important factor.

Increasing engine speed reduces knock intensity with paraffinic fuels more than with olefinc or aromatic fuels.[48] The auto-delay does not reduce with engine speed increase, knock is therefore less intense (Sec. III.C). However, the audibility of knock is also reduced with engine speed, while the damage it does to engine structures increases until at a maximum power setting preignition intervenes. High-speed knock damage may be a maximum at speeds between maximum torque and maximum power (i.e., 4500 rpm).[36]

Olefinic constituents of commercial blends confer strong high-speed knocking properties. In general gasolines with pro-preignition constituents (Sec. V) have low high-speed octane numbers.

4. *Flame Travel*

From the early days of the reduction of the chamber major axis from T-head, to L-head, to the "hemi-head," and finally arriving at the May or "Fireball" head, it has been known that a short flame travel was essential to a high-compression engine.

It can of course be said that as compression ratios increase beyond 7 : 1 the geometrical constraints indicate a short flame travel. This will most often be slightly less than the bore diameter up to roughly 10 : 1 and about 50% of the bore diameter above that. However, there is a strong proviso that at high ratios with a small head, or piston cavity, knock can occur in areas where close metal clearances may prevent normal flame travel, i.e., in the squish area, in the top-ring groove, and between the top piston land and the cylinder.[24]

The effect of a short, fast flame travel on end-gas autoignition is extremely complex. In a modern engine with a compact (low surface–volume ratio) chamber the turbulence may be sharply increased at TDC by the "pinching" of the charge from the cylinder diameter into that of the chamber. The end-gas temperature may be increased by a more nearly adiabatic flame compression, but the time in which the autoignition delay can occur is shortened.

There is a lack of information on such chambers, and flame statistics derived from research engines may be inappropriate.

5. Charge Uniformity

Mixture quality should be as uniform as possible in a multicylinder engine, otherwise it may knock in a fortuitously rich cylinder which bears more than its fair share of the load. The uniform mixture quality should be attained with the use of as little intake air heat as is practicable for good knock performance. A fact well demonstrated by the knocking of many U.S. emissions-treated cars which use excessive charge heating.

The tendency towards leaner mixtures for reasons of emissions and economy is advantageous for good knock performances, but the use of mixture enrichment at high engine load brings problems of high-speed knock.

6. Engine Temperatures

It has been a cardinal rule in engine design to initiate combustion close to the exhaust valve in order to confine the end gas to a cool area in the chamber which is supplied with coolant at as low a temperature as practicable. Intake air heating at constant compression ratio reduces knock-limited air density[53] whereas increasing end-gas temperature by raising compression ratio has almost no effect on it, because of the spark retard needed to maintain best torque. Later "ignition" means a shorter delay time for end-gas autoignition, thus compensating for the higher temperature.

Measuring end-gas temperature profiles during the cycle is not easy. It has been accomplished using a two-wavelength infrared radiation method[59] and by a benzene ultraviolet absorption method.[60] Both methods demonstrate the exothermicity of the preflame processes.

7. Exhaust Residuals

Residual exhaust gas in the fresh charge is a very variable quantity and depends on compression ratio, valve overlap, engine speed, and the intake and exhaust systems. In the guise of exhaust gas recirculation (EGR) it has been extensively studied.[61] Residuals lower flame speed and temperature but increase the thickness of wall quench layers and the charge temperature.[31]

Clearly any cyclic irregularity of residuals must reflect on combustion and end-gas conditions; little, however, is known on this point.

Active species still remaining in the residual gas may promote preflame reactions especially if the engine stagnant and the residuals are not fully mixed. Use is made of this interesting effect to effect regular ignition in a two-cycle engine with the spark switched off.[62]

III. KNOCK THEORIES

There has been constant speculation on the origin of knock from its first observations to the present day, and it can be said that there is no completely satisfactory explanation of the phenomenon even now. This is not entirely surprising because the phenomenon itself has changed from that occurring in a long chamber at a compression ratio of 4:1 to that occurring in a short compact chamber at 14:1.

It is generally agreed that knock emanates in some way from a rapid combustion of the end gas resulting in a local high-pressure region. The relocation of this pressure causes gas vibration in the chamber which may take on the nature of shock waves and may resonate with the natural frequency (gas) of the chamber.

During the thirties the various theories boiled down to two which have considerable scientific support, the detonation theory and the autoignition theory. The former postulates that, for some reason, the advancing flame front accelerates sharply as it approaches the end gas, develops sonic velocity, and consumes the end gas with great rapidity. The resultant pressure discontinuity gives rise to the gas vibrations.

The latter theory postulates that the fuel–air mixture in the end-gas region is subject to high temperatures which cause preflame reactions. These sensitize the mixture to undergo spontaneous or autoignition, perhaps in several sites simultaneously. The rapid combustion in advance of the flame front causes a pressure pulse which produces the gas vibrations.

A. End Gas

Before discussing these theories it is of advantage to examine the end-gas zone which is the seat of knock. What is striking, judged from pitting and erosion footprints, is the way in which the zone is confined to a narrow strip on the periphery of the chamber. On occasion the pitting also occurs down the top piston land and sometimes in the ring groove,[24] a condition perhaps aggravated by squish pressures.

The last observation eliminates the forward radiation theory of knock, although this has already been done spectroscopically.[63]

In the author's experiments a totally uncommercial, very heavy knock level in a 3-in. bore test engine (with a disc chamber) gave a piston "footprint" of only 0.25 in. wide at its center. The highest swirl obtainable using a 180° shrouded valve moved this footprint only marginally. Qualitative confirmation came from sampling valve tests on preflame products (peroxides and aldehydes) in which the product was plotted against sample flow. At a lower knock intensity the product concentration fell using a sample flow representing an end-gas hemisphere of 0.15 in. radius. These results indicate that with moderate to heavy knock only very approximately 2–3% of the charge is involved. Curry reports[29] that the detonation of a lead azide pellet (10% of the energy of the engine fuel–air charge) in the cylinder near TDC gave a rate of pressure rise vastly greater than did heavy knock.

The end gas seems to be a very thin sliver of gas trapped in a crevice or against the cool cylinder wall and partaking of the conditions in that region. An exception could be where two flame fronts trap end gas in the chamber center, but this must be very uncommon.

The chamber wall temperature is obviously crucial to the end-gas autoignition; but the end-gas composition, also crucial, is much less well understood because the original charge is mixed with unknown exhaust residuals, boundary layer hydrocarbons,[64] and oil mist. These factors undoubtedly add to the statistical uncertainty of knock.

B. Detonation Theory

Detonations in tubes had fascinated physicists long before engine knock was observed. The essential features were an initial flame acceleration up to a sudden development of an ultra-high-speed combustion and shock wave fed by a continuous supply of fresh combustible mixture. The details became clearer with the application of high-speed schlieren photography.[65,66] When it was shown that knock occurred at the end of flame travel in an engine, an essential link between the phenomena seemed to be broken. However, there remained the possibility that under favorable conditions the normal flame could accelerate and suddenly engulf the end gas at detonation velocity with an ensuing shock wave reflecting in the chamber at resonant frequency.

For small- to medium-size automotive engines the resonant frequency of knock, dependent on many conditions, may vary from 2–10 kHz, and the detonation velocities in the cylinder gases from 2000–7000 ft/sec.[29,67,68] Workers at NACA[67,68] using high-speed photography came to the conclusion that the knock reaction occurred at velocities of the order of 5000 ft/sec, well into the supersonic region, and hence probably associated with a detonation wave. They observed autoignition centers in various parts of the end-gas zone which appeared to initiate the detonation reactions.

In a series of brilliant studies Withrow and Rassweiler[69,70] photographed knock reactions in an L-head engine and showed autoignition in a number of separate centers well ahead of the flame. This reaction seemed to be almost instantaneous, whereas at no time did the flame front itself exceed 300 ft/sec. They used time displacement photography related to a pressure diagram to show that the gas pressure vibrations of knock could be seen as shock waves in the flame photograph[71] and were related to the engine-knocking noise frequency which in this L-head engine was approximately 3500 Hz.[72]

The knocking pressure waves cause the burned gases to reilluminate by adiabatic heating and also to reionize such that the normal ion decay has the characteristic knocking frequency imposed on it.[29] Comparing ion currents behind a true detonation wave with those measured after knock, the latter are only 10% of the former, suggesting a knocking velocity in the 500 ft/sec range.

The highest cylinder flame velocity reported is 1200 ft/sec[29] measured by Curry using ion gaps. This speed was unusual, 300–600 ft/sec was more usual. Thus there is no evidence of a normal flame accelerating to sonic velocities (2000 ft/sec) unless it is in the act of consuming the end gas at the very end of combustion, which is likely to be unobservable due to the thin layer comprising the end gas. A number of authorities[73–75] suggest that the pressure differentials associated with flame speeds only 5–10 times normal can, on collapsing, give rise to knocking vibration.

There is no evidence in all this work of tubelike detonation even in the long L-head chambers. Possibly the event is simulated just up to the threshold of the true detonation wave which collapses at the end wall in a sonic shock. Fuel mixtures, which knock readily in engines, are not necessarily detonation prone in tubes[76] and TEL hardly affects true detonation, observations which make the detonation theory even more tenuous.

C. Autoignition Theory of Knock

H. R. Ricardo first suggested the autoignition theory of knock[77] based on experiments by his colleagues using a rapid-compression machine.[78] That these were more and less readily ignitable fuels under compression was well known with the good antiknock quality fuels coming in the latter category. Although the piston compression might not autoignite the gasoline mixture, the combination of piston and flame compression would. If known partial oxidation products of the fuel (e.g., formaldehydes) were found ahead of the flame, then autoignition could be presumed.[79,80]

Using combustion in "bombs" Townsend[81] established pressure–temperature ignition diagrams for paraffinic fuels with two main regions, a high-temperature region and a so-called low-temperature peninsula (Fig. 6). In the low-temperature region ignition proceeds by a two-stage process. The first stage is a delay period up to the passage of a cool flame at a temperature of 350 °C. The second stage has a negative rate coefficient and ends with autoignition.

The high-temperature region has only one stage, a delay up to autoignition and no cool flame. Fuels such as benzene and methane show only single-stage high-temperature ignition.

A cool flame is a chemiluminescence due to the formation of an excited state of formaldehyde during the oxidative reactions of the fuel and its relaxation to the ground state.[82] In constant flow tests in a hot tube, cool flames can be stabilized, thus allowing analysis of the preflame products. Aldehydes, organic peroxides, and hydrogen peroxide are found. Burgoyne[83] showed that total peroxide concentration gave a discontinuity as the cool flame passed (Fig. 7). In a study of *n*-hexane oxidation organic peroxides were identified in the pre-cool-flame region, whereas in the cool flame a mixture of formaldehyde and hydrogen peroxide was found.[84]

Cool-flame formation varies enormously with hydrocarbon structure. Thus *n*-paraffins readily give strong cool flames, but branched-chain paraffins are more resistant. Olefins give even lower luminosity cool flames with longer induction periods. Benzene gives none, and other aromatics give hardly detectable luminosities. Attempts have been made to relate cool-flame intensities, heat releases, or induction periods to octane number without great success.[85,86] Townsend,

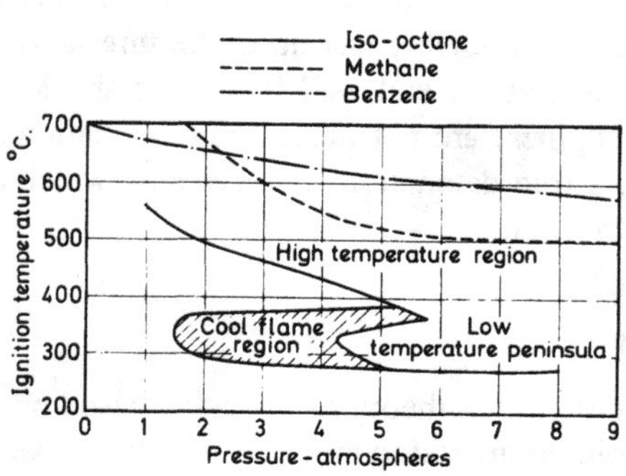

FIGURE 6. Typical ignition diagrams showing two-stage and single-stage ignition (Townsend).

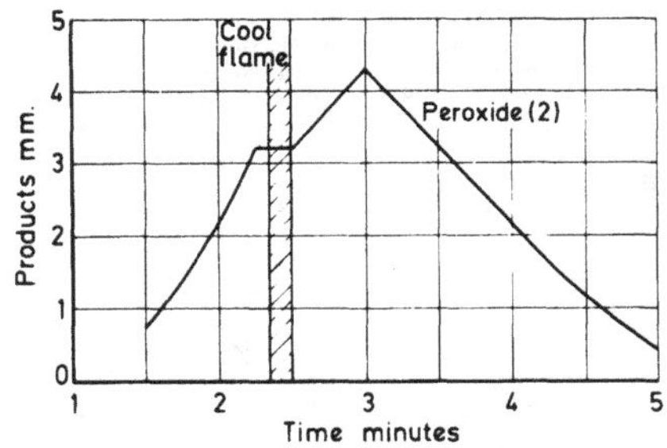

FIGURE 7. The effect of cool-flame on peroxide concentration in a flow-rig.

however, showed that the order of octane numbers was the same as the order of the minimum pressures for ignition in the "low-temperature" region for a number of paraffinic fuels.[81]

TEL does not affect the induction period to the cool flame but considerably lengthens the second stage delay to ignition.[87,88] A large body of evidence swiftly gathered favoring a complex free-radical mechanism for hydrocarbon oxidation and autoignition. This was critically reviewed by Jost[73] and compared with engine data. The similarities were impressive despite the higher temperatures and pressures of the engine and the much shorter delays.

The engine is in fact a much more convenient apparatus for the study of autoignition using optical, sampling, and additive response methods.

The relationship between RON, MON, and the molecular structure of hydrocarbons is a complex topic beyond the scope of this chapter. The purification and rating as well as the determination of blending octane numbers and lead susceptibilities of hydrocarbons has been a long-term project of the American Petroleum Institute.[89] The points of attach of free radicals on the molecules has been discussed in relation to the numbers of primary, secondary, and tertiary hydrogen atoms, with the ease of reactions declining in the same order.[90,91]

A correlation of octane numbers and total chemical structure has been presented by Boord.[92] It is an interesting idea but must await a more modern kinetic treatment.

D. Engine End Gas and Autoignition Tests

1. *End-Gas Sampling for Preflame Products*

Many workers believed that because of their instability organic peroxides must be associated with knocking reactions and therefore might provide a marker for preflame processes. Egerton used a mechanical sampling valve to obtain end-gas indication of peroxides.[79] An electromagnetic valve and improved analytical techniques allowed peroxide (Fig. 8) and aldehyde analyses to be made under

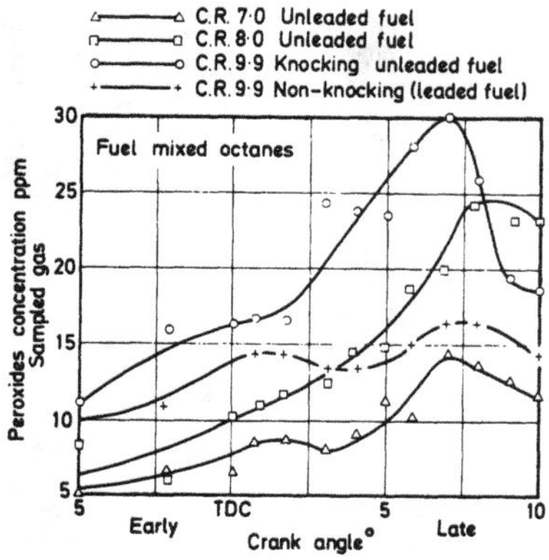

FIGURE 8. Curves of peroxide concentration in the end gas of Ricardo E6 engine.

severe end-gas conditions using an iso-paraffinic fuel.[93] The curves show clearly a two-stage nature (e.g., Fig. 7) that peroxides increase as the knock limit is approached, and that lead affects the peroxides in the second stage more than in the first.

Organic peroxides were only present in traces in the samples, almost all of the peroxide was hydrogen peroxide. Similarly formaldehyde proved to be the main aldehydic product.

Sampling tests on single-stage ignition, high-temperature fuels such as benzene and toluene gave no detectable peroxide.

Cyclic irregularity of combustion is a severe obstacle to sampling tests as the important stage only occupies a 15° crank angle.[94] However, Davis *et al.* by analyzing samples showed a close correspondence between end gases in fired engines and preflame processes in motored engines.[95]

Single-cycle "one-shot" sampling is very necessary for progress to be made. This has been achieved,[96] but further work is called for to allow end-gas compositional data to be gathered for kinetic studies.

2. *Additive Tests in Fired Engines*

Clues to the chemistry of end-gas autoignition may be obtained by observing the differential effects of additives on knock with a variety of fuels.[93] Additives may have several different effects on the prereactions of autoignition. The addition of organic radicals is proknock, adding to the chain carriers already present; thus diazomethane, which gives methylene radicals on heating, and mercury diethyl, which gives ethyl radicals, are proknocks.[91] Large molecules such as iodine are antiknock because they act as chain breakers in the gaseous phase, while other large molecules which exist as multivalency oxides of low volatility such as lead, iron, nickel, manganese, and tin are antiknocks.

TEL can, however, be a proknock in fuels that have no two-stage ignition such as benzene and cyclopentadiene because of the production of ethyl radicals.[91]

Aniline and its derivatives such as *N*-methyl aniline are good antiknocks due, it is believed, to a reaction with the chains in the first stage up to the cool flame. With "high-temperature, single-stage" fuels, it becomes a proknock.[93] A comprehensive critique of the knocking effects of ashless organic molecules has been produced by Shell.[97]

After a study of additive effects on "low-" and "high-temperature" ignition, Walsh[93,98] chose formaldehyde and nitrogen dioxide as the two most significant indicators of the ignition modes of hydrocarbons. Formaldehyde is a proknock in "high-temperature" fuels as is nitrogen dioxide. In two-stage ignition formaldehyde increases the delay to the cool flame and is a slight overall antiknock, whereas nitrogen dioxide is a mild chain breaker in the first stage but a prooxidant in the second, ending up as an overall slight proknock, Fig. 9.

The effect of these additives on mainly paraffinic or naphthenic motor fuels is in accord with their effects on two-stage ignition in the "low-temperature" region in laboratory apparatus, showing a close parallel between the chemistry of autoignition and of knock.[99]

3. *Motored Engines and Cool Flames*

It is relatively easy to demonstrate preflame reactions in motored engines by pressure indicators[100] and to show its two-stage nature.[101] Thinwise resistance thermometry can also be used to follow the exothermicity of the preflame reactions.[96]

The first visual observation of cool flames in motored engines was by Peletier *et al.* in 1939.[102] Post–World War II photomultipliers allowed cool-flame curves to be plotted accurately.[103] Figure 10 shows the limit curves, for isooctane fuel, of cool flames and bright blue flames in a Ricardo E6 engine. These luminosities are both emitted by excited formaldehyde.[104] Spectroscopic examination of cool flames from motored engines has identified the bands of formaldehyde fluorescence.[103,105]

The flame front compression lacking in motoring experiments is made up

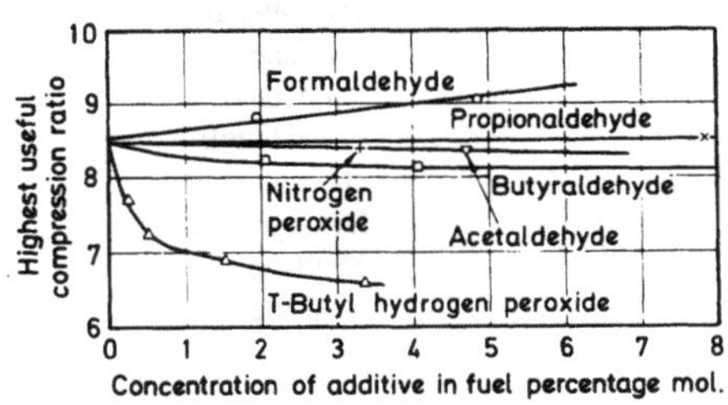

FIGURE 9. Effect of various additives on knock with paraffinic fuel.

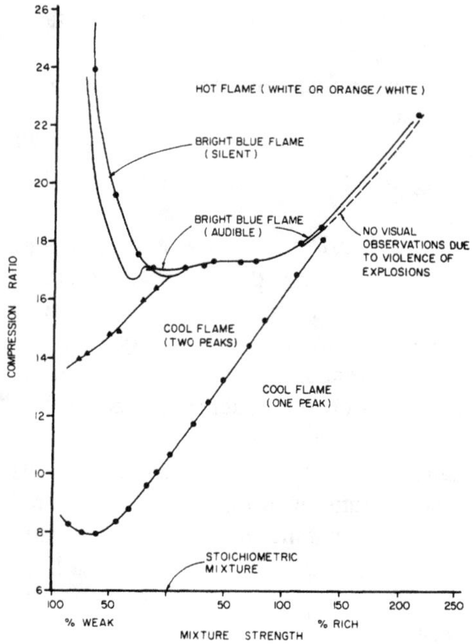

FIGURE 10. Flame characteristics of iso-octane in a motored engine.

for by a higher compression ratio such that the autoignition of isooctane occurs at seven ratios higher than knock in the same engine fired.

The cool-flame limit curves for olefins are at higher compression ratios than the equivalent paraffins and their cool-flame intensities are much lower. Some typical intensities from the E6 engine are shown in Table 1. These and other

TABLE 1. Typical Intensities from the E6 Engine

Fuel	Max. cool-flame intensity at hot-flame limit (arbitrary units)
Heptane	33
Heptene	8
Hexane	17.5
cyclo-hexane	13
Triptane	5
iso-octane	15
Di-isobutylene	0.3
Diethyl ether	30–65
Di-isopropyl ether	24
Methane	0
Benzene	0
Cumene	1.0

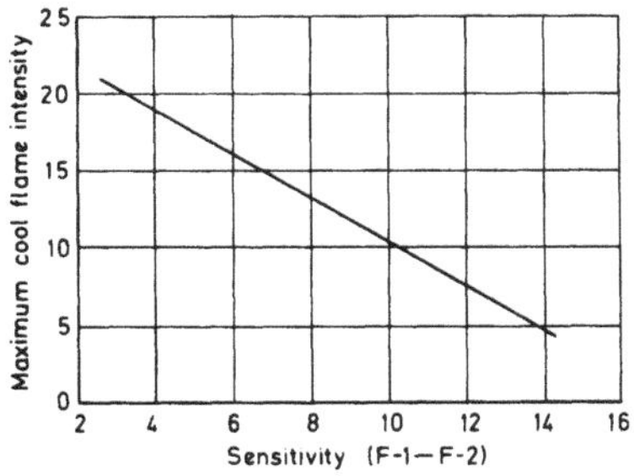

FIGURE 11. Relationship between cool-flame intensity and temperature sensitivity for a range of motor gasolines.

results led to the testing of a range of motor fuels of varying temperature sensitivities [RON (Fl)–MON (F2)] for maximum cool-flame intensity at the hot-flame limit[(103)] (Fig. 11). The good correlation suggests that sensitive fuels burn rather more by the "high"-temperature process than the "low." However, these two categories are only to be regarded as roughly descriptive, and the cool flame is just one easily observable marker in what is an extraordinarily complex kinetic problem.

If the temperature–pressure conditions for autoignition in a motored engine is compared to the pressure-calculated temperature conditions in the engine when fired, there is good correlation, and it is the basis for a calculated autoignition octane number (RON).[(106,99)]

Gas chromatographic analysis of products from cool-flame processes in static, flow, and motored engine tests on *n*-heptane fuel show a basic similarity.[(107)] Wide differences in conditions only slightly affected the oxidation chemistry. Fired engine data has been reported to give similar results,[(96)] which are in line with proposed schemes for *n*-heptane oxidation.[(108)] In the present stage of technology the rapid-compression machine offers the best means for studying the autoignition process. Pressures, temperatures, and time delays can be obtained readily for pure or blended fuels,[(109)] the data can be used as a basis for modeling the autoignition process with a view to elucidating the problem of why certain olefins and aromatics autoignite so much more readily (with a rise in temperature) than do paraffins.[(110)]

E. Antiknock Action

1. General Considerations

Since the discovery of TEL in 1921 by Thomas Midgley,[(38,111)] a vast effort has been expended to try to find a substitute, preferably ashless. Only the anilines looked possible to fit the ashless criterion, and they have an antiknock effect 1% of that of TEL.[(93)] An exhaustive research by the Shell Company[(97)] finally ended the "ashless" quest.

Some organic substances, such as tertiarybutyl acetate do, however, potentially effect TEL and are called co-antiknocks.[112]

The search for organic–metallic antiknocks was a little more successful and, as Egerton had predicted, they are all derivatives of metals with several oxides.[79]

Tetramethyllead was investigated concurrently with TEL but was turned down because of its toxicity and volatility; its volatility brought it back into favor to solve the manifold fractionation problem, which was severe with the less volatile TEL.

Various metal carbonyls were tested and iron pentacarbonyl, as well as nickel tetracarbonyl, showed a similar antiknock effect to TEL but were inadmissible for stability and toxicity reasons. Also their oxides, being abrasive, gave rapid engine wear. The last objection also condemned ferrocene, or dicyclopentadienyl iron—a very effective antiknock.

The only useful antiknock that emerged from this work was a manganese compound, methylcyclopentadienyl manganese tricarbonyl, code named AK33X, now called MMT.[113] It is a good antiknock in its own right but operates best as a supplemental additive to TEL, especially in paraffinic fuels.[114]

The provisions of the Clean Air Act of 1970 brought noble-metal catalyst technology in exhaust systems that would not operate with leaded gasolines. It has been hoped that MMT could be used to replace some of the octane numbers which had been lost with the elimination of TEL. Manganese does not have the toxic effects of lead in exhaust gases, and the body has a high tolerance to it, owing to its widespread presence in the earth's crust.[115] Unfortunately, despite manganese oxides being mild oxidation catalysts, they tend to impact on catalysts and plug the passages.[116] For this reason MMT was banned from unleaded gasoline in the United States.[117]

The situation in Europe is quite different. Lead is still in widespread use, although Germany has restricted its concentration to 0.15 g/1, and Britain will follow by 1985. The reason is not based on exhaust emissions but on alleged environmental impact.[118]

The economic penalties of unleaded petrol have received much attention. A condensed view of the U.S. and German situation has been given by Dartnell.[119]

Filtration techniques to reduce the exhaust lead emissions have been developed, and lead catchment efficiencies of 50% (70% in urban driving) have been achieved.[120]

Noble-metal catalyst technology has been developed to operate on the exhaust gases from gasolines with 0.15 g/1 lead. Halogen–lead scavengers (chlorine, bromine) play a part in catalyst poisoning and might be eliminated at low lead levels at least in high-duty European engines.[121,122]

Undoubtedly lead will continue to be used in many low population density areas of the world, and in the hydrocarbon starved future it may need to be resorted to in a widespread fashion, although probably lead resource conservation is a more cogent argument against its use than environmental impact.

2. Theories of the Antiknock Action of Lead

Two main theories exist to explain the extraordinary activity of lead. Both agree that lead monoxide is the active form of lead but disagree as to its state at the time of its action. Norrish, using kinetic (flash) spectroscopy, showed the strong presence of PbO molecules at the time of a simulated antiknock action and argued strongly for vapor phase lead oxide inhibition.[123,124] Walsh, on the other hand, on the basis of engine tests and on laboratory work with oxidation inhibitions using solid lead oxide favored a PbO "fog" theory.[125]

Although under knocking conditions an aerosol of PbO must coexist with molecular PbO, the "fog" theory is most attractive and has gained most support.[114]

That lead particles exist as a cloud of 100-Å particles in the knock zone was first propounded in 1940.[126] In 1961 Downs *et al.* showed by Tyndall scattering in a motored engine that a fog was present at TDC,[127] and 100-Å particles were sampled from both a motored and a fired engine.[128] Modern coagulation theory predicts ample time for lead aerosols to form before knock occurs.[129]

The antiknock action must probably be the surface catalytic deactivation of active species such as the HO_2 radical.[125] Mixed oxide "promotion" is a possibility among all oxide catalysts, but other than the small effect of manganese,[114] true promotion has not been observed. Walsh considered the PbO to be a nonstoichiometric defect structure, and self-promoting active red-lead monoxide has been sampled from engines running on fuel mixtures of maximum lead susceptibility,[114] although all the lead monoxides are probably antiknocks. In contradistinction with the thermally unstable iron pentacarbonyl, lead has little effect on two-stage ignition until after the cool flame.[130,103] It was believed that the cool-flame reactions decomposed the TEL to PbO, but the evidence was inconclusive.[127] It is likely that the preflame oxygenates affect the PbO particulate structure.[114]

The indications that the effectiveness of lead is related to the crystal structure of the lead oxide is in itself an argument against the molecules PbO theory.[114] Fuels with low or zero cool-flame activity such as methanol, butyl alcohol, and benzene have a low or even negative lead susceptibility in a fixed engine; whereas iron pentacarbonyl is a much more effective antiknock. Both additives must be fully decomposed at the time of knock; thus the explanation might lie between the inefficiency of chain breaking with the short-chain processes in high-temperature oxidation and the prooxidant effect of the ethyl fragments of TEL.[130,91]

Interestingly TEL, when aspirated into a fuel-free motored engine, gives a good cool flame and readily autoignites at 8 : 1 compression ratio.

F. Summary

There is no evidence at all for a classical long-tube detonation in modern engines, nor is there the certainty that end-gas autoignition on its own can cause knock. The quiet spluttering autoignition in motored engines seems to confirm this, although these are usually very late in timing. If compression ratio is raised to bring autoignition towards TDC, then the noise becomes violent and knocklike. Engine autoignition is identical to the laboratory version, but at higher temperatures and pressures with shorter time intervals.

Antiknock action is entirely explicable by the action of lead oxide surfaces on autoignition with the possibility that PbO vapor plays a minor part. The theory propounded by Curry[29] that the higher than average subsonic flame speeds can cause knock is undoubtedly true because viewed from the end-gas region they will appear in timing to be from an advanced spark, but his assertion that autoignition is unnecessary does not appear likely.

The end gas has been seen to autoignite in several places simultaneously by Male[68] and by Withrow and Rassweiler.[69] This prompted the suggestion of Echols *et al.*[131] that knock is a detonation wave in the end gas caused by a cascade process of minor autoignitions. This may be true (it would be very difficult to track the explosion by ion gaps if indeed it occurs this way), but in any case the end gas can be likened to a thin arc of compressed gaseous explosive over about 90° of cylinder circumference. It is in the form of a "shaped explosive charge" with ample energy to promote a multiply reflecting shock wave.

It would be naive to assert the truth of either theory *in toto*. Rather like Miller,[132] it is better at this stage to assert the partial truth of each.

IV. RUNNING-ON OR AFTER-RUNNING

After-running is a condition in which an engine continues idling in a sporadic and rough fashion after the ignition has been switched off and the throttle closed. It is a familiar condition in many emission-treated engines which have an excessive air intake heating and a high idle speed. Running-on (a term often used in Europe) has been investigated in a single-cylinder engine showing that only about one in five cycles fired and that very high rates of pressure rise occurred, presumably before TDC.[133] This explains the extreme engine roughness and the crank reversals sometimes seen; these could break crankshafts in early engines.

To differentiate after-run from preignition, fuel blending tests were carried out which showed a lack of correlation of after-run tendency with preignition rating, but a fair correlation with MON.[134]

Methanol, benzene (preignition prone fuels), and TEL reduce after-running, while diethyl ether and ditertiarybutyl peroxide promoted it.[134]

The correlation with autoignition is striking, although the delay of after-run is longer. The mixture is mostly free of exhaust residuals but rich in preflame reaction residuals, and temperatures are high in relation to the long delays. When the after-running exhaust is sampled, aldehydes and peroxides are found in high concentration.[134,135] Benson, whose paper[135] is a mine of information states that NO levels in his engine were five times that in normal idle exhaust, possibly a result of the high temperatures induced by the early firing. He found a correlation with RON and not with MON or fuel sensitivity, and showed that high idle speed, leaner mixture, and retarded spark gave increased after-running. A later view of the problem in relation to octane numbers and fuel hydrocarbons is given by Ingamells.[136]

To stop this objectionable phenomenon, several options exist:

1. Stall the engine using the clutch and parking brake with the manual transmission in top gear.
2. Pump the accelerator pedal to richen the mixture.
3. Use the manual choke.

None of these are recommended on the safety grounds of transmission damage or failure to restart. The best preventative is a reduction of idle speed on switching off the ignition using a solenoid to close the throttle plate, or to close the idle air supply.

After-run has some links with the hot-starting problems of autoignitions which can occur early on the compression stroke and cause stalling of, and damage to, the starter.[137]

There is some evidence that hot spots may cause after-running in low-compression engines,[138] but it has been shown that removal of combustion chamber deposits does not affect running-on.[139]

The future of the after-run problem is unpredictable with higher compression, leaner running, economy engines, and uncertain emissions laws.

V. PREIGNITION AND POSTIGNITION

A. General Remarks

Any process which effectively advances combustion from that giving best torque will cause high heat rejection because of its negative work. High heat rejection causes overheated engine parts, which have the effect of advancing combustion still further until components melt. In engines of low thermal capacity such as aircraft piston engines the destruction may be complete in seconds rather than minutes.[140] Knock in these engines gives a much smaller temperature rise,[141] which is reflected in the fact that knock damage does not involve metal heat treatment effects, whereas preignition damage does.[36]

The parts most likely to cause preignition are those least well cooled; spark plugs, exhaust valves, metallic asperities (edges of head cavities, or piston bowls), and the supplementary thermal insulation effect on these of deposits. Exhaust valve heads have been implicated in small European engines, especially if carrying calcined deposits of lubricating oil additives such as calcium and barium compounds.[28] Cooler running exhaust valves and improved lubricating oil consumption are the best remedies. Siting the valve in a compromise position between the spark and the end gas helps to avoid both the highest flame temperature region at the spark plug and the effect of the hot valve head on the end gas.[28]

Engine-build features to minimize preignition include correct heat range spark plugs, radiused metal edges, removal of asperities, and well-cooled exhaust valves, seats, and guides, with sodium-cooled valves an extreme option.

Preignition research is rendered difficult because it is very difficult to duplicate multicylinder thermal conditions in a single cylinder. In a multicylinder, a preigniting cylinder may be driven to destruction by the other (normally combusting) cylinders, whereas the same level of preignition in a single cylinder quietly stops the engine. Some multis are both preignition and knock limited.[142]

B. Effect of Engine-Operating Parameters

An immensely complicated interplay of factors determines how a hot spot arrives at its preignition temperature. Downs,[32,49] using two widely differing single-cylinder engine procedures (Sec. II.C.3), was able to obtain many useful trends for preignition with benzene as fuel. Cyclohexane was used as preignition rating = 0, and isooctane = 100.

Maximum preignition tendency on all three fuels was close to 10% richer than stoichiometric, just leaner than for maximum power.

Engine speed, owing to the effect on turbulence, flame speed, and volumetric efficiency has a complex effect,[32] but for many naturally aspirated engines preignition tendency increases with speed up to maximum power.[22,143] However, if the preigniter has a high heat capacity (exhaust valve), then a sudden drop of say 1000 rpm may bring on preignition as the time lag is increased and the preignition temperature remains high due to thermal lag.[23]

Increasing compression ratio by increasing heat input to a hot spot may promote preignition but if the hot spot is the exhaust valve, then the cooler exhaust gas may reduce the valve temperature. In general, though, deposit-induced preignition increases with compression ratio.[143]

Ignition timing advance, beyond that for MBT, hardly affected the hot-spot preignition temperature but raised the normal running temperature. When these temperatures coincide, preignition occurs so any closing of this temperature gap increases preignition tendency.

Significant changes in inlet air temperature and in coolant temperature had little effect on preignition,(49) the wire coil preigniter not being directly affected. This result may not hold for a large area, thermally close-coupled, deposit-type preigniter.

C. Preigniter Surface Effects

Preignition tests are notoriously difficult to reproduce because of changes in the surface quality and shape of the hot spot. Metal oxidation and erosion, deposit buildup and break-away as well as catalytic changes all make results unpredictable.

Tests using low heat factor thermocouple spark plugs to cause preignition in a variable compression engine had to be terminated because the preignition-limited compression ratios drifted upwards steadily with deposit buildup.(32) Spark plug central electrodes with a platinum tip ran 100 °C hotter than with a nickel–silicon tip in lean mixtures; in richer mixtures they ran at the same temperature or even lower, suggesting catalytic effects. Normal running preigniter temperatures of pure fuels do not appear to run parallel to their flame temperatures also suggesting catalysis on the preigniter surface.(32)

Higher surface ignition temperatures have been observed with methane fuel with platinum rather than base metals.(144,145) In rating fuels there is a choice between changing an engine parameter (CR or boost) to bring the preigniter normal running temperature above the preignition temperature, or to run with a cool preigniter under fixed engine conditions and add heat to it either by reducing the cooling or electrically.

The first choice runs the risk of thermal or catalytic changes due to the high normal running temperatures, the second of deposit buildup under cooler conditions. Downs,(32,49) after testing each, chose an air-cooled Nimonic thimble for boosted engines and a stiff two-part wire thermocouple coil (which was electrically heated) for naturally aspirated tests. However, the choice of wire material was important, an alloy (with 40% nickel) gave anomalous temperatures due to catalysis by nickel oxide. Kanthal Al was the only really satisfactory wire.(49)

Arrigoni used the same preigniter at a fixed temperature, assessing preignition by statistically treating pressure diagrams.(50)

There was a tendency for the larger thimble preigniter to give a preignition temperature lower than the wire coil, due either to thermal lag or surface area effects.

Catalytic "preignition" has been used in a low-compression diesel-type engine (the ignition timing is controlled by the injection timing). Platinum metal wires and platinum-coated base metal wires ignite fuels readily.[146]

Preignition from deposits built up under urban driving using leaded fuels was a major problem in the post–World War II years. The deposits consisted of lead salts and carbonaceous material which could shake loose, overheat, and glow at temperatures 350 °C. The cure was to add an organic phosphate to the fuel. Lead phosphate did not promote carbon combustion as did lead bromide.[147,148]

The European preignition problem has been ascribed to lubricating-oil ash deposits on exhaust valve heads. Lead compounds are less important, but calcium and barium salts have a large effect.

Oil consumption control is a preferred palliative, but in the long term additive metals may need revision.[28]

D. Preignition Tendency of Fuels

The temperature of a hot spot is a very complex function of the heat received from each combustion cycle and the heat lost in all four cycles. The shape, size, material, and conductive path of the hot spot affects the heat balance, as do the flame temperature, its speed, and the turbulence intensity in the region of the hot spot. The position of the hot spot in the flame path is also important.

An attempt to weigh some of these factors was made by Guibet[28] using a range of pure fuels. Flame temperatures and speeds, when compared with the observed hot-spot normal running temperatures, show that the flame speed is of more importance than the temperature. A high flame temperature and speed, such as with benzene and cumene, give a high preignition tendency, but the relationship is not strict because toluene, which is very preignition resistant, also has a moderately high flame temperature and speed. The removal of the hot spot down the flame path can effect a lowering of its temperature by up to 100%. Guibet concluded that a significant lowering of the preignition tendency of motor fuel was ruled out economically.

Downs[32,49] rated a large number of fuels for preignition and knock, finding no relationship in agreement with other workers.[43] (See Table 2.)

TEL improved the preignition resistance of nearly all fuels, assuming that tests were short and did not involve deposits or chemical modification of the surfaces. The effect of lead was, however, considerably less than it would have been on the octane ratings of the fuels.

A number of available gasoline components (pure hydrocarbons) were tested revealing that the olefins in general had the lowest resistance to preignition. On

TABLE 2. Preignition and Knock Ratings for Various Fuels

Fuel	Preignition ratings		Knock rating (E6 HUCR)[a]
	E6	"Dagger"	
Iso-octane	100	100	11
Cumene	0	0	12.5
Benzene	31	9	14.6
Methanol	0	0	15.0
Toluene	91	120	15.0
100/130 grade aviation	77	87	12.4
Cyclohexane	0	0	8.2
80 octane motor fuel	32		8.5

[a] HUCR—highest useful compression ratio.

the cyclohexane–isooctane scale, hexene-2 was −26 and *m*-xylene 131. When some of these pure fuels were blended into a mixed fuel, the resulting preignition number could be calculated from the values for the components by simple proportion.[(49)]

Viewing these results in the light of commercial fuel analyses to some extent bears out the contention[(28)] that fuel modification could only play a minor part in reducing preignition. However the observation was made that the more temperature-sensitive fuels, which autoignite more by the high-temperature single-stage process and which have weak or no cool flames, are generally the most preignition prone.[(134)]

Using the same "dopes" as were used on the autoignition process (Section III.D.2) on a preigniting engine gave some support to the theory,[(32,49)] which is discussed broadly in the content of a general theory of engine combustion.[(134)]

The large number of anomalies in the preignition behavior of fuels make any theory somewhat tenuous; thus how is the fact to be explained that benzene and ethyl benzene are highly preignition prone, while *o*-, *m*-, and *p*-xylene are not?

With the advent of high-performance thermally intensive engines, using more preignition prone fuels (aromatics, olefins, methanol) and the appearance of high-speed knock as a major problem, it may be the time to reconsider the whole area of fuel testing for abnormal combustion and seek some unification of procedures which relate more to modern problems.

E. Practical Engine Systems for Knock Reduction

1. System Constraints

Before the advent of tight NO_x control a popular emission control strategy was the use of lean running (lean burn) epitomized by the Chrysler clean air package of a lean well-prepared mixture and electronic spark advance control.[149]

During this phase a great amount of useful work on lean running was accumulated purely to allow low emissions levels. CO and NO_x could be reduced reasonably by lean burn but HC levels remained a severe problem.

The system was abandoned in the United States but more recently has evinced interest in Europe, partly for emission reasons and partly for economy in high-compression, small-capacity engines.[5]

Requirements for low-knock, high-compression engines may thus run counter to emissions standards, and whereas much of the early work used premium fuels any strategy for attaining optimum compression ratios in engines must now take into account the fuels available to them. If indeed a fuel of 105 RON was economic, then the problems would be minimal and fuel consumption at its lowest in terms of volume or weight. However, the vital fuel parameter is cost, which is related to the efficient use of crude oil, has been demonstrated by Tierrey *et al.* who argued that with the technology existing in 1975 the premixed charge spark-ignition engine with its high octane requirement is the least efficient user of crude oil.[150] The Shell Company, treating the engine and the refinery as one system, concluded that the optimum RONs are 96–97 at 0.4 g/l lead, and 92–93 in an unleaded situation.[151]

A later European view from the CONCAWE group ROFIT reached similar conclusions. Taking the base case as 96 RON with 0.4 g/l lead, the optimum with 0.15 g/l lead is 95 RON at a 2% crude oil wastage, the optimum unleaded fuel is 92 RON with a 5% crude penalty.[152]

The exhaust emission requirements of the Clean Air Act in the United States need unleaded fuel to allow noble-metal catalyst technology to be used. 91 RON unleaded gasoline became available at a small cost penalty but caused much marginal knock in the emissions treated cars. It resulted in some demand for higher RONs, although some folks now tolerate slight knock more than in the past.[153] Perhaps designers have been too afraid to approach the knock limit in the absence of any real knowledge of engine damage and of modern soundproofing of the engine compartment.

In the absence of antiknock additives and of more efficient refinery processes, 93 RON gasoline would appear to be indicated in the remaining years of the petroleum era. When coal-derived highly aromatic fuels become available, then 100 RON may again be possible, while if coal-derived methanol comes on the market, 100–110 RON may be seen depending on the necessary additive package (e.g., 5% of pentanes for starting).

It goes without saying that any big advance in engine technology will need a recalculation of the engine–refinery–crude oil system with a new optimum fuel quality.

2. Air–Fuel Mixture Preparation

It is vital in an engine running at the knock limit that the air–fuel ratio should be exactly the same in each cylinder. With carburetion and manifold distribution of the mixture, it is almost impossible to ensure that the liquid and vapors proportionate exactly.[154,155] Indeed it is often found that the air distribution itself is defective.[156] Maldistribution may cause knock in a lean running engine by bringing the air–fuel ratio in one or more cylinders down to maximum knock mixture strength. Also TEL (boiling point 200 °C) and the knock resistant fuel fractions (boiling from 100–200 °C) are partially separated from the lighter hydrocarbons (boiling from 30–100 °C) which have a lower RON called RON 100 °C or front-end octane number.[157] Under accelerating conditions, cylinders may get mostly "front-end" gasoline and knock heavily. Tetramethyllead, boiling at 110 °C, improves the situation considerably.

In principle fuel injection behind the inlet valve or into the cylinder should effect a complete cure; but pump tolerances can be quite high,[155] and it is not universally accepted as the best solution to the problem.[158] One difficulty in researching air–fuel mixtures in the past was the lack of rapid-response instruments for monitoring exhaust gases and giving direct readouts of air–fuel ratios. This deficiency has now been rectified.[159] Orthodox carburetors and simple induction systems will not allow regular combustion beyond an air–fuel ratio of 17:1.[160] To reach ratios of 22:1 it is necessary to use fully vaporized or highly atomized fuel. Numerous systems exist for this purpose, the most popular being types of sonic shock carburetors[161] represented by the Dresserator.[162] In these devices the distorted flows from orthodox throttle plates are avoided, the latter being incorporated into the sonic flow producer.

Comparison of various mixture preparation schemes have shown the sonic carburetor to give the best charge mixing.[163] It is possible to use the engine inlet valve as a sonic throttle by controlling its lift from the throttle pedal.[164]

Another popular method of attaining a homogeneous charge is to sophisticate the "hot-spot" inlet manifold. An early version was the Ethyl Corporation "Hot Box."[165] Another interesting idea is the use of a heat pipe to convey exhaust heat to a heat exchanger between the carburetor and manifold.[166]

A useful device for producing a continuous supply of homogeneous charge to an engine or to a vehicle on the dynamometer has been described.[167] A version of it is shown in Fig. 12 using hot oil to flash evaporate gasoline to such a temperature that it either recondenses into an aerosol (droplets <10 μm) or stays vaporized, depending on the rate of heat input. This equipment will indicate the true lean limit of the engine, the attainable indicated specific fuel consumption (ISFC), and its emissions potential.

Despite the body of evidence that a homogeneous charge extends lean limits it has been shown that in an engine with a quiescent chamber of relatively low compression ratio (7–8:1) a poorly atomized mixture gave an extended lean limit compared to a homogeneous mixture.[168] This result may be due to accidental stratification. However, there is some evidence from research into kerosene fuel sprays that the widest ignition limits are given by a droplet dispersion of small size and not by fully vaporized fuel.[169]

Charge preparation clearly needs further research in the context of high-compression low-knock engines. Front end octane problems are one feature of stratification, another more favorable one might be where the high RON "back end" is persuaded to occupy the end-gas position in a stagnant chamber.

3. *Ignition*

With the advent of electronic ignition systems, contactless triggering, and set-for-life spark timing, many of the problems associated with faulty initiation of combustion have disappeared. Spark plugs have been optimized for lean running with extended noses and wider gaps.[170,171] Multiple spark plugs increase burn rate and reduce flame travel so they promote economy and reduced knock. Up to five plugs have been used in emissions tests,[172] but in general only aircraft engines use two plugs. Sequential sparking is only used in stratified charge engines.

The advent of external "bolt-on" knock detectors of the accelerometer type has allowed systems to be developed which retard the spark by a feedback signal

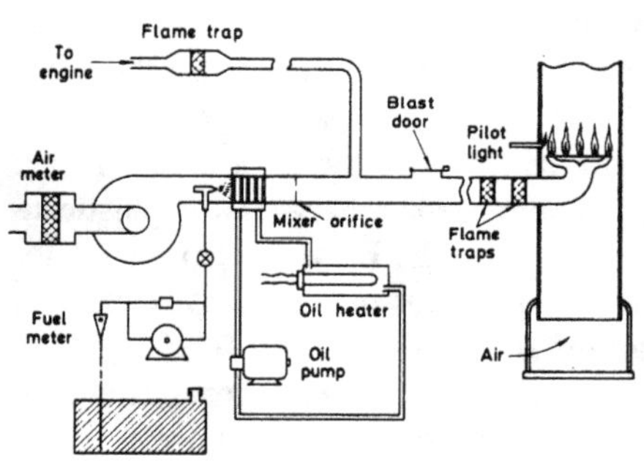

FIGURE 12. Perfect mixture generator.

at the onset of trace knock. They can allow compression ratios 1 unit higher, reduce the RON requirement by up to 10 units, and permit a saving in fuel of up to 6%, but depend on the driver's acceptance of some trace knock.[173–175]

It is not possible, as yet, to control compression ratio by a feedback signal, but the nearest approach is to control the waste-gate of a turbocharged unit so as to reduce the cylinder charge at the onset of knock. By this means the cylinder geometry can remain at an optimum while the knock-limited charge density is varied.[176]

Passive control systems, which have the engine data stored in a memory, are popular but do not have the ability to compensate for fuel quality changes. They can, however, be integrated with complex emission controls.[177]

4. Octane Requirement Increase (ORI)

Engines in typical user operation generally need a higher octane fuel to remain knock-free as miles accumulate. Thus after 10,000 miles of mixed city and freeway driving the ORI might be from 3–7 numbers. The new engine or initial octane requirement (IOR) is usually aimed at about 3–5 numbers below the specified fuel to compensate for this—a gross waste of gasoline nationally. The ORI is caused by many factors; partly deposit volume and quality, partly changes in heat transfer, along with some unknown engine factors.[178] Another very important factor is the effect of high oil consumption on deposits.

In the United States it has been reported that ORI can be higher with unleaded gasoline than with leaded,[179] whereas the experience in Europe is that there is little difference, presumably because higher engine duties there tend to burn off the carbonaceous residues of unleaded fuel.[157] Engines requiring a high IOR generally give a lower ORI than engines of low IOR. This inverse relationship holds in the United States and in Europe. It is related to the observation that high-compression, highly loaded engines give low ORIs.[180,157]

Engines operating near to, or at, the knock limit during normal driving tend to keep free from excessive deposits and give only a small ORI. Stagnant, low-turbulence chambers give high ORIs.

Production tolerance in combustion chambers allow quite a wide scatter in compression ratios, and hence in IOR engines.[178,157] Water passages also vary, allowing high temperatures in some end-gas zones. Chamber surface quality also plays some part. It is to be hoped that engine builders using experiences gained in producing diesel engines (where these factors are critical) will improve tolerances, for the cost to the purchaser in extra fuel of a needlessly low compression ratio would more than pay for the extra care in production. Betts has used the phrase "octane waste" and states the average European car wastes 3 O.N.[181]

5. Charge Motion and Combustion Chamber Design

The modern engine designer has some advantages over his counterpart in the twenties in better materials and in vastly better fuels and oils. However, in the department of combustion chamber design he still has to work with qualitative concepts of charge motion and the subsequent flame propagation pattern.

Vigorous research using high-speed cinematography,[(182)] hot-wire anemometry,[(183)] and laser doppler velocimetry (LDV)[(184,185)] is leading to a more complete understanding of the air motion in direct-injection diesel engines. This knowledge has spin-offs applicable to SI engines despite their higher speed and generally smaller size. Three main types of charge motion are important: swirl, turbulence, and squish.

1. *Swirl.* Cylinders being by definition cylindrical and inlet valves almost always offset from the axis, it is understandable that the simplest component of gas motion during the induction stroke is swirl. Its presence can be demonstrated by simple paddle wheels in motored engines, and it can be roughly averaged by counting the paddle revolutions which gives a value about one-half of the realistic figure. It is usually stated in terms of swirl ratio = charge rev./engine rev.

Scott,[(182)] using high-speed cinematography in his simulated diesel engine, measured swirl under firing conditions by tracking specific parts of the flame structure. Swirl cannot be regarded as a simple flywheeling motion as it appears to be from vane experiments. During the induction stroke its speed varies with the air velocity through the port and this gives a stratification or layering of differing swirl zones, which remain until compression at TDC. This causes them to mix and degrade into turbulence on the surfaces and discontinuites of the head.[(183)] Swirl is rarely axisymmetric in cylinders, being generally observed by LDV as an off-center vortex.

Excessive swirl leads to pumping and heat losses causing an increase in brake-specific fuel consumption (BSFC);[(19)] it also can create a charge velocity at the spark plug position which is too high and causes quenching. Ideally velocity should degrade to small scale or micro-turbulence at the plug before the spark passes.[(186,187)]

2. *Turbulence*. This is an extremely complex topic which, in this field, is only on the threshold of revealing its secrets.[(188)] Until recently designers have visualized it in terms of fluctuations of gas velocity but had no measure of it, although schlieren photography had considerably aided the imagination.[(189)]

Hot-wire anemometry has greatly advanced knowledge of turbulence in motored engines, both SI and diesel, but leaves the fired engine blank.[(186,190)]

Laser velocimetry can operate in flaming gases, but the windows necessary—one in the case of the insensitive back-scatter mode and two in the sensitive forward-scatter mode—cannot generally be fitted into a modern high-compression engine of practical design.

The idea that the degree of turbulence and its structural scale might need to be altered throughout combustion had occurred to Ricardo when he designed a later version of the "turbulent head" called the Shock Absorber head. Induction turbulence caused irregular flame initiation at the spark plug, which was controlled by thinning down the gas space under the plug until he obtained a steadier flame propagation.[191]

Using a rapid-compression machine, Matekunas[192] has shown that only with a stagnant chamber did the flame kernel remain within the spark gap and quench. With very high turbulence, particularly in lean mixtures, the kernel would commence to propagate but then die away. Even with optimum turbulence approximately 30% of the combustion period is devoted to the formation of the flame front. Clearly this vital phase is ill understood, as is the role of turbulence (if any) on the layer of end gas.

3. *Squish.* This is the name given to the action of charge expulsion from a closing piston head gap near TDC. Usually with a bowl-in piston or bath tube-in-head design, the squish is inwardly radial, but its magnitude measures less than expected.[193,194] Some engineers complain that squish, being by definition timed at TDC, comes too late and too weak to have a major influence. An ingenious Russian twin-piston design, the Kushul, allows turbulence to be generated before TDC in the main cylinder by phasing the cranks slightly such that the squish piston is in advance.[195]

Squish is a vital factor in L-head engines (Sec. E.4) being more readily controlled from the large closing area of the piston and head. However, designers consider that squish is a powerful influence in many commercial engines of overhead valve configuration because it promotes rapid flame travel well after the spark, thus allowing the flame kernel to propagate without excessive turbulence quenching.

Probably more important than the squish is the acceleration of charge swirl by simple conservation of momentum when it is forced into a chamber of smaller radius at the same time as squish occurs.

6. *Some Practical High-Efficiency Systems*

The choice of head design depends heavily on the duty required, on the cost, and on the suitability for automatic production.

For racing and exotic sports vehicles, the four-valve pent roof design, with central spark plug, has long been a favorite.[55] It has a high volumetric efficiency, a short flame travel, internally developed turbulence by mixing of the two inlet streams, and less need for squish. It is unsuitable for very high compression ratios as the combustion space becomes too thin, with flame quenching and high hydrocarbon emissions. It is, on the other hand, ideal for turbocharging where the compression ratio would be kept down to approximately 9:1.

If this engine is required for very high power, then the cylinder bore is made greater than the piston stroke (oversquare configuration). This allows a large valve area, high volumetric efficiency design with high engine speed, and modest piston speed. In this form it has an even more exaggerated Centicular shape leading to high hydrocarbon emissions and a longer flame travel.[196]

For a less costly solution the hemispherical (hemi-) head, two-valve design has much to offer. Heat losses are low as it has a low surface-to-volume ratio, induction turbulence is well maintained in the swirling charge, and there are no protuberances to act as preigniters.

The inclined valves allow easy gas flow through the ports,[196] but until recently production has been a disincentive because of the highly angled valves. A compromise solution has half of the chamber in the piston with more nearly vertical valves—a double-convex chamber.

One of the most impressive chamber studies was by Heron[53] who attempted to realize the highest efficiency, power, and compression ratio, using 100 RON fuels which at that time (1950) were becoming available. He also required a chamber which rated sensitive fuels as closely as possible to their RON, referring to this design improvement as conferring "mechanical octane numbers" (after C. F. Kettering of GM).

The research centered on a two-vertical valve, flat-head design. The inlet valve could have a shroud (to promote turbulence) or not, and the piston was either flat topped or with a central cavity giving a 55% squish area (Fig. 13). By changing pistons the compression ratio could be varied from 5:1 to 30:1.

The best overall performance, judged in terms of a knock-limited charge density index, was given at a compression ratio of 10 : 1 using both inlet valve shroud and a piston bowl with squish, the two turbulence-creating mechanisms being roughly equal in effect.

The work was performed at engine speeds of 3000 rpm since high-speed

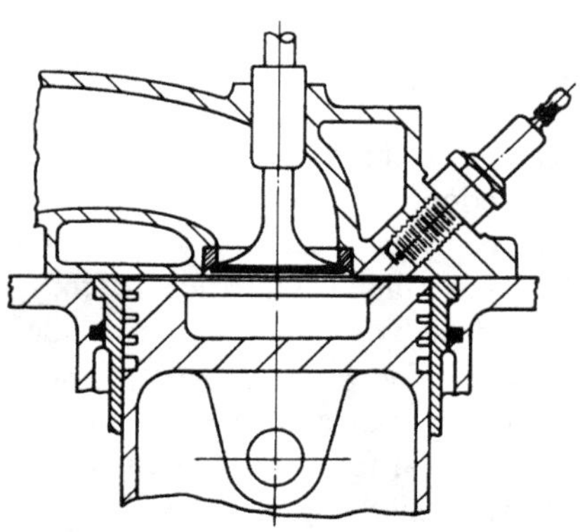

FIGURE 13. A Heron or cup-in-piston chamber.

knock was not a problem in 1950. The Heron chamber has been used on various versions. The most popular being the so-called "bowl-in-piston" design. With European performance levels it tends to need a very well designed piston to resist heat flow into the ring-pack region and to prevent high-speed knocking followed by preignition.

The bowl-in-piston high-performance engine, however, made some inroads on the natural habitat of the hemispherical head engine, namely the high-quality sports saloon because it is more easily fabricated, and is more tractable at low powers presumably due to squish.

Similar results can be obtained by transferring the combination cavity to the head with both valves in the cavity causing it to be elongated. This form is called a "bathtub" and is usually angled to the cylinder axis (Fig. 14). The angled chamber (sometimes one side tapers into the head face in which case it is a "wedge" chamber) allows a narrowed down end-gas zone. These chambers have been extremely popular operating up to a compression ratio of 11:1 on 93 RON gasoline.[197]

The lip of the "bathtub" must be carefully shaped as it can have a very significant effect on knock in the squish area opposite the plug.[24]

Despite the effect of squish and compression turbulence in the asymmetric head cavity, this chamber may need something more than normal induction turbulence to achieve a burn fast enough to avoid knock at very high compression ratios.

The next move toward higher compression economy engines was made by M. G. May.[198,199] He designed around the requirement that any new cylinder head, for small- to medium-mass-produced engines, should have two parallel

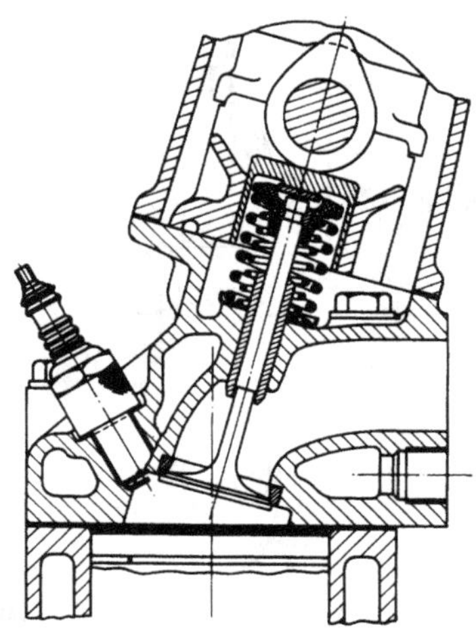

FIGURE 14. A "bathtub" chamber tending towards a "wedge."

vertical valves, a flat-tapered piston, and a "cast-in" chamber-in-head with heavy squish. To obtain compression ratios of from 13:1 to 15:1 only one valve could be included in the cavity; he preferred the exhaust valve since in that position the larger inlet valve would have a restricted flow. Not all authorities agree on this point.[(5)] The restricted inlet valve might generate very useful turbulence and in this position conforms to the traditional rule that end gases should be kept as far from exhaust valves as possible. Opposed to this is the argument that with the end gas removed by turbulent vortices, the exhaust valve merely increased flame speed and reduces the spark advance requirement.

A rough sketch of a May-type chamber is shown in Fig. 15. Induction swirl is generated clockwise by the inlet port, and on compression this concentrates in the cavity below the exhaust valve. A high squish develops at TDC over the inlet valve area and ejects through the channel into the cavity augmenting the induction swirl.

The flame front is encouraged to travel fastest along the hot wall between the valves with the net result that the end-gas region is fiercely scoured away and knock only occurs at low speed with abnormal loads.

This system is ideal for lean running and, if advantage is not taken of the potentially higher power with high compression ratio, then very considerable gains in economy can be made, otherwise only small gains are to be made.

7. *Stratified-Charge Combustion*

In the early days of his work on fuel economy, H. R. Ricardo devised the idea of stratifying the air–fuel mixture and the excess air to avoid throttling. This idea led to British Patent No. 2125,AD 1915 (Fig. 16). He was forced to

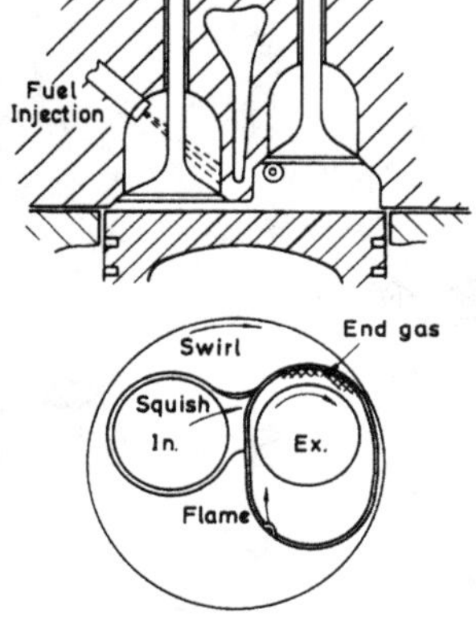

FIGURE 15. High-compression high-turbulence chamber of the May type.

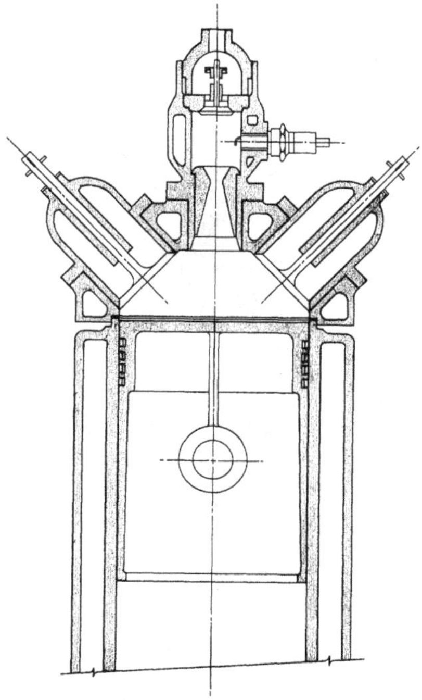

FIGURE 16. Stratified charge chamber (Ricardo 1915 patent).

separate the strata by the use of an antichamber rather than by the use of aerodynamics. Later he used the idea in a large two-cycle sleeve-valve aero-engine which operated at light load (and full load) unthrottled, but not over the middle of the range.[200]

A modern version of the same idea, but using fuel injection into the prechamber of a modified swirl-chamber diesel system has been described,[201] Fig. 17.

Looking at an ideal stratification case, clearly the charge must be in the vicinity of the spark, and the excess air in the end-gas position, thus reducing the chance of knock. Of course ideals like this never exist and little was heard about it until two separate needs rekindled interest. Firstly, a military requirement for a multifuel engine with no octane or cetane number requirements, and secondly a need for an engine to run on leaded or unleaded fuel to meet the U.S. EPA emissions requirements. The first challenge was taken up by Texaco with the TCCS system[202–204] and by the Ford Motor Company with the PROCO (FCP) engine.[205] Both chose spark-ignited, in-cylinder-injected, bowl-in-piston engines of high compression ratio (10:1–12:1) and the TCCS could also be turbocharged.

MAN in Germany also entered the fray with a spark-injected version of its

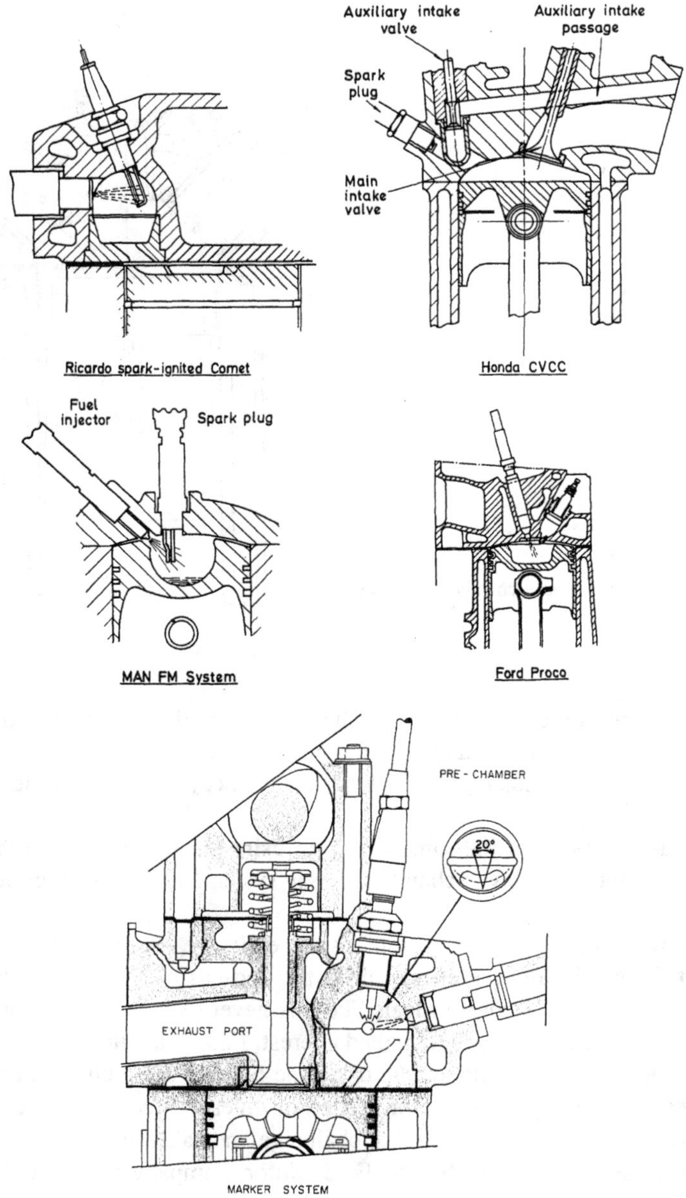

FIGURE 17. Stratified charge combustion systems.

"M" system diesel engine (Fig. 17) called the MAN-FM system. Their stratification principles differ widely but can be described simply as follows.

1. Texaco TCCS. The bowl is heavily swirled using a shrouded inlet valve and fuel is injected downstream in the swirl. It evaporates and fires off a sequentially sparking high-energy plug. Further downstream a flame front is set up into which mixture feeds until the injector closes.

In versions of this engine the injector opening fixes the spark close to TDC, reducing the time for fuel premixing to an absolute minimum. Any light or middistillate fuel can be used.[(204)]

2. Ford PROCO (which has a version for kerosene and diesel fuel called FCP). The bowl is again heavily swirled, but the fuel is injected into the eye of the swirl where it evaporates and is fired by a centrally disposed spark plug. The flame gases stay in the center because they are less dense than the air surrounding them.

3. MAN-FM. Again the bowl is swirled, but in this case the fuel is sprayed directly onto the bowl wall where it is ignited by a spark plug with an extended nose very close to the bowl. The fuel evaporates from the wall, ignites and is swirled round while the burnt mixture moves centrifugally. All three systems run at high compression ratios mainly unthrottled and are relatively knock-free, having in principle no end gas to autoignite. Probably a more cogent reason is that the necessarily late injection of fuel gives no time for the autoignition reactions to build up.

A comparative study of these and the other stratified charge engines is given in a report to the EPA.[(200)]

The requirement for lean running engines designed to overcome all three major emissions led many groups to study the divided-chamber idea of Ricardo, but high efficiency was subservient to exhaust emissions. Usually the spark plug is placed in the prechamber which is filled with rich mixture, either by a rich carburetor feeding a second inlet valve, Honda CVCC,[(206)] or by fuel injection into the prechamber Volkswagen and Porsche.[(207,208)] Very lean mixture from the main carburetor (or port injection) fill the main chamber, thus the end-gas zone is in knock-resistant lean mixture, which in any case was fired by multipled ignition sources from the prechamber throat, itself a good antiknock feature. An interesting new concept in prechamber stratified charge engines has recently emerged[(209,210)] (Fig. 17). The mixture is created by atomizing injected fuel against an axial bar through the spherical prechamber. The compression vortex stratifies the mixture, which is ignited by a spark passing into the earthed bar. The rapid combustion occurs close to TDC almost entirely within the prechamber because the piston-head clearance is very small. This engine is multifuel and essentially knock-free, it operates well in turbocharged form. It has the code name "Marker."

These engines were developed with some emphasis on emissions, so it is worth pointing out that high-compression lean-running engines have a severe problem with unburned hydrocarbons because their exhaust temperatures are low and high-octane gasoline is resistant to oxidation. Even the use of a platinum oxidation catalyst may be insufficient to meet regulations.

The Diesel Engine. Although not spark ignited, the diesel engine is in principle a system which defeats the knock barrier. Multiple ignition sources and very late injection give little time for fuel premixing and end-gas formation. There is no element of abnormal combustion in diesel knock. High peak pressures at low engine speeds excite this much lower frequency vibration within the engine structure.

Because of the expected low availability of diesel fuel in the future, there is an active development of engines running with a pilot-injection ignition of diesel fuel with a power injection of any low-cetane or even high-octane fuel such as methanol, and with no knock.[211]

VI. GENERAL CONCLUSIONS

1. To preserve private mobility in a fuel-starved future, new, small, high-compression, highly loaded spark ignition engines are needed. They must be built to higher engineering and technological standards than heretofore. They must be long-lived in order to recoup their extra cost.

2. Spark ignition fuels will always predominate over diesel fuels for private transportation, especially in small engine sizes. Spark ignition fuels will vary in quality as supplies dwindle; an option to employ antiknock additives should be retained, especially in low population density areas.

3. These engines must run close to the knock limit over their range. The driver will learn to tolerate slight low-speed knock, but a great effort in engine and fuel technology is needed to overcome high-speed knock.

4. To allow simpler, less fuel-wasting refining and to accommodate a possible collapse in high-octane and high-cetane fuel supplies, the stratified-charge engine should be researched vigorously, particularly in smaller sizes. The Texaco engine offers the greatest promise at this point in time, but swirl chamber units should not be forgotten.

5. New fuel-rating procedures for all forms of abnormal combustion in more realistic engines are needed with a fresh look at primary reference fuels.

6. The end gas still holds many mysteries which seem to have been forgotten in rather sterile disputation over knock theories. What is its precise composition, and what are its physical conditions in the thin layer which is "attached" to the relatively cool chamber wall?

7. The kinetics of fuel sensitivity need much more attention, particularly in the interplay of autoignition processes between paraffins, olefins, and aromatics in mixed fuels.

8. Much fuel economy research is frustrated by excessive emissions requirements—waivers from emissions regulations should be granted to the most promising economy developments until their prospects have been determined. Only then should some balance between economy and emissions be reached on the basis of the then-existing vehicle population and utilization.

REFERENCES

1. Ricardo, H., and Hempson, J. G. G., *The High-Speed Internal Combustion Engine,* 5th ed., Blackie, London (1968).
2. Lichty, L. C., *Combustion Engine Processes,* McGraw-Hill, New York (1967).
3. Taylor, C. F., *The Internal Combustion Engine in Theory and Practice,* 2nd ed., MIT Press, Cambridge, Massachusetts (1966).
4. Caris, D. F., and Nelson, E. E., "A New Look at High-Compression Engines," SAE Preprint 61A (1958).
5. Kuznicki, L., and Thompson, W. B., "High-Compression–Lean-Burning Engines: Signpost for the Future," *Inst. Mech. Eng.* Conference, London, 12–14 June (1979).
6. Dugald, C., *Trans. Faraday Soc.*, **22,** 338 (1926).
7. Ricardo, H. R., "The Progress of the Internal Combustion Engine and its Fuel," Melchett Lecture, Inst. of Fuel (1935).
8. Hempson, J. F. G., "The Automobile Engine 1920–1950," SAE Paper 760605 (1976).
9. Boyd, T. A., "Pathfinding in Fuel and Engines," SAE Trans. **4,** 182 (1950).
10. Campbell, J. M., Lovell, W. G., and Boyd, T. A., "Detonation Characteristics of Some of the Fuels Suggested as Standards of Anti-Knock Quality," SAE Trans. **25** (1930).
11. Report of the Empire Motor Fuels Committee. The Institution of Automobile Engineers, Vol. 18, part 1, (1924).
12. Ricardo, H. R., *Engineering,* **110,** 325 and 361 (1920).
13. Agnew, W. G., "Fifty Years of Combustion Research at General Motors," *Proc. Energy Combust Sci,* **4,** 115–155 (1978).
14. Tizard, H. T., and Pye, D. R., *Phil. Mag.*, **11,** 1094 (1926).
15. Callendar, H. L., *Engineering,* 123, 147, 182, 210 (1927).
16. Egerton, A. Smith, A. C., and Ubbelohde, A. R., *Phil. Trans. Roy. Soc. A,* **234,** 433 (1935).
17. Ricardo, H. R., and Glyde, H. S., "The High-Speed Internal Combustion Engine," Blackie, London (1941).
18. Glyde, H. S., "Experiments to Determine Velocities of Flame Propagation in a Side-Valve Petrol Engine," *J. Inst. Pet. Tech.*, **16**(85), 756–776, November (1930).
19. Alcock, J. F., "The Effect of Swirl on Petrol Engine Combustion," *The Aircraft Engineer* No. 100 (Vol. 9, No. 5), May 9 (1934).
20. Alcock, J. F., and MacLellan, G. D. S., "Thermo-centrifugal Convection," Proc. Eighth Int. Congress on Theoretical and Applied Mechanics, 1952 Istanbul, p. 385 (1953).
21. Withrow, L. L., and Bowditch, F. W., "Flame Photographs of Auto-Ignition Induced by Combustion Chamber Deposits," *SAE Trans.* **6,** 724 (1952).
22. Stebar, R. F., Wiese, W. M., and Everett, R. L., "Engine Rumble—A Barrier to High Compression Ratios? SAE Trans. **68,** 206 (1960).
23. Cornetti, G. M., De Cristoforo, F., and Gozzelino, R., "Engine Failure and High-Speed Knock," SAE paper No. 770147 (1977).
24. Henault, C., and Riviere, J. P., "Experimental Study of Pinking at High Speed," *Ingenieur de l'Automobile,* **3,** 178–198, March (1974).

25. Judge, A. W., "Testing High-Speed I.C. Engines," 3rd ed., Chapman and Hall, London. (1943), pp. 67–68; and Ref. 11, p. 77.
26. Downs, D., and Robinson, V. H., "Small Variable-Compression Research Engine," *Engineering*, December 30 (1949).
27. Rado, W. G., "Characteristics of a Plasma Generated by Combustion in a Spark Ignition Engine," *J. Appl. Phys.*, **46**(6), June (1975).
28. Guibet, J. C., and Duval, A., "New Aspects of Pre-ignition in European Automotive Engines," SAE paper no. 720114 (1972).
29. Curry, S., "A Three-dimensional Study of Flame Propagation in a Spark Ignition Engine," *SAE Trans.* **71,** 628 (1963).
30. Arrigoni, W., Cornetti, G. M., Gerbarz, G. P., Giavazzi, F., and Pozzi, V., "Chemi-ionization and Carbon-in a Spark Ignition Engine," SAE paper no. 740192 (1974).
31. Kaufman, C. W., and Gat, N., "Some Effects of Exhaust Gas Recirculation on Combustion," Spring. Mtg. Central States Sec. Combustion Inst, April (1978).
32. Downs, D., and Pigneguy, J. H., "An Experimental Investigation into Pre-ignition in the Spark Ignition Engine," *Proc. Inst. Mech. Eng.* (Auto Div.) London (1950–1951).
33. Evers, L. W., "Spark Plug Pressure Transducers for Measuring Indicated Work," SAE paper no. 780148 (1978).
34. Arrigoni, V., Cornetti, G., Gaetani, B., and Ghezzi, P., "Quantitative Systems for Measuring Knock," *Proc. Inst. Mech. Eng.*, **186,** 48–72 (1972).
35. Barton, R. K., Lestz, S. S., and Duke, L. C., "Knock Intensity of Engine Rate of Pressure Change," SAE Trans. paper no. 700006 (1970).
36. Sezzi, F., Cornetti, G., Anigoni, V., Vincenzetto, F., and Biancucci, S., "Possible Mechanisms of Piston Failure due to Detonation and Pre-ignition," *A.T.A. Jour.*, p. 23, July (1969).
37. Dunstan, A. E., and Card, S., "The Work of Midgley and Boyd Report of the Empire Motor Fuels Committee," *Inst. Auto. Eng.*, **18** (part 1), 305 (1924).
38. Midgley, T., and Boyd, T. A., various papers in *J. Ind. Eng. Chem.*, No. 7, p. 589 (1922), No. 9, p. 849 (1922), No. 9, p. 894 (1922), No. 4, p. 421 (1923).
39. Wolber, W. G., "Automotive Engine Control Sensors," SAE paper 800121 (1980).
40. ASTM, Manual for Rating Motor, Diesel and Aviation Fuels, American Society for Testing and Materials, Philadelphia, Pa., 1973–74.
41. "CRC, Revised Road Rating Techniques, Modified Borderline Method," Coordinating Research Council Inc., July (1970) with October (1975) revisions.
42. Myers, M. E., Stolsteiner, J., and Wims, A. M., "Determination of Gasoline Octane Numbers from Chemical Composition," *Anal. Chem.*, **47**(13), 2301, November (1975).
43. Marciante, A., *et al.*, "Fuel Hydrocarbon Composition and High-Speed Knock," 15th FISITA Congress, Paris, May 1974, Soc. Ing. de l'Auto, Paris (1974).
44. Cornetti, G. M., DeCristofaro, F., and Gozzelino, R., "Engine Failure and High-Speed Knock," SAE paper 770147 (1977).
45. Betts, W. E., "Knock and Engine Damage," Coordinating European Council, Int'l. Sym., Rome, June 1981, Coordinating European Council, London (1981).
46. Arrigoni, V., *et al.*, "Recent Advances in the Detection of Knock in S. I. Engines," SAE paper no. 780153 (1978).
47. Duval, A., *et al.*, "New Approach for a Better Engine Fuel Compromise," Sym. AGELF1, Rome 1979, p. 49, AGIP Petroli, Rome (1979).
48. Arrigoni, V., Cornetti, G., Gaetani, B., and Ghezzi, P., "Quantitative Systems for Measuring Knock," *Proc. Inst. Mech. Engr.* **186,** 194, London (1950–1951).
49. Downs, D., and Theobold, F. B., "The Effect of Fuel Characteristics and Engine Operating Conditions on Pre-ignition," *Proc. Inst. Mech. Eng.* (Auto Div.), 1963–1964.
50. Arrigoni, V., *et al.*, "A Quantitative System for Measuring Pre-ignition," *ATA Tech. Bull,* November (1968).

51. Rothrock, A. M., "Fuel Rating—Its Relation to Engine Performance," *SAE Trans.*, **48,** 51, February (1941).
52. Alquist, H. E., O'Dell, L., and Eward, J. C., "Correlation of Effects of Compression Ratio and Inlet Air Temperature on Knock Limits of Aviation Fuels in CFR Engine," *NACA Report*, E6, E13, Part II, June (1946).
53. Heron, S. D., and Felt, A. E., "Cylinder Performance—Compression Ratio and Mechanical Octane Number Effects," *SAE Trans.*, **4**(4), October (1950).
54. Addicott *et al.*, "Spark Knock," *Proc. Inst. Mech. Eng.*, **2A,** 182, London (1967–1968).
55. Yagi, S., Ishizuya, A., and Fujii, I., "Research and Development of High-Speed, High-Performance Small Displacement Honda Engines," SAE Paper no. 700122 (1970).
56. Harrow, G. A., *et al.*, *J. Inst. Pet.*, **49**(457), 204 (1963).
57. Mori, T., and Yamazaki, K., "Variations of the Flame Propagation Time in a Spark-Ignition Engine," *Bull. JSME*, **13**(58) (1970).
58. Halstead, M. P., Pye, D. B., and Quinn, C. P., "Laminar Burning Velocities and Weak Flammability Limits Under Engine-Like Conditions," *Combust. and Flame*, **22,** 89 (1974).
59. Agnew, W. G., "End-gas Temperature Measurement by a Two-Wavelength Infrared Radiation Method," *SAE Trans.* **69,** 495 (1961).
60. Haskell, W. W., Trumpy, D. K., and Hendrickson, C. H., "Measurement of Mixture Temperature in an Internal Combustion Engine," SAE Paper 730082 (1973).
61. Gumbleton, J. J., Bolton, R. A., and Lang, H. W., "Optimizing Engine Parameters with Exhaust Gas Recirculation," SAE Paper 740104 (1974).
62. Noguchi, M., Tanaka, Y., Tanaka, T., and Takeuchi, Y., "A Study of Gasoline Engine Combustion by Observation of Intermediate Reactive Products During Combustion," SAE Paper 790840 (1979).
63. Withrow, L., and Rassweiler, G. M., "Spectroscopic Studies of Gaseous Charges in the Gasoline Engine," *Ind. Eng. Chem.*, **25,** 923 (1933).
64. Heywood, J. B., and Kock, J. C., "Formation of Hydrocarbons and Oxides of Nitrogen in Automobile Engines," *Environ. Sci. and Tech.*, **7**(3), March (1973).
65. Payman, W., and Titman, H., *Proc. Roy. Soc.*, **A152,** 418 (1935).
66. Bone, W. A., Fraser, R. P., and Wheeler, W. M., *Phil. Trans. Roy. Soc.*, **A235,** 29 (1936).
67. Miller, C. D., Olsen, H. L., Logan, W. O., and Osterstrom, G. E., NACA Report 857 (1946).
68. Male, T., "Photography at 500,000 frames per second of combustion and Detonation in a Reciprocating Engine," 3rd Sym. on Combustion, Flame and Explosion Phenomena, p. 721, Williams and Wilkins, Combustion Institute, Pittsburgh (1949).
69. Withrow, L., and Rassweiler, G. M., "Slow-motion Shows Knocking and Non-knocking Explosions," *SAE Jour.*, **39,** 297 (1936).
70. Withrow, L., and Rassweiler, G. M., "Studying Engine Combustion by Physical Methods," *J. Appl. Phys.*, **9,** 362 (1938).
71. Withrow, L., and Rassweiler, G. M., "Engine Knock," *Auto. Engr.*, **24,** 281 (1934).
72. Grinstead, C. E., "Sound and Pressure Waves in Detonation," *J. Aeronaut. Sci.*, **6,** 412 (1939).
73. Jost, W., "Explosion and Combustion Processes in Gases," McGraw-Hill, New York (1946).
74. Ball, G. A., *5th Symposium on Combustion*, p. 366, Reinhold Co., New York (1955).
75. Bowditch, F. W., and Stebar, R. F., "Autoignition Associated with Hot Starting," *SAE Trans.*, **66,** 179, (1958).
76. Rifkin, E. B., and Sokolik, A. S., *Fuel*, **17** (1938); *J. Roy. Aero. Soc.*, **42,** 851 (1938).
77. Ricardo, H. R., "Recent Research Work on the I. C. Engine," *SAE Trans.* **17,** pt. 1, 1 (1922).
78. Tizard, H. T., and Pye, D. R., "Ignition of Gases by Sudden Compression," *Phil. Mag*, **44**(6), 79 (1922).
79. Egerton, A., Smith, A. C., and Ubbelohde, A. R., *Phil. Trans. Roy. Soc. A.*, **234,** 433 (1935).

80. Rassweiler, G. M., and Withrow, L., "Spectrographic Detection of Formaldehyde in an Engine Prior to Knock," *Ind. Eng. Chem.*, **26,** 1359 (1933).
81. Maccormac, M., and Townend, D. T. A., 5, *Chem. Soc.*, **140,** 238 (1938).
82. Emeleus, H. J., *J. Chem. Soc.* **228,** 2948 (1926).
83. Burgoyne, J. H., *Proc. Roy. Soc.*, **A 175,** 539 (1940).
84. Kahler, E. J., Bearse, A. E., and Stoner, G. G., *Ind. Eng. Chem.* **43,** 2777 (1951).
85. Barusch, M. R., Crandall, H. W., Payne, J. Q., and Thomas, J. R., *Ind. Eng. Chem.*, **43,** 2764 (1951).
86. Levedahl, W. J., and Howard, F. L., *Ind. Eng. Chem.*, **43,** 2805 (1951).
87. Sokolik, A. S., and Jantowsky, S. A., *Aita. Phys. Chim. U.R.S.S.*, **19,** 329 (1944).
88. Chamberlain, G. H. N., and Walsh, A. D., *Proc. Roy. Soc.*, **A 215,** 175 (1952).
89. "The Knock Rating of Hydrocarbons," API Res. Project 45, American Petroleum Inst.
90. Walsh, A. D., *Trans. Faraday Soc.*, **42,** 269 (1946).
91. Cramer, P. L., and Campbell, J. M., "Combustion of Hydrocarbons," *Ind. Eng. Chem.*, **41**(5), 893, May (1949).
92. Boord, C. E., Amer. Chem. Soc. (Div. Pet. Chem.), 112th meeting, N.Y. (1948).
93. Downs, D., Walsh, A. D., and Wheeler, R. W., "A Study of the Reactions that Lead to Knock in the Spark-Ignition Engine," *Phil. Trans. Roy. Soc. Series A*, **870**(243), 463 (1951).
94. Alperstein, M., and Bradow, R. L., "Investigations into the Combustion of End-Gases from Otto Engines," SAE paper no. 660410 (1966).
95. Davis, W. C., Smith, M. L., Malmberg, E. W., and Bobbitt, J. A., "Comparison of Intermediate Combustion Products Formed in Engines with and without Ignition," *SAE Trans.*, **63,** 386 (1955).
96. Luck, C. J., Burgess, A. R., Qesty, D. H., Whitehead, D. M., and Pratley, A., "A Study of the Combustion of *n*-heptane in an Engine Using a Novel High-Speed Sampling Technique," 14th Symp. (Int'l) on Combustion, Pittsburgh, Combustion Institute, Pittsburgh (1973).
97. Mackinven, R., "A Search for an Ashless Replacement for Lead in Gasoline," MK/172/74 Shell Int'l Petroleum, London (1974).
98. Walsh, A. D., "Low Temperature Oxidation," (W. Jost, ed.), Gordon & Breach, London (1961).
99. Downs, D., and Wheeler, R. W., "Recent Developments in Knock Research," *Proc. Inst. Mech. Eng.* (Auto Div.), **89** (1951–52).
100. Wolcutt, C., and Rifkin, E. B., "Pre-combustion Reactions in an Engine—Thermodynamic Analysis of Pressure Developed During Pre-flame Period," Ethyl Corp, collected tech. papers (1951).
101. Taylor, E. S., Taylor, C. F., Livengood, J. C., Russell, W. A., and Leary, W. A., *SAE Quart. Trans.*, **4,** 222 (1950).
102. Peletier, J. L., Van Hoogstraten, S. G., Smittenberg, J., and Kooijman, P. L., *Chaleur et Industrie,* **20,** 120 (1939).
103. Downs, D., Street, J. C., and Wheeler, R. W., "Cool-Flames in a Motored Engine," *Fuel,* **XXX1**(3), July (1953).
104. Sheinson, R. L., and Williams, F. W., "Chemiluminescence Spectra from Cool and Blue Flames—Electronically Excited Formaldehyde," *Comb. and Flame* **21,** 221 (1973).
105. Broida, H. P., and Levedahl, W., *Anal. Chem.*, **24,** 17776 (1952).
106. Pastell, D. L., "Pre-combustion reactions in a Motored Engine," Discussion by Walcutt, *SAE Quart. Trans.*, **4,** 571 (1950).
107. Affleck, W. S., and Fish, A., *Eleventh Symposium (Int'l) on Combustion,* The Combustion Inst., p. 1003 (1967).
108. Burgess, A. R., and Lauchlin, R. G. W., *Combust. and Flame,* **19,** 315 (1972).
109. Halstead, M. P., Kirsch, L. J., Protheso, L. S., and Quinn, C. P., *Proc. Roy. Soc.*, **A346,** 515 (1975).

110. Halstead, M. P., Kirsch, L. S., and Quinn, C. P., "Autoignition of Hydrocarbon Fuels," *Combust. and Flame,* **30,** 45–60 (1977).
111. Boyd, T. A., "Pathfinding in Fuels and Engines," *SAE Trans.* **4,** 182 (1950).
112. Newman, S. R., Dille, K. L., Heisler, R. Y., and Fontaine, M. F., "Tertiary-butyl acetate—an Octane Improver for Leaded Gasolines," SAE Fuels and Lubricants Meeting, Chicago, Paper 127U, October (1959).
113. Brown, J. E., and Lovell, W. G., "A New Manganese Antiknock," *Ind. Eng. Chem.,* **50,** 1547 (1958).
114. Graiff, L. B., "The Mode of Action of Tetraethyl Lead and Supplemental Anti-Knock Agents," SAE paper 660780 (1966).
115. Moran, J. B., "The Environmental Implications of Manganese as an Alternate Anti-Knock," SAE paper 750926 (1975).
116. Furey, R. L., and Summers, J. C., "How MMT Causes Plugging of Monolithic Converters," SAE Paper 780004 (1978).
117. In reapplication for MMT Waiver, *Federal Register,* **43,** n. 181, Monday, Sept. 18, 1978, p. 41424.
118. Turner, D., "Lead in Petrol 2—Environmental Health," *Chemistry in Britain,* **16**(6), June (1980).
119. Dartnell, P. L., "Lead in Petrol I—Energy Conservation," *Chemistry in Britain,* **16**(6), June (1980).
120. "Lead in Petrol. An Assessment of the Feasibility and Costs of Further Action to Limit Lead Emissions from Vehicles," Working party on lead in petrol, Dept. of Transport, London, U.K., July (1979).
121. Dartnell, P. L., "Formal Contribution to Technical Session 3—Economics of Environmental Protection," World Energy Conf., Munich, September (1980).
122. Acres, G. J. K., The Johnson Mattley Co., London, England, Private communication.
123. Norrish, R. G. W., Discussion, *7th Sym. on Combustion,* p. 203, Butterworths, London (1958).
124. Collear, A. B., and Norrish, R. G. W., "The Behaviour of Additives in Explosions and the Mechanism of Anti-Knock," 124: *Proc. Roy. Soc.* **A259,** 304 (1960).
125. Cheaney, D. E., Davies, D. A., Davis, A., Hoare, D. E., Prothesoe, J., and Walsh, A. D., "Effects of Surfaces on Combustion of Methane and Mode of Action of Anti-Knocks Containing Metals," *7th Sym. on Combustion,* p. 183, Butterworths, London (1958).
126. Oosterhoff, L. J., "The Action of Anti-Knock Dopes," *Rec. Trav. Chim.,* **59,** 811 (1940).
127. Downs, D., Griffiths, S. T., and Wheeler, K. W., "The Part Played by the Preparational Stage in Determining Lead anti-Knock Effectiveness," *J. Inst. Petrol,* **47**(445) (1961).
128. Zimpel, C. F., and Graiff, L. B., "An Electron Microscopic Study of Tetraethyl Lead Decomposition in an Internal Combustion Engine," *11th (Int'l) Sym. on Combustion,* Berkeley, Calif. (1966).
129. Graham, S. C., and Homer, J. B., "Coagulation of Molten Lead Aerosols," *Faraday-Chem. Soc. Sym.,* **7,** Swansea (1973).
130. Downs, D., Griffiths, S. T., and Wheeler, R. W., "Pre-flame Reactions in the Spark-Ignition Engine and the Influence of Tetraethyl Lead and Other Antiknocks," *J. Inst. Petrol,* **49**(469), January (1963).
131. Echols, L. S., Yust, B. E., and Barne, J. L., "A Review of Research on Abnormal Combustion Phenomena in Internal Combustion Engines," 5th World Petroleum Congress (1959).
132. Miller, C. D., *SAE Quart. Trans.,* **1,** 98 (1947).
133. Barrett, G. M., "Discussion of Pre-ignition in the Spark-ignition Engine," *Proc. Inst. Mech. Eng.* (Auto Div.), **141,** 141 (1950–51).
134. Downs, D., Street, J. C., and Wheeler, R. W., "A Unified Concept of Combustion in I. C. Engines," *Proc. 4th World Petroleum Congress,* Sect. VL-F, paper 5 (1955).

135. Benson, S. D., "The Influence of Engine and Fuel Factors on After-Run," SAE Paper 720085 (1972).
136. Ingamells, J. C., "Effect of Gasoline Octane Quality and Hydrocarbon Composition on After-Run," SAE Paper 790939 (1979).
137. Bowditch, F. W., and Stebar, R. F., "Auto-ignition Associated with Hot-starting," *SAE Trans.*, **66,** 179 (1958).
138. Morris, W. E., "After-running in Low-compression Engines due to Ignition by Hot Spot," *SAE Trans.*, **67,** 133 (1959).
139. Affleck, W. S., Bright, P. E., and Fish, A., "Run-on in Gasoline Engines—A Chemical Description of Some Effects of Fuel Composition," *Comb. and Flame,* **12,** 307–317, August (1968).
140. Hundere, A., and Bert, J. A., "Pre-ignition and its deleterious effects in aircarft engines," *SAE Quart. Trans.*, **2**(4), 546 (1948).
141. Anderson, J., "Pre-ignition in Aircraft Reciprocating Engines," *SAE Trans.*, **19,** 65 (1967).
142. Bowens, R. C., and Isitt, A. R., "The Observation of Automotive Pre-ignition and Knock," SAE Summer Mtg., Atlantic City (1954).
143. Sturgis, B. M., Cantwell, E. N., Morris, W. E., and Schultz, D. L., "The Pre-ignition Resistance of Fuels," *Proc. Amer. Petrol. Inst., Sect. III,* **34,** 256 (1954).
144. Coward, H. F., and Guest, P. G., "Ignition of Natural Gas–Air Mixtures by Heated Metal Bars," *J. Amer. Chem. Soc.*, **49,** 2479 (1927).
145. David, W. T., "Temperature of Flame Gases," *Nature,* **152**(278) (1943).
146. Thring, R. H., "The Catalytic Engine," *Platinum Metals Review,* **24**(4), October (1980).
147. Bowditch, F. W., and Yu, T. C., "A Consideration of Deposit Ignition Mechanism," *SAE Trans.*, **69,** 435 (1961).
148. Nebel, G. J., and Cramer, P. L., "Ignition Temperatures of Lead Compound–Carbon Mixtures," *Ind. Eng. Chem.*, **47,** 2393 (1955).
149. Chrysler "Clean Air Package," *Automotive Eng.*, **83,** 44, October (1975).
150. Tierney, W. T., Johnson, E. M., and Crawford, N. R., "Energy Conservation Optimization of the Vehicle-Fuel-Refinery System," SAE Paper 750673.
151. Van Gulick, H., "Refineries and Engines as a Single Technical System," *J. Automot. Eng.*, **6,** 11, April (1975).
152. Kohonitz, R., "The National Utilization of Fuels in Private Transport-RUFIT," *11th World Energy Conf.*, Munich, **3,** 201, September (1980).
153. Corner, E. S., Hockhauser, A. M., and Shannon, H. F., "Technical versus Customer Knock Satisfaction: Two Decades," SAE Paper 780322 (1978).
154. Sksipkin, S. P., Vorohev, V. I., and Romanchikov, P. G., "The Influence of the Distribution of the Vapour and Liquid Components of the Fuel Mixture Stream on the Non-Uniformity of Mixture Components in the Engine Cylinder," *Automob. Prom.*, **12,** March (1974); MIRA Transl. No. 74-7-190 (1974).
155. Bommeister, J., Drechler, F., and Nghia, V., "Current Problems of Mixtures Distribution in 4-str Petrol Engines," Tech. Univ. Dresden, Kraftfahrzeng-Tech., 364, December (1974).
156. Brandstetter, W. R., and Carr, M. J., "Measurement of Air Distribution in a Multi-cylinder Engine by Means of a Mass Flow-Probe," SAE Paper 730494.
157. Bell, A. G., "The Relationship Between Octane Quality and Octane Requirement," SAE Paper 750935 (1975).
158. Blackmore, D. R., and Thomas, A., *Fuel Economy of the Gasoline Engine,* p. 112, Macmillan, New York (1977).
159. Haslett, R. A., and Eidson, T. M., "Equivalence-Ratio Meter," SAE Paper 770219.
160. Robison, J. A., and Brehob, W. M., "The Influence of Improper Mixture Quality on Engine Exhaust Emissions and Performance," Western States Comb. Inst. Mtg. no. WSC1, 67-17, October (1965).

161. Ma, T. H., "Effect of Cylinder Charge Motion on Combustion," *Inst. Mech. Eng.* Conference C81-75, Cranfield, July (1975).
162. *Environmental News*, "EPA reports Dresserator control system meets interim emission standards," October 3, 1974.
163. Ford Motor Co., U.S.E.P.A. Committee on Motor Vehicle Emissions, panel of consultant on engine systems, May (1974).
164. Stivender, D. L., "Intake Valve Throttling (IVT) a Sonic Throttling Intake Valve Engine," *SAE Trans.*, **77**, 1293 (1968).
165. Eltinge, L., Marsee, F. J., and Warren, A. J., "Potentialities of Emissions Reduction by Engine Modification," SAE paper 680123 (1968).
166. Harrow, G. A., Mills, W. D., Thomas, A., and Finlay, I. C., "The Vapipe-A Practical System for Producing Homogeneous Gasoline-Air Mixtures," SAE paper 760564 (1976).
167. Lindsay, R., Thomas, A., Woodworth, J. A., and Zeschmann, E. G., "Influence of Homogeneous Charge on the Exhaust Emissions of Hydrocarbons, Carbon Monoxide and Nitric Oxide from a Multi-Cylinder Engine," SAE Paper 710588 (1971).
168. Matthes, W. R., and McGill, R. N., "Effects of the Degree of Fuel Atomization on Single-Cylinder Engine Performance," SAE paper 760117 (1976).
169. Polymeropoulos, C. E., and Das, S., "The Effect of Droplet Size on the Burning Velocities of Kerosene–Air Sprays," *Comb. and Flame*, **25**, 247–257 (1975).
170. Ryan, T. W., Lestz, S. S., and Myers, W. E., "Extension of the Lean Misfire Limit and Reduction of Exhaust Emissions of a Spark-Ignition Engine by Modifications of the Ignition and Intake Systems," SAE paper 740105 (1974).
171. Burgett, R. R., Leptich, J. M., and Sangwan, V. S., "Measuring the Effect of Spark-Plug and Ignition System Design on Engine Performance," SAE Paper no. 720007 (1972).
172. Thring, R. H., "The Effect of Varying Combustion Rate in Spark Ignited Engines," SAE paper 790387 (1979).
173. Wallace, T. F., "Buick's Turbocharged V-6 Powertrain for 1978," SAE paper 780413 (1978).
174. Currie, J. H., Grossman, D. S., and Gumbleton, J. J., "Energy Conservation with Increased Compression Ratio and Electronic Knock Control," SAE paper 790173 (1979).
175. Kraus, B. J., Godici, P. E., and King, W. H., "Reduction of Octane Requirement by Knock-Sensor Spark-Retard System," SAE paper 780155 (1978).
176. Gillbrand, P., "Knock Detector System Controlling Turbocharger Boost Pressure," SAE paper 800833 (1980).
177. Bailey, R. L., Cederquist, A. L., Hart, D. L., Florek, J. J., and Metzler, A. H., "An IIEC-2 Low Emissions Concept Car," SAE paper 780206 (1979).
178. Alquist, H. E., Holman, G. E., and Wimmer, D. B., "Some Observations of Factors Affecting O.R.I.," SAE paper 750932 (1975).
179. Forster, E. J., and Stinson, L. E., "Effects of Leaded versus Unleaded Gasolines on Stabilized Octane Requirements," Nat. Pet. Refiners Assoc., F&L 70–46.
180. "The Effect of Removing Lead from European Motor Gasolines," Cooperative Octane Requirement Committee Project Report, October (1972).
181. Betts, W. E., "Improved Fuel Economy by Better Utilization of Available Octane Quality," SAE paper 790940 (1979).
182. Scott, W. M., "Looking in on Diesel Combustion," SAE SP-345 (1970).
183. Brandt, F., Reverencic, L., Cartellieri, W., and Dent, J. C., "Turbulent Air Flow in the Combustion Bowl of a D.I. Diesel Engine and its Effect on Engine Performance," SAE paper 790040 (1979).
184. Witze, P. O., "Application of Laser Velocimetry to a Motored Engine," Third Int'l Workshop on Laser Velocimetry, Purdue Univ., July (1978).
185. Hutchinson, P., Morse, A., and Whitelaw, J. H., "Velocity Measurements in Motored Engines-Experience and Prognosis," SAE paper 780061 (1978).

186. Dent, J. C., and Salama, N., "Turbulence Structure in the Spark Ignition Engine," Inst. Mech. Engrs. Conf., Cranfield, C83/75, p. 23, July (1975).
187. Dent, J. C., and Salama, N., "The Measurement of Turbulence Characteristics in an Internal Combustion Engine Cylinder," SAE paper 750886 (1975).
188. Hinze, J. O., "Turbulence," 2nd ed., McGraw-Hill, London (1975).
189. Lyn, L. S. T., and Valdemanis, E., "The Application of High-Speed Schlieren Photography to Diesel Combustion Research," *J. Photo Science,* **10** (1962).
190. Lancaster, D. R., Krieger, R. B., Sorenson, S. C., and Hull, W. L., "Effect of Turbulence on Spark-ignition Engine Combustion," SAE paper 760160 (1976).
191. Ricardo, H. R., "Cylinder Head Design," *The Automobile Engineer,* July-August (1929).
192. Matekunas, F. A., "Ignition Studies in a Rapid Compression Machine," *G.M. Res. Publ.*, GMR 2681 (1978).
193. Alcock, J. F., and Scott, W. M., "Some More Light on Diesel Combustion," *Proc. Inst. Mech. Eng.* (Auto Div.), **5,** 179 (1962–1963).
194. Woods, W. A., and Ghirlando, R., "Radial Flow in an Engine Cylinder Near the End of Compression," *Inst. Mech. Eng.*, Conference, C82/75 Cranfield, July (1975).
195. Beale, N. R., and Hodgetts, D., "The Cranfield Kushul Engine," *Inst. Mech. Eng.*, Conference C90/75 Cranfield, July (1975).
196. Barnes-Moss, H. W., "A Designer's Viewpoint," *Inst. Mech. Eng.*, Conference publ. 191, C343-73 (1973).
197. Gillbrand, Per, "The Saab-Scania Fuel Economy Engine," *Inst. Mech. Engrs.*, Conference on land transport engines, Paper C6-77, January (1977).
198. May, M. G., U.S. Patent 4000722, January 4, 1977.
199. May, M. G., "Lower Specific Fuel Consumption with High Compression Lean Burn Spark Ignited 4-stroke Engines," SAE paper 790386 (1979).
200. Ricardo and Co., "A Study of Stratified Charge for Light-Duty Power Plants," vol. 1, p. 127, Publ. no. EPA 460/3-75-011A, R.T.P., N.C. 27711.
201. Overington, M. T., and Haslett, R. A., "A New Stratified Charge Engine Based on the Ricardo Comet Design," *Int. Mech. Eng.*, "Stratified Charge Conference," London, C253/76, November (1976).
202. Barber, E. M., Reynolds, B., and Tierney, W. T., "Elimination of Combustion Knock-Texaco Combustion Process," SAE Summer Mtg., Indian, paper 473, June (1950).
203. Tierney, W. T., Mitchell, E., and Alperstein, M., "The Texaco Controlled-Combustion System—A Stratified Charge Engine Concept—Review and Current Status," *Inst. Mech. Eng.*, Conference on Power Plants and Future Fuels, London, January (1975).
204. Lewis, J. M., and Tierney, W. T., "United Parcel Service Applies Texaco Stratified Charge Engine Technology to Power Parcel Delivery Vans—Progress Report," SAE paper 801429 (1980).
205. Scussel, A. J., Simko, A. O., and Wade, W. R., "The Ford Proco Engine Update," SAE paper 780699 (1978).
206. Date, T., Yagi, S., Ishuzuya, A., and Fuju, I., "Research and Development of the Honda EVCC Engine," SAE paper 740605 (1974).
207. Brandstetter, W. R., Decker, G., Schafer, H. J., and Steinke, D., "The Volkswagen PC1 Stratified Charge Concept—Results from the 1.6 litre air-cooled engine," SAE paper 741173 (1974).
208. Gruden, D., "Combustion and Exhaust Emissions of an Engine Using the Porsche Stratified-Charge Chamber System," SAE paper 750888 (1978).
209. U.S. Patent No. 3,970,053 (1976) Fiat (Turin) and C. L. Goodacre (London).
210. Goodacre, C. L., "Clean Automobiles on Leaded Gasoline," SAE paper 810778 (1981).
211. Holmer, E., Berg, P. S., and Bertilson, B. L., "The Utilization of Alternative Fuels in a Diesel Engine Using Different Methods," SAE paper 800544 (1980).

7

Mechanical Friction and Lubrication in Automobiles

KENNETH C. LUDEMA

I. INTRODUCTION

Friction and lubrication, like liberty, are not fully appreciated until there is none. Friction is defined as the resistance to motion due to sliding, and all machinery consume frictional energy because their parts slide against one another. Lubrication can control friction in some of these instances but very often, such as in engines, most mechanical energy is lost by shearing liquid lubricant films.

Friction is most popularly thought of as the resistance to sliding of dry, solid surfaces over each other. The coefficient of friction, μ, is defined primarily for this case and is numerically defined as the force which causes the sliding, F, divided by the force or load that presses the two surfaces together, W. Values of μ for dry surfaces range from about 0.05 to 10 or more, but the usual range for metals is between 0.1 to 0.7.

A value of μ is often also available in handbooks for lubricated surfaces. It usually ranges from 0.01 to values in excess of 0.5. Strictly, in full fluid film lubrication, the resistance to sliding is the viscous shear in the fluid film rather than due to contact of solid surfaces.

Friction has meaning in more cases than the above. For example, the resistance to rolling offered by roller element bearings and by tires is called rolling friction. The resistance to the flow of fluids along solids is referred to as fluid friction, and finally some types of resistance to the deformation of solids is referred to as internal friction.

KENNETH C. LUDEMA • Department of Mechanical Engineering and Applied Mechanics, University of Michigan, Ann Arbor, Michigan 48109.

Several types of friction exist in the mechanical parts of automobiles, and these will be enumerated below. It is probably most surprising that lubrication is not done to reduce friction but to prolong the life of machinery. Neither is lubrication done primarily to eliminate wear but to control a form of catastrophic surface failure known as scuffing. The several types of wear and the principles of lubrication are given below. In the sections that follow, the discussion will focus upon the mechanical parts, made of common metals in moderate precision, that are found in automobiles.

II. PRINCIPLES OF LUBRICATION AND WEAR

Lubrication ameliorates the damage that occurs to dry surfaces when sliding. All solid surfaces, however carefully made, retain some roughness. For example, crank shafts are ground to a surface finish of about 10 to 15 microinch specification. This refers to the average height of the microscopic bumps or asperities left by grinding and in fact by all surface manufacturing processes. Lapping and honing form smoother surfaces to a finish of about 2 to 8 microinches, but asperities remain.

When two dry surfaces come into contact they actually contact upon the asperities and produce, for the most part, plastic flow in the asperities and in the substrate below the base of the asperities. With sliding, the oxides and adsorbed water film, each about 2 to 7 nm thick, are partially pushed aside from the minute contact spots, exposing a fraction of bare metal on each asperity. When uncovered regions of touching asperities coincide, there is adhesion (bonding, welding) with a joint strength near that of the weaker of the two contacting metals. This adhesion together with a lesser adhesion between metals, oxides, and water vapor is well proven and constitutes the major resistance to sliding. The effect that sliding has on the metal is to severely plastically deform the contacting asperities.

When sliding takes place at a speed greater than 50 cm/sec (100 ft/min) the rate of plastic flow in ductile asperities, and the attendant heating of the asperities is high enough, relative to the rate of heat dissipation into the substrate, to produce a significant temperature rise. The energy input rate is the product, μFV. Still higher speeds, such as the surface speed of grinding wheels, produce visible evidence of heating. Grinding wheel surface speeds are about 30 m/sec (6000 ft/min or 68 mph) and the grinding of ordinary steel produces sparks which are particles heated to above 660 °C (1200 °F).

The effect of surface temperature rise is twofold. First, the thin adsorbed water layer and other "contaminants" are evaporated and are no longer available to reduce bare metal contact. Secondly, heating softens metal and allows severe surface distortion and exposure of large areas of bare metal surface. The latter is a self-regenerating mechanism, often leading to very high friction and sometimes seizing of machine parts to each other.

The basic function of a lubricant interposed between two metal surfaces is to reduce the stress concentration upon asperities by separating the surfaces and by distributing the load over the entire "contact" region. This reduces the plastic flow and heating of asperities considerably. However, viscous shear in the fluid film produces heating as well. By way of comparison of dry sliding with lubricated sliding, let us take the example of a plate of steel, 3 cm thick and 45 cm square, (~45 kg) sliding on a smooth surface at normal walking speed of about 125 cm/sec (~250 ft/min). When the value of μ is 0.25, a force of about 11 kgf is required to slide the plate. However, if the plate is separated from the smooth surface by a fluid film of 0.014 mm (5.5×10^{-4} in.) and of 6 centipoise viscosity (e.g., SAE30 oil at 240 °F, and 7 centistokes) the force required to move the plate at 125 cm/sec is also 11 kgf. The equation governing the conditions in a fluid film between two surfaces is

$$F = \frac{\eta A V}{h} \tag{1}$$

where F is the viscous (sliding) resistance force, η is the fluid viscosity, V is the sliding speed, A is the wetted area over which shear occurs, and h is the fluid film thickness.[(1)]

Note in the above example that the viscous friction force can be less than dry friction force ($F = \mu W$) at low sliding speed but higher at high sliding speed. The conditions were chosen to give the same friction force, which implies also that the same amount of heat will be generated in both cases. The major difference between the two cases is that heat must be conducted away through the bearing material in dry sliding whereas in the lubricated case the heat is generated in the fluid and is largely removed by circulating the fluid through the bearing. Thus even where the viscous losses are high, e.g., in the above example, but at 1000 cm/sec, good fluid circulation will prevent excessive heating of bearings.

In keeping with the modern need to reduce frictional losses in automobiles, it is useful to determine which of the variables in Eq. (1) can be changed to reduce F. Obviously, it is necessary to reduce η, A or V, or increase h, but there are practical limits to these quantities. The magnitidue of A is controlled by design changes, V may be controlled by design and operating conditions, and η is a matter of choice of lubricants. However, film thickness, h, is controlled by many variables and is not simply "designed in," except in bearings in which fluid is pumped into the lubricated conjunction through a hole in either solid surface. Bearings using this principle are known as hydrostatic bearings because the film thickness is not controlled by the motion of the solid surfaces. This design is best seen in distinction to hydrodynamic bearings, which will be described below.

Bearings have developed over the last 200 or more years without benefit of a pump to supply a steady flow of lubricant. If successful lubrication required a flow of pressurized lubricant, bearing technology would never have developed for lack of a successful lubricant pump! Rather, bearings became increasingly useful over the years as substances were found that could be put into bearings to improve their performance. Animal fat was found to be effective until vegetable oil, mineral oil, and sperm oil became available. The industrial revolution brought increased precision and improved materials into the art of bearing-making. As more and more machines were built, the art of bearing making expanded in scope but remained an art.

Only 100 years ago the first science of bearings began and that by the observation of the annoying events in a bearing test program. Beauchamp Tower, an English engineer, had designed a simple test stand to test the load-carrying capability of a simple shaft and journal bearing. The shaft rotated, partly immersed in oil. The journal bearing was loaded downward against the shaft. It had a hole through, such that the pair could be oiled from outside. When the shaft was rotated, oil emerged from the hole "and made a mess."[(2)] The hole was plugged with a wooden plug, but the pressure was sufficient to force the plug out of the hole. A pressure gauge was then fitted to the oil hole. The pressure exceeded the capacity of the gauge, which was 200 psi, whereas the mean pressure between the shaft and bearing was 100 psi.

These results came to the knowledge of Professor Osborne Reynolds at Manchester University, England. He modeled the geometry of Tower's machine as a long shaft of radius R that rotated and "carried" oil from the oil supply upward to the bearing. The bearing radius exceeded R by about $R \times 10^{-3}$. Viscous resistance prevented easy removal of oil from the shaft so that some oil passed through the upper load-carrying region. The higher the shaft speed the more oil flowed through the upper region and the thicker the fluid film became between the shaft and journal. This type of bearing is known as a hydrodynamic bearing because its load-carrying ability depends upon its speed.

Professor Reynolds published an equation for fluid film thickness in 1886, which has been revised and is simply expressed as follows

$$\frac{h}{R} = \frac{2.47\eta VL}{P} \tag{2}$$

where R is the shaft radius, P is the load, L is the length of shaft ($>R$), and the constant 2.47 is appropriate for English units on the variables. This equation served as the basis of journal bearing design for over 50 years. With this equation designers calculated the conditions required to comfortably maintain a value of film thickness, h, greater than the combined heights of asperities on the shaft and journal surfaces.

As machines were built to higher precision, bearing capacity increased, partly because asperities "became" smaller and partly because alignment and other dimensions were well controlled. With the development of airplanes it soon became apparent, however, that Reynolds equations led to an over-design of bearings by as much as 10 times. The reason is that the Reynolds equation assumes the shaft and journal materials to be rigid, thus concentrating the load on the bearing surfaces in a small region. When elastic deformation of each surface is accounted for, the two surfaces conform partially, thus increasing the area of close proximity. Equations that take into account both the hydrodynamics and the elastic state of the solids are equations of elastohydro-dynamics (EHD). One of the early equations of EHD is,[3]

$$\frac{h_{min}}{R} = \frac{0.88(\alpha E)^{0.6}\left(\dfrac{\eta_0 V}{ER}\right)^{0.7}}{\left(\dfrac{W}{LER}\right)^{0.13}} \tag{3}$$

where E is the effective Young's Modulus. This equation introduces the quantity, α, which is the pressure–viscosity index. The quantity, α, reflects the observation in modern machinery that the viscosity, η_0, of oil at atmospheric pressure is not the viscosity of the lubricant in service but it increases because of the fluid pressure in the contact or load-carrying region. This behavior of fluids was not known to be significant by Reynolds.

The above equations have companion equations for predicting the viscous drag or bearing friction, but they are rather complex and give the friction force only over the region of the calculated film. This may be only about 30° of arc at most, which is less than 10% of the bearing length or area as shown in Fig. 1. In this region we may combine Eqs. (1) and (2) to gain a perspective on the variables that influence the EHD viscous friction force:

$$F = \frac{\eta VA}{h} \quad \text{and} \quad h_{min} \propto \frac{\alpha^{0.6}\eta_0^{0.7}}{W^{0.13}}$$

or assume

$$h_{ave} \propto \eta^{0.7}/W^{0.13} \tag{4}$$

then

$$F_i \propto \frac{\eta VAW^{0.13}}{\eta^{0.7}} = \eta^{0.3}VAW^{0.13}$$

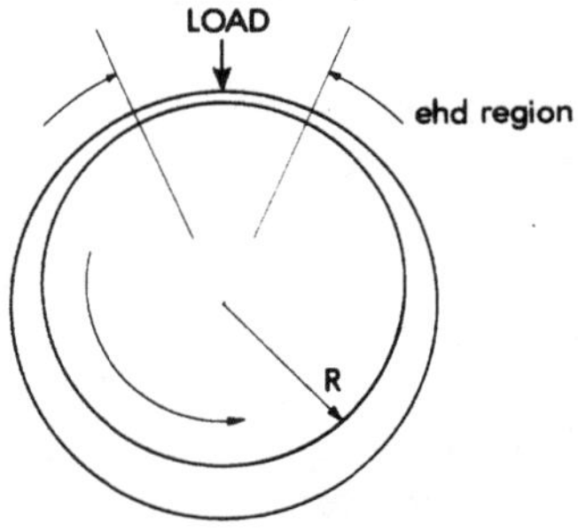

FIGURE 1. Shaft in bearing.

Outside of the EHD film region the average thickness of the fluid film is half the bearing clearance, and

$$F_0 = \frac{2\eta VA}{c} \tag{5}$$

Thus it may be seen that a reduction in viscosity has a larger effect outside of the EHD region than in it.

The total friction force of the shaft-bearing pair is $F_t = F_i + F_0$. The magnitude of F_i/F_0 is strongly dependent on the design of the bearing and the surface condition of the bearings. With the aid of experimental data shown in Fig. 2, by McKee,[4] we may estimate F_i/F_0. Figure 2 gives the value of μ for a lubricated shaft-bearing pair as a function of $ZN'/P'\delta$, where Z is the fluid viscosity, N' is the shaft turning speed, P' is the ratio of applied load to area of the bearing, and δ is the clearance ratio in the bearing (clearance is the difference between shaft diameter and bearing diameter). To the right of the

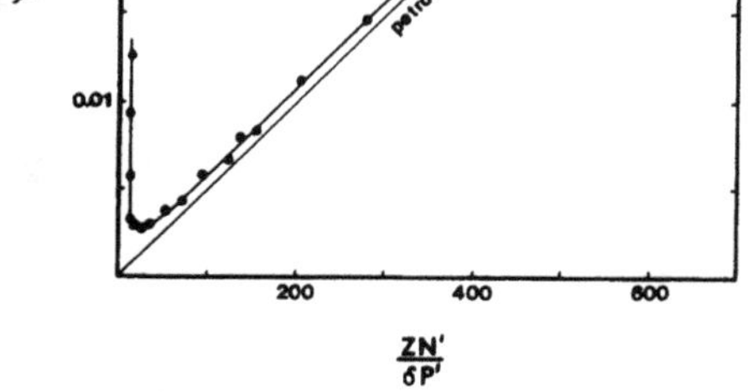

FIGURE 2. Friction for a journal bearing. Z = centipoises; N' = rev/min; P' = load length × dia(psi); δ = clearance ratio × 10^{-3}.

minimum there is full fluid film lubrication, where $F_t = F_0 + F_i$. To the left the fluid film is too thin in the EHD region to prevent solid "contact" between the shaft and the journal. Thus to the left the total friction force is made up of $F_0 + F_i + F_{sc}$, where the latter subscript is for "solid contact."

The fluid film thickness at the minimum of the curve in Fig. 2 has been experimentally measured to be about 10^{-5} in. ($\sim 2.5 \times 10^{-5}$ cm) in well-made bearings. If $c = 10^{-3}$ in., then we can approximate $F_i/F_0 \sim 10$ because the viscous drag per unit area in the EHD region is 100 times that outside, but the EHD region is less than 10% the total area of the bearing. By Eq. (3), a 27-fold increase in shaft speed beyond that at minimum will increase h by a factor of 10. At that point $F_i/F_0 \sim 1$. Where there are time-varying loads, such as in automobile engine parts, the average ratio F_i/F_0 can be taken to be near 0.5. With this ratio it is possible to estimate the importance of varying the parameters, particularly in Eqs. (4) and (5).

The above-mentioned finding, of the minimum in μ being associated with a film thickness of 10^{-5} in., is reasonably compatible with discussions in the literature which suggest that $h_{\min}$ be greater than 3σ where σ is the composite roughness of the "contacting" surfaces. For smaller values of h the surface roughness may strongly influence the flow patterns of the fluid film, which produces conditions outside of those assumed in the analysis. In any case some lubricated surfaces such as gear teeth and camshafts appear to survive even where the calculated value of h is less than σ. Perhaps the analytical difficulties in predicting viscous drag are compounded by the same difficulties.

The equations of EHD are used to design lubricated machinery, particularly weight efficient machinery. There are limits to the practical applicability of parts designed from the equations apart from the accuracy in calculation of h. The most important limit is the case of surfaces that have a load applied when there is insufficient velocity to maintain an adequate fluid film. Such is the case when stopping and starting engines and drive trains.

Starting is a more severe condition than stopping because time is required for film thickness to change. This is a particularly severe problem in the contact area between the engine camshaft and slider followers. One solution to the problem might be to increase the load-bearing area. With a cam this means either a long cam or a profile design producing a low rate of rise. Another solution might be to increase the viscosity of the lubricant, which would provide protection against overload for all rubbing pairs. However, an increase in viscosity merely increases the viscous resistance or drag in close-fitting journal bearings.

An interesting development occurred in the early 1950s to improve the reliability of sliding surfaces particularly for cases of excessive film thinning. Complex chemical compounds were added to engine oil, which adsorb to the metal oxides of bearing surfaces. The exact physical nature of the adsorbate has never been determined. The action of the adsorbate is that of a very thin film of very high viscosity, and yet it appears to degrade with use and needs constant replenishment. The most common protective compound (also referred to as boundary additives, etc.) in use today is zinc-dialkyl-dithio-phosphate (ZDDP). It contains zinc, sulfur, and phosphorous, all of which in separate studies have been found to be effective in friction reduction. The final effect of the ZDDP and similar compounds is that engine parts (and gears) can readily accommodate the absence of fluid film conditions during start-up and they can endure short-term overloads. Thus engines can safely be made with smaller bearings and can use lubricant of lower viscosity than if protective compounds were not available. This might suggest that improved protective compounds should be developed to reduce engine friction, except that these compounds are usually somewhat corrosive. The most severe effect of corrosive species is upon the fatigue properties of the multimetal shell bearings on the crankshaft. Again, a compromise must be reached between short-term overload protection and long-term degradative potential of protective compounds.

Some solid materials are useful lubricants. The two most common are graphite and MoS_2. Less common, in automobile engines at least, are Teflon and other long-chain polymers.

Graphite and MoS_2 have gained their reputation as effective preventers of galling, scratching, and other forms of surface damage in hot forging and similar operations. Hot working is usually done at temperatures where liquid lubricants burn or decompose, leaving the sliding surfaces unprotected. Graphite and MoS_2 are each best for certain applications. Their effectiveness derives from their unique hexagonal layer lattice structure as shown in Fig. 3.[5] The bonds between the layers are relatively weak, allowing shear to take place at low stress. The

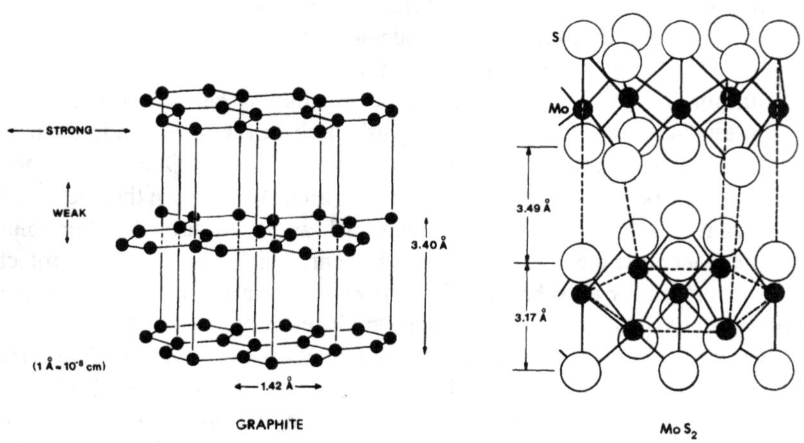

FIGURE 3. Lattice structure of graphite and MoS_2.

shear stress is not strongly dependent on the forces applied perpendicularly to sliding planes. Thus, for large values of W, the value of F is relatively small, yielding a value of μ as low as 0.01. At very small values of W however, the value of μ may be as high as 0.6 for MoS_2 and 0.48 for graphite.

The bond strength in the hexagonal planes of MoS_2 and graphite are very high, which makes the ends of the planes very hard and abrasive. It is this property that causes wear of machinery when MoS_2 or graphite is added to liquid lubricant. However, some wear of machinery may be tolerable if the solid lubricant prevents galling and seizing.

III. FRICTION IN AUTOMOTIVE COMPONENTS

The largest frictional energy loss in automotive components operating at a steady speed is the engine. The details of the engine will be discussed in Sec. IV. The energy balance in engines varies widely in automobiles and depends on the type of service. For example, the combustion energy distribution and use when driving the mixed urban and highway course of the EPA (U.S. Environmental Protection Agency) is divided, 33% into exhaust gas heat, 29% into cylinder and crankcase cooling, and 38% into indicated power (average cylinder pressure multiplied by piston area and speed).[(6)] It would have been more consistent to give a balance in terms of brake power, but the useful and surprising conclusion is that only 12% of the combustion energy is finally used to propel the automobile. This is a loss of 88% of the input energy to heat, which is little better than the 100% loss when an automobile is standing still with the engine running!

Another view of the entire automobile is shown in Fig. 4 for a passenger car with a 348-$in.^3$ engine, moving at 50 mph, showing that energy is lost by internal pumping of gases, by friction in various locations, and by driving the accessories.[(6)] It is difficult to reduce the internal friction in an engine. However, by the elimination of some of the accessories, by the elimination of the automatic transmission and the hypoid ring-pinion final reduction gear, and by reducing tire friction the size of the engine could be reduced by one-third. This obviously reduces engine friction by one-third as well. In addition, all of the conventional vehicle components are capable of duty far more stringent than carrying passengers along a roadway at 50 mph and are proportionately less efficient than they could be.

All of the power transmitted along the drive axle shafts is applied to the drive wheels except for a very small friction loss at the oil seals and rolling element bearings on the drive and axle shafts. The drive wheels must apply a driving force to overcome the small retardation forces due to brakes which do not retract completely, due to the rolling resistance of the tires and due to

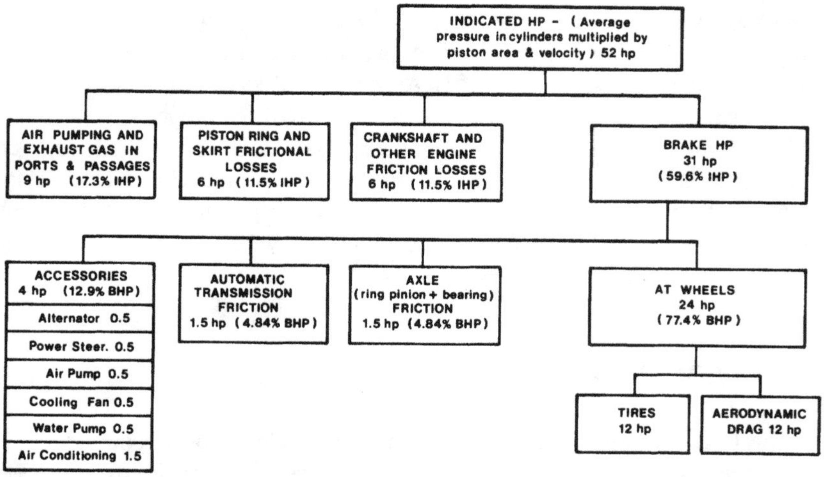

FIGURE 4. Frictional losses in a 348-in.^{3} gasoline engine passenger car at 50 mph.

aerodynamic losses from vehicle motion through air. The rolling resistance of tires is about 20 lb per 1000 lb load and is strongly affected by inflation pressure but hardly affected by speed. Brake drag will range up to 20 lb per vehicle, and aerodynamic drag increases logarithmically with the vehicle speed.

IV. FRICTION IN ENGINES

About 30 ft lb of torque is required to rotate the crank of a large automobile engine. Few drivers experience this torque directly, but they know that continuous cranking of a large engine with the starting motor should be avoided. In fact, less than 5 min of cranking will fully discharge a 150-A h, 12-V battery. This consists of a power dissipation of about 20 Kw or with an efficiency of 50%, a power usage of 10 Kw. Cranking turns the engine at about 125 rpm.

About 10 ft lb of torque is required to overcome mechanical friction in large engines, the remaining 20 ft lb is required to compress air in one or two cylinders at one time. The mechanical friction is the focus of this chapter, but the fluid friction arising from either turbulence and/or flow resistance of air exhaust gases, oil, and water in and near the engine is mentioned for perspective.

Mechanical friction exists at very many locations in an engine. It is very difficult to make friction measurements in engines precisely. The only precise way to do so is to instrument each of the components.

Easier methods have been attempted, such as to turn an unfired engine and to measure the fuel consumption with and without a load. Each method introduces

TABLE 1. Friction Losses in Various Parts of Engines

Bearing	Number	Diameter (in.)	Length (in.)	Wetted area	Surface speed (fpm) for 4000 rpm engine speed	Fluid film thickness h (in.)	Relative friction factor × 10^7	% total
Crankshaft main	5	2.5	1.25	49	× 2618	× 5.5 × 10^{-4}	= 23.8	19.5
Crankshaft rod	8	2.25	0.875	49.5	× 2356 (avg)	× 5.5 × 10^{-4}	= 21.2	17.4
Wrist pins	8	0.75	1	19	× 400	× 5 × 10^{-4}	= 1.52	1.2
Pistons	8	3	3	226	× 3500 (max) 1750 (avg)	× 2 × 10^{-3}	= 19.8	16.2
Piston rings	8	3	0.75	56	× 3500 (max) 1750 (avg)	× 2.3 × 10^{-4} (avg)	= 43.0	35.2
Cam mains	5	1.5	0.75	18	× 700	× 10^{-3}	= 1.26	1.0
Cam profiles	16	1.5	0.75	57	× 700	× 10^{-4} × $\frac{1}{4}$ turn (neg. $\frac{3}{4}$ turn)	= 9.98	8.2
Cam followers	16	1	0.75	38	× 150	× 10^{-3}	= 0.57	0.5
Rocker arm	16	1	1	50	× 20	× 10^{-3}	= 0.1	0.1
Valve stems	16	0.25	1.5	19	× 100	× 2 × 10^{-4}	= 0.95	0.8
								100%
Rocker ends			small					
Distributor/oil pump gear			small					
Distributor			small					
Oil pump			large (1–2 hp)					
Fuel pump			small					

uncertainties due to changes in loading of engine parts, gas flow, and lubricant temperature from that in a fired engine.

For perspective, all of the sliding elements in an engine are listed in Table 1, which permits easy comparison of the relative importance of friction losses in each part. The table also gives, for each part, a number which is its relative friction factor. This is the product of the shear rate of the fluid film in each bearing and the wetted area of the bearing, assuming a uniform film thickness without EHD considerations. The above friction factor multiplied by the viscosity of the fluid gives the viscous drag force in Eq. (1). In the table, to provide a reasonable magnitude of sliding speed and contact condition a large V-8 engine, turning at 4000 rpm, is used.

V. FUEL ECONOMY

From Table 1 it is readily apparent that considerable energy could be saved first of all by using engines of less excess capacity than in a 350-in.^{3} engine, and further by reducing the friction at several locations in the engine, mainly the crankshaft and the piston rings. As mentioned above, there are solutions to the large energy consumption of several other components, as given in Table 2.

In engines the proper method of reducing friction will depend on where the part is operating relative to the minimum safe fluid film thickness, as illustrated

TABLE 2. Solutions for Various Friction Losses in Automobiles

Dragging brakes	Improve retraction of brakes
Rolling loss of tires	Increase inflation pressure, use natural rubber which has low damping, and use very stiff belts in the tire
Air pumping in the engine	Redesign the parts and passageways
Aerodynamic losses around the body	Revise body design
Hypoid gear in differential housing	Use lower viscosity oil, or use spur gears or chain drive (available on some front-wheel drive automobiles)
Automatic transmission	Include a lock-up clutch and minimize oil turbulence
Accessories	Beyond the scope of this chapter

in Fig. 2. Recall that in components of fixed clearance, such as the crankshaft and bearings on the crankshaft, there is a time-varying "whirling" and oscillating load on each bearing. Thus a film-thinning region moves around in the bearing while the lubricant in the remaining major arc of the bearing performs no useful function except to be available for a load-carrying excursion. There is therefore a continual viscous drag over the entire 360° in the bearing clearance, which could be reduced by using a low-viscosity lubricant. In principle, a larger clearance and smaller bearing would serve the same purpose but with some loss of bearing reliability. These same overall principles apply to other bearings and the pistons in the engine but not to piston rings and the cam-lifter interfaces.

The latter sliding surfaces are supported by an EHD film, for which Eq. (4) may be recalled. It can be seen that a change in viscosity is a less effective means of changing the viscous friction forces in piston rings–cylinder and cam-lifter pairs. Of the two, the piston ring–cylinder pair consumes much more frictional energy, so it must be further studied. In fact, a solution to the cam-lifter pair exists in the use of rollers in the place of sliders.

Piston ring friction can be reduced by at least two design changes, but the primary sealing function of the rings might then be compromised. These are to decrease the pressure of the ring against the cylinder wall and by designing the ring surface to maintain a thick EHD film between the ring and cylinder. These factors can only be discussed in general terms because there is no detailed or formal methodology of piston ring design available.

A common principle to all sliding surfaces in an engine is that they all suffer some damage when starting, when overloaded, or when the fluid film is inadequate. Piston rings stop and start twice per turn of the crank, which makes the piston ring and cylinder pair the most crucial design problem in the engine. In fact, the design activity of engines is supported by considerable amount of testing of engines on dynamometers and engines in the hands of the general public. Frictional energy loss cannot, in general, be predicted accurately but in most components it is not possible to have minimal friction loss and maximum component life without imposing limits on how the engine shall be operated. That may come.

REFERENCES

1. Cameron, A., *Principles of Lubrication*, Chapter 1 and 2, Wiley, New York (1966).
2. Cameron, A., *Principles of Lubrication*, p. 271, Wiley, New York (1966).
3. Cameron, A., *Principles of Lubrication*, p. 209, Wiley, New York (1966).
4. McKee, S. A., and McKee, T. R., *Trans. Amer. Soc. Mech. Eng.*, 161 (1929).
5. Buckley, D., *Surface Effects in Adhesion, Friction, Wear and Lubrication*, p. 573, Elsevier, Amsterdam (1981).
6. "Strategy for Energy Conservation through Tribology", (O. Pinkus and D. F. Wilcock, eds.), Res. Comm. on Lubrication, Amer. Soc. Mech. Eng., New York (1977). Fig. 4 is a composite of figures from pp. 26 and 68 in this reference.

8

Tire Rolling Resistance and Vehicle Fuel Economy

S. K. CLARK

I. INTRODUCTION

Interest in fuel conservation and the goal of more energy efficient vehicles has generated considerable interest in the phenomenon of the rolling resistance of pneumatic tires. It is generally recognized that the pneumatic tire represents one of the major loss mechanisms for the engine output of a vehicle, the other mechanisms being aerodynamic loss, transmission and drive train inefficiency, and the power needed for acceleration of the vehicle.

The importance of tire rolling resistance clearly lies in its influence on vehicle fuel economy. This is difficult to determine quantitatively, since field test results are subject to large fluctuations due to varying atmospheric conditions and engine map characteristics. However, extensive work at General Motors Proving Grounds[(1)] has shown that the slope of fuel consumption, change with change in total vehicle retarding force, i.e., drag of all four tires, is approximately 0.02 gal/100 miles per pound of tire drag for a typical passenger car powered by a gasoline internal combustion engine. This is illustrated in Fig. 1.

The slope of this effect seems to be independent of driving schedule, but may, of course, vary for different classes of vehicles and different types of engines.

Assuming a fleet of 10^8 vehicles in the United States, each operating 10^4 miles per year, one concludes that an improvement in tire rolling resistance of 0.25 lb per tire, easily achieved by small changes in inflation pressure, would result in a fuel saving of 4×10^6 barrels of gasoline per year.

S. K. CLARK • Department of Mechanical Engineering and Applied Mechanics, University of Michigan, Ann Arbor, Michigan 48109.

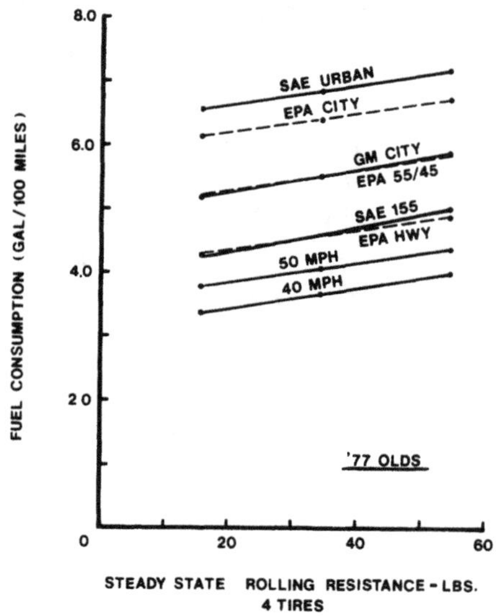

FIGURE 1. Influence of rolling resistance on vehicle fuel consumption.

Assuming the same vehicle fleet, but with fuel efficient tires exhibiting a decrease in rolling resistance of 2.5 lb per tire, the fuel savings will amount to 40×10^6 barrels, or 7×10^6 metric tons, of gasoline per year.

A rule of thumb which seems to be generally true for passenger car vehicles is that a five-to-one ratio exists between improvements in tire rolling resistance and improvements in fuel economy, i.e., a 10% improvement in tire rolling resistance will yield a 2% improvement in fuel consumption.

Both savings are significant and make a more thorough knowledge of tire rolling resistance well worth the effort.

II. FUNDAMENTALS OF TIRE ROLLING RESISTANCE

The forces acting on a free rolling tire under constant velocity and straight line motion are shown in Fig. 2. The F_r is the tire rolling resistance, which arises from both the tangential force F_r at the tire road interface, which is *not* necessarily caused by highway friction, and the shift of the vertical center of pressure, which is caused by tire material internal losses.

Most data on the rolling resistance of a tire is given in terms of the force F_R, although other measures such as power loss or efficiency clearly could be used.

Under conditions of free rolling or steady-state driving, where external torque applied to the wheel is not large, it is conceded that the hysteretic com-

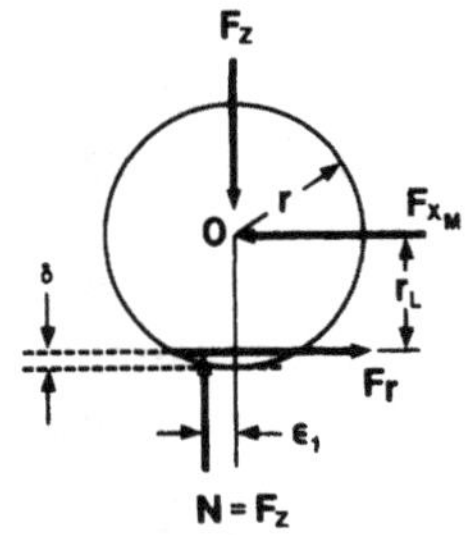

FIGURE 2. Free-body diagram of a rolling tire on a flat surface.

ponent of the total loss is the dominant one, making up something more than 90% of the total mechanical losses in the tire. Aerodynamic drag of the tire itself and friction between the tire and roadway are minor components. This has been demonstrated by Bowden and Tabor.[(2)]

Starting from rest at ambient temperature, the temperature of the rolling tire begins to rise as the vehicle starts in motion due to the fact that the hysteretic losses inside the material generate heat. Due to the poor thermal conductivity of rubber, it takes a relatively long time for the tire to come into temperature equilibrium. For example, a normal passenger car under typical loads requires about 30 min to reach steady-state temperature beginning from rest. Eventually this equilibrium is attained. This higher temperature is not uniform throughout the tire body but varies according to tire geometric and material design. Due to mixing, the air inside the tire cavity will have a higher average temperature than at the beginning of the rolling process. This results in a higher pressure than the

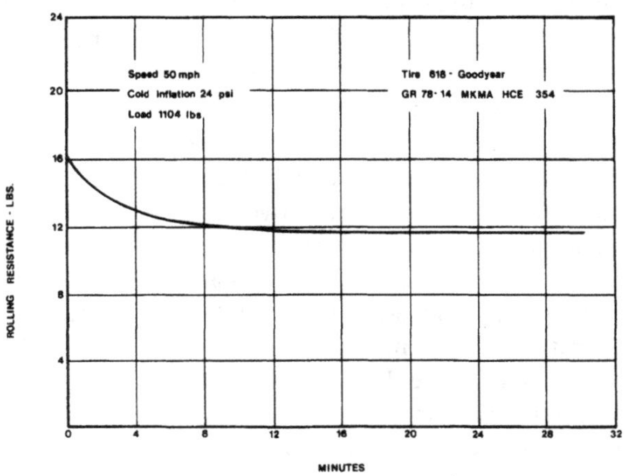

FIGURE 3. Typical rolling resistance versus time variation.

original inflation pressure of the tire, due to the increased temperature of the trapped air. This rise in pressure causes a reduction in the tire deflection. The hysteretic loss properties of the tire materials are temperature sensitive and they, coupled with the decreased deflection, both influence the equilibrium state rolling loss value. A typical plot of rolling resistance versus time is shown in Fig. 3 for a modern passenger car tire.

From this discussion it may be seen that an unambiguous description of the rolling loss of a pneumatic tire requires the complete specification of its operating characteristics. For example, one must specify temperature state and inflation state, both of which depend not only on the length of time of running but also on ambient temperature and on the load carried.

III. TEST PROCEDURES

The measurement of tire rolling resistance is not a particularly simple one since it is necessary to measure a small force, say 12 lb such as shown in Fig. 2, in the presence of a large load on the tire, such as 1000 lb. This means that whatever force-measuring device is used must be capable of exerting large vertical forces while at the same time measuring very small horizontal forces. For this reason most tire rolling resistance measurements are made indoors on a cylindrical test drum, typically used in the tire industry for durability and qualification testing, because only indoors can the ambient effects such as temperature change, road surface conditions, and speed be accurately controlled. All of these effects are present in outdoor testing and tend to distort the results of tire rolling resistance measurements.

By far the most common method for the determination of tire rolling resistance is the use of measurements of the force or torque needed to rotate a 67-in. steel drum while the tire is loaded against it. The simplest technique used in drum measurements is to measure the torque necessary to rotate the wheel when the tire is just in the skim position, i.e., in contact with a load of 10 lb or so, followed by the torque necessary to rotate the drum when the tire is in its fully loaded position. The difference in these two readings tends to cancel out any bearing losses and gives the torque necessary to rotate the drum because of the loaded tire against it. This is the tire rolling resistance. While this method requires a good quality torque transducer, it is not particularly difficult to implement. Other methods that are commonly used for drum measurements are the power input to the drive system of the drum, which is treated in exactly the same way as the torque measurement, and in which a so-called skim or initial reading is taken followed by a reading of the tire in the fully loaded state. An additional method, not now widely used, involves warming the tire up against the drum and then allowing the drum to coast freely down to rest. The time necessary for this to happen, or the time necessary for the drum to reach some predetermined lower speed, is taken as a measure of the tire rolling resistance.

Finally, on some occasions the tire axle is instrumented and the force tangent to the drum surface is actually measured by some form of strain gauge transducer. This method, while also accurate, requires additional data reduction and is dependent upon very high quality and carefully aligned force transducers. It is not widely used.

A variety of methods have been developed for measuring the rolling resistance of a tire under the more realistic conditions of operating on a flat roadway. These methods generally have demonstrated substantially less precision than is possible on indoor tests but have the advantage of giving a more realistic picture of the tire operating conditions, since they are done on a flat surface, as well as involving realistic road roughness effects. A discussion of the most popular methods follows.

A. Trailer Tests

In this method one or two tires are used on a loaded trailer which is towed behind a vehicle using normal road surfaces. The load or force necessary to tow the vehicle is measured and used as a measure of the tire rolling resistance. This method suffers from the difficulty that both the acceleration of the vehicle and road gradients cause large forces in the trailer, forces which mask the relatively small rolling resistance. For this reason such trailers are difficult to instrument and use.

B. Fuel Consumption

In this method the vehicle is equipped with tires of a particular type and run over fairly long distances while measuring the fuel consumed during the trip. If the distances are long enough and if the same route is used, random effects may eventually be cancelled out by the statistical nature of their occurrence. However, this method suffers from the difficulty that during long-time tests ambient conditions, such as temperature or humidity, may change and these in turn effect the engine efficiency. In addition, temperature effects themselves change the tire rolling resistance, and so such fuel consumption tests are often done using pairs of vehicles and more complicated experimental designs in order to cancel out unwanted effects.

C. Coast-Down Tests

Coast-down tests are sometimes used as measurements of tire rolling resistance since they can be performed rather quickly, and thus they avoid the long-term ambient changes associated with direct fuel consumption measurement. This test method consists of mounting a set of tires on the vehicle, bringing it to a fixed speed, and then either depressing the clutch or putting the vehicle in neutral and measuring the total distance through which it rolls before coming to rest. One of the difficulties with coast-down methods is that the stopping distance is affected not only by tire rolling resistance but also by aerodynamic resistance and drive-train friction from bearing and differential losses. The aerodynamic loss is highly velocity sensitive and is difficult to extract from the tire rolling resistance portion. In addition, the drive-train friction may not necessarily be independent of velocity. For these reasons coast-down methods, while superficially attractive, have proven difficult to implement in a large-scale way.

IV. INFLUENCE OF TIRE CONSTRUCTION VARIABLES

Conant[3] has studied the influence of tire construction variables in detail. Some of his conclusions are summarized in this section.

A. Tire Type

Radial tires have become the standard for passenger car use because of their excellent wear, traction, and rolling resistance characteristics. They seem to excel in each category because strains imposed by footprint loading are concentrated in the sidewalls, and thus they undergo a minimum of unnecessary deformation in the contact patch area of the tread.

The ability of the radial-ply tire to deform to the flat road with less tread distortion or shuffle, together with the virtual elimination of the sidewall interply shear, probably accounts for the considerable reduction in rolling resistance of radial tires.

Figure 4 illustrates the superior fuel economy of radial tires over belted-bias and bias tires in a statistically significant test. The comparison is on a practical basis since the tires are production and the results are in terms of actual savings.

Radial truck tires show about the same gain in fuel economy over bias-ply tires as do passenger car tires. In one study this gain was 5–7% in a comparison of 16R-20 to 16-20 10-ply rating tires on army vehicles over a variety of terrain surfaces. Comparisons were made on the basis of ton–miles per gallon.

B. Tire Size

Upsizing of tires generally decreases the rolling resistance coefficient if operating factors are held constant.[4] The effect is due mostly to the reduced

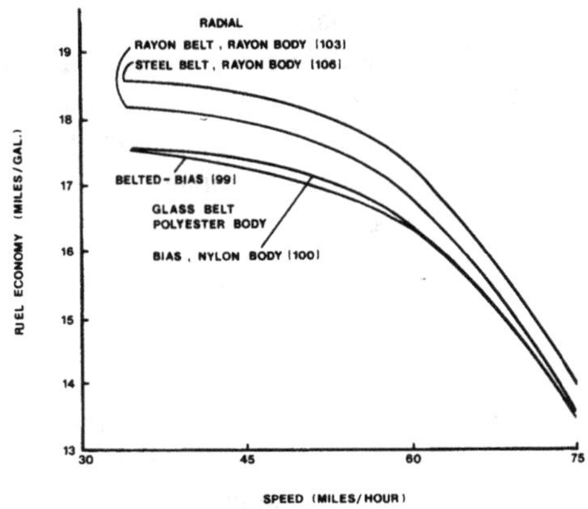

FIGURE 4. Tire type, cord material, and speed effects on fuel economy.

tire deflection needed to carry the same load. This applies to increase in either rim diameter or section width and of course is beneficial to fuel efficiency. Table 1 shows recent data taken on a variety of passenger car tires at the same load.

There is little analytical information on the influence of tire size and geometry on rolling resistance, but two studies throw some light on this question. The earlier, by Evans,[5] shows that for a solid tire the rolling resistance is approximated by

$$F_R = K\frac{W^2}{D}\left(\frac{h}{W}\right)^{1/2} \tag{1}$$

TABLE 1. Equilibrium Rolling Resistance for Tires of Various Sizes under the Same Load

Tire	Mfgr.	Load/rated load	Equilibrium rolling resistance, lb
C78-14	Goodyear 2P + 2F/2P	840/1050	11.44
E78-14	Dunlop 2P + 2F/2P	840/1190	10.03
F78-14	Dunlop 4P	840/1280	9.55
H78-14	Goodyear 2P + 2F/2P	840/1510	9.63
GR78-15	Goodrich 2P + 2S/2P	1104/1380	12.49
LR78-15	Goodrich 2P + 2S/2P	1104/1680	11.97

where K is a constant depending on material characteristics, W is the load on the tire, D is tire diameter, h is section height, and W section width.

Trends in experimental measurements seem to confirm Eq. (1), although in many cases the effects of geometry are masked by the changes in other characteristics which accompany changing geometry. Nevertheless, the conclusions seem to be that for a fixed load:

1. increasing tire diameter reduces rolling resistance and
2. decreasing aspect ratio reduces rolling resistance.

A similar study was carried out by Clark[6] for pneumatic truck tires, and in this study it was concluded that replacement of two conventional 10 : 00–20 truck tires by a single wide-base tire could result in measurable fuel savings. It was also shown that increasing tire diameter reduced rolling resistance at a given load. A typical result from this study is shown in Fig. 5. It was also shown that for multiwheeled vehicles appreciable increases in efficiency could be obtained by increasing the load carried per tire, both by increases in section width and in inflation pressure. Typical data of this type is shown in Fig. 6.

Finally, confirmation of the trends indicated by these computations has been provided by measurements taken at Delft Technical University, Netherlands. These are shown in Fig. 7.

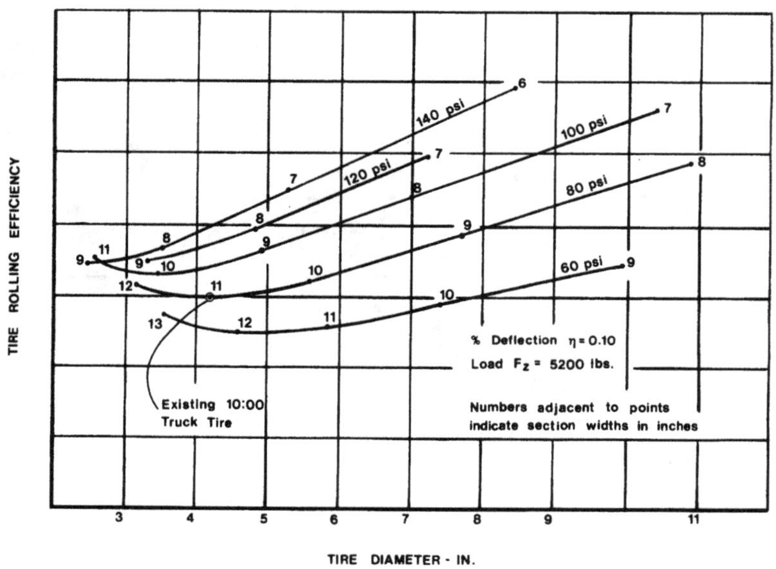

FIGURE 5. Influence of tire diameter at constant load and aspect ratio.

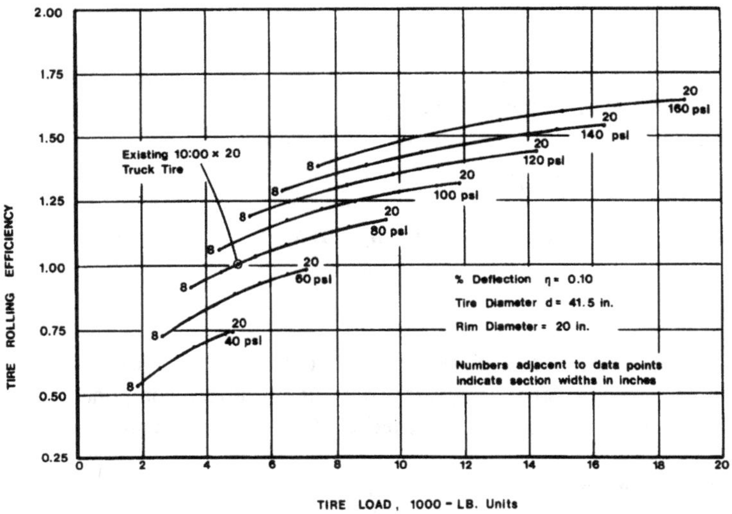

FIGURE 6. Influence of tire pressure, load, and section width at constant tire diameter.

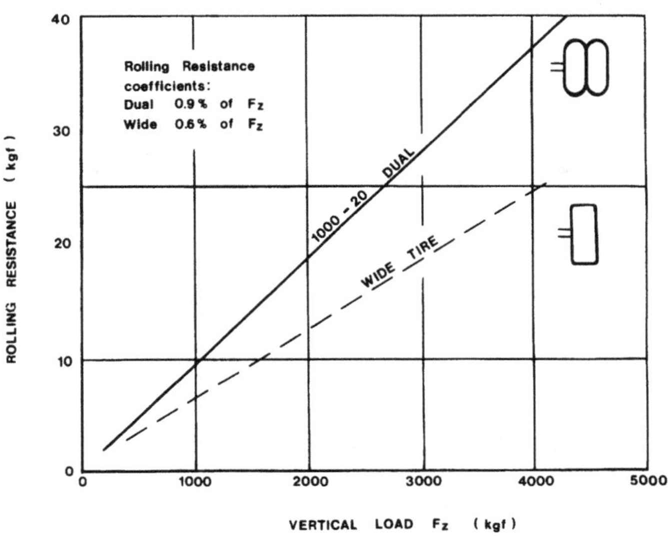

FIGURE 7. Rolling resistance versus vertical load for truck tires.

C. Cord Material

Cord properties also affect rolling resistance, both through losses within the cords by their own hysteresis and by the effect of their modulus on rubber deflection. Figure 8 shows Aramid to be better than the higher modulus steel in a belt and better than lower modulus nylon or rayon in the body. Since the tests were at 50 mph, the mass of the steel may have been important to inertial distortion at the footprint exit. More than hysteretic losses need to be considered.

D. Polymer

Tire rolling resistance is obviously related to the viscoelastic (i.e., temperature and frequency dependent) properties of its rubber components. Many investigators have either assumed or tried to determine the nature of the relationship.

The tread is generally acknowledged to be the largest contributor to viscoelastic losses, and its duty cycle is seemingly a combination of radial compression, bending, and scrubbing. A direct approach, therefore, is to construct tires with treads having a range of viscoelastic properties and to relate these properties to rolling losses of the tires. One such study on radial tires found better correlation to be given by dynamic damping coefficient (loss modulus) of the tread compound than by its dynamic spring rate or rebound. In this study seven tread polymers were compounded to the same hardness. Natural rubber had the lowest rolling

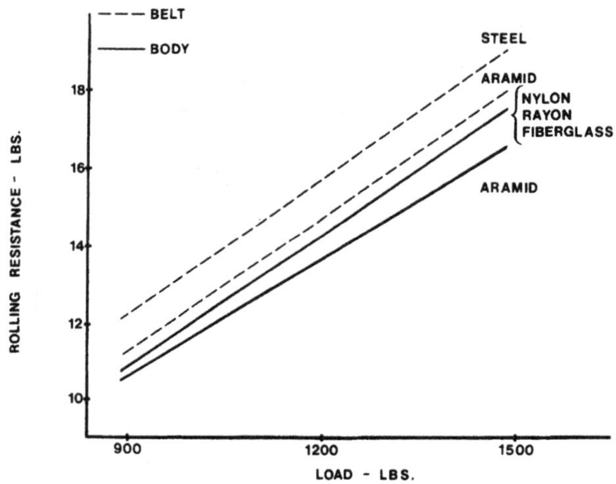

FIGURE 8. Effect of cord material on rolling resistance.

ranked at 100, the IIR (butyl) was 73, while IR (polyisoprene) and BR (butadiene) were at 99. Rankings were unchanged by variations in load and pressure, and very slightly by warm-up.

The contribution of the rubber in the tread and that in the body may not be additive. Rather, there seems to be a pronounced interaction between the two which can cause the power loss to be greater than anticipated when different polymers are included. This possible interaction has not been fully considered by investigators who have tried to allocate energy consumption to specific parts of a tire by experiments that presumed to separate the effects.

V. EFFECT OF OPERATING VARIABLES

Since all rubber compounds exhibit temperature sensitive loss characteristics, rolling resistance of a pneumatic tire depends on its operating temperature, which in turn is controlled by its inflation pressure, load, and the length of time which the tire has run.

There has been considerable interest in the influence of load and running time on the rolling resistance of pneumatic tires, and Figs. 9 and 10 present data on this subject for two common sizes of passenger car tires. Here the tire rolling resistance is plotted as a function of load carried for the case where the cold inflation pressure is set prior to the beginning of the test. Plots are given at various values of time ranging from the initial value up to the equilibrium value. These figures show that there is a nearly linear relationship between the tire rolling resistance and the load for the equilibrium case. This is illustrated by means of a dashed line extending from the data points to the zero load condition.

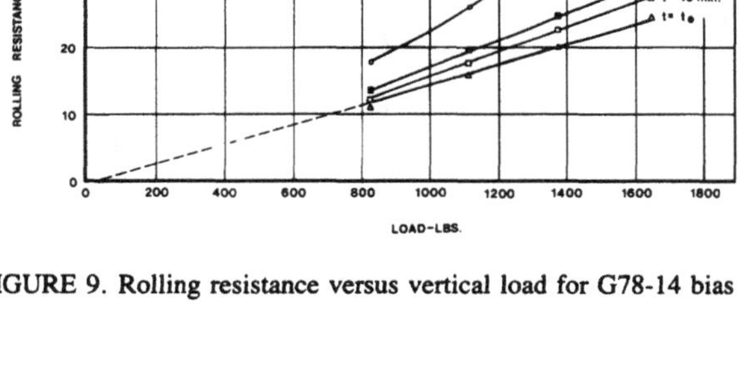

FIGURE 9. Rolling resistance versus vertical load for G78-14 bias tire.

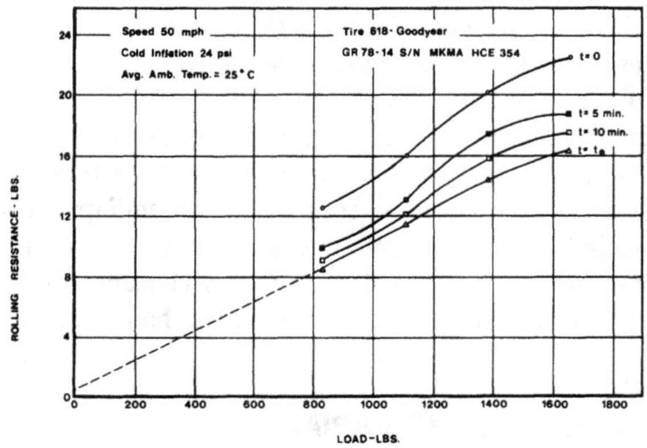

FIGURE 10. Rolling resistance versus vertical load for GR78-14 radial tire.

The linear nature of the equilibrium rolling resistance as a function of load is apparently fortuitous, but is well known and has led to the common and very useful concept of the coefficient of rolling resistance, which is defined as the rolling resistance divided by the load carried. Using the data of Figs. 9 and 10, the coefficient of rolling resistance may be replotted as a function of load and time and is shown in Figs. 11 and 12 for the same two tires.

The coefficient of rolling resistance is a convenient concept since it allows one to compare various tires for use on the same vehicle. The load carried by a tire will be the same on a given vehicle in a given tire position, so a comparison of the rolling resistance coefficients will show which tire is the most efficient for each application.

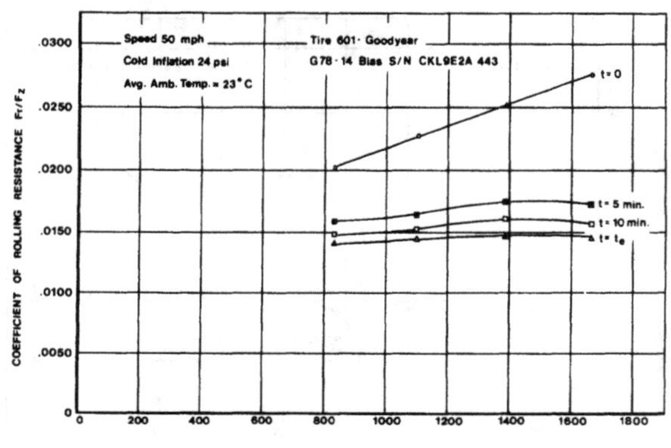

FIGURE 11. Coefficient of rolling resistance versus vertical load for G78-14 bias tire.

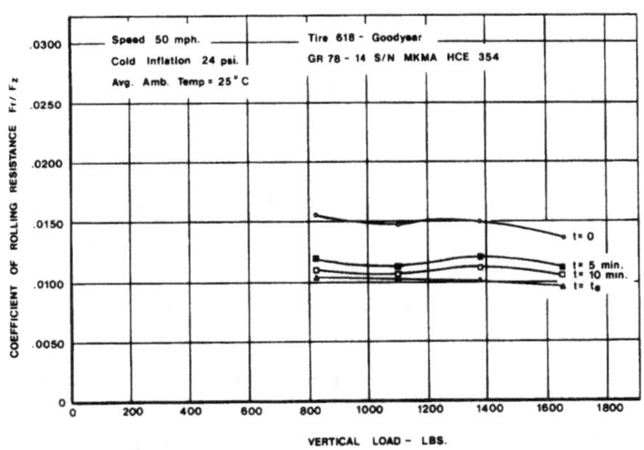

FIGURE 12. Coefficient of rolling resistance versus vertical load for GR78-14 radial tire.

Examination of data taken at fixed tire load and variable initial inflation pressure shows that the tire rolling resistance decreases as the inflation pressure is increased. This is caused primarily by the reduced deflection of the tire when running under a higher inflation pressure as compared with lower. The rolling resistance is nearly linear with the reciprocal of initial inflation pressure under conditions of capped air, steady-state running, and constant load. The two tires of Figs. 9–12 were tested under a variety of initial inflation pressures and the rolling resistance values are plotted as a function of the reciprocal of inflation pressure in Figs. 13 and 14, again for various times so that the influence of running time may be illustrated.

The question of speed effect on rolling resistance has been one of consid-

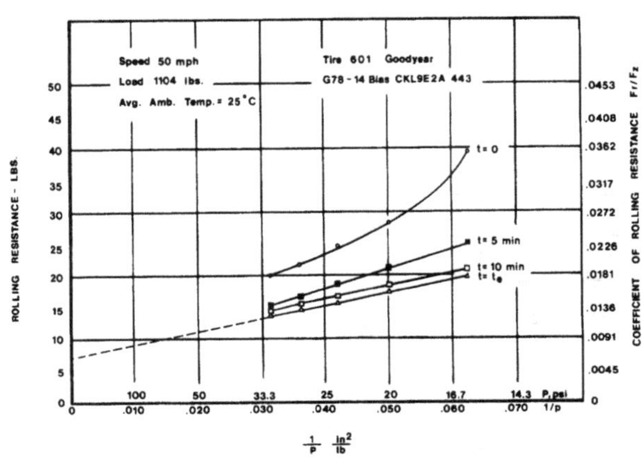

FIGURE 13. Rolling resistance and coefficient of rolling resistance versus reciprocal of inflation pressure for G78-14 bias tire.

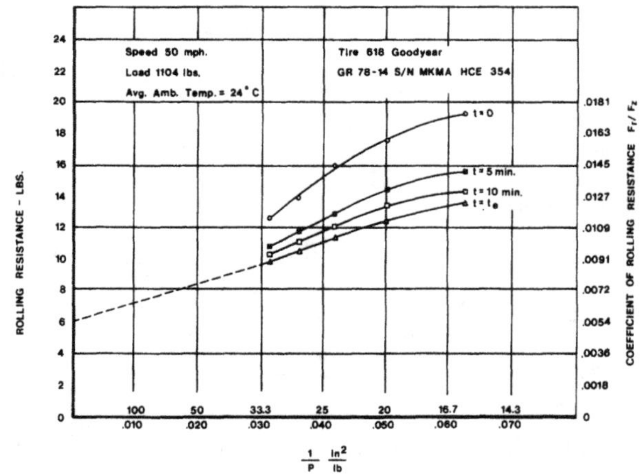

FIGURE 14. Rolling resistance and coefficient of rolling resistance versus reciprocal of inflation pressure for GR78-14 radial tire.

erable uncertainty in the earlier published literature. Some measurements have been reported showing rather marked effects of speed on rolling resistance, while other measurements show very little effect. A recent series of measurements on a single set of tires carried out by a number of the major American tire manufacturers showed that the rolling resistance of the sample tires was nearly independent of speed within the speed ranges normally encountered on American highways today. Figure 15 shows the mean value of rolling resistance for five

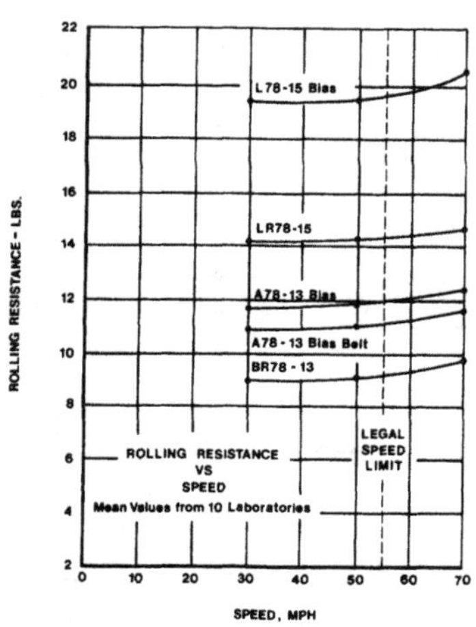

FIGURE 15. Rolling resistance versus speed for a group of modern passenger car tires.

different common American passenger car tires at three different speeds as measured over nine different laboratories ranging from a 64-in. diameter cylindrical drum to a flat surface.

Using the linear relationship between load, reciprocal of pressure, and tire rolling resistance and assuming that rolling resistance is independent of speed, then it may be shown[(4)] that the load-inflation pressure rolling resistance relation for a given tire is expressed by

$$F_r = F_{r_0}\left(\frac{F_z}{F_{z_0}}\right)\left[1 + c_p\left(\frac{p_0}{p} - 1\right)\right] \tag{2}$$

This is an expression for rolling resistance at any load and pressure as a function of a baseline value of rolling resistance F_{r_0} at the baseline conditions of load F_{z_0} and pressure p_0. This expression contains only one constant, c_p, characteristic of the tire and expressing the sensitivity of that tire's rolling resistance to inflation pressure. Hence, the use of the symbol c_p denoting a pressure coefficient is appropriate.

This means that the entire load-pressure–rolling resistance map of the tire can be determined if one value of the rolling resistance F_{r_0} is known at one load F_{z_0} and one inflation pressure p_0, provided that the constant c_p is also known for the tire. This constant may be determined most easily by fixing the load F_{z_0} at its baseline value and measuring the tire rolling resistance over several pressures.

Values calculated using Eq. (2) are compared with measured values in Table 2. It is seen that agreement is generally quite good.

TABLE 2. Measured and Predicted Equilibrium Rolling Resistance Using Eq. (2)[a]

	500 lb	1000 lb	1500 lb	2000 lb	2500 lb	3000 lb
20 psi	6.25	13.80	21.10	—	—	—
	(6.89)	(13.78)	(20.68)	—	—	—
35 psi	5.25	10.70	15.90	—	—	—
	(5.28)	(10.56)	(15.84)	—	—	—
50 psi	4.95	9.50	13.90	18.50	22.80	—
	(4.63)	(9.27)	(13.90)	(18.53)	(23.17)	—
65 psi	4.85	8.80	12.90	16.80	20.80	24.80
	(4.29)	(8.57)	(12.86)	(17.14)	(21.40)	(25.70)
80 psi	4.75	8.70	12.20	16.27	20.30	24.40
	(4.07)	(8.13)	(12.20)	(16.27)	(20.30)	(24.40)

[a] 67-in. drum; 8.75 R-16.5 light-truck tire; c_p = 0.325; Abs. avg. error = 0.33 lb (Calculated values are in parentheses).

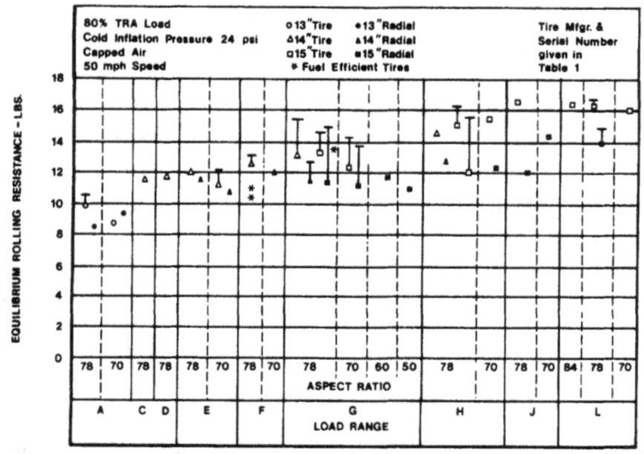

FIGURE 16. Rolling resistance versus tire load rating for a typical selection of passenger car tires.

VI. CURRENT STATUS

The present status of passenger car tire rolling resistance is well summarized in Fig. 16, where coefficients of rolling resistance at equilibrium are presented for a large variety of modern American passenger car tires. This shows that radial tire designs are somewhat more efficient than bias tire designs, but that on the whole the present average coefficient of rolling resistance in the United States is in the range of 0.011 to 0.012. It also shows traces of the scale effect, i.e., small tires generally exhibit somewhat higher rolling resistance coefficients than do larger ones. This effect is now generally accepted as being inherent in most test observations.

VII. FUTURE STATUS

While the present understanding of the rolling resistance of pneumatic tires is greatly improved over that of the past, considerably more work remains to be done. At this time the rather considerable advances which have been made in reducing tire resistance have been through:

1. lighter tires with less material in them as a result of lower overall vehicle weights;
2. higher tire inflation pressures, as some ride quality is sacrificed for fuel economy; and
3. more use of radial tire designs, coupled with lower hysteresis materials.

Efforts still need to be made to understand and exploit the optimum composite geometries of cord rubber materials for the tire service.

REFERENCES

1. Private communication, General Motors Proving Grounds, Milford, Michigan. Courtesy of Richard Moore.
2. Bowden, F. P., and Tabor, D., *Friction and Lubrication,* Wiley, London (1956).
3. Conant, F. S., "SAE Course in Tire Rolling Losses," Society of Automotive Engineers, Inc., Detroit, Michigan, June (1978).
4. Clark, S. K., "A Handbook for the Rolling Resistance of Pneumatic Tires," University of Michigan Research Report 013658-I-T, prepared for the Dept. of Transportation, Transportation Systems Center, Cambridge, Mass. (1978).
5. Evans, I., "The Rolling Resistance of Wheel With a Solid Rubber Tyre," *Brit. J. Appl. Phys.,* **5,** 187 (1954).
6. Clark, S. K., and Dodge, R. N., "The Influence of Tire Geometry on the Rolling Resistance Efficiency of Commercial Vehicle Tires," University of Michigan Report No. UM-013663-4-I, prepared for Dept. of Transportation, Transportation Systems Center, Cambridge, Mass. (1976).

9

Properties of Sheet Molding Compounds

GEORGE S. SPRINGER

I. INTRODUCTION

The use of lightweight fiber-reinforced plastics (FRP) in automotive applications offers a promising avenue towards achieving better fuel economy through weight reduction. In addition to their high strength-to-weight ratios, FRP materials offer other potential advantages over sheet metals, such as a wide choice of thickness and shape, reduced cost, corrosion resistance, and good vibration damping properties.

Owing to their favorable performance characteristics, lightweight composite materials have been gaining wide use in passenger cars (Fig. 1). However, FRP materials have been used mostly as body panels where the material is not subjected to heavy loads. In order to take full advantage of the weight reduction possibilities of FRP materials, they should be used also as load-bearing members. Such applications require a comprehensive data base. For this reason in recent years several investigators have measured the properties of glass fiber-reinforced sheet molding compounds (SMC), which are the materials most widely employed in automotive applications. In this chapter a summary is given of the engineering properties of SMC materials and of the effects of service environment on these properties. Attention is focused on SMC materials since they can be processed in the high volumes necessary in the automotive industry. Continuous fiber composites that require long process time (e.g., graphite–epoxy composites) are not included in this survey.

In presenting the results emphasis is placed on the main features and char-

GEORGE S. SPRINGER • Department of Aeronautics and Astronautics, Stanford University, Stanford, California 94305.

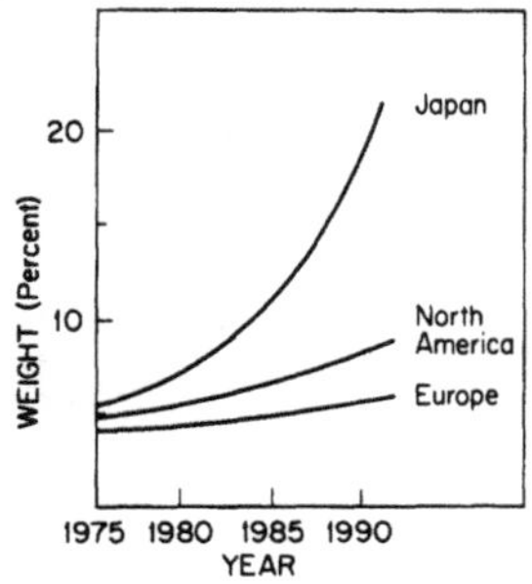

FIG. 1

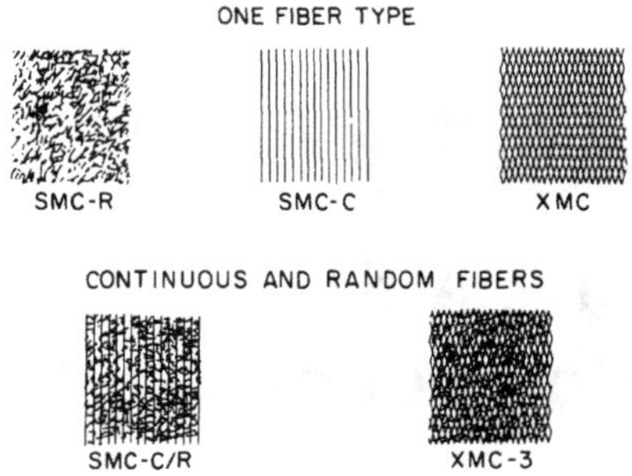

FIG. 2

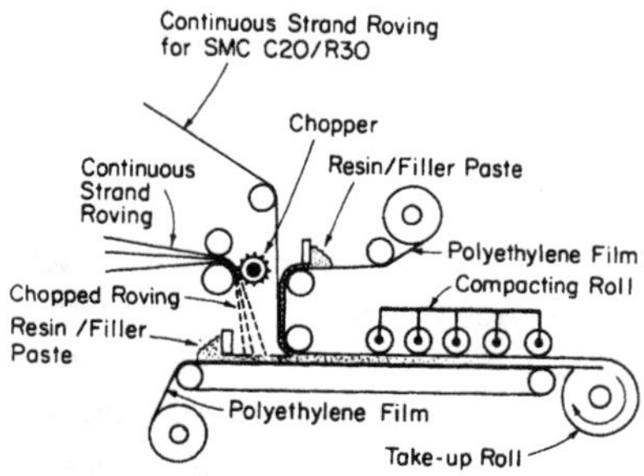

FIG. 3

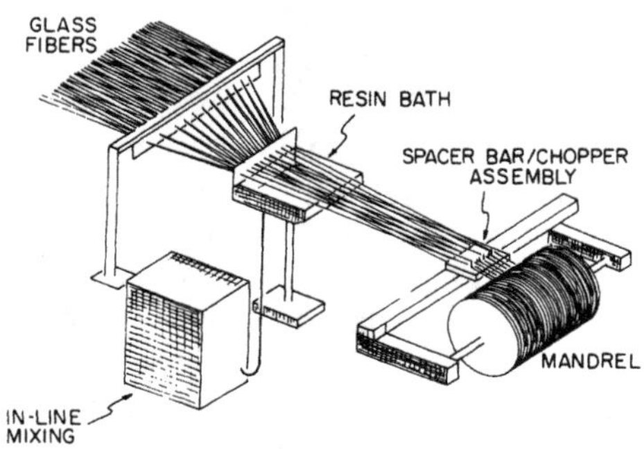

FIG. 4

FIGURE 1. The use of plastics in passenger cars (Ref. 1).

FIGURE 2. Illustration of glass fiber reinforced sheet molding compounds.

FIGURE 3. Schematic of process used to manufacture glass fiber reinforced SMC composites (Owens Corning Fiberglas System).

FIGURE 4. Schematic of process used to manufacture XMC composites (PPG Industries System).

acteristics of the data. Readers interested in details of the material behavior are referred to the appropriate references quoted in the text, figure captions, and table headings.

II. MATERIALS AND PROCESSING

Sheet molding compounds consist of polyester (or, less frequently vinylester or epoxy) resins reinforced with glass fibers. A given material may contain randomly oriented chopped fibers (designated as SMC-R), continuous fibers (SMC-C and XMC), or a combination of chopped and continuous fibers (SMC-C/R and XMC-3), as illustrated in Fig. 2. Numbers added after the letters R and C indicate the weight percent of chopped and continuous fibers, respectively. XMC contains 75% glass fibers by weight. The continuous fibers provide added strength and stiffness in the direction of maximum load. The random fibers provide strength and stiffness in the direction perpendicular to the fibers. The formulations and densities of different types of materials are given in Tables 1 and 2.

SMC materials are processed on machines illustrated in Figs. 3–5. The materials are compounded on machines shown in Fig. 3 (used for all types of SMC materials except XMC-3) and in Fig. 4 (used for XMC-3). The compounded material is allowed to set at room temperature for 5–10 days. The required part is then manufactured by compression molding in metal dies (Fig. 5). The die temperature may range from 130–165 °C and the die pressure from 3–14 MPa. The process time may vary from 1–3 min. The exact conditions depend upon the material, the thickness, and the shape of the part.

The material formulation as well as the control of the compounding and molding processes affect significantly the properties of the finished product. Care must be exercised, therefore, during the manufacturing process to avoid undesirable and unacceptable variations in material properties.

III. STATIC PROPERTIES

In this section the tensile, compression, shear, and flexural properties of SMC materials are reviewed.

A. Tensile Strength and Modulus

The room temperature tensile strength, tensile modulus, tensile strain at failure, and Poisson's ratio are summarized in Table 3. Materials which contain continuous fibers (XMC-3 and SMC-C20/R30) have different properties in the

TABLE 1. Material Formulations and Densities of SMC Materials[a]

Material	Ingredient	Type	Weight %	Density kg/m³
XMC-3	Continuous glass fibers— ±7.5°, X-pattern	PPG XMC strand type 1064	50	
	2.54 cm chopped glass fibers	PPG XMC strand type 1064	25	
	Resin	PPG selectron RS-50335 isophthalic polyester	21.5	1970
	Monomer	Styrene	2.4	
	Thickener	PPG selectron RS-5988	0.8	
	Catalyst	TBPB	0.2	
	Mold release	Zinc stearate	0.1	
SMC-C20/R30	Continuous glass fibers—aligned	OCF 433AB roving	20	
	2.54 cm chopped glass fibers	OCF 433AB roving	30	
	Resin	OCF-E980 polyester	32.3	
	Filler	Calcium carbonate	16.1	1810
	Mold release	Zinc stearate	0.8	
	Thickener	Magnesium oxide	0.5	
	Catalyst	TBP	0.3	
	Inhibitor	Benzoquinone	Trace	
SMC-R25	2.54 cm chopped glass fibers	E-glass (OCF 951 AB)	25	
	Resin	Polyester (OCF E-920-1)	29.4	
	Filler	Calcium carbonate	41.8	
	Internal release	Zinc stearate	1.1	
	Catalyst	Tertiary butyl perbenzoate	0.3	1830
	Thickener	Magnesium hydroxide	1.5	
	Pigment	Mapico black	0.8	
SMC-R50	2.54 cm chopped glass fibers	OCF 433AB	50	
	Resin	OCF-E980 polyester	32.3	
	Filler	Calcium carbonate	16.1	
	Mold release	Zinc stearate	0.8	1870
	Thickener	Magnesium oxide	0.5	
	Catalyst	TBP	0.3	
	Inhibitor	Benzoquinone	Trace	
SMC-R57	Formulated epoxy resin	Epoxy sheet molding compound (Gulf 1057)	43	

TABLE 1. (*Continued*)

Material	Ingredient	Type	Weight %	Density kg/m³
	1.27 cm chopped glass fibers	E-glass (OCF-495)	57	1740
SMC-R65	2.54 cm chopped glass fibers	E-glass (PPG 518)	65	
	Rigid resin	Polyester (PPG 50271)	16	
	Flexible resin	Polyester (PPG 50161)	16	1820
	Thickener, etc.		3	
EA-SMC-R30	2.54 cm chopped glass fibers	E-glass (OCF 956)	28	
	Resin	Polyester	19.9	
	Filler	Calcium carbonate	41	1830
	Thickener	Balance	11.1	

[a] From PPG-PPG Industries, OFC-Owens Corning Fiberglas (Refs. 2,3).

directions along and perpendicular to the fibers (Fig. 6). Therefore, properties for these materials are listed separately for the longitudinal and transverse directions. For materials containing both continuous and chopped fibers, the amount of chopped fibers also affects the tensile properties, as shown in Fig. 7.

Both the ultimate tensile strength and the tensile modulus depend on the material formulation. In addition, the environment has a marked effect on these properties. Generally, both the ultimate tensile strength and the tensile modulus decrease at elevated temperatures (Fig. 8) and during exposure to fluids com-

TABLE 2. VE-SMC-R50 Paste Formulation

Component	Parts
A-side	
XD-9013.03	10
TBPB	1
Camelwite	92
Zinc stearate	3
B-side	
Derakane[a] 470-45	100
Maglite D	50
Camelwite	100
Pump 10-14/1 by weight A/B	
Max. viscosity = 6000 cps	
(90F, RVT, #4 spindle, 20 rpm)	

[a] Registered trademark of Dow Chemical Company.

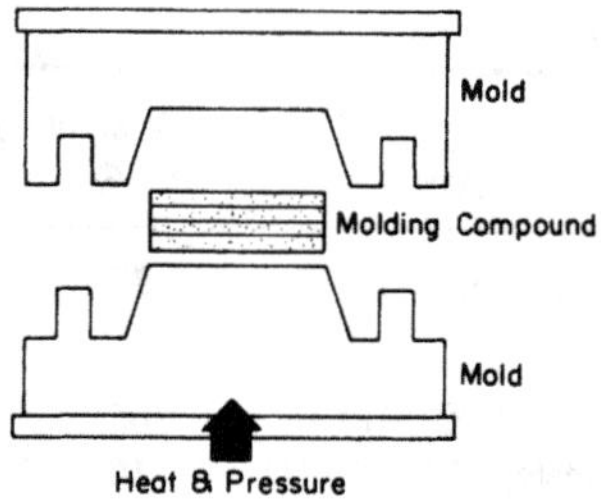

FIG. 5

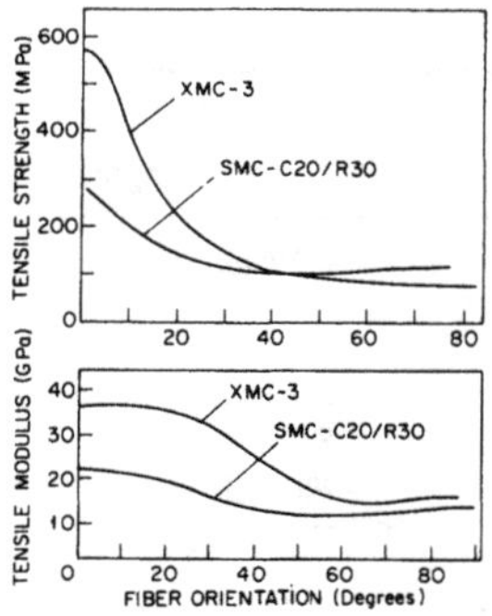

FIG. 6

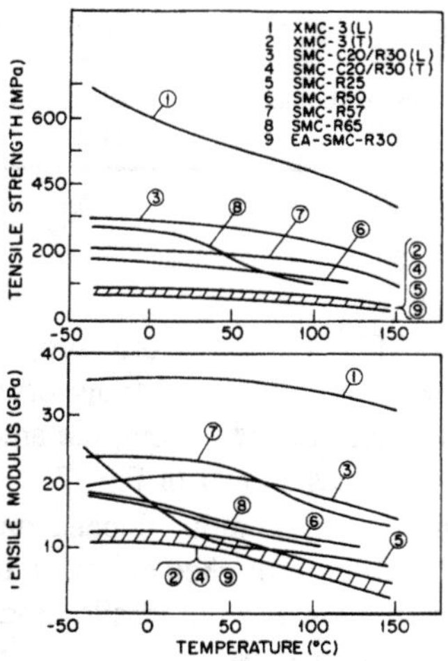

FIG. 7

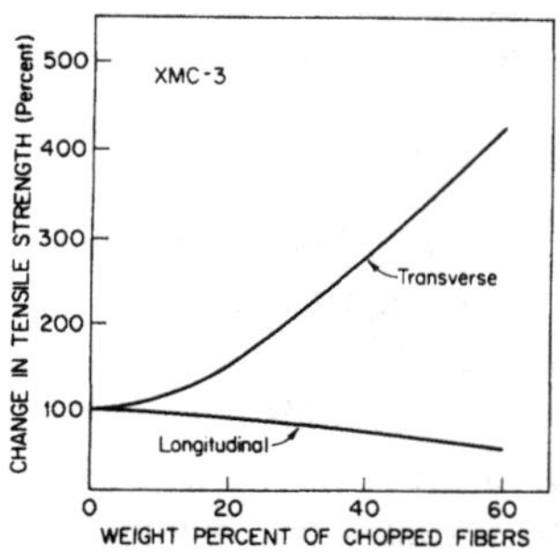

FIG. 8

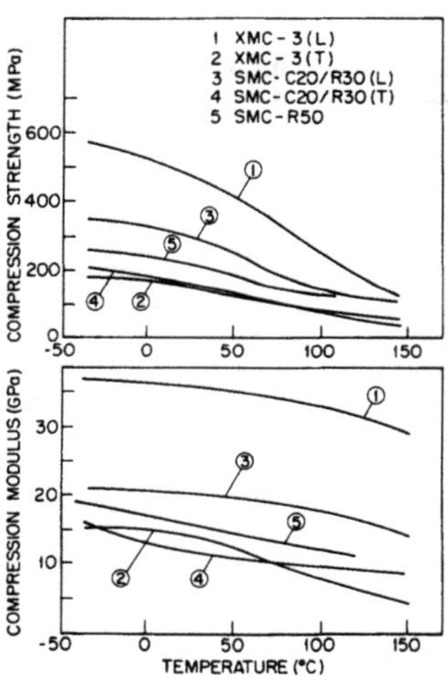

FIG. 9

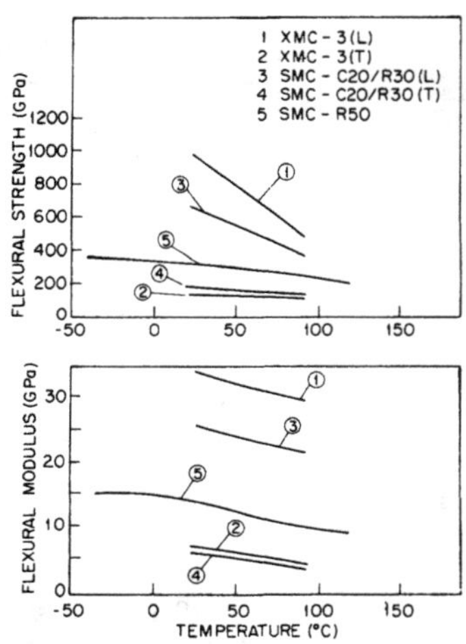

FIG. 10

TABLE 3. Room Temperature Ultimate Tensile Strength (S_t), Tensile Modulus (E_t), Tensile Failure Strain (ε_t), and Poisson's Ratio (ν_t)

Material[a]	S_t(MPa)	E_t(GPa)	ε_t(%)	ν_t
XMC-3(L)	561	35.7	1.66	0.31
XMC-3(T)	70	12.3	1.54	0.116
SMC-C20/R30(L)	289	21.4	1.73	0.3
SMC-C20/R30(T)	84	12.4	1.58	0.18
SMC-R25	82	13.3	1.34	0.25
SMC-R50	164	15.8	1.73	0.31
SMC-R57	160	16.5	—	—
SMC-R65	227	14.7	1.67	0.26
EA SMC-R30	30	8.7	1.43	0.30
VE-SMC-R50	165	7.0	—	—
VE-SMC-C40/R10(L)	426	—	—	—
VE-SMC-C40/R10(T)	57	—	—	—
VE-XMC-3(L)	648	—	—	—
VE-XMC-3(T)	74	—	—	—

[a] L, longitudinal; T, transverse direction (Refs. 2–6).

monly encountered in automotive applications (Tables 4 and 5). The decrease in properties depends on the temperature, the type of fluid, and the length of exposure. Interestingly, under some conditions there is a slight (~10%) increase in both the tensile strength and the tensile modulus. The increase is probably due to plasticization of the material.

B. Compression Strength and Modulus

The room temperature compression strength, compression modulus, and the strain at failure of different types of materials are listed in Table 6. Both the

FIGURE 5. Schematic of compression molding of SMC composites.

FIGURE 6. The effect of fiber orientation on the tensile strength and tensile modulus of XMC-3 and SMC-C20/R30 composites (Ref. 3).

FIGURE 7. The effect of the amount of chopped fibers on the tensile strength of XMC-3 composites. Total fiber content by weight = 75% (Ref. 7).

FIGURE 8. The effect of temperature on the tensile strength and tensile modulus. (L, longitudinal; T, transverse direction) (Refs. 2, 3, 5).

FIGURE 9. The effect of temperature on the compression strength and compression modulus. (L, longitudinal; T, transverse direction) (Ref. 3).

FIGURE 10. The effect of temperature on the flexural strength and flexural modulus. (L, longitudinal; T, transverse direction) (Ref. 3).

TABLE 4. Tensile Strength Retained (%) after Immersion in Different Fluids for 30 and 180 Days

Fluid	XMC-3	SMC-C20/R30	SMC-R25		SMC-R50		VE-SMC-R50	
	30 days[a]	30 days[a]	30 days	180 days	30 days	180 days	30 days	180 days
Water, 23 °C	89(L)	103(L)	—	—	—	—	—	—
	107(T)	73(T)	—	—	—	—	—	—
Humid air, 23 °C, 50% r.h	—	—	100	90	100	100	100	105
Humid air, 93 °C, 50% r.h	—	—	95	95	100	90	105	102
Humid air, 23 °C, 100% r.h	—	—	90	80	95	80	105	75
Humid air, 93 °C, 100% r.h	—	—	95	55	95	65	95	55
Salt water 23 °C	95(L)	90(L)	95	65	105	80	102	80
	107(T)	87(T)	—	—	—	—	—	—
Salt water, 93 °C	—	—	70	45	105	50	85	53
No. 2 diesel, 23 °C	—	—	90	90	98	98	103	105
No. 2 diesel, 93 °C	—	—	95	90	98	98	101	102
Motor oil, 23 °C	95(L)	100(L)	95	80	95	95	95	102
	110(T)	108(T)	—	—	—	—	—	—
Motor oil 93 °C	—	—	90	80	90	97	105	102
Antifreeze, 23 °C	95(L)	78(L)	95	80	95	95	95	105
	110(T)	108(T)	—	—	—	—	—	—
Antifreeze, 93 °C	—	—	75	30	85	30	105	45
Gasoline, 23 °C	97(L)	101(L)	90	90	100	100	100	95
	108(T)	96(T)	—	—	—	—	—	—
Gasoline, 93 °C	—	—	75	70	95	70	105	85
Transmission fluid, 23 °C	99(L)	82(L)	—	—	—	—	—	—
	120(T)	110(T)	—	—	—	—	—	—
Brake fluid, 23 °C	97(L)	97(L)	—	—	—	—	—	—
	93(T)	109(T)	—	—	—	—	—	—

[a] L, longitudinal; T, transverse direction (Refs. 3, 4).

TABLE 5. Tensile Modulus Retained (%) after Immersion in Different Fluids for 30 and 180 Days[a]

Fluid	SMC-R25		SMC-R50		VE-SMC-R50	
	30 days	180 days	30 days	180 days	30 days	180 days
Humid air, 23 °C, 50% r.h	105	110	100	90	98	95
Humid air, 93 °C, 50% r.h	120	110	90	80	95	90
Humid air, 23 °C, 100% r.h	100	95	90	80	95	90
Humid air, 93 °C, 100% r.h	120	110	85	80	90	90
Salt water, 23 °C	90	95	90	80	95	90
Salt water, 93 °C	110	90	85	65	90	85
No. 2 diesel, 23 °C	110	115	90	90	95	95
No. 2 diesel, 93 °C	120	95	95	90	90	90
Motor oil, 23 °C	95	110	80	90	90	95
Motor oil, 93 °C	110	115	90	90	95	95
Antifreeze, 23 °C	90	110	85	80	90	95
Antifreeze, 93 °C	85	85	80	50	90	75
Gasoline, 23 °C	95	90	85	85	95	90
Gasoline, 93 °C	80	80	88	60	85	75

[a] From Ref. 4.

strength and the modulus depend on the material composition, on the fiber orientation, and on the temperature (Fig. 9). As expected, the strength and the modulus are highest along the fiber direction of composites containing continuous fibers (XMC-3 and SMC-C20/R30).

C. Shear Strength and Shear Modulus

Room temperature values of in-plane shear strength, in-plane shear modulus, short-beam shear strength, and short-beam shear modulus are given in Table 7.

TABLE 6. Room Temperature Compression Strength (S_c), Compression Modulus (E_c), and Compression Failure Strain (ε_c)

Material[a]	S_c(MPA)	E_c(GPa)	ε_c(%)
XMC-3(L)	480	37	1.36
XMC-3(T)	160	14.5	1.38
SMC-C20/R30(L)	306	20.4	2.50
SMC-C20/R30(T)	166	12.2	1.74
SMC-R25	183	11.7	2.16
SMC-R50	225	15.9	—
SMC-R65	241	17.9	1.81

[a] L, longitudinal; T, transverse direction (Refs. 2, 3).

TABLE 7. Room Temperature in Plane Shear Strength (S_{LT}), In-Plane Shear Modulus (E_{LT}), Short-Beam Shear Strength (S_s), and Short-Beam Shear Modulus (G_s)[a]

Material	S_{LT}(MPa)	E_{LT}(GPa)	S_s(MPa)	G_s(GPa)
XMC-3	91.2	4.47	55	—
SMC-C20/R30	85.4	4.09	41	—
SMC-R25	79	4.48	30	5.0
SMC-R50	62	5.94	25	7.0
SMC-R65	128	5.38	45	—
VE-SMC-R50	—	—	35	4.0

[a] From Refs. 3, 4, and 8.

In most cases there is a significant decrease in strength and in modulus at elevated temperature (Table 8) and during exposure to humid air and to different types of liquids (Tables 9 and 10). As in the case of tensile properties, shear properties also increase slightly under some conditions. Again, this increase is caused by plasticization of the material.

D. Flexural Strength and Modulus

The flexural strengths and flexural moduli of different types of SMC materials are presented in Table 11. These properties are highest in the longitudinal directions of XMC-3 and SMC-C2/R30 composites and are about five times higher in the longitudinal than in the transverse direction. The strength and the modulus decrease with increasing temperature (Fig. 10). For XMC-3 composites

TABLE 8. Losses in In-Plane Shear Strength (S_{LT}), Shear Modulus (E_{LT}), and Ultimate Shear Strain (ε_{LT}) When the Temperature is Raised from 23 °C to 93 °C[a]

	Loss (%)		
	S_{LT}	E_{LT}	ε_{LT}
XMC-3	38	48	13
SMC-C20/R30	40	44	0.7
SMC-R50	—	22	—

[a] From Ref. 3.

TABLE 9. Short-Beam Shear Strength Retained (%) after Immersion in Different Fluids for 30 and 180 Days[a]

Fluid	SMC-R25		SMC-R50		VE-SMC-R50	
	30 days	180 days	30 days	180 days	30 days	180 days
Humid air 23 °C, 50% r.h	105	120	110	110	100	103
Humid air 93 °C, 50% r.h	120	110	98	110	105	120
Humid air 23 °C, 100% r.h	110	110	90	95	95	95
Humid air 93 °C, 100% r.h	102	95	92	85	95	95
Salt water 23 °C	110	100	95	95	99	98
Salt water 93 °C	95	65	80	35	95	75
No. 2 diesel 23 °C	115	120	90	75	95	110
No. 2 diesel 93 °C	125	125	103	110	100	105
Motor oil 23 °C	100	125	85	95	90	115
Motor oil 93 °C	110	120	90	110	95	120
Antifreeze 23 °C	105	110	90	100	95	105
Antifreeze 93 °C	75	55	50	15	85	50
Gasoline 23 °C	95	95	80	95	98	99
Gasoline 93 °C	70	75	85	85	75	75

[a] From Ref. 4.

TABLE 10. Short-Beam Shear Modulus Retained (%) after Immersion in Different Fluids for 30 and 180 Days[a]

Fluid	SMC-R25		SMC-R50		VE-SMC-R50	
	30 days	180 days	30 days	180 days	30 days	180 days
Humid air 23 °C, 50% r.h	110	125	105	100	105	115
Humid air 93 °C, 50% r.h	125	115	95	95	110	120
Humid air 23 °C, 100% r.h	100	115	85	85	90	95
Humid air 23 °C, 100% r.h	115	120	95	85	105	105
Salt water 23 °C	110	100	85	90	100	95
Salt water 93 °C	85	80	65	50	90	95
No. 2 diesel 23 °C	125	120	95	95	105	115
No. 2 diesel 93 °C	110	115	95	95	110	110
Motor oil 23 °C	105	125	75	90	90	110
Motor oil 93 °C	105	125	85	95	90	115
Antifreeze 23 °C	105	115	75	90	90	110
Antifreeze 93 °C	80	75	54	15	85	55
Gasoline 23 °C	95	90	80	95	105	110
Gasoline 93 °C	60	85	70	70	80	75

[a] From Ref. 4.

TABLE 11. Room Temperature Flexural Strength (S_f) and Flexural Modulus (E_f)

Material[a]	S_f(MPa)	E_f(GPa)
XMC-3(L)	973	34.1
XMC-3(T)	139	6.8
SMC-C20/R30(L)	645	25.7
SMC-C20/R50(T)	165	5.9
SMC-R25	220	4.8
SMC-R50	314	14.0
SMC-R65	403	5.7

[a] L, longitudinal, T, transverse direction (Ref. 3)

the strength and the modulus also depend on the weight percent of chopped fibers (Fig. 11).

E. Notch Sensitivity

The notch sensitivities of three types of SMC materials were investigated by Riegner and Sanders.[(3)] The reduction in tensile strength (due to circular holes or "notches") of XMC-3, SMC-C20/R30, and SMC-R65 materials are shown in Fig. 12. The solid lines in this figure represent results obtained by matching the three-parameter model of Pipes *et al.*[(3,9)] to the data.

Riegner and Sanders also measured the notch strength of SMC-R25 materials.[(3)] This material did not fail through the circular holes. Apparently, the weakening effects of the holes were less than the strength of the material itself.

IV. FATIGUE

Most of the data available on the fatigue behavior of SMC materials are for tests performed in tension–tension mode. A typical test record is shown in Fig. 13. In this figure the measured loads and deformations have been converted to stresses and strains, respectively.

Fatigue life (S–N) data for tension–tension fatigue are presented in Figs. 14–16. In general, the fatigue strengths follow the same trends as the static strengths. The material that has the highest static strength has also the highest fatigue strength. A good indicator of the relative fatigue performance of different types of materials is the maximum stress the materials can sustain without failure for one million cycles. These data are presented in Table 12. As can be seen, the maximum stress depends not only on the material but also on the temperature.

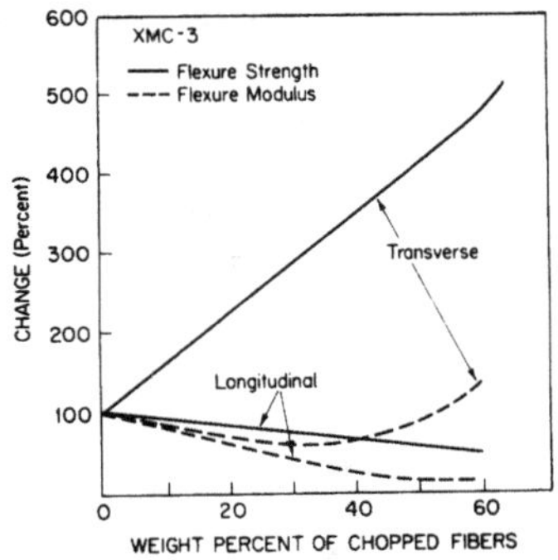

FIG. 11

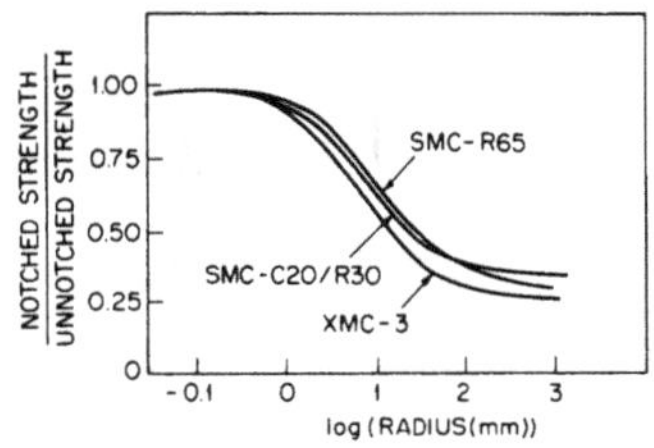

FIG. 12

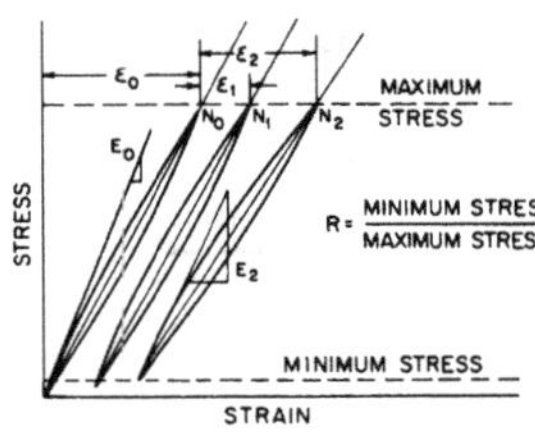

FIG. 13

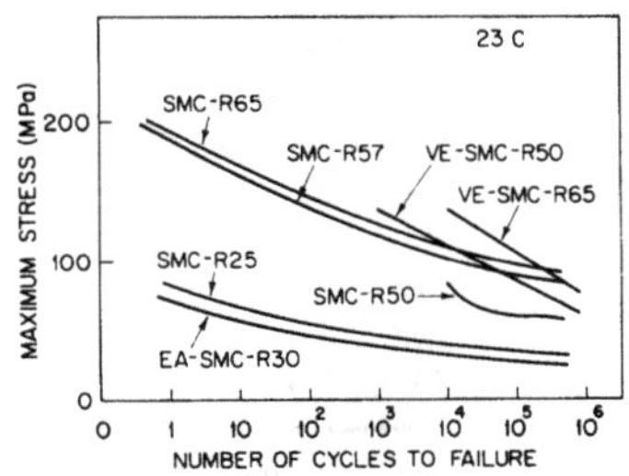

FIG. 14

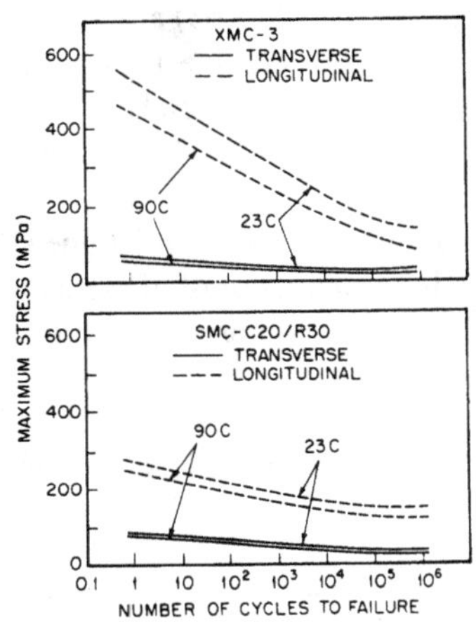

FIG. 15

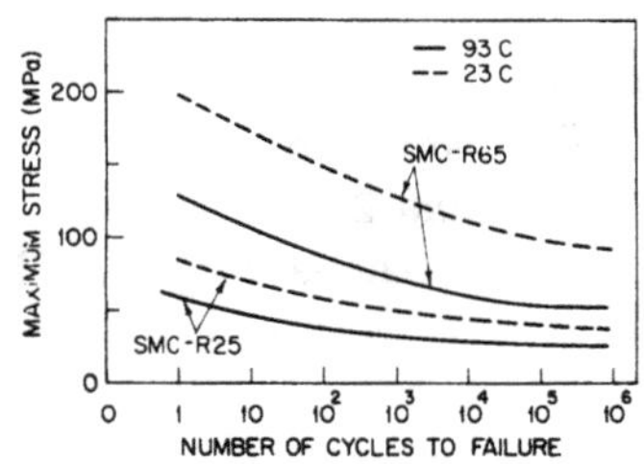

FIG. 16

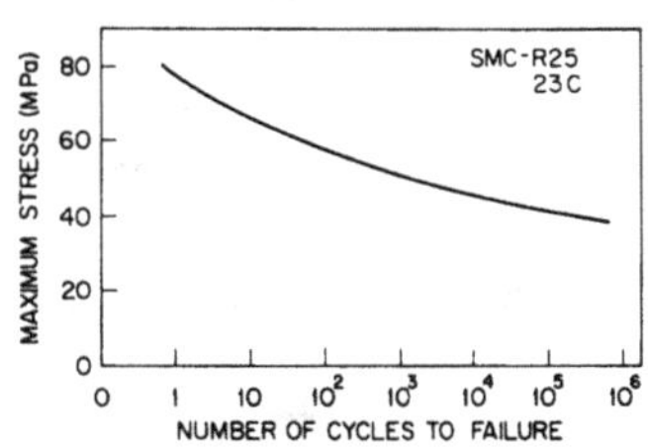

FIG. 17

FIGURE 11. The effect of the amount of chopped fibers on the flexural strength and flexural modulus of XMC-3 composites. Total fiber content by weight = 75% (Ref. 7).

FIGURE 12. Changes in tensile strengths due to different diameter circular holes ("notches") (Ref. 3).

FIGURE 13. Typical result of fatigue test and definition of symbols used in the presentation of the fatigue results.

FIGURE 14. Tension–tension fatigue results. $R = 0.05$ (except for VE-SMC-R50 and VE-SMC-R65) (Refs. 2, 5, 6).

FIGURE 15. Tension–tension fatigue results. $R = 0.05$ (Ref. 3).

FIGURE 16. Tension–tension fatigue results. $R = 0.05$ (Ref. 2).

FIGURE 17. Compression–tension fatigue results. $R = -1$ (Ref. 2).

TABLE 12. Fatigue Stress for Survival to One Million Cycles

Material[a]	Percent static ultimate tensile stress	
	23 °C	90 °C
XMC-3(L)	23	18
XMC-3(T)	27	26
SMC-C20/R30(L)	45	44
SMC-C20/R30(T)	52	43
SMC-R25	49	55
SMC-R50	38	41
SMC-R57	42	—
SMC-R65	31	28
EA-SMC-R30	44	—

[a] L, longitudinal; T, transverse direction (Refs. 2, 3).

The effect of temperature on fatigue life is further illustrated in Figs. 15 and 16. For chopped fiber composites (SMC-R25 and SMC-R65) an increase in temperature from 23–93 °C results in about a twofold decrease in the fatigue strength of the material (Fig. 16). The fatigue strengths of materials containing continuous fibers (XMC-3 and SMC-C20/R30) seem to be affected less by changes in temperature (Fig. 15) than by the orientation of the fibers (Fig. 16).

A limited amount of data is available for compression–tension fatigue (Fig. 17). Data generated with SMC-R25 composites indicate that most of the damage is done during the tension portion of the cycle.[(2)] The compression portion of the cycle seems to contribute less to the damage.

Results for flexural fatigue are presented in Fig. 18. For SMC-R25, SMC-R57, and SMC-R65 composites at one million cycles the stress levels are about 0.4, 0.35, and 0.25 times the static flexural stress.[(2)] These values are about

FIGURE 18. Flexural fatigue results (Ref. 2).

FIGURE 19. Creep strain ($\varepsilon_N - \varepsilon_0$) during tension–tension fatigue tests as functions of load (percent of ultimate tensile strength). $R = 0.05$ (Refs. 2, 3).

FIGURE 20. Modulus decay during tension–tension fatigue tests as a function of load (percent of ultimate tensile strength). $R = 0.05$ (Ref. 3).

FIGURE 21. Modulus decay during tension–tension fatigue tests as a function of load (percent of ultimate tensile strength). $R = 0.05$ (Refs. 2, 4).

FIGURE 22. Creep of XMC-3 at 70, 50, and 30% of static ultimate tensile strength (L, longitudinal; T, transverse direction) (Ref. 3).

FIGURE 23. Creep of SMC-C20/R30 at 70, 50, and 30% of static ultimate tensile strength. (L, longitudinal; T, transverse direction) (Ref. 3).

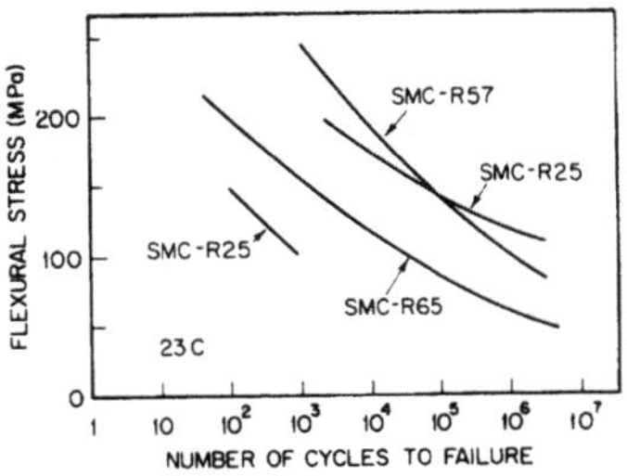

FIG. 18

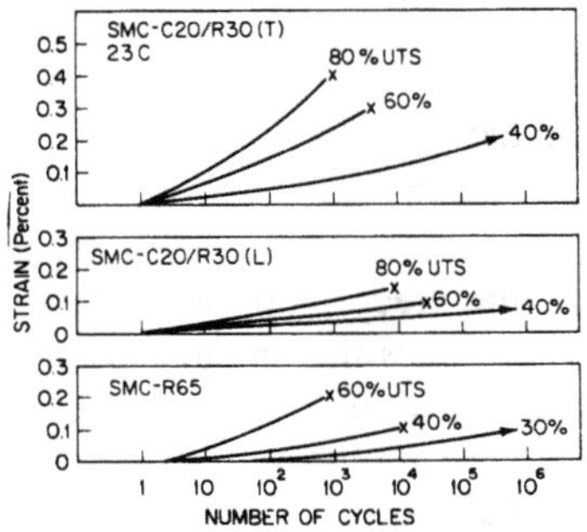

FIG. 19

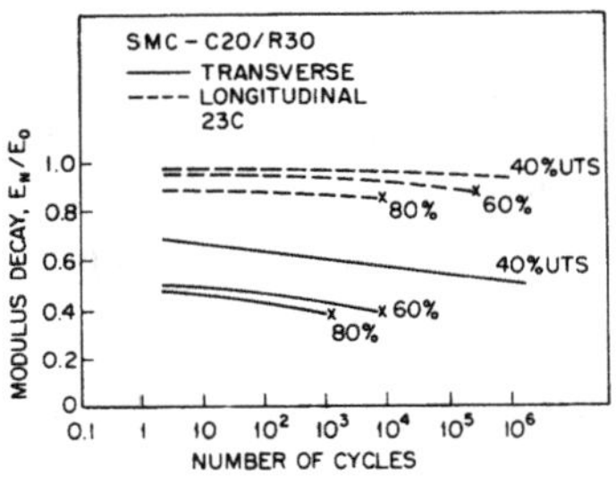

FIG. 20

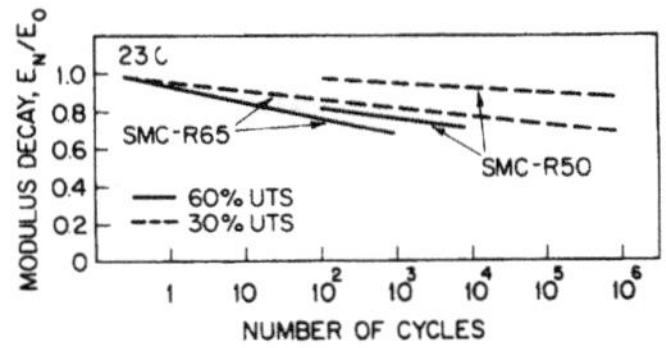

FIG. 21

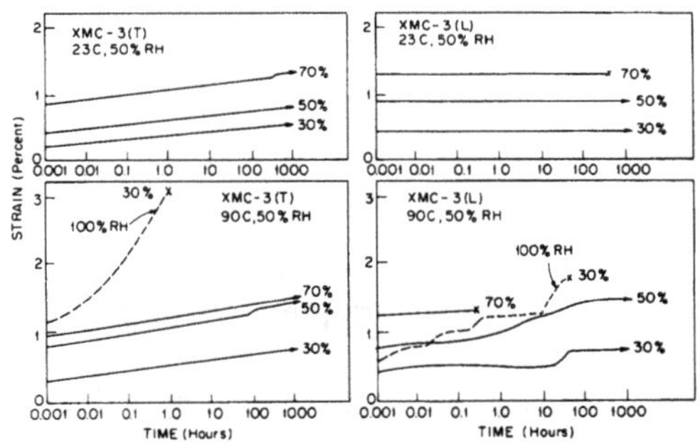

FIG. 22

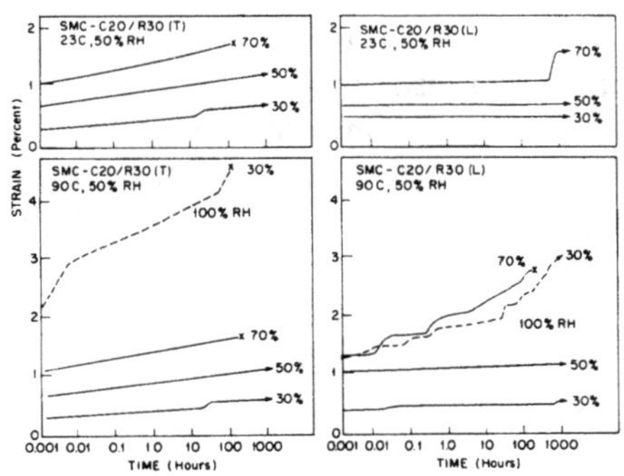

FIG. 23

twice those reported by Maaghul and Potkanowicz[10] for SMC-R25 and SMC-R65 composites.

Room temperature creep during fatigue tests is shown in Fig. 19. As expected, the creep increases with increasing load. Furthermore, for SMC-C20/R50 composites the amount of creep is about twice as large in the direction of the fibers as in the transverse direction.

In those applications where stiffness of a part is an important factor, changes in modulus during fatigue must be taken into consideration. Sheet molding compounds may decrease in modulus during fatigue cycling (Figs. 20 and 21). At room temperature the modulus decay is small (less than ~20%) except in the transverse direction of SMC-C20/R30 composites. Hence, in most automotive applications the decay in modulus would not play an important role.

Heimbuch and Sanders[2] evaluated the effects of test frequency on fatigue behavior. For SMC-R25 and SMC-R57 composites test frequencies from 5–20 Hz did not change markedly the fatigue life.

Heimbuch and Sanders[2] also examined the effects of notches on the fatigue characteristics of SMC-R25 and SMC-R65 composites by drilling 0.6-mm diameter holes into the samples. These holes did not have a pronounced effect on the fatigue behavior, suggesting that these materials are not notch sensitive in fatigue.

V. CREEP

The results of static creep tests are presented in Figs. 22–28. The curves are average values. There is considerable scatter in the actual data. An arrow at the end of a curve indicates that the specimen did not fail at the end of the test, while a cross indicates specimen failure.

As expected, the strain increases with load, temperature, relative humidity, and time. The increase in strain with time is not uniform. Step "jumps" occur in strain at random times. Because of these unpredictable jumps, the strain cannot be described by simple viscoelastic models.

Heimbuch and Sanders[2] also investigated the stress rupture of SMC-R25, SMC-R57, and SMC-R65 composites in air at 23, 60 and 90 °C and at 50% and 100% relative humidities. Owing to the large scatter in the data, the effect of the environment on stress rupture cannot be ascertained from the results of these tests.

VI. ADHESIVE-BONDED SINGLE-LAP JOINTS

In this section behavior of adhesive-bonded single-lap joints exposed to automotive-relative fluids is discussed. The results presented were obtained with

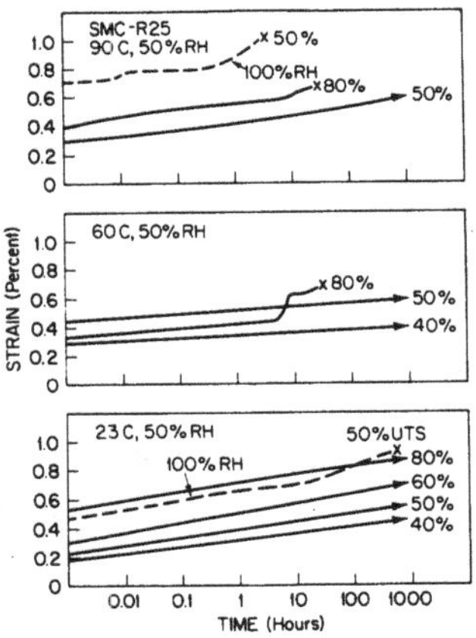

FIG. 24

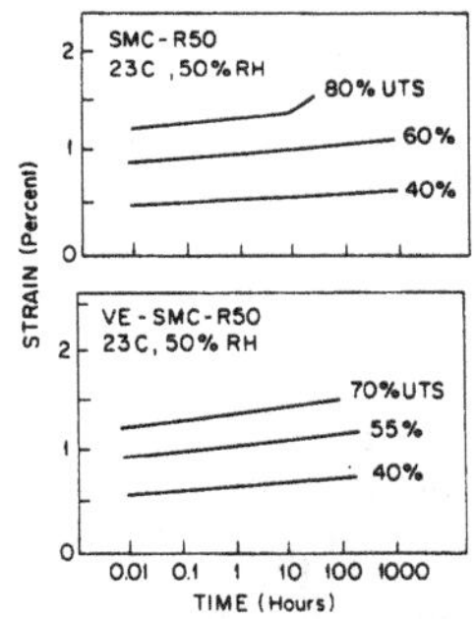

FIG. 25

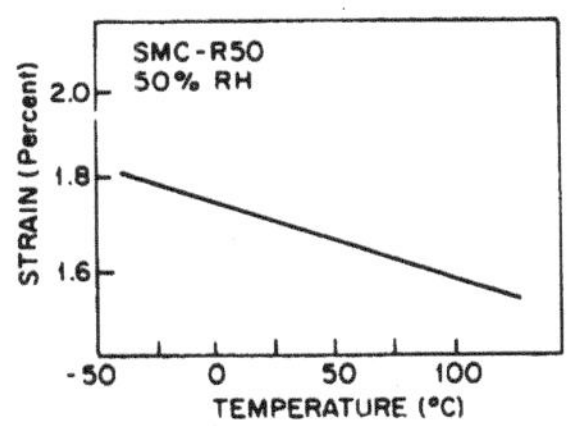

FIG. 26

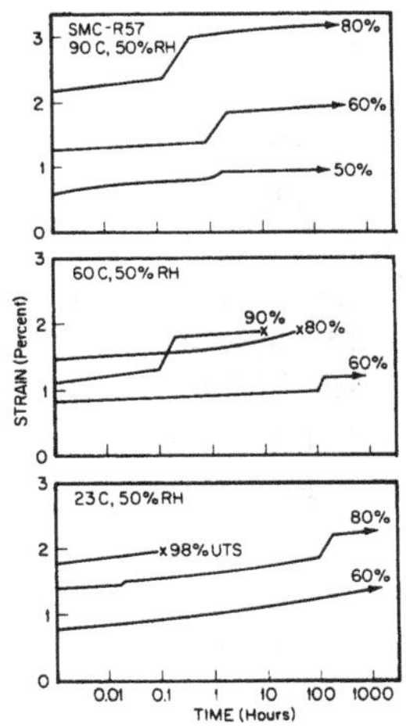

FIG. 27

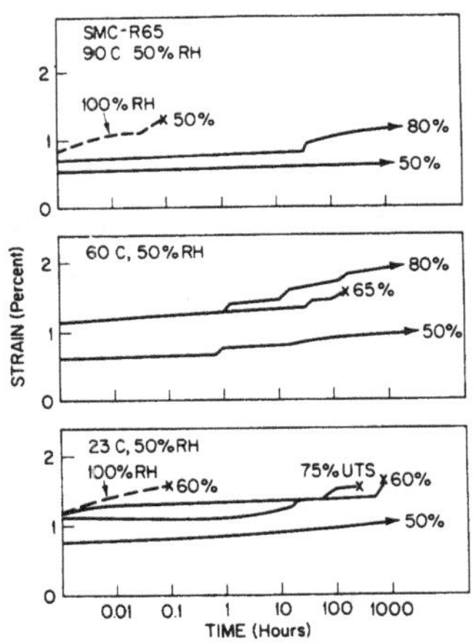

FIG. 28

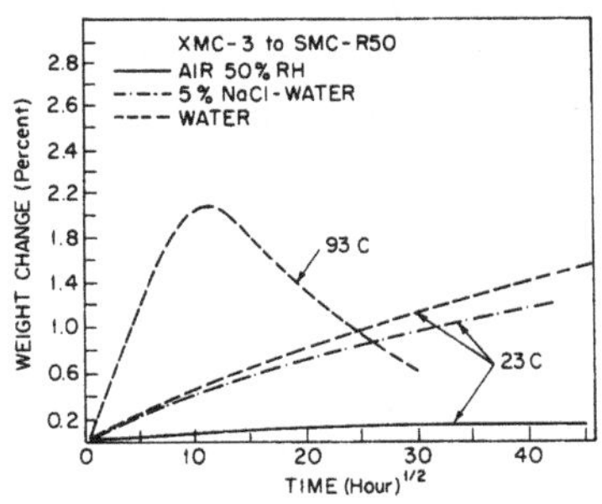

FIG. 29

FIGURE 24. Creep of SMC-R25 under different loads (percent of static ultimate tensile strength) (Ref. 2).

FIGURE 25. Creep of SMC-R50 and VE-SMC-R50 under different loads (percent of static ultimate tensile strength) (Refs. 5, 6).

FIGURE 26. Strain (elongation) of SMC-R50 at failure as a function of temperature (Ref. 5).

FIGURE 27. Creep of SMC-R57 under different loads (percent of static ultimate tensile strength) (Ref. 2).

FIGURE 28. Creep of SMC-R65 under different loads (percent of static ultimate tensile strength (Ref. 2).

FIGURE 29. Moisture absorption of adhesive-bonded XMC-3 to SMC-R50 single-lap joints.

adherends bonded with a two-part urethane adhesive, characterized in detail by Wang, Sanders, and Lindholm.[11]

A. Moisture Absorption Characteristics

Typical moisture absorption data obtained with XMC-3 to SMC-R50 joints are given in Fig. 29. Data for SMC-R50 to SMC-R50 joints exhibit similar trends. At 23 °C both XMC-3 to SMC-R50 and SMC-R50 to SMC-R50 joints seem to approach asymptotically the same maximum moisture content when immersed in the same fluid. During a two-month test period, maximum moisture is reached only in air. In water and in 5% NaCl–water mixture the maximum moisture contents are not attained. The maximum moisture values can be estimated by extrapolating the data, giving 0.18, 1.5, and 2.0% for air, salt water, and water, respectively.

At 93 °C (immersion in water) the maximum moisture level is not approached asymptotically. Here the weight increases for about the first 100 h and then decreases at a rapid rate. This indicates that the material deteriorates during exposure. At 93 °C both bonded and unbonded test specimens behave similarly, suggesting that degradation is mostly in the composite and not in the adhesive.

Joints loaded up to 30% of their strength did not show appreciable change in their moisture absorption characteristics.

B. Lap Shear Strength

Lap shear strengths of adhesive-bonded single-lap joints are given in Table 13. Changes in baseline strength and modulus during environmental exposure are illustrated in Table 14. Neither the strengths nor the moduli change significantly when the joints are exposed to room temperature fluids. In some cases the strength improves slightly (10–15%) during environmental conditioning. The beneficial effects of fluid and temperature are likely due to plasticization. The

TABLE 13. Baseline ("As Received") Lap Shear Strengths of Adhesive-Bonded Single-Lap Joints

	Strength (MPa)	
	23 °C	93 °C
XMC to SMC-R50	6.55	3.89
SMC-R50 to SMC-R50	6.11	2.12
SMC-R25 to SMC-R25	3.83	—

TABLE 14. Changes in Lap Shear Strength (S/S_B) and Modulus (E/E_B) of Adhesive-Bonded Single-Lap Joints after 30 Days of Environmental Exposure at 23 °C[a]

	SMC-R25 to SMC-R25		SMC-R50 to SMC-R50	
Fluid	S/S_B	E/E_B	S/S_B	E/E_B
Air	1.00	1.00	1.00	1.00
Motor oil	0.89	1.05	0.85	0.92
Transmission fluid	0.92	1.01	0.91	0.91
Gasoline	0.95	0.80	1.20	0.77
Salt water	0.97	0.72	0.98	0.78
Brake fluid	0.95	0.87	0.96	0.85
Antifreeze	0.96	1.15	0.81	0.81

[a] B, baseline value (Ref. 11).

strength of joints immersed in hot (93 °C) water and salt water for 30 days decrease by a factor of two. Loading (up to 30% of the baseline strength) during exposure does not seem to affect the strength.

The joints may fail by delamination of the composite or by separation of the adherent. In these tests most failures occurred by delamination of the adherent. Separation of the adhesive was predominant only at higher (93 °C) temperatures.

C. Fatigue

Wang *et al.*[11] conducted fatigue life tests on SMC-R25 to SMC-R25 and SMC-R50 to SMC-R50 single-lap joints. During the tests the stress levels were 30, 50, 70, and 90% of the static shear strength. Prior to the fatigue tests the specimens were soaked for 30 days at room temperature in the following liquids: 50% by weight salt water, motor oil, transmission fluid, and gasoline. The ranges of data are shown in Fig. 30. The data are not shown separately for specimens immersed in the different fluids because the fluids did not have a significant effect on the fatigue life.

The residual strengths and moduli were also measured for those specimens surviving for one million cycles.[11] Cyclic stressing at 30% of ultimate strength did not degrade appreciably either the strength or the modulus; in general, both the strength and the modulus retained at least 80% of their initial values.

D. Creep

Creep deformations of adhesive-bonded single-lap joints under static and cyclic loadings are shown in Figs. 31 and 32. These figures illustrate the effects

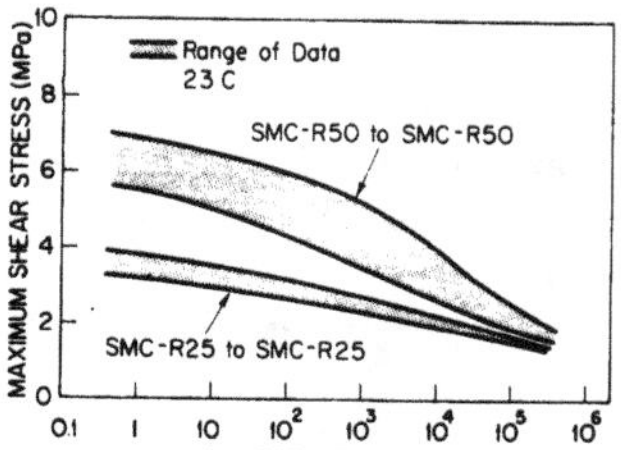

FIG. 30

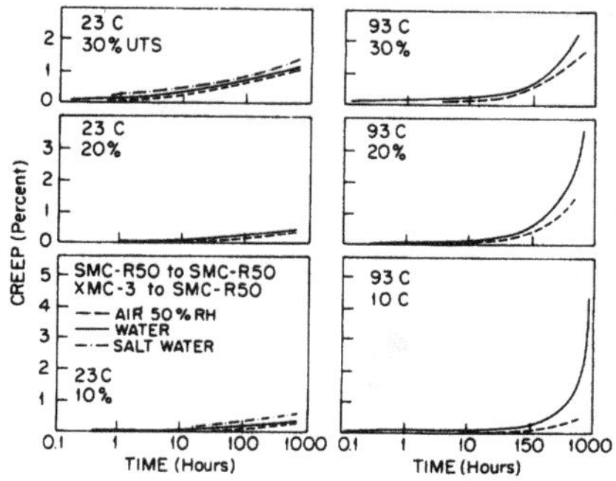

FIG. 31

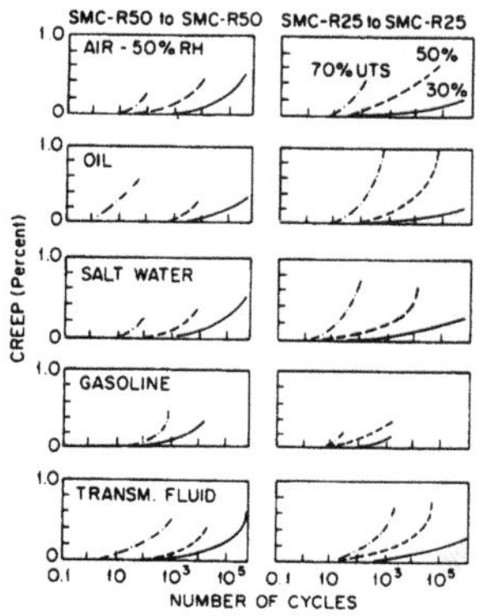

FIG. 32

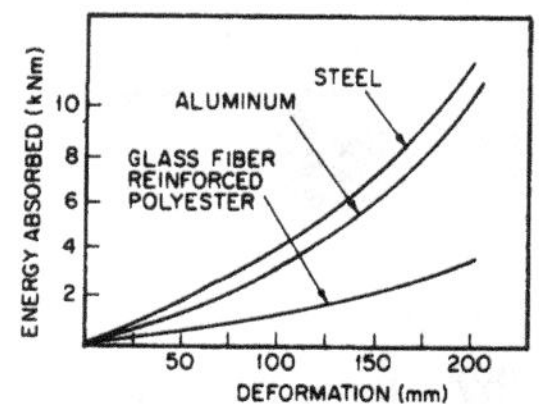

FIG. 33

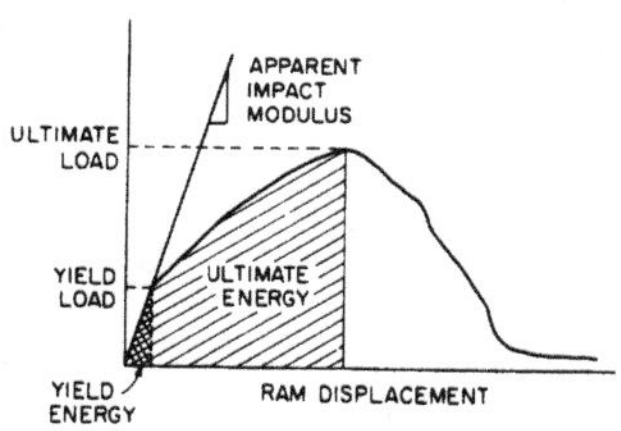

FIG. 34

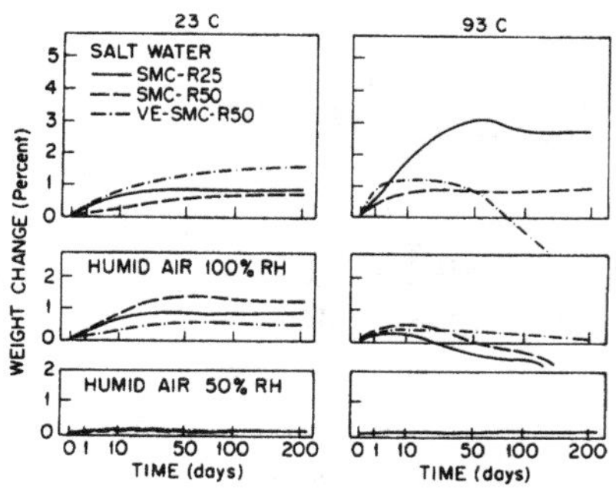

FIG. 35

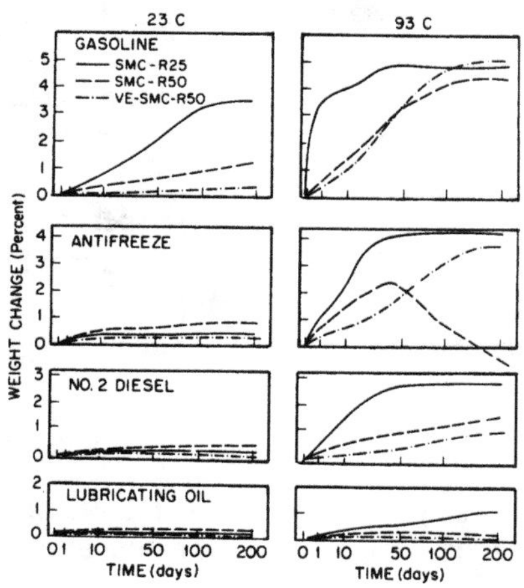

FIG. 36

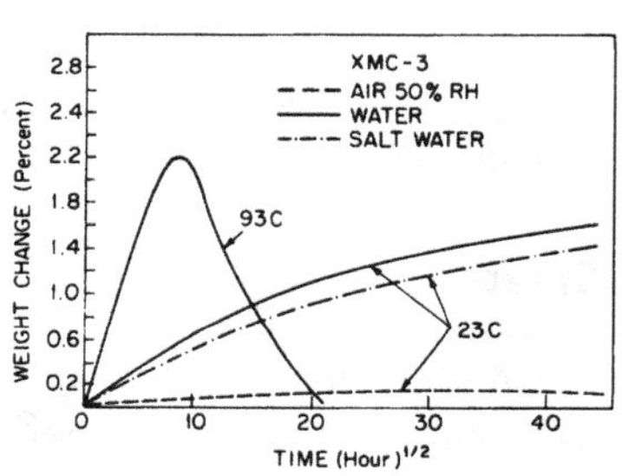

FIG. 37

of material, fluid, temperature, and applied load on creep behavior. The type of material used in forming the joints has smaller effect on creep than does the type of fluid, the temperature, and the applied load. The creep is least in air and is higher in water, in salt water, and in hydrocarbons. The creep also increases with temperature and with applied load. For example, at 23 °C none of the XMC-3 to SMC-R50 or SMC-R50 to SMC-R50 joints failed during static creep. In air at 93° C only one of the joints failed, this occurring at 30% load level. During water immersion (at 93 °C) all but three coupons failed before the end of the 715-h test. During cyclic creep, only joints with 30% load survived for one million cycles. At higher loads none of the joints survived for one million cycles (Fig. 32).

VII. VIBRATION DAMPING

Materials used in commercial applications should damp out noise and vibrations. It is important, therefore, to know the vibration damping properties of SMC materials. These properties may be characterized by two parameters, the loss factor and the storage modulus. These parameters are obtained by exciting the material with forced sinusoidal oscillations and by measuring the input stress and output strain.[3] The loss factor is the tangent of the phase angle between the stress and the strain, and is equal to the ratio between the energy dissipated and the energy stored in the material. The storage modulus is the in-phase component of the ratio of input stress to output strain.

Room temperature values of loss factors and storage moduli are listed in Table 16. The loss factor is insensitive to the exciting frequency in the range of 0.1–10 Hz. For the SMC materials tested the loss factors are an order of mag-

←

FIGURE 30. Maximum shear stress of adhesive-bonded single-lap joints (SMC-R50 to SMC-R50 and SMC-R25 to SMC-R25) during tension–tension fatigue (Ref. 11).

FIGURE 31. Creep of adhesive-bonded single-lap joints (SMC-R50 to SMC-R50 and XMC-3 to SMC-R50) immersed in air, water, and 5% NaCl–water mixture under different loads (percent of static ultimate tensile strength).

FIGURE 32. Creep of adhesive-bonded single-lap joints (SMC-R50 to SMC-R50 and SMC-R25 to SMC-R25) during tension–tension fatigue under different loads (—30% UTS, - -50% UTS, -·-70% UTS) (Ref. 11).

FIGURE 33. Energy absorbed by different materials during impact against a fixed barrier. Impact speed 24 km/h (Ref. 13).

FIGURE 34. Typical output of impact test using Rheometrics impact tester (Ref. 3).

FIGURE 35. Weight change during immersion in humid air and in saturated salt water (Ref. 4).

FIGURE 36. Weight change during immersion in different types of hydrocarbons (Refs. 4, 14).

FIGURE 37. Weight change of XMC-3 immersed in humid air, water, and in 5% NaCl–water mixture (Ref. 4).

TABLE 15. Loss Factor and Storage Modulus at 23 °C and Maximum Changes in These Parameters When the Temperature Is Increased from 23 °C to 120 °C

Material[a]	Loss factor at 23 °C		Storage modulus at 23 °C (GPa)	Max. change loss factor	Percent storage modulus
	0.1 Hz	10 Hz			
XMC-3(L)	0.028	0.025	36	+129	−7
XMC-3(T)	0.063	0.053	—	+355	−54
SMC-C20/R30(L)	0.034	0.029	—	+ 80	−5
SMC-C20/R30(T)	0.051	0.049	—	+204	−44
SMC-R25	0.037	0.035	4	+471	−48
SMC-R65	0.039	0.034	8	+241	−30
Steel	~0.001	—	—	—	—

[a] L, longitudinal; T, transverse direction (Refs. 3, 13).

TABLE 16. Maximum Changes in Dynamic Properties during 1000 Hours of Soak

Fluid	Material[a]	Maximum change (%)	
		Loss factor	Storage modulus
Distilled water 22 °C	XMC-3(L)	+53	0
	SMC-C20/R30(L)	+38	0
	SMC-R25	+193	−20
	SMC-R65	+180	−7
Distilled water 50 °C	XMC-3(L)	+88	0
	SMC-C20/R30(L)	+112	0
	SMC-R25	+210	−10
	SMC-R65	+247	−12
Salt water 23 °C	XMC-3(L)	+55	0
	SMC-C20/R30(L)	+50	0
	SMC-R25	+117	−4
	SMC-R65	+178	−7
Motor oil 23 °C	XMC-3(L)	0	0
	SMC-C20/R30(L)	0	0
	SMC-R25	+25	0
	SMC-R65	+24	0
Antifreeze	XMC-3(L)	0	0
	SMC-C20/R30(L)	0	0
	SMC-R25	+29	0
	SMC-R65	+22	0
Gasoline 22 °C	XMC-3(L)	0	0
	SMC-C20/R30(L)	0	0
	SMC-R25	178	−15
	SMC-R65	30	0

[a] L, longitudinal; T, transverse direction (Ref. 12).

TABLE 17. Impact Properties of SMC Composites. Temperature 23 °C, Impact Velocity 1 m/sec[a]

Material	Load (N)		Apparent modulus (N/m)	Displacement (mm)		Energy (J)	
	Yield	Ultimate		Yield	Ultimate	Yield	Ultimate
XMC-3	1.5×10^3	7×10^3	1.2×10^6	1.5	9.0	1	30
SMC-C20/R30	1.2×10^3	5×10^3	1×10^6	1.5	9.0	1	30
SMC-R25	1.2×10^3	5×10^3	1×10^6	1.5	9.0	1	30

[a] From Ref. 3.

nitude higher than for steel. Thus SMC materials would damp out vibrations more effectively than steel.

The effects of temperature and soaking in different types of liquids are illustrated in Tables 15 and 16. An increase in temperature (from 23–120 °C) increases the damping and reduces the stiffness. Soaking in liquids has similar effects. Soaking for 1000 h considerably increased the damping of chopped fiber composites (SMC-R25 and SMC-R65), while their stiffness decreased slightly. The damping characteristics of continuous fiber composites (XMC-3 and SMC-C20/R30) change little in the fiber direction. It is noteworthy that both temperature- and moisture-induced changes in the vibration properties appear to be reversible.[(12)]

VIII. DYNAMIC IMPACT

Fiber-reinforced composites used in automotive applications should be able to resist impact damage. Unfortunately, glass fiber-reinforced polyester composites absorb less energy during impact than metals. This is illustrated in Fig. 33, where a comparison is made between the energies absorbed by different types of materials during impact against a fixed barrier. At a given deformation metals absorb about twice as much energy as the composite.

Impact resistances of composites may be evaluated using the Rheometrics High Rate Impact Tester.[(3,5)] In tests with this device a hydraulic ram impacts a clamped plate of material and the load versus displacement is recorded, as shown in Fig. 34. From the data several parameters can be calculated, including yield and ultimate load, apparent impact modulus, yield and ultimate displacement, and yield and ultimate energy absorbed. Typical values of some of these parameters are given in Table 17. The data in this table are a summary of the results of Riegner and Sanders,[(3)] who reported data for different impact velocities (from 0.5–10 m/sec) and for different temperatures (23 and −35 °C). The data do not show clear trends. In general, neither changes in impact velocity nor changes in temperature had a significant effect on the load, modulus, displacement, and energy values given in Table 17.

IX. MOISTURE ABSORPTION

Glass fiber-reinforced organic matrix composites absorb moisture when exposed to humid air or to liquids. The weight changes of different types of SMC composites exposed to typical automotive service environments are presented in Figs. 35–37. The weight change M is defined as

$$M = \frac{\text{wet weight} - \text{dry weight}}{\text{dry weight}} \times 100\%$$

The data show that, in general, when the dry material is submerged in the fluid the weight at first increases then levels off for some length of time. Both the initial rate of weight increase and the value at which the weights level off depend on (a) the material, (b) the temperature, and (c) the environment (relative humidity of air or the type of liquid used). The data also show that in some instances the weight does not remain constant after it reaches a level value but keeps either increasing or decreasing. This suggests that under some conditions the moisture transport is by a non-Fickian process. One reason for the non-Fickian behavior may be that moisture transfer through the resin does not proceed by a process that can be described by Fick's law. Another plausible explanation of the observed non-Fickian absorption process is as follows. Owing to the moist, high temperature environment, microcracks develop on the surface and inside the material. Moisture rapidly enters the material, causing an increase in weight. As the cracks grow larger, material, most likely in the form of resin particles, is actually lost. In fact, such material loss is frequently observed after a few hours of exposure to the moist environment. As long as the moisture gain is greater than the material loss, the weight of the specimen increases. Once the weight of the lost material exceeds the weight of the absorbed moisture, the weight of the specimen decreases. Of course, when the material is lost, the measured weight change no longer corresponds to the moisture content of the material.

The foregoing results were all obtained with unstressed specimens. However, as was indicated in Sec. VI.A, the moisture absorption characteristics of stressed and unstressed SMC materials do not differ appreciably.

X. THERMAL EXPANSION

The dimensional changes of the material during temperature cycles must be taken into account in the design process. The thermal expansion coefficient values reported by Heimbuch and Sanders[2] and by Riegner and Sanders[3] are

TABLE 18. Thermal Expansion Coefficient α at Room Temperature

Material[a]	α(μm/m °C)
XMC-3(L)	8.7
XMC-3(T)	28.6
SMC-C20/R30(L)	11.3
SMC-C20/R30(T)	24.6
SMC-R25	23.2
SMC-R50	14.8
SMC-R65	13.7

[a] L, longitudinal; T, transverse direction (Refs. 2, 3).

reproduced in Table 18. As expected, the thermal expansion coefficients are lowest along the fiber direction of composites containing continuous fibers (XMC-3 and SMC-C20/R30). The coefficient is high in the transverse direction of these materials and also for SMC-R25 composites. More comprehensive data on the variation of the thermal expansion coefficient with temperature are not yet available.

XI. CONCLUDING REMARKS

As this survey indicates, the numerous studies made in recent years provide much of the needed information on the mechanical and thermal properties of glass fiber-reinforced SMC composites. For many materials we now have the extensive data bases required for the design of practical engineering systems. There are still areas in need of further exploration. In particular, information is needed on problems in which the viscoelastic behavior of the material predominates. Hopefully, the continuing efforts in this area will soon provide information which will lead to an understanding of these problems.

REFERENCES

1. Charlesworth, D., "Review of Plastics Applications Within BL Cars," in *Worldwide Applications of Plastics*, pp. 73–85, Society of Automotive Engineers, SP-482, February (1981).
2. Heimbuch, R. A., and Sanders, B. A., "Mechanical Properties of Chopped Fiber Reinforced Plastics," in *Composite Materials in the Automotive Industry*, pp. 111–139, American Society of Mechanical Engineers, December (1978).
3. Riegner, D. A., and Sanders, B. A., "A Characterization Study of Automotive Continuous and Random Glass Fiber Composites," Report GMMD79-023, General Motors Corporation, Manufacturing Development, GM Technical Center, Warren, Michigan (1979).

4. Springer, G. S., Sanders, J. A., and Tung, R. W. "Environmental Effects of Glass Fiber Reinforced Polyester and Vinylester Composites," *J. Compos. Mater.*, **14,** 213–232 (1980).
5. Denton, D. L., "Mechanical Properties Characterization of an SMC-R50 Composite," 34th Annual Technical Conference, Reinforced Plastics/Composites Institute, The Society of the Plastics Industry, Section 11-F (1979); also SAE paper 790671 (1979).
6. Enos, J. H., Erratt, R. L., Francis, E., and Thomas, R. E., "Structural Performance of Vinylester Resin Compression Molded High Strength Composites," 34th Annual Technical Conference, Reinforced Plastics/Composites Institute, The Society of Plastics Industry, Section 11-E (1979).
7. Ackley, R. H., and Carley, E. P., "XMC-3 Composite Material Structural Molding Compound," 34th Annual Technical Conference, Reinforced Plastics/Composites Institute, The Society of Plastics Industry, Section 21-D (1979).
8. Adams, D. F., and Walrath, D. E., "Iosipescu Shear Properties of SMC Composite Materials," Department of Mechanical Engineering, The University of Wyoming, Laramie, Wyoming (1981).
9. Pipes, R. B., Wetherhold, R. C., and Gillespie, J. W., "Notched Strength of Composite Materials," *J. Compos. Mater.*, **13,** 148–160 (1979).
10. Maaghul, J., and Potkanowicz, E. J., "HMC - A High-Performance Sheet Molding Compound," 31st Annual Technical Conference, Reinforced Plastics/Composites Institute, The Society of Plastics Industry, Section 7-C (1976).
11. Wang, T. K., Sanders, B. A., and Lindholm, U. S., "A Loading Rate and Environmental Effects Study of Adhesive Bonded SMC Joints," Report GMMD80-044, General Motors Corporation, Manufacturing Development, GM Technical Center, Warren, Michigan (1980).
12. Gibson, R. F., Yau, A., and Riegner, D. A., "The Influence of Environmental Conditions on the Vibration Characteristics of Chopped-Fiber-Reinforced Composite Materials," Presented at AIAA/ASME/ASCE/AHS 22nd Structures, Structured Dynamics and Materials Conference, April (1981).
13. Seiffert, V. W., "Review of Recent Activities and Trends in the Field of Automobile Materials," in *Worldwide Applications of Plastics,* pp. 1–6, Society of Automotive Engineers, SP-482, February (1981).
14. Loos, A. C., Springer, G. S., Sanders, B. A., and Tung, R. W., "Moisture Absorption of Polyester-E Glass Composites," *J. Compos. Mater.*, **14,** 142–154 (1980).

10

Aerodynamics of Road Vehicles

THOMAS MOREL

I. INTRODUCTION

Vehicle aerodynamics concerns the effects arising due to motion of the vehicle through, or relative to, the air. Its importance to road vehicles became apparent when they started to achieve higher speeds. The automobile as we know it came onto the scene in the last decade of the nineteenth century. Its beginnings roughly coincided with the advent of powered flight, and perhaps for this reason, it became of interest to aerodynamicists right from the start. One of the first attempts to apply aerodynamic principles to road vehicles was the streamlining given to the first holder of the land speed record, a car named Jantaud driven by Gaston Chaseloup-Laubat (Fig. 1). This vehicle held the record several times, culminating with 93 km/hr (58 mph) achieved in 1899.

The early interest in vehicle drag continued and an early paper published in 1922 by Klemperer[(1)] (apparently the very first paper on the subject) already reported actual wind tunnel studies on several then current automobile shapes and also on one low-drag shape. The work was done in the wind tunnel of the Zeppelin Company which was involved in the development and construction of dirigibles. The influence of dirigibles showed in the shape of the low-drag model, which had a streamlined teardrop shape and no wheels. The drag coefficient of this model was 0.15, and it is interesting to note that this value is still among the lowest obtained for a vehicle-shaped body near the ground.

The thrust of the original work during the first few decades of the twentieth century was toward the reduction of drag to increase the top speed of road vehicles. That objective became less important with the steady improvements and increasing power of automobile engines, and with the increasing efficiency

---

THOMAS MOREL • Integral Technologies Incorporated, Westmont, Illinois 60559.

FIGURE 1. First land speed record holder Jantaud, driven by Gaston Chaseloup-Laubat.

of the whole power-train system. Although the fuel costs were never negligible, efficiency improvement was not considered a pressing issue and, consequently, aerodynamics was often subjugated to other objectives. The original need for it, related to top speed, remained an issue somewhat longer with racing cars (but even there the increasing engine power shifted the emphasis) and land speed record cars. Examples of the latter are the Goldenrod,[2] holder of the world speed record for wheel-driven automobiles set in 1965 with 664 km/h (413 mph), and the rocket-propelled Blue Flame,[3] record holder in the unlimited class with 1001 km/h (622 mph) set in 1970. The drag coefficient claimed for the pencil-shaped Goldenrod is $C_D = 0.12$. This is the lowest value reported for any full-scale driveable vehicle. Practical road vehicles are much less slender and so comparison to Goldenrod as the ideal is probably not appropriate.

With the more recent emphasis on fuel economy, aerodynamics has regained some of its former importance. The need for increased fuel economy, dictated by the realities of the marketplace, has led to a climate in which aerodynamics is being asked to contribute its share to the attainment of the highest possible efficiency of the road vehicle as a system. This renewed interest shows up in the number of meetings organized in recent years, at which developments in the area of vehicle aerodynamics have been discussed.[4–10]

There is a commonly held belief that the principles of low-drag vehicle design are well understood. It is, after all, fairly widely known that the lowest drag body at subsonic speeds is one with a teardrop shape, already considered by Klemperer. An optimized teardrop in free flight (away from the ground) is usually assumed to have a drag coefficient as low as 0.04, but, somewhat surprisingly, the actual proportions of an optimum teardrop and their dependence on Reynolds number and yaw angle are in fact not that well established. Also, it is in fact not necessary to have the body taper at the end to a point, but it may be abruptly terminated well before the point would be reached, with no increase in drag as was determined experimentally in the 1930s. This idea is credited to Kamm, and the resulting shape is referred to as Kamm-back. These subtleties notwithstanding, it is a fact that current production automobiles have drag coefficients an order of magnitude larger than tear-shaped bodies in free flight and several times larger than shapes like Klemperer's. They range from

0.8 for the open-roof early automobiles to today's typical cars averaging 0.45, to the best current designs quoted at approximately 0.35. The major reason for the difference between the ideal and the current state of the art is that the vehicle shapes are subject to many constraints, which will be discussed in the next section.

Most wind tunnel investigations of the flow around road vehicles have concentrated on the measurements of forces and moments, while the details of the flow have not been studied in any great detail. This is mostly because of the complexity of vehicle shapes, which is the direct consequence of the practical constraints which influence and dictate various aspects of vehicle shape. An important consequence of the complexity of the vehicle shapes is that the resulting flow fields are also complex. They are turbulent, three-dimensional, they have numerous regions of separated flow, typically have strong streamwise vortices in the wake, and are influenced by the proximity of the ground. Some of these features are graphically illustrated in the sketch in Fig. 2.

As a result of this complexity, the work in the area of vehicle aerodynamics was not well focused until recently, being fragmented into many parallel investigations often involving the development of a particular vehicle, with not enough emphasis being given to the general discipline. An additional contributing factor was that aerodynamics had not been considered an essential and integral part of vehicle design, but had been included into the process after the fact. The vehicle shape was designed by the stylists, and the aerodynamicists who followed tried to make the most out of what they were given. The final decision whether to incorporate any changes proposed by aerodynamicists rested with the stylists, and this situation did nothing to encourage fundamental research in aerodynamics.

The literature related to vehicle aerodynamics is very extensive. No effort to compile all of these numerous publications is being made here, but interested readers may consult a paper by McDonald[11] which includes a list of over 100 references. In addition to these studies one may consider the large amount of

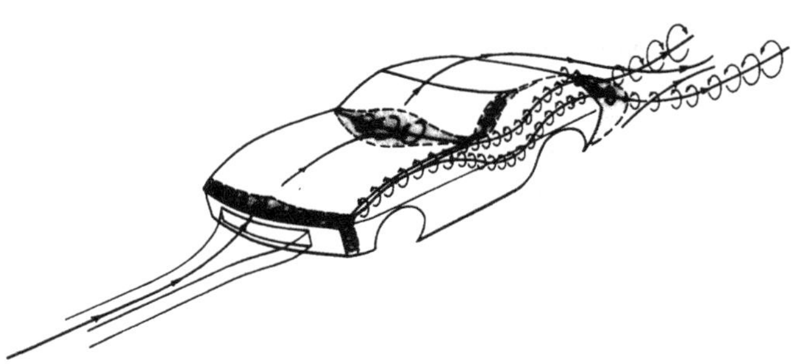

FIGURE 2. Schematic view of some of the complex features of vehicle flows (from Ref. 8).

work done on bluff or nonstreamlined body aerodynamics related to other fields. Its relevance to road vehicles was reviewed during a symposium at the General Motors Research Laboratories in 1976.[8] Most of the available bluff body work may be classified into several categories: fundamental—concerning simple shapes studied to provide basic understanding of separated flows; aeronautical and ballistic—concentrating on airplanes, projectiles, and missiles; and architectural—related to flows over buildings and structures. Since all this literature is available, how much can be learned and transferred from it to road vehicles?

The last category, architectural aerodynamics, is not particularly relevant as it deals with bodies placed directly on the ground, immersed in boundary layers whose thickness is often larger than the body height. The resulting flow fields are very different from road vehicle flow fields. The other two categories have more in common with road vehicles, and thus they might be expected to be a valuable source of information and ideas. As will be discussed later in a section dealing with the mechanisms of drag generation, the great majority of that literature deals with two-dimensional and axisymmetric bodies, whose flow fields are by and large not too relevant to the complex three-dimensional flows of road vehicles. In this context it is worthwhile to bring out one of the differences between the aeronautical and vehicle aerodynamics, which should be kept in mind when viewing the presented results, and this concerns the coordinate system. In consequence of the positive contact of the vehicle with the road, all forces and moments are referred to body axes, rather than to the relative wind direction as is done for free-flying objects.

The objective of this review is to attempt to address a number of key questions concerning road vehicle aerodynamics:

1. How does aerodynamics fit into the context of a practical vehicle?
2. How much does aerodynamics contribute to fuel consumption?
3. How is drag generated?
4. What is the current state of the art of low-drag design?
5. What is the future outlook for low-drag design and what is the practical lower limit of C_D for production vehicles?

These questions guided the outline of the review, with separate sections devoted to each of them.

II. BASIC CONSIDERATIONS INFLUENCING VEHICLE SHAPE

As pointed out in the introduction, current production automobiles have drag coefficients an order of magnitude larger than tear-shaped bodies in free flight and several times larger than idealized shapes like Klemperer's. An important reason for this discrepancy is that vehicle shapes are subject to many practical constraints from the point of view of intended vehicle purpose and use, vehicle safety, maintenance, and cooling and product identity. Taking these constraints in turn, the vehicle has to provide a reasonably spacious passenger compartment which dictates its overall size and means for a convenient entry from outside, and it must also provide engine and luggage compartments. All of these combine to largely dictate the overall shape, and they also add to body details, such as door handles, which increase drag. The result is graphically illustrated in Fig. 3 taken from Hucho.[12] For a typical current automobile the overall proportions of the "enveloping box" are about 3:1.33:1 (length:width:height). Part of this constraint includes the wheels, which must be left partly unshielded. From the safety point of view the passenger compartment must be configured to permit adequate visibility frontward, backward, and to the sides. The vehicle must provide maximum protection in accidents, which means the shape should incorporate adequate crush zones. The windshield may not be inclined below some 30° because of light refraction limits which reduce visibility. External mirrors providing rear and side view also add to overall drag.

For ease of maintenance there should be good access to the engine compartment, as well as to the many components laid out along the underbody. This tends to preclude the use of a smooth underbody shield that could be quite beneficial from the drag point of view. Another reason against the use of a shield is that a number of components such as brakes, transmission, and the exhaust system need to be exposed to the airstream for cooling reasons. Finally, exterior vehicle appearance is an important factor in customer appeal. It is this aspect which motivates vehicle designers to provide each vehicle with its own identity, often while maintaining a family theme common with the rest of the product line. In an effort to achieve this objective some of the principles of low-drag design are subjugated to the appearance.

As a consequence of these constraints, and others not mentioned, the designer–aerodynamicist does not start with a clean sheet of paper, but rather with

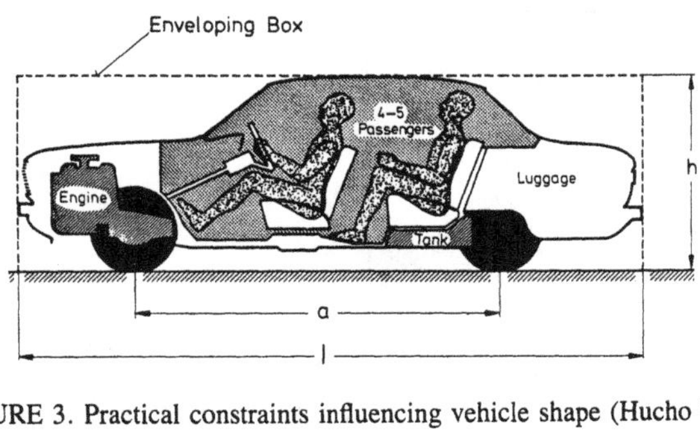

FIGURE 3. Practical constraints influencing vehicle shape (Hucho [12]).

a rough-cut vehiclelike shape conforming to practical requirements. The challenge facing him is to design a final vehicle shape so that the drag is at a minimum, which inevitably will be substantially higher than that of an idealized smooth teardrop shape. On the other hand the gap between the current average drag coefficient of 0.45 and the drag coefficient of an idealized shape of 0.15 is very large. As will be pointed out in subsequent sections, this gap is wide enough to permit significant reductions in C_D within the practical constraints discussed above.

III. EFFECT OF DRAG ON FUEL ECONOMY

The motion of a road vehicle is subject to three different forces—tire rolling resistance, aerodynamic drag, and gravity—which oppose the tractive force propelling the vehicle. In the simple case where the vehicle is traveling on a level road at a constant speed, only the first two forces are present. The first of them, the rolling resistance, depends on the tire construction and inflation pressure, and it is proportional to the vehicle weight. Its magnitude is fairly constant, increasing only slowly with vehicle speed as well as its shape. The drag, by contrast, depends strongly on the vehicle speed. Its magnitude is proportional to the vehicle frontal area and to the square of the air speed. At very low speed the drag is negligible compared to the rolling resistance, but at high speeds the drag rises rapidly and eventually it dominates the total resistance.

The fact that drag increases rapidly with speed is well known, and it is responsible for the perception that drag is important only at high speeds. This is not really true. A yardstick commonly used to address this question and provide some indication as to the speed at which the drag *does* become important is the so-called crossover speed. This is the steady-state speed at which the aerodynamic drag equals the rolling friction (Fig. 4). For today's automobiles this crossover has been estimated to occur around 50 km/h (30 mph) (Hucho *et al.*[13]) and

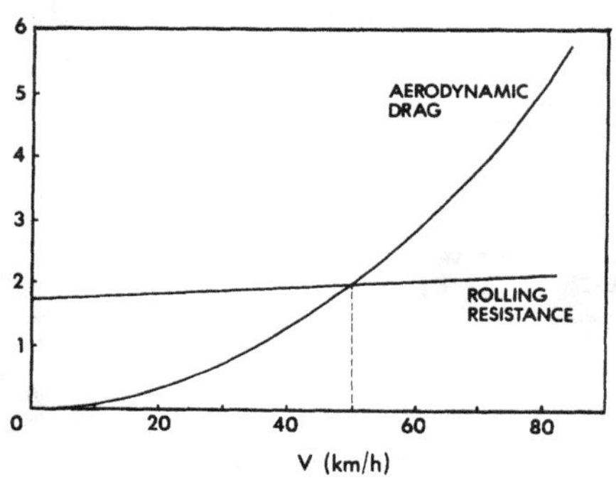

FIGURE 4. Aerodynamic drag versus rolling friction; the intersection of the two curves defines the "crossover speed."

with the trend towards lighter vehicles and reduced tire rolling resistance the crossover speed will tend to be even lower on future road vehicles. This means that even at speeds attained in city and suburban driving, 50% of the total resistance at constant speed travel is due to drag. It follows that drag is not an exclusively high-speed highway phenomenon. We shall return to this point later, when discussing cycle-averaged energy consumption of the federal EPA city driving schedule.

A. Fuel Consumption Due to Drag

Let us now address directly the question of importance of drag. There are several ways this question may be approached, one being the crossover speed discussed above. This is a simple and effective way which brings out the level of vehicle speeds at which drag starts to dominate. However, it treats drag only in comparison to the rolling resistance, which varies from vehicle to vehicle with weight and tire characteristics. A common way to quantify the importance of drag is by estimating the percentage of drag contribution to the overall resistance over a typical driving schedule. This approach is used very often to estimate improvements in fuel consumption that might be expected for a given reduction in drag. This measure is again relative, and it does not isolate the true contribution drag makes to the fuel consumption.

To obtain that information, one ought to start from the energy that is required to overcome the drag alone. We shall now follow this avenue, and calculate the energy lost to drag at an arbitrary speed in terms of fuel consumption, expressed in terms of liters/100 km. Starting with energy expressed in Joules (calculated as drag force times 100 km):

$$E_A = C_D A \tfrac{1}{2} \zeta V^2 \cdot (10^5 \text{ m}) \tag{1}$$

where A is the frontal area in m^2, air density $\zeta = 1.177$ kg/m^3 and V is vehicle velocity in m/sec. The resulting energy loss is shown in Fig. 5, where each line corresponds to a fixed value of the product C_DA. The values of C_DA used in Fig. 5 span the range from the high values of some past cars to the expected lowest ones. The highest value of 1.4 refers to the large automobiles of a few years ago with drag coefficients near 0.6. Current compacts with frontal areas approximately 2 m^2 and C_D around 0.45 have C_DA of close to 0.9. The lowest line $C_DA = 0.5$ refers to a very low-drag minicar (or a sports car) with $A = 1.66$ m^2 and a very low-drag coefficient of 0.3. This range of C_DA indicates the extremes of the drag energy loss: the large car consuming three times as much fuel to overcome drag than an optimized minicar.

Up to this point no assumptions had to be made, but in order to translate the energy into L/100 km one now has to make some assumptions about the

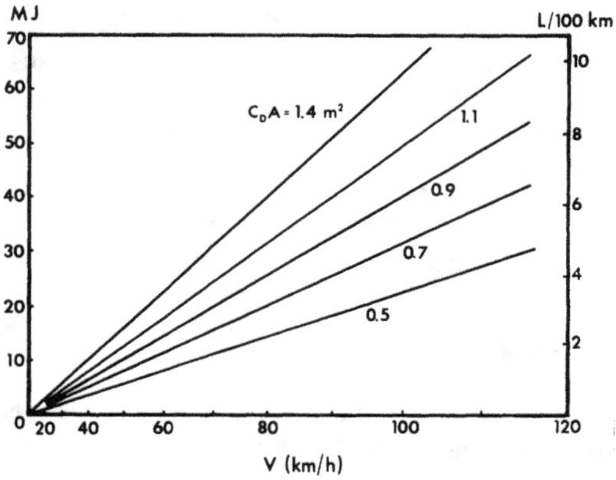

FIGURE 5. Energy and fuel consumption as a function of speed for various values of C_DA.

overall efficiency of energy conversion with which the energy contained in fuel may be used to propel a vehicle, i.e., the efficiency of the entire engine–driveline system. Taking tentatively $\eta = 0.2$ as a typical value, and accepting 32.2 MJ/L to be the heating value of gasoline, one may write for the fuel consumption needed to overcome drag

$$C_A = E_A/(0.2 \times 32.2) \qquad \text{(L/100 km)} \tag{2}$$

This relation between C_A and E_A was used to construct the scale on the right-hand side of Fig. 5. It should be stressed that any other values of efficiency and fuel heating value could be used, and in that case, the scale would have to be changed to reflect these values. The only reason for including the right-hand scale is to provide a quick estimate of the magnitude of the fuel consumption. The left-hand side scale is independent of any assumptions and truly reflects the *energy* consumed.

B. Drag Amplification by Wind

All of the above concerned a strictly windless environment, for which one can use the drag coefficients measured in wind tunnels at zero yaw, and take the air speed to be equal to the vehicle speed. In the presence of wind, such a simple procedure is not appropriate as will be shown below. Instead, one has to provide for the actual values of the drag coefficient and air speed, and in the analysis that follows we shall lump these two effects into an "effective" drag coefficient. Once the effective drag coefficient has been determined, it may be used to replace the zero-yaw coefficient in Fig. 5.

The wind direction does not necessarily coincide with the direction of travel. From Fig. 6 one finds that as a result, the vehicle is exposed to air coming at

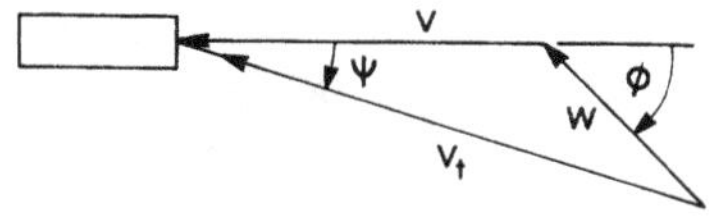

FIGURE 6. Vector diagram of vehicle and wind velocity.

an apparent yaw angle ψ, which is not the same as the wind angle ϕ. It is related to it by

$$\psi = \sin^{-1}[W \sin \phi / V_t] \tag{3}$$

where the total apparent air speed V_t is given by

$$V_t^2 = V^2 + 2VW \cos \phi + W^2 \tag{4}$$

The change in the apparent air speed is not the only effect produced by the wind. Another effect is a change in the drag coefficient with the effective yaw angle ψ. One part of that change comes about from the fact that in road vehicle practice the relevant drag force is the projection of the total force vector into the direction of travel. Thus even for a sphere, which has the same total force irrespective of the wind direction, the drag force resolved along a fixed axis varies as $\cos \psi$, giving zero drag at $\psi = 90°$ and negative drag at angles larger than that. This is in contrast to aeronautical practice where the relevant drag force is the projection into the air flow direction. The drag coefficient of a road vehicle in yaw has the same gross characteristics as for the sphere, but its details are more complicated. Typically, the drag coefficient first rises at small yaw angles, to fall off rapidly beyond a maximum (Fig. 7), which for conventional sedans lies near $\psi = 30°$–$35°$ (Bowman[14] and Cogotti *et al.*[3]), near $\psi = 25°$ for station wagons (Cogotti *et al.*[15]) and for trucks around $\psi = 20°$ (Mason and Beebe[16]). A 30–50% drag overshoot at the maximum is not uncommon. The magnitude

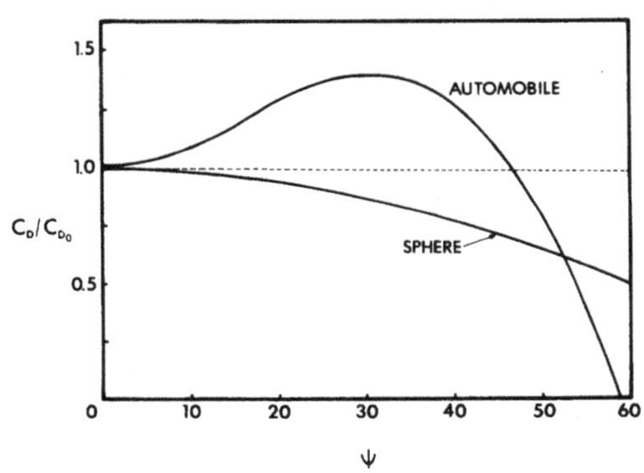

FIGURE 7. Dependence of drag coefficient on yaw angle for a typical automobile and for a sphere.

of the drag overshoot depends on a number of parameters of the body shape, but among them the strongest is likely to be the body length to "diameter" ratio. The longer the body, the larger will be the body area exposed to the airstream in yaw. Thus for a sphere with length to diameter ratio of 1 the overshoot is 0, while for a long truck its value easily reaches 50%. The curve in Fig. 7 may be approximated by a cosine curve, in a form similar to that used by Bowman, as

$$C_D(\psi)/C_{D0} = 1 + \tfrac{1}{2}B[1 - \cos(\pi\psi/\psi_1)] \tag{5}$$

where B is the maximum relative overshoot and ψ_1, is the location of the maximum.

The combined effect of the apparent air speed and of the yaw effect on the drag coefficient may be expressed in terms of an effective drag coefficient. In order to do that, one has to make some assumption about the probability distribution of the wind angle ϕ with respect to the direction of travel. For a long-distance truck traveling mainly on a straight-line highway between two truck terminals in an area with prevailing winds, the distribution may be approximated by a double-peaked delta distribution, at two angles separated by 180° corresponding to the two directions of travel on the highway. For a general-vehicle population a more appropriate distribution is an equal probability for all wind angles. Let us consider that particular case and write

$$\frac{C_{D\text{eff}}}{C_{D0}} = \frac{1}{2\pi}\int_0^{2\pi} \frac{C_D(\psi)}{C_{D0}}\left[1 + 2\frac{W}{V}\cos\phi + \left(\frac{W}{V}\right)^2\right] d\phi \tag{6}$$

where ψ is given by Eq. (3). The ratio on the left-hand side is a function of the wind speed ratio W/V and of the assumed values of B and ψ_1. A parametric study was run to determine the range of $C_{D\text{eff}}/C_{D0}$ for a series of values of W/V and B, with a fixed value of $\psi_1 = 30°$. The results are presented in Fig. 8. The most obvious conclusion one can immediately draw from the figure is that the presence of wind always magnifies the average drag force, and effect increases roughly quadratically with increasing wind speed. The presence of a typical drag overshoot found on road vehicles at moderate yaw angles compounds the direct wind effect, i.e., the increase in relative air velocity, roughly doubling it.

In order to use the results of Fig. 8, one has to assume a typical value for the wind speed. A good source for that value is the meterological service, which monitors and compiles wind speeds at regular intervals. Thus, for example, one finds for the Detroit area that the average wind speed (yearly average) is more than 16 km/h (10 mph).[17] The actual wind speed close to the ground where vehicles operate is less than this value. However, the mean wind speed underestimates the true wind effect, which is magnified by wind unsteadiness and

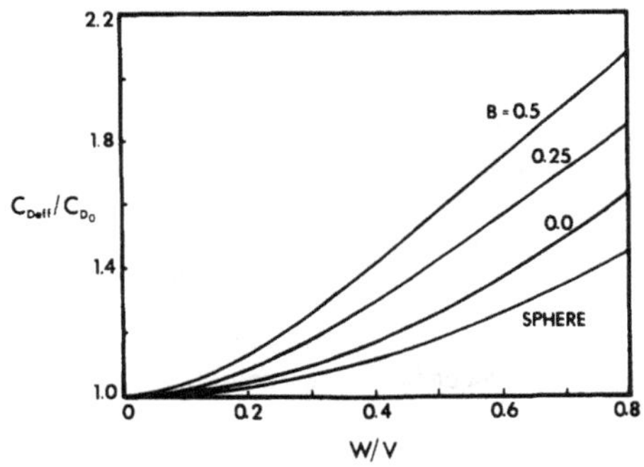

FIGURE 8. Drag amplification by wind. (Note that the curve for $B = 0$ means $C_D(\psi) = C_{D0}$, while for a sphere $C_D(\psi) = C_{D0} \cos \psi$.)

turbulence due to nonlinearity of drag (drag increases as the square of total velocity V_t). Thus taking the meteorological mean wind data as representative of the wind encountered by vehicles is probably a reasonable approximation. Taking this value as the wind speed and $B = 0.25$ as a representative value for the drag coefficient overshoot, one can obtain an estimate of the effect of wind on vehicle drag: at $V = 50$ km/h (30 mph) the effective drag coefficient is about 20% higher than the zero-yaw baseline C_{D0}, at $V = 30$ km/h (20 mph) it is about 45% higher, and at $V = 22$ km/h (14 mph) it is more than 70% higher.

In consequence, the unavoidable presence of wind increases the contribution of aerodynamic drag to the overall losses. This is particularly true at lower speeds. As a result, the crossover speed at which aerodynamic drag predominates is in fact even lower than the zero-yaw (no-wind) considerations would indicate, i.e., below 50 km/h (30 mph).

C. EPA City–Highway Driving Schedules

All the previous discussions related only to constant speed operation, but actual operation also involves periods of acceleration, braking, and idling. The EPA highway driving schedules have been created to approximate the real driving situation and we can employ them as representative of typical operation. These cycles are prescribed in terms of vehicle speed given in 1-sec intervals, and they are graphically represented in Fig. 9. As may be seen from these figures, the average vehicle speed in the EPA cycles is 77.6 km/h for the highway schedule and 38.4 km/h for the city schedule. These arithmetic averages are not the proper ones to use in estimates of the average drag force; for that, one has to integrate the drag force with respect to distance traveled (because $E_A = \int \text{Drag}\, ds$) and this yields

$$V_{\text{ave}} = \left[\frac{1}{S}\int V^2\, ds\right]^{1/2} \tag{7}$$

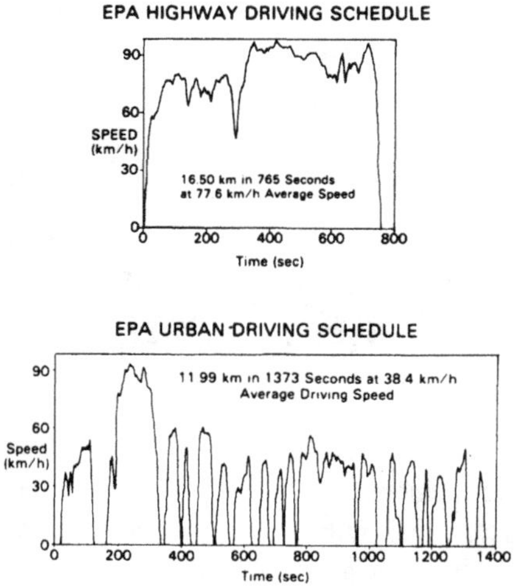

FIGURE 9. EPA driving schedules (Sovran and Bohn [18]).

where S is the distance traveled in the driving schedule. Such integration of the EPA cycles has been performed by Sovran and Bohn,[18] and it yielded 81.9 km/h (50.9 mph) for the highway schedule, and 53.3 km/h (33.1 mph) for the city schedule. It follows that even during the city schedule the average vehicle speeds are fairly high and the aerodynamic drag is important. The paper of Sovran and Bohn delves quite deeply into the analysis of the variable-speed schedules and extracts a lot of useful information about the distribution of the total energy supplied to the vehicle into its acceleration, aerodynamic drag, and rolling resistance. The amount of energy that could in principle be recovered by regenerative braking during EPA schedules is also given.

D. Fuel Economy Potential of Drag Reduction

In order to assess the fuel economy potential of drag reduction, let us consider one particular example. Taking a compact automobile with frontal area of 2 m^2 (roughly the size of Chevrolet Citation) and a 0.45 drag coefficient typical of today's automobiles, one can calculate the energy needed to overcome aerodynamic drag over the distance of 100 km on the two EPA schedules. For this purpose let us adopt the mean velocities calculated by Eq. (7) for the two schedules, i.e., 81.9 km/h and 53.3 km/h, respectively. Using these values in Eqs. (1) and (2) one obtains

$$C_A = 1.8 \text{ L/100 km} \qquad \text{city schedule}$$

$$C_A = 4.25 \text{ L/100 km} \qquad \text{highway schedule}$$

Including now into the consideration the effect of wind with W = 16 km/h, and maximum drag coefficient overshoot at yaw of 25%, one finds from Fig. 8 that the fuel consumption due to aerodynamic drag is augmented by 17% for the city schedule and by 8% for the highway schedule. Incorporating this drag increase, we find C_A = 2.11 and 4.59 L/100 km, respectively. The combined city–highway fuel consumption (55/45 percent combination) is then calculated to be 3.23 L/100 km. A typical compact automobile weighing 1200 kg (2700 lb) may be expected to have a fuel consumption of around 10 L/100 km (24 mpg) on the combined EPA schedule. For it a *10% reduction in aerodynamic drag* would produce a fuel consumption reduction of 0.32 L/100 km or *3.2% improvement in fuel economy*. This result, although fairly typical, is exactly valid only for the example used here, as the actual percentage improvement per 10% drag reduction depends on a number of parameters, the most important among them being the ratio of C_DA to vehicle weight and the tire rolling resistance. As for the total fuel economy potential of drag reduction, one may assume that the now typical C_D = 0.45 could be reduced to levels around 0.30 or by about 33%. This drag reduction translates into a very significant 11% decrease in fuel consumption on the EPA combined schedule, and much more than that at highway speeds.

It should be mentioned that the stated fuel economy benefit will not be fully realized by simply reducing the drag coefficient of an *existing* vehicle. This is because changing the drag changes the vehicle's energy requirements. Presuming that the original vehicle had a power train perfectly matched to the original energy requirements (load), the lower-drag version of that vehicle will not be as well matched, and the power train will not operate at its maximum efficiency. This is because the effectively oversized engine will run at excessive speeds, at which it will have to be throttled more than necessary. Thus the full economy potential of reduced drag will be realized only if the power train is rematched to the new load. This aspect is treated in more detail, for example, by Janssen and Emmelmann.[19] For a newly designed low-drag vehicle this problem does not exist, because its power train may be expected to be matched as closely as possible to the actual load characteristics including the low aerodynamic drag.

In summary, the aerodynamic drag on current vehicles exceeds rolling resistance at speeds lower than the average achieved even in the EPA city schedule. With the present trend towards lighter vehicles and better tires, the relative importance of the aerodynamic drag will be accentuated, unless parallel reductions in C_DA are made.

IV. MECHANISMS OF DRAG GENERATION

As will be shown in the next section, it is important to give careful attention to small body details, because the cumulative effect of a number of such details can be quite significant. Following that strategy, a competent aerodynamicist can reduce significantly, by trial and error, the drag of almost any vehicle supplied by the stylists. However, to step beyond the state of the art and reduce the drag even more substantially, the following two questions must be addressed:

1. What is the practical lower limit of C_D for production vehicles?
2. What are the key design criteria one should observe to achieve this goal?

To answer these questions one first has to gain more understanding of the fluid mechanics involved in drag generation.

The purpose of this section is to review what is known about the mechanisms by which drag is generated on bluff bodies and, in particular, on road vehicles. We shall discuss what is known about the nature of road vehicle flow fields and provide some rudiments of fluid mechanics relevant to them. It is quite important to be familiar with the state of the art in understanding the drag mechanisms, because it represents a consolidated body of knowledge of the basics underlying the whole vehicle aerodynamics subject area. This knowledge provides the foundation needed for the analysis of experimental data, and it can provide ideas for new key experiments. It is central to our ability to make assessments of the most profitable avenues for drag reduction, and of the capabilities and applicability of various theoretical or numerical approaches. For these reasons this section, dealing with drag mechanisms, constitutes a substantial part of the whole review.

A. Components of Drag

A technique that is useful in analysis of aerodynamic drag is a division into its components. The purpose of the division is to gain conceptual simplification of the problem which promotes understanding of how drag is generated. An important benefit is that this permits focusing attention on those areas showing the greatest promise from the point of drag reduction. The first step that one can make is to divide the total drag between internal and external contributions, the internal being due to losses in internal passages, most prominently in the cooling system. Of the two, the latter one is by far the larger, accounting for more than 90% of the total, and to it most of the following discussion will relate. The internal drag will be dealt with separately later.

The next step in the breakdown is to separate the external drag into two parts: the pressure drag and the skin friction drag. These two parts may be evaluated as integrals over the entire body surface, the pressure drag being the integral of static pressure over projections of the surface normal to the direction of travel, and the friction drag being the integral of surface shear stress over surface projections parallel to the direction of travel. In terms of force coefficients, one writes

$$C_p = (\text{pressure drag})/(A \cdot q) \tag{8}$$

$$C_f = (\text{friction drag})/A \cdot q) \tag{9}$$

where A is the frontal area of the body and q is the dynamic pressure, $\frac{1}{2}\rho V^2$. The ratio of C_p to C_f may be taken as a convenient criterion to decide whether a body is "streamlined" or "bluff," for example

$$\text{Streamlined body: } C_p/C_f < 1 \tag{10}$$

$$\text{Bluff body: } C_p/C_f \gg 1 \tag{11}$$

Road vehicles are, according to this definition, bluff bodies. Due to the presence of large separated flow regions, which are unavoidable on practical shapes, their pressure drag is always much larger than the friction drag. By contrast, missiles, airplane fuselages, or submarine hulls would qualify as streamline bodies, their ratio C_p/C_f being on the order of unity or less. The friction drag will be discussed in more detail later, but let us mention at this point that for road vehicles its magnitude is less than 10% of the total drag.

Since the exterior pressure drag is responsible for more than 80% of the total drag (the remaining 20% being due to the internal drag and skin friction drag), it seems worthwhile to attempt to divide it further into separate parts. From here on the division is less straightforward, but one way is to divide the body into the upper shell and the underbody, and then further divide the upper shell into a forebody and an afterbody. This type of division is not ideal as it suggests an independence between the flow fields of the three resulting parts, while in fact there are strong links between them. Nevertheless, the division is quite useful conceptually in many instances, for example, when discussing small changes in geometry affecting localized separated regions.

B. Forebody Drag

1. Single-Body Configurations

The division of the upper body shell into a forebody and an afterbody is not a precise one; it is mainly an illustrative aid. In general, the body is divided into these two parts at its maximum cross section. In the case of a van or a station wagon the location of the maximum cross section is ambiguous, but that presents no difficulty. In fact, the presence of a constant area section between the two parts makes the two flow fields more independent.

The pressure distribution over a forebody surface always has regions where the pressure is higher than the ambient pressure, and regions where it is lower. The high-pressure regions are near stagnation points and on most concave surfaces, while the low-pressure regions are found along convex surfaces and corners. If the flow over the forebody is attached everywhere, it may be approximated to a good accuracy by an inviscid flow. In such a case, one can show that the forebody pressure force, the integral of pressure over the whole forebody, is always negative (it pulls the body forward), tending to zero for bodies having a long parallel-sided section between the forebody and the afterbody (Fig. 10). This result is independent of the details of the forebody shape (Morel[20]).

In real-flow situations the forebody pressure force is often positive due to the presence of flow separation. This separation may be of three basic types: (1) separation on convex surfaces which have a curvature large enough so that the flow cannot follow the contour and breaks away, (2) localized separation on strongly curved concave surfaces where the flow breaks away to bridge the region of strong curvature, and (3) three-dimensional separation of skewed surface flows on surfaces that deviate substantially from axisymmetric or two-dimensional shapes.

The strategy concerning the minimization of forebody drag of a single-body configuration seems fairly straightforward. It calls for the elimination of all regions with excessive curvature, and control of shape skewness, which lead to flow separation. From the point of view of the effects of forebody flow on the afterbody flow, the strategy is more involved. At a minimum, one ought to try to minimize the curvatures around the maximum body cross section and thus the local pressure minimum there. Beyond that, there are indications that some gains may be made on road vehicles by shaping the forebody in such a way that the external flow is channeled sideways around the vehicle, rather than over the top, as will be discussed in a later section. A further increase in complexity is

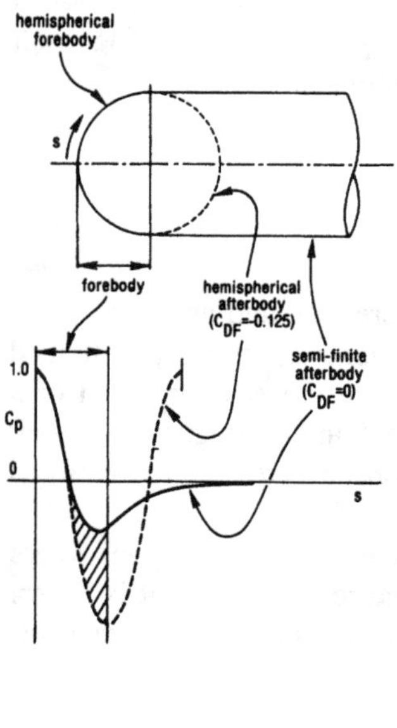

FIGURE 10. Pressure distributions on a hemispherical forebody followed by two different afterbodies. Note the excess suction peak near the maximum cross section for the finite afterbody, the shaded area, which gives rise to a negative forebody drag coefficient (Morel [20]).

encountered when yaw characteristics (crosswind operation) are included and we have already seen that the drag coefficient of road vehicles is typically larger in yaw than at zero yaw. The resulting three-dimensional flows are an order of magnitude more complex, a subject that has not been addressed to date in the context of road vehicles. On the other hand, a considerable amount of work has been done on this subject in the aeronautical engineering related to airplane fuselages and missiles. Many of the concepts proposed in that context may be of value in relation to streamlined road vehicles in crosswinds. An excellent reference in this regard is the extensive review of Peake and Tobak.[(21)] This subject is rather important conceptually, as it bears on the optimization of vehicle shapes for the whole range of yaw angles of practical interest (from 0° to around 15°).

2. *Tandem Configurations*

One of the more intriguing topics in bluff-body aerodynamics is the interaction of two bluff bodies placed in tandem. The intriguing facet of this topic is that the flow pattern and drag of a tandem configuration cannot be easily predicted from the known flow characteristics of the two individual bodies that form it. The reason for this is that the rear body is exposed to a flow strongly perturbed by the front body, and there is also some upstream influence from the rear body on the front one. The most relevant examples here are the flow over a tractor–trailer combination, or over two vehicles traveling close behind one another. Another example would be two neighboring buildings. When the bodies are not directly connected, one is most interested in the forces acting on each of them. When they are connected, as in the case of a tractor and trailer, one is also interested in the forces on the system as a whole.

Much work in this area has been done on truck–trailer combinations, often in connection with wind deflectors (or fairings) mounted on truck cabs. An example of such work is shown in Fig. 11, taken from Mason and Beebe.[(16)] The flow fields of tractor–trailer combinations are three-dimensional and very complex, which makes analysis of such experiments often quite difficult. Because of that, studies have been made of simpler axisymmetric configurations. One of them is that of Roshko and Koenig,[(22)] concerning an axisymmetric cylinder preceded by a concentric thin circular disk. The study concentrated on the forebody drag of this combination (defined as the force on the disk plus the force on the front face of the circular cylinder), and the optimum disk diameter and distance ahead of the cylinder were sought at which the forebody drag was minimum. The cylinder alone had a forebody drag coefficient of 0.75 when the circumferential edge was sharp. Rounding the edge by a radius equal to one eighth of the cylinder diameter reduced the forebody drag coefficient to zero, provided the boundary layer on the front face was tripped. The Reynolds number

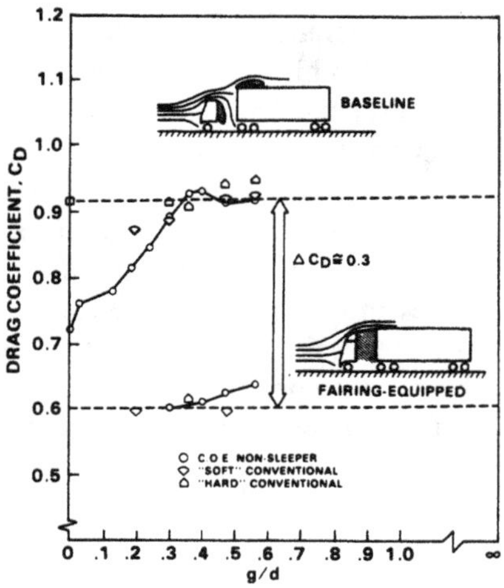

FIGURE 11. Effect of a cab-roof deflector on drag as a function of tractor–trailer gap. The upper curve is for the baseline tractor, the lower one is for the tractor with a deflector; g is the gap between the tractor and the trailer, d is the equivalent diameter of the trailer. (Mason and Beebe [16]).

in this experiment was 100,000–800,000. When an optimized disk was placed ahead of the cylinder with the sharp edge, with disk-to-cylinder diameter ratio of 0.75 and gap-to-cylinder diameter ratio of 0.375, an amazing drag reduction was achieved from $C_{DF} = 0.75$ down to $C_{DF} = 0.02$ (Fig. 12). This value is almost the same as for a well-rounded forebody with no separation—but in this case there was a substantial separated region present. One lesson learned from this study was that for tandem bodies, where some flow separation is unavoidable, the overall drag depends very strongly on the nature of the flow separation and reattachment: on the optimum configuration the flow separated from the front disk reattached itself neatly at (or perhaps a very small distance just aft of) the

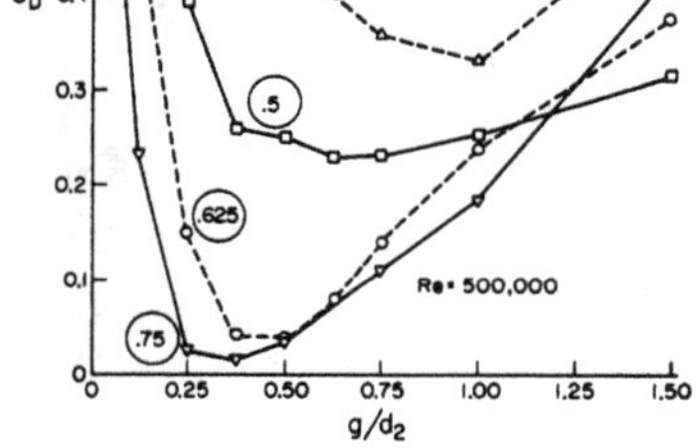

FIGURE 12. Forebody drag of an axisymmetric cylinder preceded by a concentric disk, with disk-to-cylinder diameter ratio and the gap between the two as parameters (Roshko and Koenig [22]).

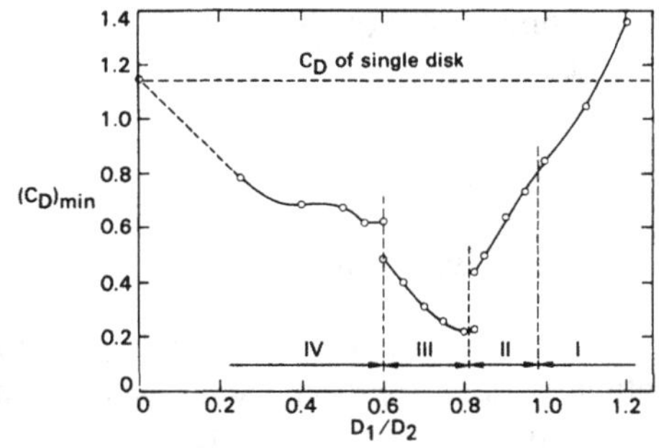

FIGURE 13. Drag coefficient of two disks in tandem as a function of the ratio of the front disk diameter to the rear disk diameter D_1/D_2 (Morel and Bohn [(23)]).

front edge of the rear cylinder. Another point was that the high forebody drag of a nonoptimum forebody (with separation occurring at the maximum body cross section) may be reduced by shielding that forebody by another bluff body—provided the two are matched.

Another study concentrating on optimization of tandem bodies was that of Morel and Bohn[(23)] concerning two circular disks in tandem. In this case the drag coefficient of an isolated disk was $C_D = 1.15$, of which the forebody drag contribution (front face of the disk) to the total drag was $C_{DF} = 0.71$. Figure 13 shows the effect of adding a second disk with diameter D_1 ahead of the original one (D_2) on the total drag of the configuration. Each point on the curve represents the minimum drag obtained at the optimum distance between the disks $(L/D_2)_{\text{opt}}$ (Fig. 14). The lowest drag was obtained with $D_1/D_2 = 0.75$ and $L/D_2 = 0.54$ with a value of $C_D = 0.21$. Subtracting the drag on the rear face of the downstream disk (base drag), one finds that the forebody drag coefficient of the combination was reduced to $C_{DF} = 0.03$, a value close to that obtained by Roshko and Koenig. It is interesting to note that even a disk slightly larger than the original one reduced the overall drag if placed at the right distance ahead. Four different flow regimes were identified, separated in Figs. 13 and 14 by vertical lines. Of these, regime 1 and the lowest-drag regime III were characterized by steady flow, while the other two regimes were characterized by

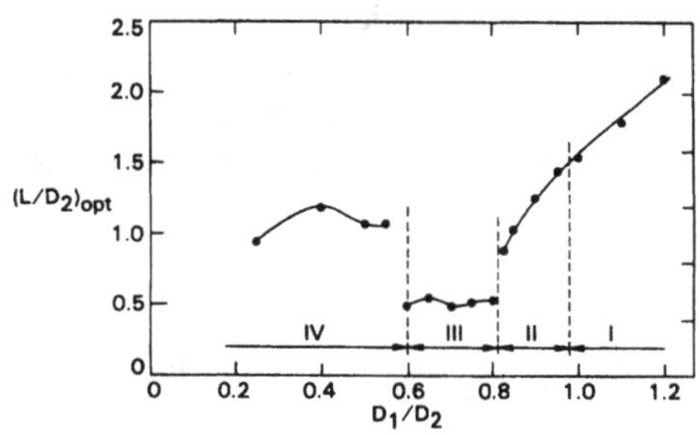

FIGURE 14. Optimum gap size for two tandem disks as a function of D_1/D_2 (Morel and Bohn [(23)]).

oscillating unsteady flows at L/D_2 = 0.5–1.0, most likely caused by cavity oscillations.

C. Afterbody Drag

1. Flow Separation

While the strategy for low drag on forebodies was the avoidance of separation, such an approach cannot be used on afterbodies: to avoid separation, the afterbody would have to be too long to be practical. Consequently, flow separation, usually quite massive, takes place on vehicle afterbodies.

When the flow separates from the afterbody, it may do so in two fundamentally different ways. One is a two-dimensional (or axisymmetric) separation, and the other one is a three-dimensional separation. Of these the former is much better known and understood. It describes the situation where boundary layers developing along a surface in an adverse pressure gradient reach a point where the velocity gradient in the direction perpendicular to the surface is reduced to zero. At that point the flow breaks away from the surface to relieve the pressure gradient, and a reverse flow sets in between the surface and the separated stream. By continuity requirements, the separated stream must close up somewhere downstream, enclosing a recirculating flow region. If the recirculating region is small and localized, the flow reattaches on the body surface itself, and the region is referred to as a "bubble." This happens typically for localized separated regions induced by abrupt convex or concave curvatures or steps, usually on the forebody (Fig. 15). For an afterbody separation the usual result is a large recirculating region enveloping the rest of the afterbody and closing up in the fluid downstream. In either case the recirculating flow regions are never exactly steady, but they display random or periodic oscillations with magnitudes that sometimes are quite substantial. This unsteadiness is concealed in the usual portrayal of separated regions in terms of time-averaged streamlines, but it is always there.

The three-dimensional separation is of a much more complicated nature. It occurs on nonaxisymmetric bodies and also on symmetric bodies inclined with respect to the flow. According to a definition proposed by Wang,[24] the three-dimensional separation can be of two basic types—closed and open. A closed separation region is one where the surface separation line divides the flow into two regions, each originating from a different stagnation point. The line separates the main stream from the recirculating flow much like the separation point does

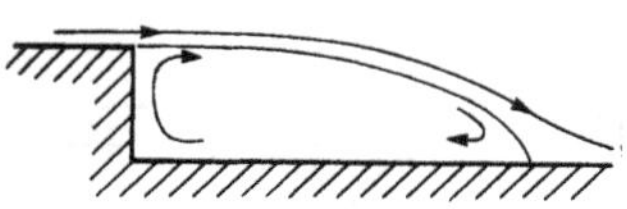

FIGURE 15. Localized flow separation in the form of a recirculating bubble.

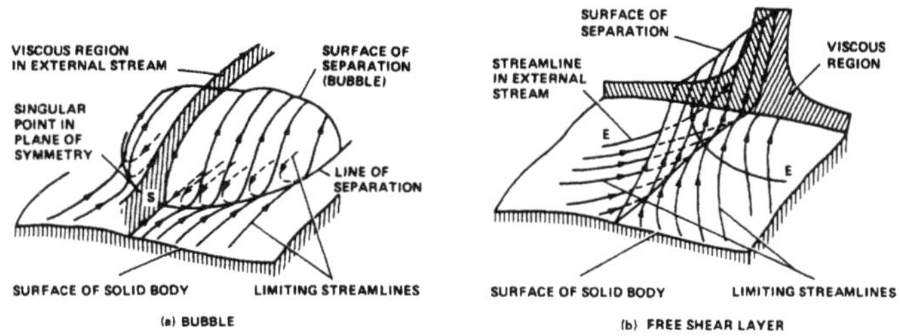

FIGURE 16. Two types of three-dimensional flow separation (Maskell [77]).

in the case of a two-dimensional separation [Fig. 16(a)]. In open separation the flow on both sides of the separation line comes from the same stagnation point [Fig. 16(b)]. The dominant feature of the open separation is that the flow leaving the surface along the separation line forms longitudinal trailing vortices. Examples of open separation are the flows over the side edges of finite-span airfoils[25] (Fig. 17) and over the leading edges of delta wings (Fig. 18) at angles of attack below stall. An interesting feature of open separation is that it is steady, as there is no recirculating flow region; this aspect has a considerable importance in some aeronautical applications (for example, in the design of the supersonic wing for the Concorde airplane).

Understanding the complexities of the three-dimensional separation is bound to be very important to low-drag vehicle design, because reliance on conventional wisdom about separation based to a large degree on two-dimensional reasoning may be quite misleading. The well-documented fact that wakes of road vehicles contain strong streamwise vortices is an indication that three-dimensional separations of the more complex, open, type are present on typical road vehicles (Fig. 2). Some very useful information related to three-dimensional separation may be found in Peake and Tobak[21] and in Landahl.[26]

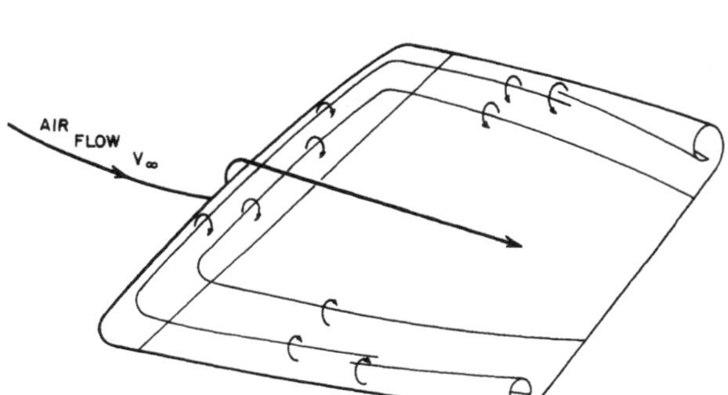

FIGURE 17. The classical conception of vortex lines in the flow past a thin wing of large aspect ratio (Thwaites [25]).

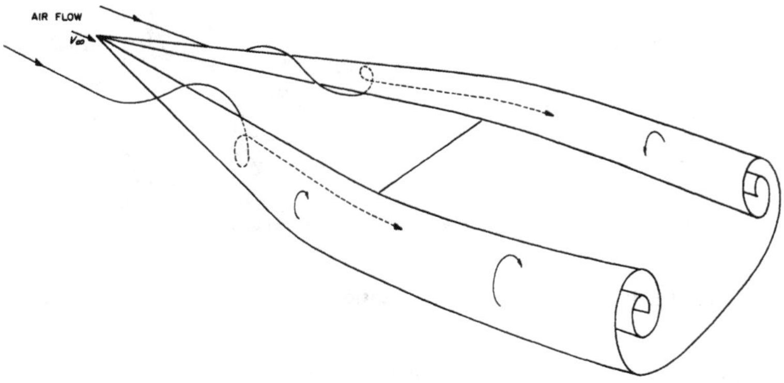

FIGURE 18. Conception of the flow past a thin delta wing (Thwaites [26]).

2. *Critical Geometries*

There is a large amount of information in the literature about the pressure drag of various bluff bodies, and also about the mechanisms by which this drag is generated. Considering this wealth of information, one would hope that it can be developed into building blocks from which at least the trends of flow pattern and force variation with changes in body geometry may be predicted. The objective would be progress to a point where educated guesses can be made to guide experimental programs. However, while an evolution towards a capability for better prediction of the flow behavior of bluff bodies is undoubtedly taking place, there will always be a need to check bluff-body flows experimentally through systematic perturbation of the geometrical parameters, in order to find the optimum configuration for any given purpose.

One of the reasons why a continued need for systematic experiments can be expected is the existence of what can be called "critical geometries," meaning cases where drag exhibits a local maximum with respect to some geometrical parameter of a body shape. The existence of a maximum, i.e., the fact that the variation of C_D with the parameter does not follow a monotonic trend, creates a difficulty. The implication is that one is faced with two opposing trends and that, at some *unknown* critical value of the parameter, the trends change. A survey of the bluff-body literature will reveal few carefully documented cases of such critical geometries. It is probable that a number of others may have been encountered but not studied in depth or reported for a variety of reasons, e.g., being unexpected and therefore suspect, or being of no interest to the purpose of the particular investigation and thus simply avoided.

Mair[27] observed an unexpected drag increase while investigating the drag-reducing capability of circular disks placed concentrically in the near-wake of a

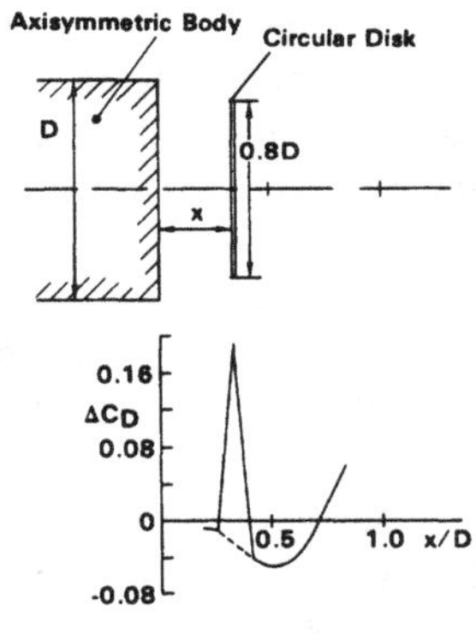

FIGURE 19. Effect of a circular disk placed in the near wake of an axisymmetric body. Note the critical behavior at $x/D = 0.3$.

body of revolution with a blunt end at $\mathrm{Re}_D = 150{,}000$. He found the effect of the added disks beneficial in general, particularly so for a disk with $d/D = 0.8$ placed at about $x/D = 0.5$ (Fig. 19). However, when the disk that gave the largest drag reduction was moved from its optimum position towards the base, a new flow regime was formed which was highly unsteady and produced a large drag increase in a relatively narrow range of x/D.

Critical geometries may also be found on smooth-shaped bodies as in the case noted by Mair[28] in his study of the effect of boat-tailing on the drag of axisymmetric bodies, at $\mathrm{Re}_D = 460{,}000$. On three of eight boat-tails tested, he noted critical boat-tail lengths at which the drag had a local maximum (length of each boat-tail was varied by simply cutting off the end). Figure 20 shows the drag curve of one of the three "critical" boat-tails, on which a local drag maximum was observed for $L/D = 0.8$.

One case of a critical geometry concerning road vehicles was documented by Janssen and Hucho.[29] Their experiment involved changes in the angle of the slanted rear portion of the roof of a car and their effect on drag. They

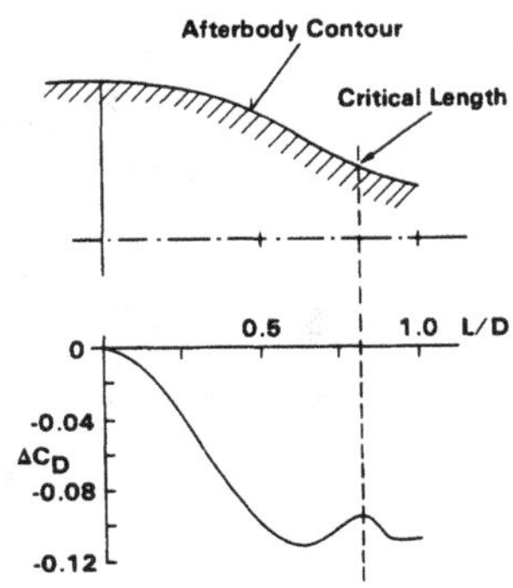

FIGURE 20. Boat-tailing on an axisymmetric cylinder, with critical behavior occurring around $L/D = 0.8$.

observed that for a small range of roof angles (25°–35°) the curve of overall drag exhibited a large overshoot (Fig. 21). They also observed a change in the extent of afterbody flow separation in the critical range, and documented it by the two sketches included in Fig. 21. They observed that the upper separation line was at the top of the slanted surface for angles more than 32°, but that it moved to the bottom for angles less than 28°; in between these two angles the point of separation was seen to pulse randomly from the top to the bottom and vice versa. (A more detailed study of this particular critical geometry will be discussed in the next section.)

What do these examples have in common? From the practical point of view the existence of a local drag maximum means that for critical geometries a small change of the critical geometrical parameter in either direction leads to a drag reduction. It follows that it should be feasible in many cases to avoid the high-drag region altogether, once its existence is established. In fluid–mechanical terms the common denominators seem to be the presence of flow separation and the existence of two competing flow patterns. Excessive unsteadiness is often, but not always, present as well. Also, all of the critical geometries identified here are associated with afterbody flows.

It is instructive to note that most of these examples of critical geometries, if not all, were encountered by chance with no prior expectation of the critical behavior. The fact that the drag overshoots are often unsuspected, coupled with the complex and difficult-to-analyze types of flow from which they result, means that one has to rely on systematic experimental studies for identification of critical geometries and for description of their behavior. Finally, the above examples

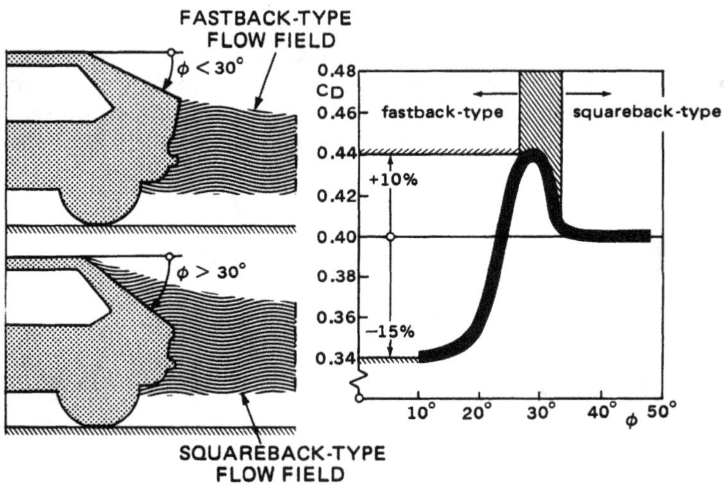

FIGURE 21. Effect of rear-roof slant angle on drag coefficient (Hucho [12]).

demonstrate that critical geometries can generate sizable excess drag, and this should be of concern to designers of road vehicles.

3. *Afterbody Shaping*

Road vehicles are only one category of bluff bodies of practical importance, and there is a large body of literature that deals with bluff bodies in general and also with other applications. The reason for discussing this literature in this section is that much of it relates to afterbody flows and to drag reduction by afterbody shaping. The applicability of the general bluff-body literature to road vehicles was reviewed by Maull[(30)] and Mair.[(31)] They found that the great bulk of published works on general bluff bodies related to simple two-dimensional shapes, such as circular cylinders and airfoils with blunt ends and also to a lesser extent to simple axisymmetric bodies. The topic attracting the greatest attention was base flows, meaning separated flow adjacent to blunt body ends, and the various means for reduction of the base (afterbody) drag. The main emphasis has been on the increase of base pressure, and a number of techniques for doing this have been proposed and tried. Several of these have been shown to be very successful in reducing the base drag of two-dimensional bodies. Examples are splitter plates—placed along the body centerline into the near wake to isolate the two separated shear layers from each other; base bleed—low-speed ejection of air into the separated region at the body end; ventilated cavities—hollowed out base area with perforated walls; trailing edge notches or serrations—whose purpose is to break up vortex shedding; and boat-tailing—contouring the body sidewalls inward in the rear.

Analyzing all of these drag reduction techniques Mair[(31)] concluded that the reason for the effectiveness of most of them came about due to their effect on weakening or elimination of vortex shedding (by interference with its formation), the regular shedding being well recognized to be responsible for a large share of the base drag of two-dimensional bodies. However, since wakes of axisymmetric and three-dimensional bodies exhibit at most a very small degree of periodic shedding in the near wake, most of these methods are not likely to be useful for them. In fact, base pressure of these bodies is already much lower than that of two-dimensional bodies, most of the difference being due to the absence of vortex shedding. Mair documents several examples of how much less (order of magnitude) effective these methods are for axisymmetric bodies. One of the more surprising twists in the application of two-dimensional techniques to axisymmetric geometries arises with ventilated cavities, which are among the most effective ones for two-dimensional shapes. When applied to an axisymmetric cylinder aligned with the stream (Morel[(32)]), they were found to provide a minor drag reduction at small cavity depths, while generating an exceedingly large drag increase for cavity depths larger than half the cylinder diameter (Fig. 22). This surprising large drag increase has not yet been understood.

The only drag-reducing technique used successfully on two-dimensional bodies, which was identified by Mair[31] as having a potential for three-dimensional bodies, is boat-tailing. This technique is also the only one of those enumerated earlier which does not depend on weakening of wake periodicity to produce a drag reduction. Rather, it is a form of streamlining, whereby the body end is contoured inward in order to recover some of the low base pressure and to reduce the base area. Most of the work done on boat-tailing was for two-dimensional airfoils or axisymmetric cylinders. Of the work done on axisymmetric boat-tails, the only ones that appear to be available in the literature are in the already-mentioned paper of Mair[28] and a thesis by Bostock.[33] These results are discussed in depth in Mair.[31] There certainly were many more studies done in connection with ballistics and for airplane fuselages, but information about them does not seem to be easily accessible.

One of the more popular schemes for increasing the base pressure of road vehicles is the base bleed. One of the first studies aimed at application to road vehicles was that of Sykes,[34] who studied the effect of bleed on drag of an axisymmetric cylinder away from and near ground. The base bleed was introduced through a central hole whose diameter was varied between 20 and 78% of the cylinder diameter. A moderate drag coefficient reduction of 0.035 was achieved with the largest bleed hole at dimensionless bleed rates (ratio of bleed flow rate to the product of free-stream velocity and base area) of 0.06 (Fig. 23). As can be seen from the figure, the base bleed is most effective when the bleed area is large and the bleed rate is fairly small. At the optimum point on the curve $d/D = 0.78$, the bleed velocity was about 10% of the free-stream velocity and thus the total thrust generated by the bleed air was negligible. Somewhat larger drag reductions were obtained by placing a porous screen over the bleed hole

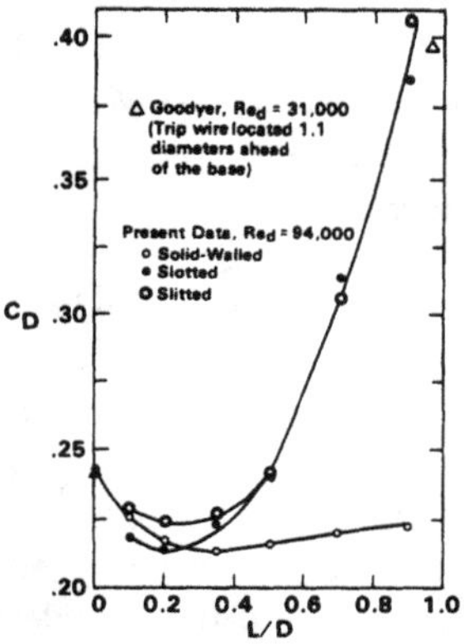

FIGURE 22. Effect of base cavities on the drag of an axisymmetric cylinder for solid-walled and perforated cavities (Morel [32]).

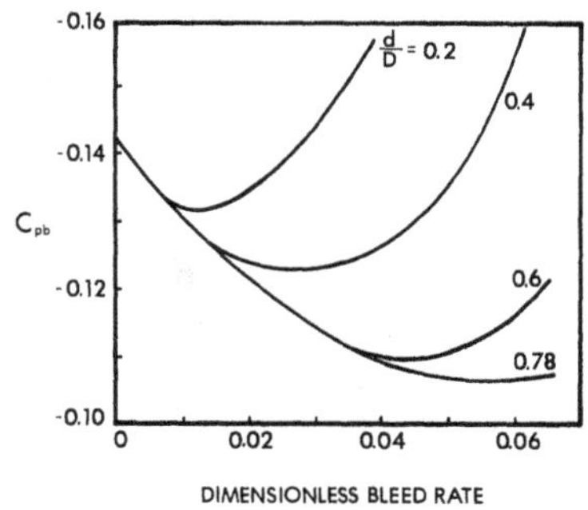

FIGURE 23. Effect of base bleed rate on base pressure coefficients for various base orifice diameter ratios. The larger the orifice diameter ratio, the slower the bleed air velocity for the same bleed rate.

to make the bleed air velocity uniform. The experiment of Sykes was repeated by Przirembel[(35)] who came to essentially the same conclusions. Since the amounts of air needed for base bleed are small, the concept may seem practical. However, examining the problem one finds that flow rates available for base bleed, such as the passenger compartment ventilation air, exhaust gases, and engine cooling air, are insufficient. This means that an auxiliary pumping system would have to be introduced to supply the bleed air. The practicality of such a system was analyzed by Mair[(31)] who concluded that the resulting drag reduction would be more than offset by the pumping and intake momentum losses. As a result, base bleed does not appear to be a promising drag reduction technique.

In addition to the drag reduction methods mentioned so far, many others have been proposed. These include boundary layer control by suction or injection, guide vanes at locations with large curvatures, movable surfaces, and even protuberances[(36)] (Fig. 24).

Most of these proposals turn out on closer examination not to be practical,

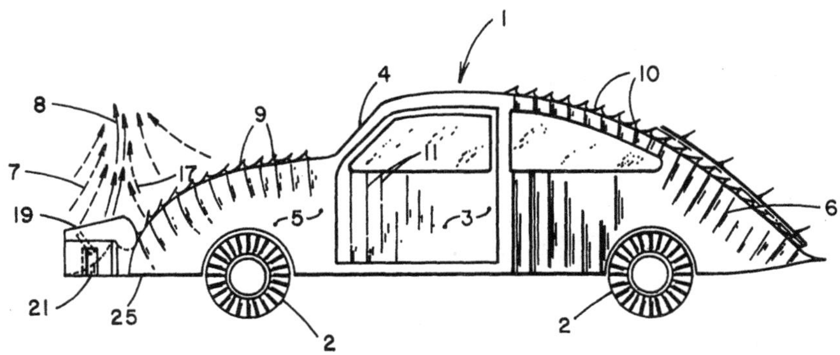

FIGURE 24. Sketch of a vehicle equipped with surface protuberances proposed for drag reduction in a recent patent (Ref. 36).

because they either do not provide any improvement when the system as a whole is considered, or are excessively complex and bulky, or the objective they seek to achieve may more easily be accomplished by better shaping of the vehicle. Apparently the lone exception to the "rule" that no improvement is achieved with mechanical devices when the system as a whole is considered, is the "edge suction effect" as reported by Heskestad.[37]

Road vehicles come in a variety of body styles, ranging from sedans through fastbacks and hatchbacks to station wagons. (The last style also includes vans, trucks, and buses.) Going from one style to another is a form of afterbody shaping, and it turns out that this shaping can have an important effect on drag. If one looks at fastbacks, hatchbacks, and station wagons, one observes that they mainly differ by the angle of the roof slant. As noted in connection with Fig. 21, Janssen and Hucho[29] observed a drag overshoot on hatchback automobiles for slant angles around 30°. This finding stimulated a basic study by Morel[38] of the effect of slant angle on an axisymmetric cylinder, on which the base slant was varied over the range from 90° (vertical base) to 20° (Fig. 25). The results showed that there were clearly two flow regimes. In regime I the drag was slowly increasing with decreasing slant angle up to a critical angle at 43°. At that angle the flow abruptly changed to regime II and the drag coefficient more than doubled. Further study showed that in regime I the flow separation of the base was quasiaxisymmetric (closed). In regime II the separation was open as on delta wings, with streamwise vortices sweeping past the surface, generating suction peaks responsible for the increased drag. A follow-up study on a vehiclelike model showed essentially the same drag behavior near the ground and in the free stream (Fig. 26), and the lift coefficient also displayed evidence of two distinct flow regimes (Fig. 27). The magnitude of the drag overshoot found in these experiments provided the stimulation to probe this critical geometry further. A series of experiments was run on the effects of free-stream turbulence (FST), rounding of the upper (roof) edge, aspect ratio of the slanted surface, and the effectiveness of spoilers. The effect of FST (Fig. 28) with 6% intensity was to shift the critical slant angle to a slightly higher value, but otherwise the shape of the drag curve was unchanged (Morel[39]). Rounding of

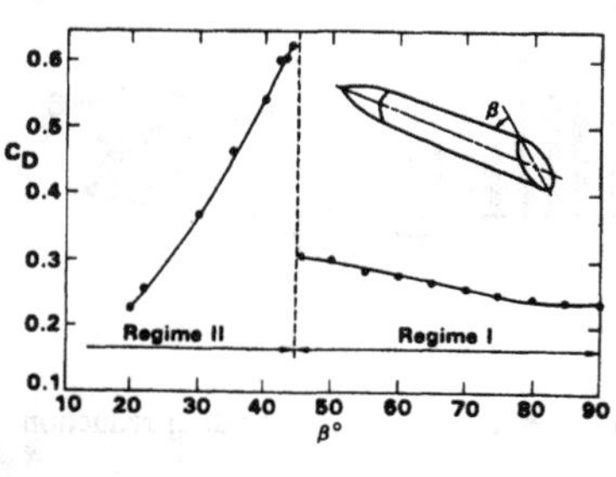

FIGURE 25. Drag coefficient of an axisymmetric cylinder with a slanted base. The discontinuity in the drag curve indicate the existence of two different flow regimes (Morel [40]).

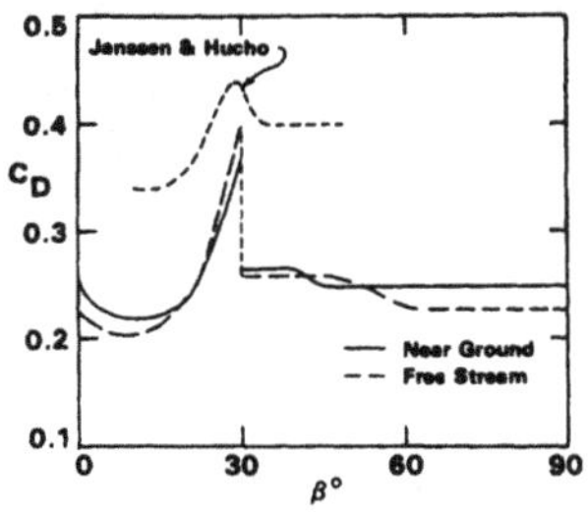

FIGURE 26. Drag coefficient of a vehiclelike body with a slanted base as a function of the slant angle. Figure compares data taken in free stream and near the ground, and it also includes the data from Fig. 21 taken on an automobile model (Morel [40]).

the upper edge, on the other hand, had a more profound effect, shifting the critical angle to a substantially higher value (Fig. 29). The strongest effect of all parameters was attributable to the aspect ratio of the slanted surface (width-to-length of the slanted surface). It was found that by making the aspect ratio close to unity the overshoot could be greatly increased over that shown in Fig. 26, which corresponds to aspect ratio of 1.5. In this case the ratio of the drag maximum at the critical angle to the minimum drag at small slant angles was 3.5 in the free stream and 2.3 near the ground, and the critical angle shifted to larger slant angles.

It is clear from these results that, when designing a hatchback, care should be exercised not to choose an angle close to the critical. It should be stressed that the critical angles quoted here are not to be taken at exact values. The angle depends on the precise body proportions and is certain to be influenced by forebody flows. The important fact to remember is the existence of a critical angle, and that it tends to occur in the range of slant angles favored by hatchback car designers. To avoid it one ought to explore the possibility that a lower drag may be obtained, with a steeper angle for a hatchback or with a shallower angle (less than 20°) for a fastback. Each choice has its advantages and disadvantages: the steep angle leads to a closed separation region which tends to deposit soil

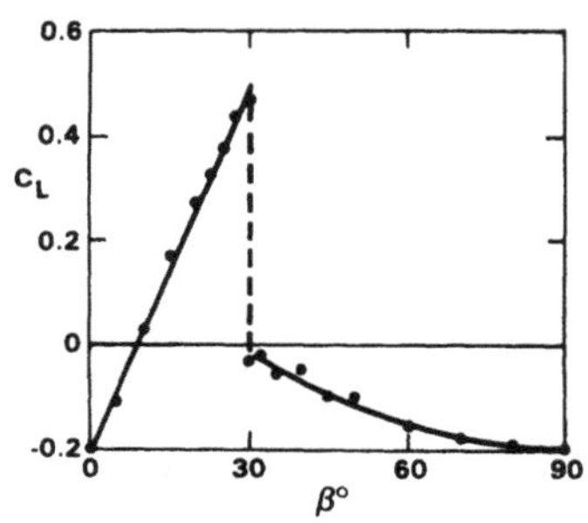

FIGURE 27. Lift coefficient of the vehiclelike body near the ground, $h/D_{eq} = 0.12$, (Morel [40]).

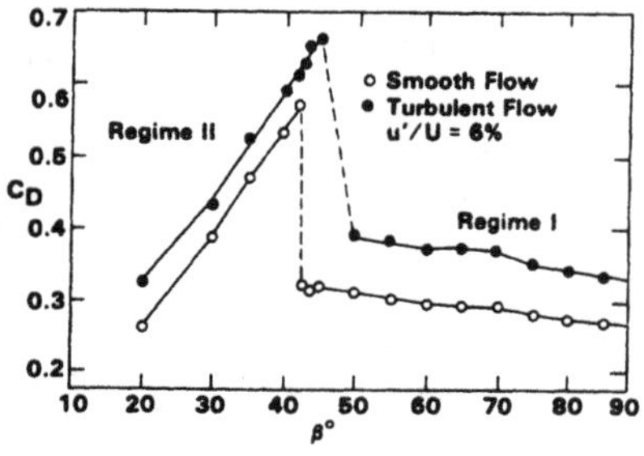

FIGURE 28. Effect of free-stream turbulence on the drag of an axisymmetric cylinder with a slanted base (Morel [40]).

on the rear window, the shallow angle has no problem with cleanliness, but provides poorer visibility. Fortunately, it turns out that proper use of spoilers can help the situation. Depending on the actual slant angle, i.e., on the particular flow regime in the slanted surface flow, it is sometimes advantageous to attach a spoiler at the top of the roof (slant angles in the approximate range of 25°–45°), while for shallower angles (15°–25°) spoilers attached at the bottom of the slanted surfaces are more effective.

The results obtained with the simple models (Fig. 26) confirm what has been known about station wagons—despite their bluff shape, station wagons are known to have a lower drag than sedans. The trends of Fig. 26 indicate that station wagons also have a lower drag than hatchbacks and are bettered only by very shallow fastbacks. In this connection one should mention some interesting observations of Ahmed and Baumert[40] who studied the streamwise vortices in the wakes of fastback and station wagon models. They found that the vortices in the wake of the station wagon are weaker than those behind the fastback—and that they have the opposite sense of rotation! It seems plausible that at some angle intermediate between the fastback (22° slant angle) and the station wagon (0°), there will be no streamwise vortices in the wake. This would roughly correspond to the drag minimum in Fig. 26, indicating that the drag is related to the strength of streamwise vortices shed into the wake. This is an important

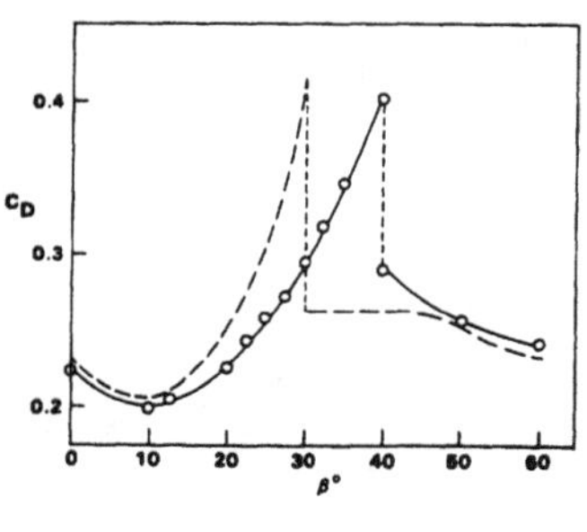

FIGURE 29. Effect of rounding the upper edge of the vehiclelike body (corresponds to the roof-rear window intersection), – – –sharp-edged case (Morel [40]).

implication, which is worth further study. Additional information on the effect of slanted bases may be found in the review of Bearman.[41] An interesting contribution to the subject is the finding of Maull[42] that the critical angle and the amount of drag overshoot depends very strongly on the angle of body incidence; positive incidence greatly enhances the effect, while negative incidence attenuates it. The problem of high drag due to base slant is also encountered on cargo planes with upswept fuselage tails, and a good reference on the subject is Peake and Tobak.[21]

D. Vortex Drag

It is generally accepted that the presence of a lift force contributes to drag. The reason for that is abundant experimental evidence that body shape changes which reduce lift simultaneously tend to reduce drag. That part of the overall drag which is attributable to lift is often quoted to be up to 10% of the total drag, and sometimes even larger than that. Although this drag force is not the dominant part of the total drag, it stands out in one respect: it may be entirely eliminated, while most other drag components cannot. Thus the drag attributable to lift is especially promising from the point of view of drag reduction.

It is a common practice to refer to this drag as "induced drag." However, this terminology has some fairly specific implications not applicable to road vehicles, and so its use is potentially misleading. The concept of induced drag comes from aeronautics. Its origins lie in the work of Lanchester[43] and Prandtl[44] who developed it in connection with airplane wings of finite span (two-dimensional wings have no lift-related drag). Based on an idealized inviscid model of the flow over the wing and in its wake, it was proposed that the wake flow behind the wing has a downward component (downwash) which induces an apparent angle of attack on the airfoil itself. As a result, the lift force on the wing is tilted backward by this angle, giving rise to a projection into the streamwise direction, the "induced drag." For small angles of attack this drag is proportional to the square of the lift, i.e.,

$$C_{Di} \sim C_L^2 \tag{12}$$

and this formula proved to be quite accurate when applied to wings of large and moderate aspect ratio at small angles of attack.

This success led to the tendency to apply the concept of induced drag [and Eq. (12)] to road vehicles, including sometimes even the constant of proportionality. That practice is incorrect, and this may be seen once one observes the differences between road vehicles and airfoils, and once one considers the assumptions used to arrive at Eq. (12). The most important assumption used was that the flow over the airfoil is attached everywhere, as shown in Fig. 17. For airfoils with moderate or large aspect ratio this tends to be a good approximation, not valid only close to the side edges, i.e., on a very small portion of the surface. Road vehicles have an aspect ratio (width to length) of much less than one, and so the assumption of attached flow is violated to a great degree. Consequently, one should not expect the standard induced drag formula to be applicable to road vehicles.

Since road vehicles have small aspect ratios, let us consider experimental and theoretical results generated at NASA for wing planforms with small aspect ratios on the order of unity. An important contribution to the understanding of the lift and drag generation on such planforms was made by Polhamus.[45,46] He studied the generation of lift on delta wing airfoils with sharp leading edges on which flow separates and rolls up to form streamwise vortices in the manner shown in Fig. 18. The work of Polhamus was subsequently extended by Lamar[47] to wings which have flow around the side edges (rectangular airfoils). On these airfoils the flow on the bottom surface is driven by pressure over the side edge and onto the top surface, where it separates and rolls up into streamwise vortices. These vortices pass close to the upper surface and create a region of low pressure under them, which generates a force perpendicular to the surface which has both a lift and a drag component. This lift, termed vortex lift, augments the potential flow lift that would exist if the flow remained attached. The vortex is nonlinear, increasing roughly as the square of the angle of attack. Its magnitude relative to the potential lift grows with decreasing aspect ratio (larger portion of the upper surface is exposed to the streamwise vortices) and with increasing angle of attack. The model indicates that the drag associated with the lift rises approximately linearly with the total lift.

$$(C_D)_L \sim \alpha C_L = \alpha(C_{L\text{pot}} + C_{L\text{vortex}}) \tag{13}$$

where α is the angle of attack. Since the potential lift varies as α and the vortex lift as α^2, the total lift coefficient must vary as a power between 1 and 2. Consequently, $(C_D)_L$ varies as a power between 2 and 3 of α, which expressed in terms of C_L gives

$$(C_D)_L \sim C_L^n \tag{14}$$

where n is between 1.5 and 2.

As for road vehicles, they are thick small-aspect-ratio bodies, and so one cannot apply to them directly either the Lanchester–Prandtl theory, or the Polhamus–Lamar model, but the latter model is conceptually more applicable than the first. Consequently, one may expect that the lift-related drag on road vehicles should vary as C_L raised to a power less than 2. However, before applying such an expression to road vehicles, one should realize that some of the potential lift on a road vehicle is due to the presence of ground (or due to the presence of its "image"), which introduces a further complicating factor. As a result, there is no guarantee that reducing lift to zero, say by changing the body attitude, will produce the minimum drag. That minimum could still occur at a somewhat different attitude at which the lift still has some small nonzero value.

E. Effect of Ground Proximity

One of the distinguishing features of road vehicles is that they travel near the ground. The typical vehicles ground clearance, expressed in terms of an "equivalent diameter"

$$D_{eq} = (4A_f/\pi)^{1/2} \tag{15}$$

is on the order of $0.1D_{eq}$, which is small enough to cause significant changes in the flow field. Perhaps the easiest way to visualize the effect of ground is to picture the effect of a mirror image in a free-flight situation. This is not a perfect analogy, because the axial velocity along the dividing horizontal plane is allowed to "slip," while in the real situation the magnitude of this velocity equals the relative vehicle velocity with respect to the ground. Notwithstanding this, the image concept is quite instructive. As elucidated by Thwaites[(25)] (page 527) for a symmetric finite-thickness nonlifting body, if one considers the flow around the image, one finds that the streamlines are deflected to go around it. As a result, the real body above it experiences an upwash over its front part, and a downwash over its rear part. With respect to the curved streamlines of this modified free stream, and real body appears to have an effective camber bowed downward, producing a force directed towards the image.

This simple picture portrayed by Thwaites describes the case of a finite-thickness nonlifting body placed not too close to the ground. If we extend the discussion to bodies that have lift in a free stream on account of body camber or angle of attack, we find that the presence of the ground can have the following two effects. When the lift is negative (towards the ground), then the streamline pattern produced by the image around the real body will be similar to that produced by thickness alone as already discussed. As a result, the negative lift force will be amplified as the distance to the ground is decreased. However, if the lift is positive, the streamline pattern is just the opposite, and it will produce an effective camber bowed upward. The result in this case will be a continually increasing lift force as the ground plane is approached. In this latter case the two effects of angle of attack and the finite thickness (which is always present) oppose each other, and so the final trend of lift with ground clearance depends on the balance of the two.

A confirmation of these arguments may be found in the work of Saunders,[(48)]

who used a linearized two-dimensional model to study the flow over wings in ground effect. He concluded that as the ground is being approached, lift goes towards negative values due to thickness effects, while lift due to incidence gets amplified. The case of a very strong ground effect was addressed by Widnall and Barrows,[(49)] who studied a wing of finite span in very close proximity to the ground. In this case the thickness and lifting effects do not decouple. The lift coefficient is found to be only a function of the wing lower surface and planform. An extension of this work was done by Tuck[(50)] who considered a two-dimensional body with a vehiclelike cross section very close to the ground (Fig. 30). He was able to obtain a numerical solution, but experienced difficulties in choosing a physically meaningful trailing edge condition. Such condition (essentially a Kutta condition) is needed to fix the rear stagnation point location and thus the circulation around the body. He noted, however, that choosing as his condition that the circulation be zero gave a negative lift force due to the Venturi effect in the gap under the body—a consequence of the finite body thickness. The problem of rear-end stagnation point was overcome by Fackrell[(51)] who considered a two-dimensional body with a blunt base on which flow separation points were fixed (Fig. 31). The base was considered fully separated and the base pressure was assumed to be uniform, with an arbitrarily set magnitude. The presence of the wake was included in the model. The predictions of the model are given in Fig. 31, which shows that as the ground is approached, both the positive and negative lift forces are accentuated.

All of the above discussion and results referred to inviscid flow analyses. In real flow situations the flow develops boundary layers and, if the body is smooth, there is some freedom in the location of the rear separation line, and the effects of these have to be studied experimentally. Especially important is the boundary layer development along the bottom surface, which is subjected to very severe pressure gradients when the ground clearance becomes small. One of the earliest experimental studies was that of Fink and Lastinger,[(52)] on small-aspect-ratio cambered wings with aspect ratio as low as unity. To avoid ground

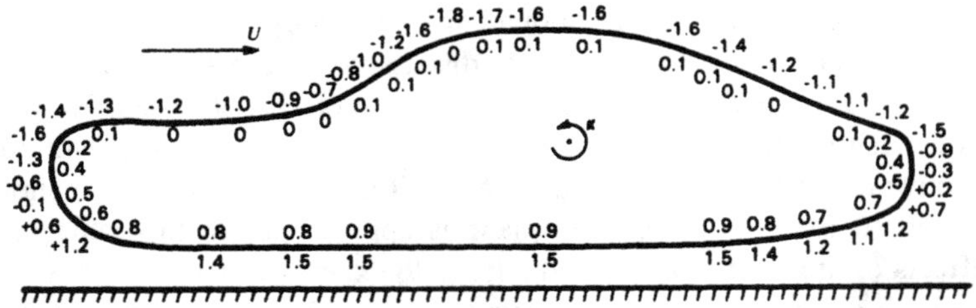

FIGURE 30. Potential flow around a two-dimensional profile with a automobilelike cross section (Tuck [(51)]).

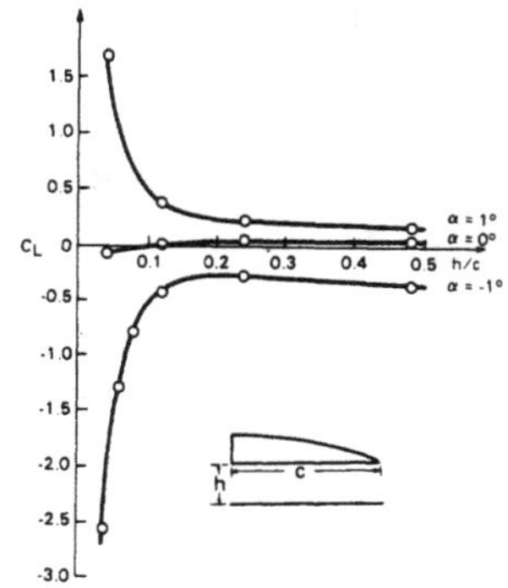

FIGURE 31. Numerical predictions of the variation of lift coefficient with ground clearance. Two-dimensional calculations for a shape with fixed rear separation points (Fackrell [51]).

boundary layer buildup, they used an image wing instead of a ground plane. Their results confirmed the general theoretical conclusions that both positive and negative lift due to angle of attack are amplified as ground is approached. All other experimental investigations reported since then have used a solid ground plane in their studies. Examples are Stollery and Burns[53] who tested a well-rounded teardrop-shaped model, Waters[54] with elliptical and semielliptical wings of very small aspect ratios, and Fackrell[51] with a three-dimensional rounded model whose longitudinal section was the same as that of his two-dimensional theoretical model discussed above. All of these studies reported trends in agreement with the theoretical models. The major exception was in the behavior very close to the ground, where the first two references (but not the last one) reported a trend reversal in cases where the lift was decreasing with decreasing ground clearance. In those cases the lift eventually turned around and started to increase. An example of such behavior may be seen in Fig. 32 showing results obtained on a sports car model without wheels. The initial lift decrease is due to flow acceleration under the body (Venturi effect) and resulting low pressure on the underbody. When the ground clearance is finite but small, viscosity effects effectively impede the flow through the gap and force a part of it to pass over the top, where it produces a low pressure and a positive contribution to the lift. This mechanism is responsible for the eventual lift reversal. By contrast, in an inviscid flow the lift reversal occurs only when the body actually touches the ground.

As we have seen, the presence of ground has a strong effect on the lift force. How does this lift impact drag? Does the lift generated by the presence of the ground have the same relationship to drag as lift produced by body attitude or camber? This question is very important conceptually, as it relates to the key question of overall shaping for minimum drag. This is a subject that has not received too much attention. The only reference to it may be found in the work of Tuck,[55] who suggested that if there is no overall circulation around the

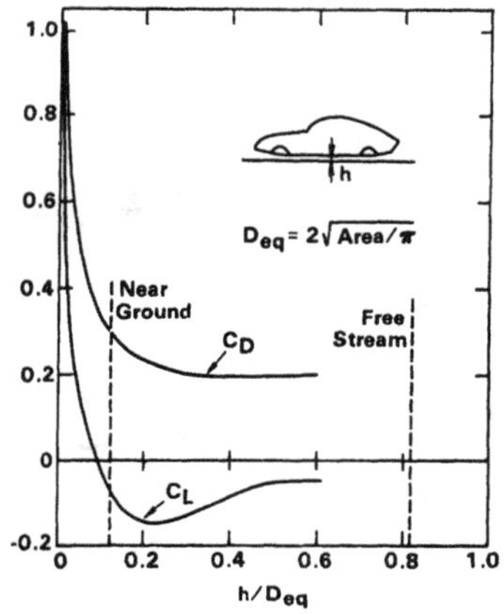

FIGURE 32. Lift and drag of a sporty automobile model without wheels as a function of ground clearance.

longitudinal body section, then even if the body experiences a negative lift (arising presumably from the thickness effect), there will be no lift-related drag. The experimental data of the references cited in this section may be interpreted as being supportive of that view, but there really are no clear-cut experiments that have been designed to test this hypothesis, and the available data are too scattered to give any firm conclusions.

The idea that all lift, regardless of its source, contributes to drag was adopted by Morelli,[56] who undertook the design of a low-drag shape based on this premise. Using a simple two-dimensional theoretical model describing only the camberline of the body (the body thickness effects were not considered), he determined camberlines producing zero lift and zero pitching moment near the ground. The rest of the body shape, local thickness, and cross section were chosen fairly arbitrarily requiring only that their variation in the stream direction be gradual (Fig. 33). Wind tunnel models designed using this theoretical approach were found to yield very low drag coefficients of about 0.05 at ground clearances characteristic of road vehicles. Although many of the assumptions used are not very realistic, e.g., the use of a two-dimensional theory and the arbitrariness in choosing the body thickness and cross section, this attempt is an indication of

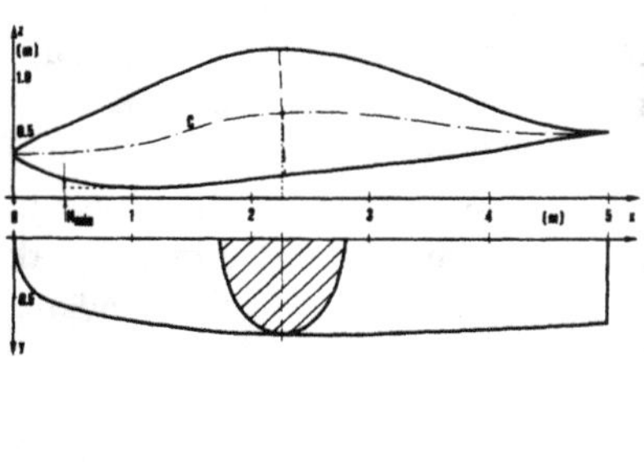

FIGURE 33. Low drag shape developed theoretically, based on the assumption that drag is minimized by lift-free design (Morelli [56]).

approaches that one might want to follow in the future; once more understanding of the lift and drag relationship is generated.

Almost all experimental work related to road vehicle aerodynamics is done using a fixed ground plane, although it is agreed that properly one should be using a moving belt to represent the on-road situation. The main reason for this is that the moving belt introduces a very substantial complication to the test technique. Also, it is the opinion of most practitioners of road vehicle aerodynamics that for the ground clearances typical of road vehicles, there is almost no difference in drag between the moving–nonmoving ground results (see discussion in Ref. 8, pp 120–123). This opinion is supported by the results of Fackrell[(51)] and Bearman,[(41)] and it is in line with the general insensitivity of drag to ground clearance (Fig. 32). One likely exception to this "rule" may be experiments concerning the effect of the increasingly popular front underbody dams, which have small ground clearances. Therefore, caution should be exercised in extrapolating fixed ground plane wind tunnel results related to dams to on-road operation.

With regard to lift force measurements, lift is as sensitive to the moving floor as it is to ground clearance (Fig. 32). As pointed out by Bearman,[(41)] although it is possible to reduce the boundary layer thickness δ^* by suction or other means in an effort to represent more accurately the on-road situation, the boundary layer will always possess nonzero values of $d\delta^*/dx$. The effect of $d\delta^*/dx$ will be to produce a slight positive incidence on the floor near the ground, which may result in a measurable increase in lift with a stationary ground. Consequently, lift coefficients obtained at small ground clearances in conventional wind tunnels with fixed ground planes are subject to a greater uncertainty than drag coefficients.

Before leaving the subject of ground proximity, a remark is due concerning the yaw effect. Typical road vehicles have almost always a positive lift, ranging from C_L of around zero to more than 0.5. When yawed, vehicles (and vehicle-shaped bodies) in ground proximity tend to show a marked increase in lift (Carr[(57)]). That is true even for symmetric shapes that have zero lift away from the ground at *all* yaw angles. The reason for that increase must be that vehicles are typically elongated in the direction of travel, and so in yaw they show a wider projection over which the air has to pass. The drag tends to increase in yaw as well. It is not clear whether the drag increase is mainly due to the lift increase, or due to the flow separation off the lee side of the vehicle. However, judging from the large magnitude of the drag increase (often 50% and more), the second mechanism is likely to be the dominant one.

F. Free-Stream Turbulence

Free-stream turbulence (FST) is the name given to the background level of random, three-dimensional velocity fluctuations present in every fluid stream.

Its characteristics, e.g., its intensity and spectral composition, depend on how the FST was generated and on the history of the flow up to the point of interest. FST is present in the natural wind, and all bodies exposed to the wind experience its effects. Turbulence is also generated in the wakes of obstacles. Road vehicles are subjected to turbulence generated by the natural wind and also by flows over other vehicles. Even research wind tunnels, carefully designed to generate smooth uniform flow, have small residual levels of turbulence.

One consequence of the presence of turbulence in a stream is that it produces unsteady buffeting loads. In addition, there is also a complex interaction between the turbulence and the flow around the body which leads to changes in the mean flow field. It is this interaction and its effect on the time-mean aerodynamic forces which are important in the present context, and these have been recently reviewed by Bearman and Morel.[58] The findings most pertinent to road vehicle aerodynamics are summarized below.

There are three basic mechanisms by which FST interacts with the overall mean flow: accelerated transition to turbulence in shear layers, enhanced mixing and entrainment into these shear layers, and distortion of FST itself by the mean flow. The overall effect of FST is often the result of more than one of these basic mechanisms.

Turbulence is usually described in terms of its intensity and scale. The intensity is most often defined as the ratio of the root mean square velocity fluctuation along the air stream direction u' to the mean flow velocity U. The turbulence originates in shear layers, which are generally present somewhere upstream. It is customary to describe the magnitude of the intensity in relative terms such as "high," meaning more than 10% or so, found, for example, in the natural wind; "low," meaning less than about 0.5%; or "very low," meaning less than 0.1%, which corresponds to levels found in low-turbulence wind tunnels. FST whose intensity is on the order of 1 or 2%, usually gets no label at all. The turbulence intensity encountered on the road varies widely depending on wind and vehicle speeds and on the presence of other vehicles. Nevertheless, it may be argued that an intensity of about 5% may be taken as typical (Bearman[41]). By contrast, wind tunnels used for road vehicle aerodynamics have FST levels around one-half of one percent, or 10 times less.

The length scale of turbulence, L, is a quantity usually used to describe various portions, or the whole, of the power spectrum of turbulent fluctuations. The most commonly used length scale is the longitudinal integral length scale, L_x, which is the most easy to obtain experimentally (usually as the area under the curve of autocorrelation of u'). More important than the absolute scale itself is its ratio to some characteristic dimension of the flow feature under study. For an isolated shear layer the most appropriate dimension is its thickness, while for the flow field around a bluff body as a whole a more appropriate choice would be some characteristic dimension, D, of the entire body. When the length scale is very large with respect to the body dimension, $L_x \gg D$, FST will appear to the local flow as a correlated unsteady mean flow of varying magnitude and direction. The two turbulent fields (free stream and local) will not interact and the gross effect of FST can be estimated as a succession of quasisteady flow conditions with different mean velocities and directions. This will lead to an increase in drag and other force coefficients due to their dependence on the square of the air velocity. Because of this decoupling of free stream and local fields, very large scale FST is sometimes referred to as "passive" turbulence. If the length scale of turbulence is very small, $L_x \ll D$, the turbulence will tend to decay rapidly in intensity and be less effective, except in small localized regions where small-scale turbulence is being distorted by the mean flow and amplified. The turbulence whose scale is of the same order of magnitude as the body size is the most interesting, because its field can interact with the local flow features to actually produce changes in the mean flow patterns. The length scales of turbulence encountered on the road, be it those produced by wind (wind length scale is proportional to the height above the ground) or by other vehicles, fall into this last scale range.

Both intensity and length scale must be considered when the effects of FST are studied. It is therefore natural that many investigations have attempted to correlate experimental results in terms of parameters involving both quantities usually in the form

$$u'/U \, (L/D)^n \tag{16}$$

This parameter reflects the fact that FST effects always increase with increasing value of its intensity. However, the exponent of the length scale is less clear-cut. Depending on the mechanism involved, the exponent may be negative (typically around -0.2) for FST effects to which small scales are important, such as transition of laminar shear layers to turbulence. In other cases it is found that positive exponents on the order of unity correlate the best experimental data, presumably because it is the large-scale turbulence which is essential. And in a number of other flow situations it is found that the length scale is not important at all ($n = 0$). In all of these cases, however, it should be stressed that the use of the parameters defined by Eq. (16) should be restricted to FST whose scales are comparable to the characteristic flow dimension.

The magnitude of changes in drag which result from the effects of FST on the nature of the mean flow can be substantial. One example, comparing the effect of base slant angle on drag for smooth and turbulent streams, was already discussed earlier (Fig. 28). Another one concerns the transition to turbulence on a two-dimensional cylinder and its effect on flow separation and drag in the critical Reynolds number range (Fig. 34). Increasing the turbulence intensity leads to earlier and earlier transition and, as the data show, to a very significant

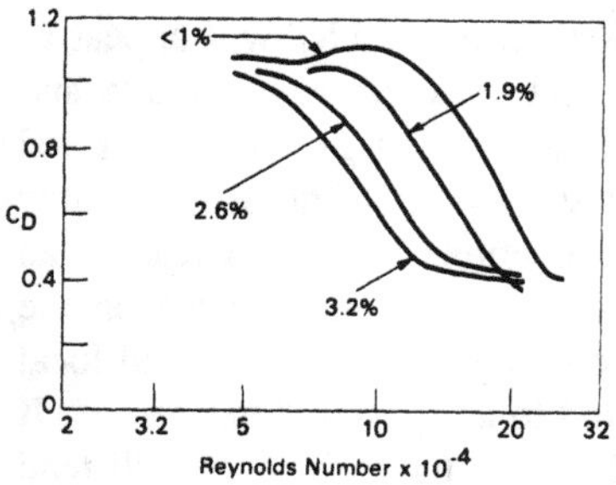

FIGURE 34. Effect of free-stream turbulence on drag of a two-dimensional circular cylinder (Fage and Warsap [59]).

(60%) reduction in drag. Even bodies with fixed location of flow separation may be affected by FST. This may be seen in Fig. 35 showing data for the base pressure coefficient on square and circular plates, which is seen to decrease (which means the drag increases) as a function of (u'/u) (L/D). The opposite trend is found with a cube exposed to a turbulent flow (Fig. 36), where base pressure is found to increase substantially with increasing turbulence intensity (Martin[63]). In this particular study no systematic dependence of the data on turbulence length scale was found over a range of $L_x/D = 0.1$ to 1.2.

The major conclusion that one can draw from this evidence is that the effect of FST is not simply to raise the "effective" Reynolds number of the flow as is often supposed. In reality, the effect is much more complex and difficult to predict. The resulting changes in the drag coefficient can be quite large and they can go in either direction: drag increase and decrease. Because of this, it is advisable to test road vehicles in streams that simulate the intensity and length scale of FST anticipated in actual on-road situations.

The effect of FST on drag has been addressed in the context of road vehicle aerodynamics by Buckley *et al.*[64] and Cooper and Campbell,[65] who concluded

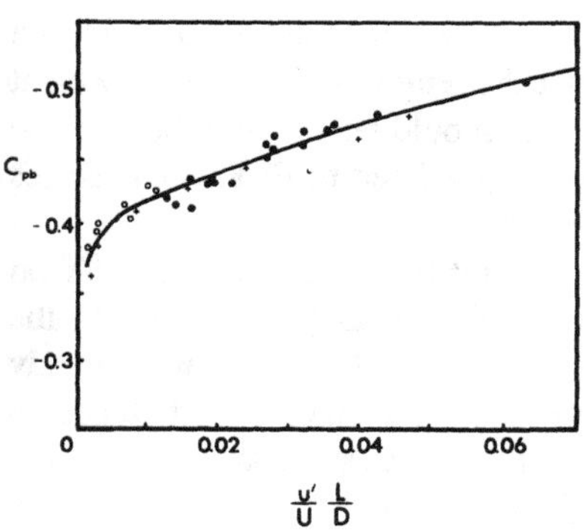

FIGURE 35. Base pressure coefficient of square plates and circular disks in turbulent flow; ○ Schubauer and Dryden [60], ● Bearman [61], × Humphries and Vincent [62]; + "smooth" flow value. (Bearman and Morel [58]).

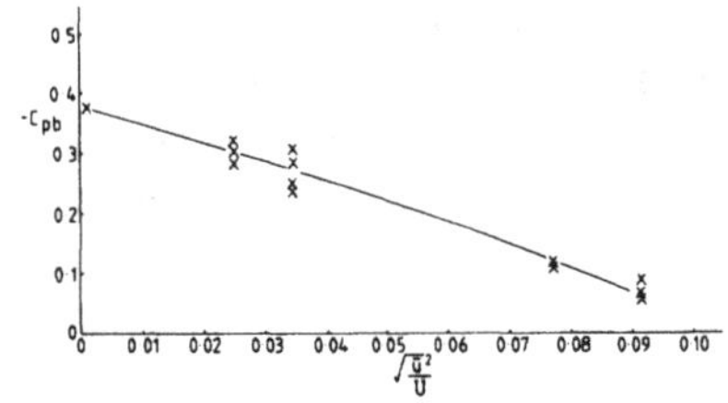

FIGURE 36. Base pressure coefficient on a cube in turbulent flow (Bearman [41]).

that the presence of FST in the actual road environment cannot be neglected. Cooper and Campbell attempted to develop a theory that would permit the extrapolation (correction) of data taken in smooth wind tunnel streams to the road environment. Their theory is based on the quasisteady approximation to the effects of turbulence, which neglects any interaction between FST and the mean flow around the vehicle. As already discussed, this approximation is valid only for $L_x/D \gg 1$, and this applies only to low-frequency wind gusts. The true turbulence has scales comparable to the dimensions of road vehicles and complex interactions affecting the mean flow patterns may be expected to take place. These are not accounted for by quasisteady analyses such as that of Cooper and Campbell.

G. Smaller Drag Components

1. *Skin Friction Drag*

In addition to pressure forces air flow generates tangential frictional stresses along the body surfaces. For a bluff body, skin friction drag is much less than the pressure drag. Its magnitude depends on the boundary layer thickness and, through it, on the Reynolds number: it decreases with increasing Reynolds number. By contrast, the pressure drag is practically Reynolds number independent, and so the total body drag has only a weak Reynolds number dependence. The magnitude of the friction drag on a smooth body may be estimated in the following way. Consider a typical vehicle which is box shaped with overall dimensions of length to width to height of 3 : 1.33 : 1. For such a shape the ratio of the total surface area parallel to the direction of travel to the frontal area is about 10. Taking a typical value of the skin friction coefficient to be 0.002 in the Reynolds number range of interest, we find that the friction drag contributes about 0.02 to the overall drag coefficient. This value is about 5% of the drag coefficient of a typical road vehicle. It may be noted that the drag coefficient of an optimized teardrop shape away from the ground, which is almost entirely due to the friction drag, is 0.04 (Scibor-Rylski[66]), or twice as high than the above estimate. The difference is due to the presence of some pressure drag on the teardrop and also due to its larger ratio of skin friction area to the frontal area.

A typical road vehicle is not smooth but has many small surface imperfections such as window recesses, door crevices, etc. These imperfections may be considered as surface roughness and their drag estimated by taking a larger value of the skin friction coefficient. If this definition of skin friction drag were adopted, the friction drag contribution to the total drag would be much higher, probably somewhere between 10 and 15% for current designs.

2. *Internal Drag*

Part of the airflow passes through the vehicle interior for purposes of cooling various components and for ventilation. Of these, the amount of air required for ventilation is quite small and its effect is negligible. The large flow needed for cooling the engine is much more significant. In order to move air through the vehicle interior, it has to be either driven by the pressure difference between the inlet and outlet, or it is driven by a fan. In its passage through the vehicle interior the air loses its momentum due to resistance in the heat exchanger and in the internal ducting. The first of these losses is unavoidable, because some pressure loss in a heat exchanger is necessary for efficient heat transfer. The second loss is strictly parasitic and it is the consequence of wall friction and pressure drop in bends and sudden expansions.

Wind tunnel testing of actual production automobiles has shown that there is a difference in drag coefficient between the case when the radiator is masked and when it is open to permit air throughflow (with the fan not running). With cooling flow permitted, the drag coefficient is found to increase by 0.01 to 0.06 (Hucho[(12)]). The wide range of values is likely to be due to differences in engine size (larger engines need more cooling air), radiator, grill, and passage design with different pressure drop characteristics and also due to differences in the local pressure on the outlet side. The majority of vehicles tested was found to have drag coefficient increases around 0.03 due to radiator flow, and Hucho[(12)] expressed the opinion that values as low as 0.01 could be standard in properly designed systems.

Well-designed systems will require that attention be given to three main elements. One is the optimum design of high-efficiency–low-pressure-drop heat exchangers. Another is the proper channeling of air through the interior, avoiding constrictions that increase drag. And the final consideration is the proper choice of air inlet and outlet, perhaps in combination with underbody dams that create stagnation pressure ahead of them and a low pressure behind. On the other hand it may be more advantageous in some cases to design air intake and grill so that the underbody dam is not needed, reducing the drag of the system as a whole.

3. Protruding Elements

On every road vehicle one finds a number of small protruding elements such as external mirrors, lights, radio antennas, hood ornaments, door handles, wipers, luggage racks, bumpers, window moldings, drip rails, and a number of different underbody details. Each of these elements has a localized flow field associated with it which produces drag additional to that of the basic body. This drag often referred to as parasitic or interference drag. The latter name arises from the fact that the protruding elements are often located in areas where the local stream velocity is substantially larger than in the free stream, being accelerated by the presence or "interference" of the basic body. The drag produced by the protruding elements can add up to a drag coefficient increment on the order of 0.05 (assuming their frontal area being equal to 5% of vehicle frontal area and their drag coefficient, including the effect of velocity augmentation, being on the order of unity). This represents about 10% of the drag of today's cars and a more significant portion of the drag of future low-drag cars unless a conscious effort is made to reduce it.

4. Rotating Wheels

Because of practical considerations, vehicle wheels have to be partially exposed. Their bluff shape contributes a moderate amount to the total drag, and their relative contribution is likely to grow as improvements are made on the basic vehicle shapes. Almost all wind tunnel testing is conducted with stationary wheels. In a real situation the wheels spin in the wheel wells against the direction of the air flow, and this produces an additional source of drag that is not included in the wind tunnel tests. According to data collected in the Pininfarina wind tunnel in Italy with rotating and nonrotating wheels, the effect of wheel rotation for a passenger car is to add a small drag coefficient increment of 0.005. The flow around a rotating wheel is very complex. Description of the flow pattern associated with it may be found in the work of Fackrell and Harvey[(67)] for an isolated wheel case, and for a discussion of a rotating wheel inside a wheel well, see Scibor-Rylski.[(66)]

5. Cavity Flows

All of the discussion so far considered the basic body shape to be closed. When the side windows are opened, the drag is usually observed to increase. There seem to be some exceptions, though, as shown by data of Kurtz[(68)] who measured the drag of 48 vehicles with their front windows closed and open. Although for most of the cars the drag coefficient at zero yaw increased due to opening of the windows, for four of them it decreased (on one of them by 4%). At nonzero yaw the opening of the windows was found to be more detrimental than at zero yaw. Similarly, opening the roof or removing the top on a convertible leads to a significant drag increase, and removing the whole top of a convertible leads to a very large increase (Janssen and Hucho[(69)]). A related topic is the aerodynamics of pick-up trucks with open loading area. This subject was studied experimentally by Goetz[(70)] who showed that opening of the loading area leads to a large drag increase. Goetz investigated the use of a system of vertical screens to break up the flow pattern on the loading area, but although that does reduce the drag, that arrangement is not too practical. It is sometimes mentioned that opening the rear gate on an empty pick-up truck would provide a drag reduction, because the vertical gate interferes with the flow which is going over the cab and reattaches onto the loading area. However, no data supporting this plausible claim were found in the open literature.

V. STATE OF THE ART IN LOW-DRAG DESIGN

A. Aerodynamic "Tuning" of Vehicle Shapes

The process of vehicle shape design usually starts in the styling studios which are given the task to produce a design consistent with the vehicle "mission." The prescribed mission involves given parameters such as the interior space (number of passengers), weight, category, and market (family, sports car, specialty, etc.). The styling studio produces a first design which complies with the general constraints described in Sec. II and which has a certain styling theme, whose purpose is to give the vehicle its identity.

This first design is then tested in the wind tunnel where aerodynamicists look for ways to reduce its drag by making modifications to a clay mock up of the vehicle. In their effort the aerodynamicists rely on their experience with previous vehicles and they concentrate on making small changes in individual model details. Quite small changes are often found to have a disproportionate effect on drag, and by systematic changes being made to many details, guided by force measurements, smoke, tufts, and surface oil and ink techniques, fairly significant overall drag reductions may be achieved. This type of aerodynamic input to vehicle design is sometimes referred to as aerodynamic tuning or optimization and it is the most common approach to aerodynamics today. Its basic premise is that the original styling theme should be preserved as much as possible, and only small body-shape changes not altering vehicle appearance can be allowed in the effort to reduce the drag.

Typical areas where aerodynamicists look for possible improvements in drag are shown in Fig. 37. They concentrate on rounding corners to minimize

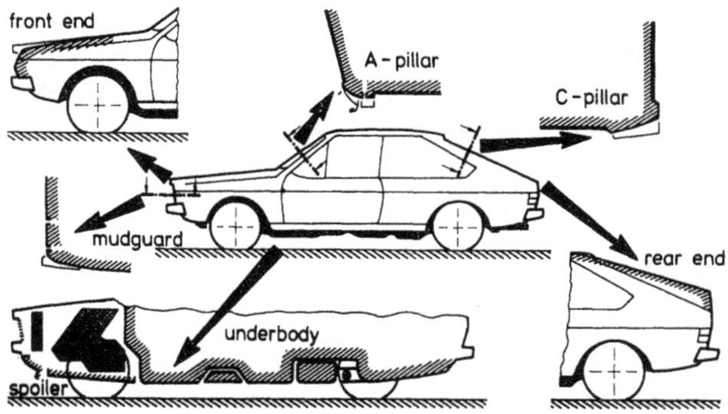

FIGURE 37. Most common areas where aerodynamicists seek drag reduction by aerodynamic tuning (Buchheim *et al.* [71]).

flow separation on the front portion of the vehicle, reduction of drag along the underbody, and on the formation of the separated flow region in the back. One typical example taken from Hucho *et al.*[13] concerns the optimization of the headlight area (Fig. 38). It shows that a drag coefficient reduction of 0.05 can be achieved by encapsulating the headlight details by a rounded fairing. The most interesting aspect of Fig. 38 is that an almost identical drag reduction was achieved by fairly moderate rounding of the edges. This finding is typical of efforts to eliminate flow separation by rounding, which is known to occur at

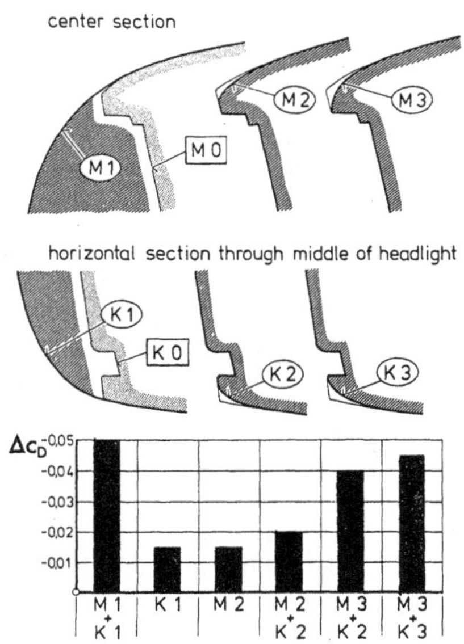

FIGURE 38. Optimization of the details around a headlight (Hucho *et al.* [13]).

some value of the rounding radius beyond which no additional benefits can be accrued, because the flow is already attached. The radius at which this happens decreases with increasing Reynolds number, and so radii should be optimized at proper Reynolds numbers representative of the on-road conditions. Another example of optimization, which was already discussed earlier, concerns the slant angle of the rear window on hatchbacks and fastbacks (Figs. 21 and 25–28). There the objective was to choose the proper slant angle so that a favorable flow pattern would form over the slanted surface. More examples of the optimization of several automobiles may be found in Hucho *et al.*[(13)]

Since aerodynamic tuning has as one of its objectives the preservation of vehicle appearance, it is clear that it cannot be used to arrive at really low-drag vehicles, and in that sense it is limited. That is its weakness, but at the same time it is also its strength, because it permits a useful drag reduction even in situations where the vehicle appearance is given the main priority.

B. Empirically Determined Principles of Low-Drag Design

Aerodynamic tuning (optimization) is a technique inherently tied to particular vehicles, and the precise forms of body details, values of optimum radii, and angles are not directly transferable to other vehicles. Nevertheless, in the process of designing a number of vehicles, aerodynamicists tend to develop their own set of design rules by observing the differences between low-drag cars and the high-drag ones. Surveying the literature, one finds mentions of guidelines for low-drag vehicle design based on this type of experience. The consensus of these empirically established guidelines on what the main elements of low-drag design are appears to be the following list of items:

Vehicle Front

1. All forward-facing corners should be rounded and this includes the leading edge of the hood, fender edges, fairing of headlights, well-rounded side edges of windshield, and a flush A-pillar with no flow separation, and also a rounded top of the windshield.
2. In the plan view, the vehicle front should have some taper towards the front to divert the oncoming stream partially sideways; this reduces the drag and also the lift, since there is less flow acceleration over the top.
3. From the side view, the hood should be sloping, and the grill should be inclined backwards some 30° from the vertical (Carr[73]).

Whole Vehicle

1. The whole vehicle should have a slight (1° or 2°) negative attitude (nose down).
2. Lift should be minimized to reduce the drag related to lift.

Vehicle Rear

1. Sides should be boat tailed.
2. Roof should be curved down at the back end.
3. Hatchback end should be of the fastback type with a shallow 15°–20° angle and equipped with an optimized rear spoiler at the lower edge of the slanted surface.
4. Notchback should have a rear deck high enough and long enough to permit the roof flow to reattach; the rear edge of the deck should be sharp.

Underbody

1. Underbody should be shielded by an underbody pan, or at least all elements should be hidden in underbody surface depressions.

In addition to these main elements there is a number of smaller items which are also known to be beneficial:

1. optimized engine cooling system with an efficient heat exchanger, well-designed internal passages and, most of all, well-chosen intake and outlet openings;
2. shallow windshield angle;
3. underbody dams, which deflect the air to reduce underbody flow (this may prove not to be beneficial when underbody pans are used);
4. well-rounded top edge of front fenders,
5. minimum of surface protrusions and imperfections such as mirrors, roof drip rails, window moldings, door handles, etc.; also minimum gaps around openings;
6. flush wheel covers and fender skirts for rear wheels; and
7. the underbody surface behind the rear wheels angled upwards.

The above list of low-drag guidelines is the result of a great many investigations and the conclusions that could be drawn from them. It represents a significant part of what can be called the state of the art, and it is already there waiting to be incorporated into vehicles designed with aerodynamics in mind right from the start.

VI. FUTURE OUTLOOK FOR LOW-DRAG DESIGN

A. Lower Limits of Aerodynamic Drag

The present interest in vehicle aerodynamics is not just a temporary, passing phenomenon. It is to be expected that aerodynamics will become a permanent discipline integral to the vehicle design process, part of the effort to increase energy efficiency of the total vehicle system. And in the future, when alternate powerplants such as battery-operated electric motors will become common, aerodynamics will have to play its part to extend the range and permit higher speed operation of these typically underpowered (small power-to-weight ratio) vehicles.

Once we accept the permanence of the aerodynamics discipline, it is natural to ask what is its long-range outlook? In particular, what is the lower limit of the drag coefficient for practical vehicles? The average current value of C_D is often quoted to be around 0.45 (see, e.g., Janssen and Emmelmann,[9] Carr, [72] and Buchheim *et al.*[71] The question of the future lower limit of C_D has been addressed in all of the above three papers, as well as in some earlier ones (Hucho [12]). The consensus of these papers is that a value of $C_D = 0.15$ is achievable with scale models which are completely smooth, but have wheels. Furthermore, the opinion is that a variety of shapes is capable of achieving these low values. A Volkswagen-developed racing car (ARVW) is reported to have that same low $C_D = 0.15$ on a drivable vehicle, with cooling flow and other details included (Buchheim *et al.*[71] However, the ARVW is not exactly a practical design, presenting a relatively large frontal and platform area for a small useful interior space. For more conventional designs with larger ratios of interior to exterior space (and thus lighter weight) the outlook is less optimistic. The opinion of many is that the lower limit of C_D for practical cars is somewhere just below 0.30, and this should hopefully include good yaw characteristics with at most a small drag increase at yaw. At this level the vehicles are expected to be practical while maintaining a good degree of identifiability. Since targets in technology tend to move, as do ideas about what is practical and identifiable, one can expect that the ultimate minimum may perhaps be closer to $C_D = 0.25$.

B. Use of Computers for Aerodynamic Design

So far, all or almost all of the applied and basic work done in vehicle aerodynamics has been experimental. The reason for the predominance of experiments has been that the flow fields involved are very complex. They are three-dimensional, with many small but significant details, separated and turbulent, all of which combine to make analytical or numerical work very difficult. Of course, the same ingredients also create, difficulties for experimental work. The main problem with experiments does not lie in making the actual measurements; that part is fairly straightforward. The real problem arises from analysis of the results, where it is exceedingly difficult to separate effects arising on one part of the body from those on another part. As a result, it is often quite difficult to draw conclusions from an experiment done on one body, which would be transferrable to other bodies.

Really important progress could be made if numerical simulations could be used to give support to and to guide the experiments. The attributes of numerical solutions are known: they permit parametric studies to be executed rapidly, and the detailed information they provide allows investigations of cause-and-effect relationships. In view of these advantages, what are the prospects for numerical simulations today and in the future? A good indication of the capabilities and difficulties of numerical simulations may be gleaned from the work of Hirt and Ramshaw.[(73)] Their paper is a good review of the numerical issues encountered when attempt is made to calculate three-dimensional flow fields using Navier–Stokes equations, including boundary conditions, description of geometry, accuracy, stability, spatial resolution, and computational time and cost. As a demonstration of capabilities, they presented a solution for a three-dimensional flow over a simplified truck–trailer combination. The flow was assumed to be laminar with Reynolds number based on trailer width of 140. The results are shown in Fig. 39 in terms of velocity vectors. What is impressive in this demonstration is that one can see the flow details not only on the surface of the vehicle, but also throughout the flow. Similar information (albeit only qualitative) can be obtained using smoke visualization in a wind tunnel, but in a numerical solution, one can take cuts through arbitrary planes very quickly and generate a much more comprehensive picture of the flow. In addition to the streamline

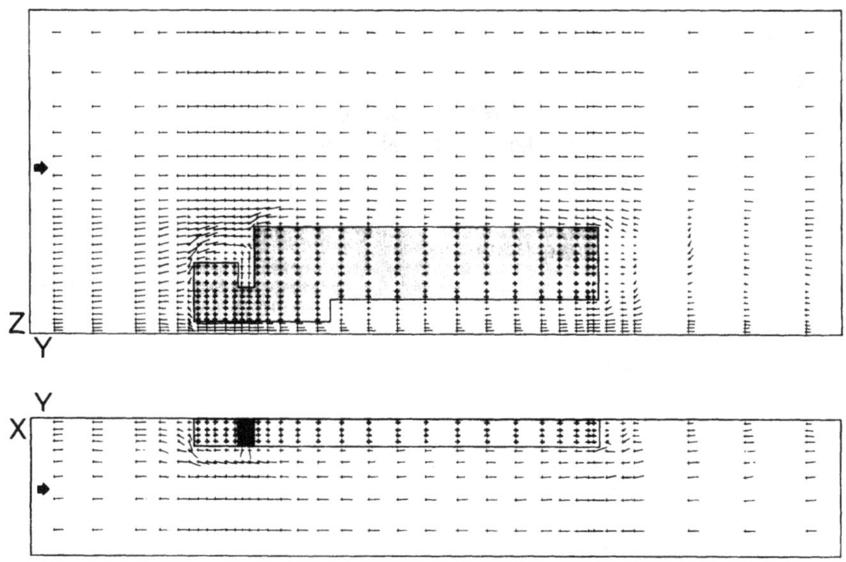

FIGURE 39. Computed velocity vectors around a crude tractor–trailer model, Navier–Stokes solution for Re = 140. The side view shows vectors in the symmetry plane of the model, the plan view shows velocity vectors in a horizontal plane intersecting the tractor cab and trailer at midcab height; top edge of this view is the plane of symmetry of the flow (Hirt and Ramshaw [(74)]).

pattern, one also has the information about velocities and pressure throughout the flow.

These are, of course, the types of results that make numerical simulations attractive. However, let us look at the difficulties with the numerical approach. The greatest problem of the simulation lies in the grid resolution. The just-mentioned tractor–trailer flow was described in terms of its values at discrete nodes—38 in the flow direction, 30 in the vertical, and 11 in the transverse directions. This amounts to over 12,000 grid points, a number large enough to give problems to large computers like CDC 7600, and the resulting computational time on the CDC 7600 was 90 min. However, even with this number of grid points, the flow was insufficiently resolved. The extent of the external flow was effectively restrained within a "duct" only three times the trailer height and five times the trailer width. At the same time the details close to the body surface were only crudely described, the smallest grid spacing corresponding to about 0.3 m in full size, i.e., much larger than the boundary layer thickness and many smaller body details. With such a coarse resolution one cannot even try to use turbulence modeling for the turbulent flow existing in close proximity of the body surfaces. Away from the body surface, the flow can be handled as laminar or even inviscid, since the Reynolds number based on trailer width is on the order of 5×10^6. However, due to numerical problems the presented calculations could not do better than Re = 1400, (30,000 times less). Consequently, the predicted flow field is considerably modified from the real one by the effects of high viscosity. This may be seen in Fig. 40 which shows a solution for Re = 1400, and where large differences are seen in details above the tractor, below the trailer, and in the wake. Additional problems would arise when solving for flow in yaw because one could not use symmetry, and the required number of grid points would double. Also, the "duct" approximation would not be appropriate resulting in an increase in grid points and problems with boundary

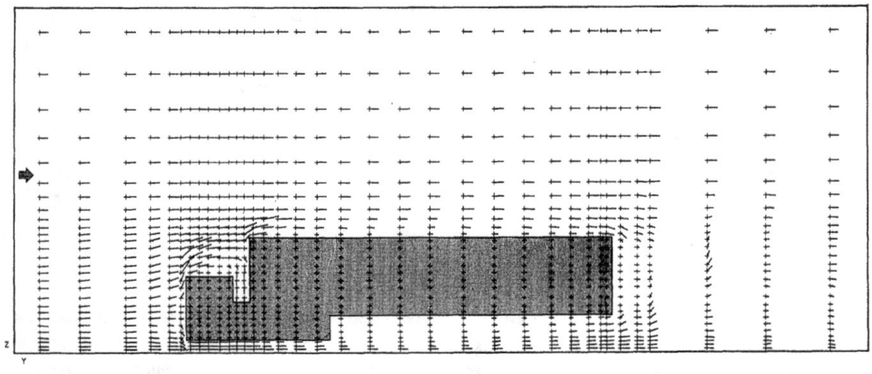

FIGURE 40. Computed velocity vectors in the symmetry plane of the flow around a tractor–trailer, Re = 1400 (Hirt and Ramshaw [74]).

conditions. Finally, any attempt to simultaneously resolve the external flow and the details of the flow near surfaces (described by widely disparate length scales) would require a number of grid points at least two orders of magnitude larger than that used by Hirt and Ramshaw. Altogether, some one million grid points appear to be required for adequate description of real-vehicle flows. This puts a tremendous requirement on computers, not only in terms of memory, but also in terms of their speed, since computational time rises faster than linearly with the number of grid points. Extrapolating the trend in the development of large-scale computers into the future (Fig. 41) it may be seen that computers needed to do this job, which would have to be some three orders of magnitude more powerful than today's machines, cannot be expected to arrive for many years to come. In summary, it is fairly clear that prospects for simulation of complete road vehicles by the straightforward solution of Navier–Stokes equations are not very good.

What is more likely to happen is that calculations will be made of simplified flow fields, using inviscid flow solutions based on body-surface distribution of singularities, so-called panel methods, as is being done in the aircraft industry. This approach to the problem was reviewed by Landahl,[(26)] who discusses some of the work done by Palko[(74)] and Kandil *et al.*[(75)] An example of the latter work is shown in Fig. 42 which displays the separated wake flow behind a delta wing. Flow structures of this type may be formed, for example, behind fastback automobiles as discussed in Sec. IV. Landahl also reviewed other promising types of models, based on discrete vortex elements. These models are used for separated flows, whereby discrete vortices representing vorticity present in a boundary layer are released into the fluid at the point of flow separation. This method is inherently time dependent and it can capture quite well the unsteady nature of separated flows. Originally developed for two-dimensional flows, the method has been extended to axisymmetric and three-dimensional flows (Leonard[(76)]). One weakness of the method is that is requires as an input prescribed locations of flow separation.

Both types of methods are very attractive and they conceivably may, when

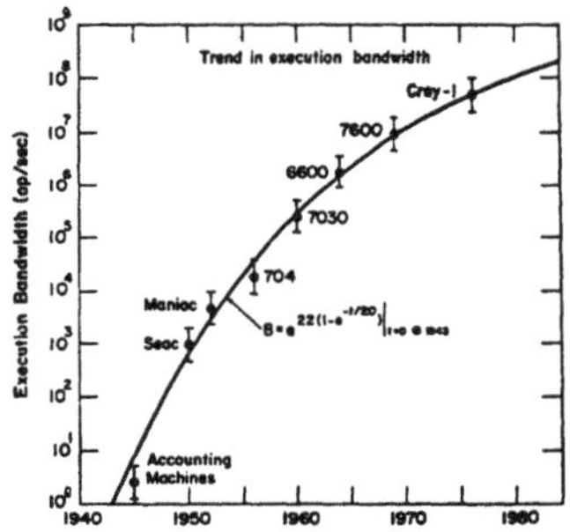

FIGURE 41. Projected rate of growth of capabilities of the largest computers (Hirt and Ramshaw [(74)]).

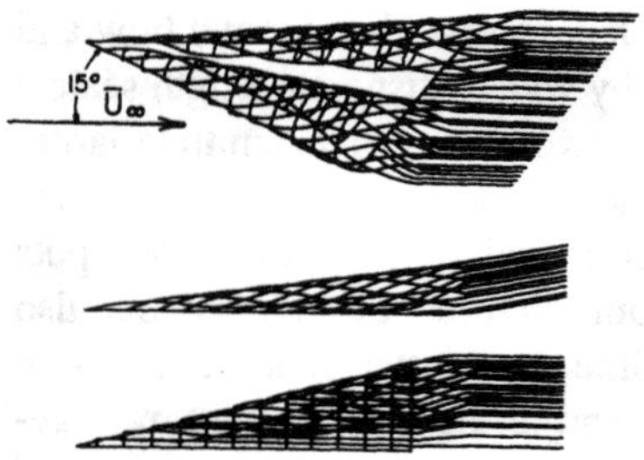

FIGURE 42. A typical solution of the wake development for a delta wing calculated with a 12 × 12 lattice (Kandil *et al.* [76]).

used together, form a powerful method for calculation of external flows. To link them one needs to solve the boundary layer equations together with the pressure distribution from the panel methods, providing the location of flow separation needed for the vortex methods. Since the flow is being solved only in the vortical part of the flow field, there is no need to be concerned with the far field of the flow; the implicit assumption is being made that it is infinite in extent. Similarly, there is no essential problem with calculation of nonzero-yaw cases. One drawback of such a combined technique is that it would require a very large computer to describe a typical road vehicle. The number of individual panels and vortices needed would be very large. On the other hand the technique could provide meaningful answers even with the present computers when applied to simple shapes. However, it should be stressed that this is only a conjecture, because the feasibility of such panel-boundary layer-discrete vortex combination has yet to be demonstrated.

C. Strategy for Achieving the Lower Limits of Drag

In order to achieve the projected lower limit of drag with $C_D = 0.25$, it will be necessary to step beyond the method of aerodynamic tuning. An obvious place to start is to integrate the work of aerodynamicists and of stylists into a single team effort, in order to bring aerodynamics to bear on vehicle design from inception.

The next step is to provide sufficient understanding of the main issues of vehicle aerodynamics. It is likely that the profitable approach is to follow three parallel avenues:

1. Utilize accumulated empirical knowledge about the principles of low-drag design and of the effects of changes of small details.
2. Complement these by making use of the current understanding of the mechanisms of drag generation; since there are still too many gaps in what is known about drag mechanism, these will have to be bridged by carefully formulated experiments and parametric studies; a greater

use should be made of simplified models which allow easier perception of trends and of mechanisms involved.

3. Make increasingly larger use of numerical simulations; simulations of whole vehicle flow fields are not likely to be feasible for years to come, but simulations of simplified flows could very well dovetail with experiments directed at drag mechanisms and provide support and guidance to them.

VII. OTHER AERODYNAMIC EFFECTS

The main topic of this review was the effect of aerodynamics on fuel economy, which comes through the drag force. However, the drag is not the only aerodynamic effect that is of interest in vehicle design. Of the other forces and moments generated aerodynamically, one is interested also in the lift force. Lift coefficients on today's road vehicles range between 0.0 and 0.7 or more (based on the frontal area). Taking $C_L = 0.7$ and frontal area of 2 m^2, this translates into a force of about 600 N at 100 km/h, or about 5% of the weight of a compact car. This means that lift can produce a measurable reduction in the tractive force at high speed. There is another reason why efforts to reduce lift are made, and that is that flow fields generating large lift forces tend to produce excessive drag as well; this aspect has been discussed in Sec. IV.

Side force and yawing movement are of concern from the point of view of vehicle stability under crosswind conditions. The location of the center of pressure with respect to center of gravity affects the response of vehicles to steady and gusty crosswinds and to transient situations generated by passing a large vehicle, or when emerging from an underpass or a tunnel.

Other important aerodynamic effects include noise generation around the windshield–side-window area, affecting passenger's comfort, and dirt deposition on the side windows and on the rear windows of station wagons and hatchbacks. Included in this list should also be engine cooling by ram air passing through the radiator, as well as cooling of important components such as brakes, oil pan, transmission, and the exhaust system. An optimization of body shapes in these areas is a routine part of wind tunnel testing, as these effects are recognized to be vital parts of vehicle reliability and key to customer acceptance.

VIII. SUMMARY

Despite some 80 years of work in vehicle aerodynamics, many important questions still remain unanswered, and the technology is still more art than science. One of the reasons is that the road vehicle flow fields are very complex and difficult to analyze. Not all trends are well understood, and practitioners still encounter many unexpected phenomena.

The drag of current automobiles is quite high, with the mean value for the drag coefficient being around 0.45. Based on the current knowledge of principles of low-drag design, it is projected that the minimum for a drag coefficient of a practical vehicle is approximately 0.25, and so there is a fairly substantial gap between what is being done and what could be achieved. The value 0.25 is still well above the minimum for a well-streamlined vehicle with wheels, which was demonstrated to be near 0.15. The difference is due to practical constraints such as the size of the passenger compartment, safety features, engine cooling, ease of maintenance, product identity, etc.

As far as the importance of aerodynamic drag to fuel economy is concerned, the drag already exceeds rolling resistance at speeds below 50 km/h (30 mph); in the presence of wind and FST, this speed is even lower. This is partly because the wind and FST increase the effective air velocity. In addition, since the drag of a typical road vehicle increases with yaw angle, and wind introduces effective yaw, there is yet a further drag increase. For a typical wind velocity magnitude of 16 km/h (10 mph), these two effects combine to produce a 17% drag increase (above windless values) for the EPA city schedule and an 8% increase for the highway schedule. Evaluating the benefit accrued by a drag reduction on a compact automobile (frontal area 2 m^2, mass 1200 kg), a reduction from $C_D = 0.45$ to 0.30 (a realistic low-drag value) translates into about 11% fuel economy improvement on the combined EPA city–highway schedule and substantially more at typical highway speeds. This large fuel economy improvement comes at a relatively small cost, which makes it especially appealing.

The common method of application of aerodynamics to road vehicles today is through aerodynamic "tuning" of shapes developed by stylists. This process often bypasses the wealth of information on principles of low-drag design, obtained empirically by observations of differences between high- and low-drag cars. Once the efforts of stylists and aerodynamicists will have been integrated, this information will be available for designing vehicles with aerodynamics in mind from inception. Beyond that, it will be necessary to improve our understanding of the mechanisms of drag generation, and the progress in this area will come from careful experiments on simplified body shapes and increasingly from numerical simulations.

One area where changes should be made in the present test techniques is the inclusion of FST into any comprehensive test matrix. This is because FST has been shown to have the potential to alter, often significantly, the drag of bluff bodies. This change is quite unpredictable and can be positive or negative; thus testing road vehicles in streams with properly simulated FST is advisable.

In closing, it is perhaps useful to address the sometimes voiced fears that low-drag vehicles will tend to lose their identity. This does not need to happen, as there is a sufficient cushion between the streamlined body with $C_D = 0.15$ and the practical lower limit of $C_D = 0.25$, which will provide the leeway needed by stylists. One also should not underestimate the capabilities of stylists to differentiate vehicles by quite subtle changes, for example, in window shapes and their skillful use of paint, stripes, and chrome to produce an effect appealing to the eye. The potential for fuel economy improvements through drag reduction, with its added virtue of low cost, is significant enough to warrant an integrated approach to aerodynamic design and styling.

REFERENCES

1. Klemperer, W. "Air Resistance Studies on Automobile Models," *Zeitschrift fur Flugtechnik und Motorluftschiffahrt,* **13,** 201–206 (1922).
2. Korff, W. H., "The Aerodynamic Design of the Goldenrod—To Increase Stability, Traction and Speed," SAE paper 660390 (1966).
3. Torda, T. P., and Morel, T., "Aerodynamic Design of a Land Speed Record Car," *AIAA Journal of Aircraft,* **8,**(12), 1029–1033 (1971).
4. Proceedings of *"Road Vehicle Aerodynamics,"* (A. J. Scibor-Rylski, ed.), City University, London, November (1969).
5. Proceedings of *"Advances in Road Vehicle Aerodynamics,"* (H. S. Stevens, ed.), BHRA (1973).
6. Proceedings of *"2nd AIAA Symposium on Aerodynamics of Sports and Competition Automobiles,"* (B. Pershing, ed.), AIAA, May (1974).
7. Proceedings of *"Colloquium on Industrial Aerodynamics,"* (C. Kramer and H. J. Gerhardt eds.), Aachen, Part 3, (1974).
8. Proceedings of *"Aerodynamic Drag Mechanisms of Bluff Bodies and Road Vehicles,"* (G. Sovran, T. Morel, W. T. Mason, eds.), Plenum, New York (1978).
9. Proceedings of a Symposium on *"Aerodynamics of Transportation,"* (T. Morel and C. Dalton eds.), ASME (1979).
10. Proceedings of the *"4th Colloquium on Industrial Aerodynamics,"* (C. Kramer and H. J. Gerhardt, eds.), Aachen, June (1980).
11. McDonald, A. T., "A Historical Survey of Automotive Aerodynamics," in *Aerodynamics of Transportation,* (T. Morel and C. Dalton, eds.), ASME, pp. 61–69, (1979).
12. Hucho, W. H., "The Aerodynamic Drag of Cars—Current Understanding, Unresolved Problems and Future Prospects," in *Aerodynamic Drag Mechanisms,* (G. Sovran, T. Morel, and W. T. Mason, eds.), pp. 7–40, Plenum, New York (1978).
13. Hucho, W. H., Janssen, L. J., and Emmelmann, H. J., "The Optimization of Body Details—A Method for Reducing the Aerodynamic Drag of Road Vehicles," SAE paper 760185 (1976).
14. Bowman, W. D., "Generalizations on the Aerodynamic Characteristics of Sedan Type Automobile Bodies," SAE paper 660389 (1966).
15. Cogotti, A., Buchheim, R., Garrone, A., and Kuhn, A., "Comparison tests Between Some Full-Scale European Automotive Wind Tunnels—Pininfarina Reference Cars," SAE paper 800139 (1980).
16. Mason, W. T., and Beebe, P. S., "The Drag Related Flow Field Characteristics of Trucks and Buses," in *Aerodynamic Drag Mechanisms,* (G. Sovran, T. Morel, and W. T. Mason, eds.), pp. 45–90, Plenum, New York (1978).

17. Local Climatological Data, Annual Summary, Detroit, Michigan, Metropolitan Airport Department of Commerce (1975).
18. Sovran, G., and Bohn, M. S., "Formulae for the Tractive-Energy Requirements of Vehicles Driving the EPA Schedules," SAE paper 810184 (1981).
19. Janssen, L. J., and Emmelmann, H. J., "Aerodynamic Improvements—A Great Potential for Better Fuel Economy" (SAE paper 780265 (1978).
20. Morel, T., "Theoretical Lower Limits of Forebody Drag," Aeronaut. J., 23–27, January, (1979).
21. Peake, D. J., and Tobak, M., "Three-Dimensional Interactions and Vortical Flows with Emphasis on High Speeds," AGARDograph No. 252 (1980).
22. Roshko, A., and Koenig, K., "Interaction Effects on the Drag of Bluff Bodies in Tandem," in *Aerodynamic Drag Mechanisms,* (G. Sovran, T. Morel, and W. T. Mason, eds.), pp. 253–273, Plenum, New York (1978).
23. Morel, T., and Bohn, M., "Flow Over Two Circular Disks in Tandem," *ASME Journal of Fluids Engineering,* **102,** 104–111 (1980).
24. Wang, K. C., "Separation of Three-Dimensional Flow," in *Reviews in Viscous Flows,* Lockheed Georgia Company, Report LG77ER004, pp. 341–414, (1977).
25. Thwaites, B., *Incompressible Aerodynamics,* pp. 294, Clarendon Press, Oxford (1960).
26. Landahl, M. T., "Numerical Modeling of Blunt-Body Flows—Problems and Prospects," in *Aerodynamic Drag Mechanisms,* (G. Sovran, T. Morel and W. T. Mason, eds.), pp. 289–302, Plenum, New York (1978).
27. Mair, W. A., "The Effect of a Rear-Mounted Disk on the Drag of a Blunt-Based Body of Revolution," *The Aeronautical Quarterly,* **XVI,** 350–360 (1965).
28. Mair, W. A., "Reduction of Base Drag by Boat-tailed Afterbodies in Low-Speed Flow," *The Aeronautical Quarterly,* **XX,** 307–320 (1969).
29. Janssen, L. J., and Hucho, W. H., "Aerodynamische Formoptimierung des Typen VW-Golf and VW-Scirocco," in *Colloquium on Industrial Aerodynamics,* Aachen, Part 3, pp. 46–69 (1974).
30. Maull, D. J., "Mechanisms of Two and Three-Dimensional Base Drag," in *Aerodynamic Drag Mechanisms,* (G. Sovran, T. Morel, and W. T. Mason, eds.), pp. 137–153, Plenum, New York (1978).
31. Mair, W. A., "Drag-Reducing Techniques for Axisymmetric Bluff Bodies," in *Aerodynamic Drag Mechanisms,* (G. Sovran, T. Morel and W. T. Mason, eds.), pp. 161–178, Plenum, New York (1978).
32. Morel, T., "Effect of Base Cavities on the Aerodynamic Drag of an Axisymmetric Cylinder," *The Aeronautical Quarterly,* **XXX,** 400–412 (1979).
33. Bostock, B. R., Slender Bodies of Revolution at Incidence, Ph.D. Dissertation, University of Cambridge (1972).
34. Sykes, D. M., "The Effect of Low Flow Rate Gas Ejection and Ground Proximity on Afterbody Pressure Distribution," in *Proceedings of Road Vehicle Aerodynamics,* (A. J. Scibor-Rylski, eds.), City University, London, November 1969, Paper No. 3.
35. Przirembel, C. E. G., "The Effect of Base Bleed/Suction on the Subsonic Near-Wake of a Bluff Body," in *Aerodynamics of Transportation,* (T. Morel and C. Dalton, eds.), ASME, pp. 43–52 (1979).
36. Drews, H. F. P., "Propelled Apparatus having Surface Means for Developing Increased Propulsion Efficiencies," U.S. Patent 4,180,290, December (1979).
37. Heskestad, G., in *Aerodynamic Drag Mechanisms,* (G. Sovran, T. Morel, and W. T. Mason, eds.), p. 184, Plenum, New York (1978).
38. Morel, T., "The Effect of Base Slant on the Flow Pattern and Drag of 3-D Bodies with Blunt Ends", in *Aerodynamic Drag Mechanisms,* (G. Sovran, T. Morel and W. T. Mason, eds.), pp. 191–217, Plenum, New York (1978).

39. Morel, T., "Aerodynamic Drag of Bluff Body Shapes Characteristic of Hatch-Back Cars," SAE paper 780267 (1978).
40. Ahmed, S. R., and Baumert, W., "The Structure of Wake Flow Behind Road Vehicles," in *Aerodynamics of Transportation,* (T. Morel and C. Dalton, eds.), ASME, pp. 93–103 (1979).
41. Bearman, P. W., "Bluff Body Flows Applicable to Vehicle Aerodynamics," *ASME Journal of Fluids Engineering,* **103,** 265–274 (1980).
42. Maull, D. J., "The Drag of Slant-Based Bodies of Revolution," *The Aeronautical Journal,* June, 164–166 (1980).
43. Lanchester, F. W., *Aerodynamics,* Constable, London (1907).
44. Prandtl, L., "Tragflugeltheorie," Nachr. Ges. Wiss, Gottingen, 107 and 451 (1918).
45. Polhamus, E. C., "A Concept of the Vortex Lift of Sharp-Edge Delta Wings Based on a Leading-Edge-Suction Analogy," NASA TN D-3767 (1966).
46. Polhamus, E. C., "Predictions of Vortex-Lift Characteristics by a Leading-Edge Suction Analogy," *Journal of Aircraft,* **8,** 193–199, April, (1971).
47. Lamar, J. E., "Some Recent Applications of the Suction Analogy to Vortex-Lift Estimates," NASA TMX 72785 (1976).
48. Saunders, G. H., "Aerodynamic Characteristics of Wings in Ground Proximity," *Canadian Aeronautics and Space Journal,* June 185–192 (1965).
49. Widnall, S. E., and Barrows, T. M., "An Analytic Solution for Two and Three-Dimensional Wings in Ground Effects," *Journal of Fluid Mechanics,* **41,** 769–792 (1970).
50. Tuck, E. O., "Irrotational Flow Past Bodies Close to a Plane Surface," *Journal of Fluid Mechanics,* **50,** 481–491 (1971).
51. Fackrell, J. E., "The Simulation and Prediction of Ground Effect in Car Aerodynamics," Imperial College, London, Aeronautics Report 75-11 (1975).
52. Fink, M. P., and Lastinger, J. L., "Aerodynamic Characteristics of Low-Aspect Ratio Wings in Close Proximity to the Ground," NASA TN D-926 (1961).
53. Stollery, J. L., and Burns, W. K., "Forces on Bodies in the Presence of the Ground," 1st Symposium on *Road Vehicle Aerodynamics,* City University, London, (A. J. Scibor-Rylski, ed.), November (1969).
54. Waters, D. M., "Thickness and Camber Effects of Bodies in Ground Proximity," in Symposium on *Advances in Road Vehicle Aerodynamics,* BHRA, (H. S. Stephens, ed.), pp. 185–205 (1973).
55. Tuck, E. O., "Matching Problems Involving Flow Through Small Holes," in *Advances in Applied Mechanics,* **15,** 89–158 (1975).
56. Morelli, A., "Low Drag Bodies Moving in the Proximity of the Ground," in *Aerodynamics of Transportation,* (T. Morel and C. Dalton, eds.), ASME, pp. 241–248 (1979).
57. Carr, G. W., "Aerodynamic Lift Characteristics of Cars," *Proc. Inst. Mech. Eng.* (Auto Division), **187,** 333–347 (1973).
58. Bearman, P. W., and Morel, T., "Effect of Free Stream Turbulence on the Flow Around Bluff Bodies," *Progress in Aerospace Sciences,* **20(2–3),** 97–123 (1983).
59. Fage, A., and Warsap, J. H., "The Effects of Turbulence and Surface Roughness on the Drag of a Circular Cylinder," ARC R&M No. 1283 (1929).
60. Schubauer, G. B., and Dryden, H. L., "The Effect of Turbulence on the Drag of Flat Plates," NACA Report 546 (1935).
61. Bearman, P. W., "An Investigation of the Forces on Flat Plates Normal to a Turbulent Flow," *Journal of Fluid Mechanics,* **46,** 177–198 (1971).
62. Humphries, W., and Vincent, J. H., "Experiments to Investigate Transport Processes in the Near Wakes of Disks in Turbulent Air Flow," *Journal of Fluid Mechanics,* **75,** 737–749 (1976).
63. Martin, L. J., "The Effect of Turbulence on the Flow Around A Cube," M. Sc. Dissertation, Department of Aeronautics, Imperial College (1977).

64. Buckley, F. T., Marks, C. H., and Walston, W. H., "Analysis of Coast-Down Data to Assess Aerodynamic Drag Reduction on Full Scale Tractor-Trailer Trucks in Windy Environments," SAE Paper 760850 (1976).
65. Cooper, K. R., and Campbell, W. F., "An Examination of the Effects of Wind Turbulence on the Aerodynamic Drag of Vehicles," Proceedings of the *4th Colloquium on Industrial Aerodynamics,* (C. Kramer and H. J. Gerhardt, eds.), Aachen, June pp. 283–294 (1980).
66. Scibor-Rylski, A. J., *Road Vehicle Aerodynamics,* pp. 106, Wiley, New York (1975).
67. Fackrell, J. E., and Harvey, J. K., "The Aerodynamics of an Isolated Road Wheel," Proceedings of The Second AIAA Symposium on *Aerodynamics of Sports and Competition Automobiles, (B. Pershing, eds.), AIAA,* pp. 119–125, May (1974).
68. Kurtz, D. W., "Aerodynamic Design of Electric and Hybrid Vehicles: A Guidebook," Jet Propulsion Laboratory Publication 80-69 (1980).
69. Janssen, L. J., and Hucho, W. H., "The Effect of Various Parameters on the Aerodynamic Drag of Passenger Cars," in *Advances in Road Vehicle Aerodynamics,* (H. S. Stevens, ed.), BHRA, pp. 223–253 (1973).
70. Goetz, H., "Schuttguttransport, Verschmutzung und Abgasgeruch bei Kraftfahrzeugen-Auswirkung und Aerodynamische Abhilfemassnahmen," in *Colloquium on Industrial Aerodynamics,* (C. Kramer and H. J. Gerhardt, eds.), Aachen, Part 3, pp. 97–108 (1974).
71. Buchheim, R., Deutenbach, K. R., and Luckoff, H. J., "Necessity and Premises for Reducing the Aerodynamic Drag of Future Passenger Cars," SAE Paper 810185 (1981).
72. Carr, G. W., "Alternative Routes to a Low-Drag Automobile," in *Aerodynamics of Transportation,* (T. Morel and C. Dalton, eds.), ASME, pp. 105–112 (1979).
73. Hirt, C. W., and Ramshaw, J. D., "Prospects for Numerical Simulation of Bluff-Body Aerodynamics," in *Aerodynamic Drag Mechanisms,* (G. Sovran, T. Morel and W. T. Mason, eds.), pp. 313–350, Plenum, New York (1978).
74. Palko, R. L., "Utilization of the AEDC Three-Dimensional Potential Flow Computer Program," NASA SP-405, pp. 127–143 (1976).
75. Kandil, O. A., Mook, D. T., and Nayfeh, A. H., "New Convergence Criteria for the Vortex-Lattice Models of the Leading Edge Separation," NASA SP-405, pp. 285–292 (1976).
76. Leonard, A. "Simulation of Three-Dimensional Separated Flow with Vortex Filaments," *Lecture Notes in Physics,* Springer-Verlag, Berlin (1977).
77. Maskell, E. C., "Flow Separation in Three Dimensions," RAE Aero Report 2565 (1955).

11

Power-Train Matching and Fuel Economy Projection Methods

THOMAS R. STOCKTON

I. INTRODUCTION

The fuel economy level established for a particular existing or hypothetical vehicle must be determined strictly on the basis of some prescribed driving cycle and testing procedure. The advent of legislation specifying the corporate average fuel economy (CAFE) that a car manufacturer must meet can legally limit the mix of car lines that can be sold in a given model year. Such legislation is a year-by-year changing situation, but fuel economy ratings mandated by federal law are considered permanent. Planning and developing the appropriate car product lines several years in advance places major emphasis on projecting fuel economy, with experimental programs to support fuel economy improvement technology.

The wide variety of factors that influence fuel economy fall into broad categories such as:

1. General vehicle specifications and characteristics
2. Legislative requirements
3. Manufacturer's product acceptance standards
4. Manufacturer's marketing objectives
5. Power-train component efficiencies
6. Engine displacement
7. Transmission and final drive ratios

Power-train matching is a technique for optimizing vehicle fuel economy through the process of determining the best overall combination of power-train parameters

THOMAS R. STOCKTON • Scientific Research Laboratories, Ford Motor Company, Dearborn, Michigan 48121.

compatible with all of the given constraints. This intricate interaction among the power-train parameters can be analyzed most easily through a comprehensive computer program. Computer fuel economy projections, however, are most generally used to evaluate the effects of parameter changes. Absolute fuel economies are usually derived by differential adjustments to known baseline test results because of the uncertainty of the computer projections.

This chapter deals primarily with the basic power-train and vehicle parameters, including:

1. Engine fuel consumption throughout the operating speed and load range
2. Engine displacement
3. Prescribed driving cycles
4. Vehicle performance, gradeability, and maximum speed
5. Transmission and driveline efficiencies
6. Gear ratios and shift schedules

These will be discussed in detail. Improvements in engine fuel consumption and emissions effects, and vehicle rolling resistance are discussed in other chapters.

II. PARAMETERS INFLUENCING FUEL ECONOMY

A. Engine Size and Thermal Efficiency

During the age of large vehicles and low cost fuel in the United States and many other countries, the range of engine options offered included large V-8's down through smaller six cylinders, with a (displacement–weight) ratio of 180 CID/ton to 160 CID/ton. Buyers selected the smaller engines when fuel economy was a more important consideration than power. With the vehicle downsizing that followed the oil crisis beginning in the 1970s, with a major emphasis on high fuel economy, both the average vehicle weight and the displacement-to-weight ratio have been significantly reduced.

The displacement ratios have essentially leveled off at approximately 100 CID/ton for naturally aspirated engines, where further displacement reductions would not generally improve fuel economy but would degrade performance and power.

The smallest engine displacement capable of meeting acceptable vehicle performance targets usually results in a fuel economy compromise over the EPA and other driving cycles. Best brake-specific fuel consumption (BSFC) at any particular power level occurs at some one combination of engine speed and torque, which generally falls below the maximum torque condition which the smallest engine demands.

Figure 1 shows the fuel consumption characteristics of a typical four-cylinder engine as a function of BMEP (brake mean effective pressure) and engine speed. Superimposed upon islands of constant BSFC in LB/BHPH (pounds of

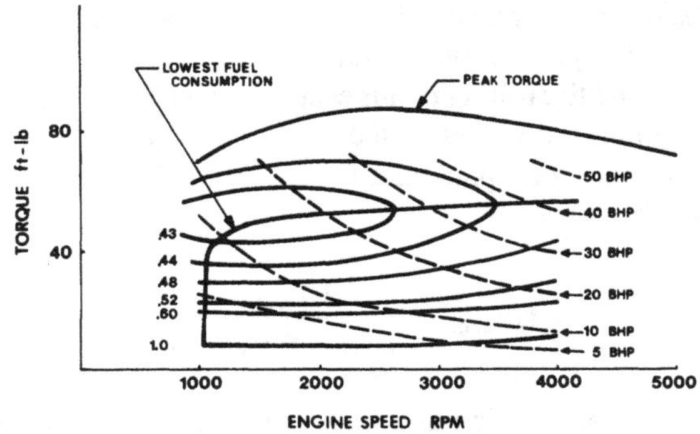

FIGURE 1. 1.6-liter engine brake-specific fuel consumption.

fuel per brake horsepower hour), a line representing lowest fuel consumption throughout the power range is located.

The minimum total of engine losses, including cycle losses, induction, pumping losses, and mechanical friction losses favors high BMEP and low rpm. Not only is engine thermodynamic efficiency improved but pumping losses across the throttle restriction are significantly reduced. Low rpm minimizes engine friction from the bearings, pistons, valve gear, and oil pump. Accessory losses from the cooling fan, alternator, and others also are reduced. Practical engine designs, however, provide air–fuel mixture enrichment at and near full throttle position so as to allow the engine to develop maximum power capability and to provide good throttle response. On carbureted engines power enrichment is controlled by manifold vacuum level and/or an accelerator pump actuated by movement of the accelerator pedal.

For best fuel economy the engine displacement must be sized adequately to avoid or minimize operation in the power enrichment zone over the prescribed driving cycle. The role of the transmission is thus vital in providing the optimum gear ratios and gear shift points to follow the best BSFC combinations of engine speed and torque as closely as practical.

B. Transmission Systems

1. Transmissions—Manual

The four-speed front-wheel drive (FWD) manual transaxle, typical of modern worldwide practice for small cars, is a low cost, highly efficient ratio changing gear box, which integrates the final drive function with the transmission. The ratios are best expressed as an overall engine-to-drive-wheel relationship since distinct transmission output speeds and axle ratios are less meaningful when combined as one system. In conventional rear-wheel drive (RWD) vehicles the transmission ratios always relate to and include one ratio of unity as direct drive. Overall ratios are the product of separate transmission ratios and axle ratio.

Wheel speed and tire rolling radius establish road speed, which is the starting point for determining the values of the various power-train parameters for optimizing fuel economy for a particular driving cycle. The required engine brake horsepower then becomes

$$HP_{Eng.} = \frac{HP_{Vehicle}}{E_{Trans} \times E_{Final\ Dr.}}$$

The importance of transmission and final drive efficiencies is apparent. Figure 2 shows efficiency characteristics for a typical four-speed manual transmission. Generally, the efficiency in all gears is approximately 95% at medium and heavy load and falls off to approximately 80% at light load.

Since the EPA driving cycle includes appreciable operation at light input torque loads, the fuel economy projections should be based upon an efficiency map rather than on a flat efficiency number. Final drive efficiencies are discussed later in this chapter.

The ideal transmission will cause the engine to operate along the best economy line shown in Fig. 1 and also permit rapid increase in ratio as required for acceleration demands. A manual transmission, however, must compromise the ideal because it must employ a limited number of discreet gear ratios. These ratios, including final drive ratio, and the steps between ratios are selected to affect the best overall compromise between economy, performance, maximum speed, and driveability. Excessive number of ratios, or close steps between ratios, could result in too frequent shifting. Also, the interruption of power during the shift interval factors into the performance, and excessive operation of the accelerator pump adversely effects economy. Addition of a fifth speed to a four-speed transmission can achieve an approximate 3.5% improvement in average

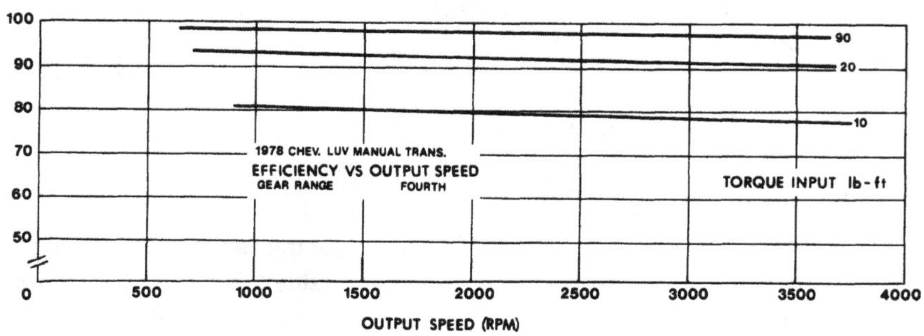

FIGURE 2. Manual 4-speed transmission efficiency.

fuel economy. A sixth speed would result in only a lesser gain and is not attractive for passenger car usage because of mechanical complexity and shift business.

Two shift schedules must be taken into consideration; first for maximizing full throttle performance, and second for conformance with legislated schedules which must be followed during the federal test procedure to establish EPA fuel economy ratings. For vehicles marketed in the United States, the EPA prescribes the shift schedule, which is 15–25–40 mph four-speed transmissions, which results in an EPA rating below the optimum.

Figure 3 illustrates that maximum performance is achieved when the transmission is upshifted where the falling torque (BMEP) or acceleration curve for a gear ratio intersects the curve for the next higher ratio. Figure 4 illustrates how the optimum shift points are established to meet the accelerations during the metro portion of the EPA driving cycle. This cannot be used, however, during the federal test procedure.

2. *Transmissions–Automatic*

The three-speed automatic transmission with open torque converter dominated automotive application in the 1950–1980 period. The hydrodynamic torque converter, comprised of an impeller, turbine, and reactor, possesses ideal characteristics including smooth operation, excellent driver feel from torque multiplication, and good efficiency throughout most of its operating range. A slip loss, however, exists at road load driving conditions. Other transmission losses include the oil pump, gear box windage, and viscous drag of friction clutches or bands. These losses result in a 12–15% EPA average fuel economy penalty versus manual transmissions. In the next generation of conventional-type automatic transmissions, the addition of converter lock-out or bypass plus more gear ratio coverage from wide ratio three speeds or four speeds, in combination cause the engine to operate at higher load factor for improved fuel economy. The fuel economy penalty versus manual transmissions is generally reduced by half. The evolution of transverse front-wheel drive also contributed an improvement in

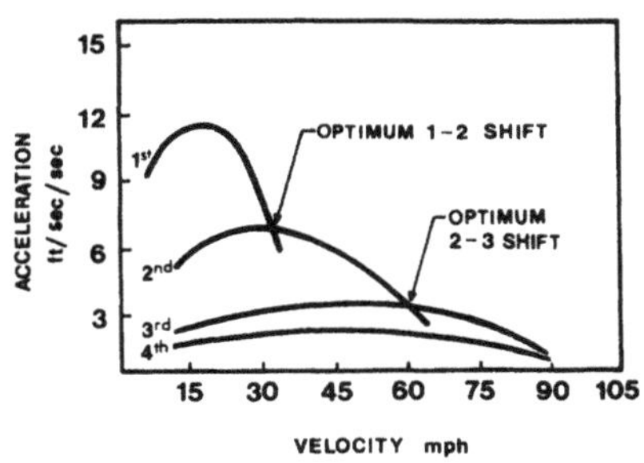

FIGURE 3. Optimum manual transmission shift schedule for performance.

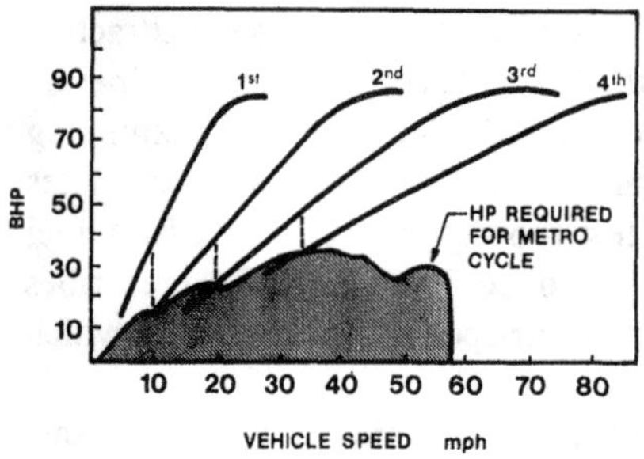

FIGURE 4. Optimum manual transmission shift schedule for fuel economy.

final drive efficiency. For computer modeling, the FWD system shown in Fig. 5 is considered typical of the new generation automatic transaxle configurations.

The capacity of a torque converter is usually expressed by an input size factor represented by Eq. (1).

$$K = \frac{N}{\sqrt{T}} \tag{1}$$

where K is the input size factor, N is the input rpm, and T is the input torque, in ft lb.

The K factor level is generally stated at stall conditions and is convenient in setting engine stall rpm at full throttle. The K factor along with torque ratio and efficiency are plotted against speed ratio at a constant input torque.

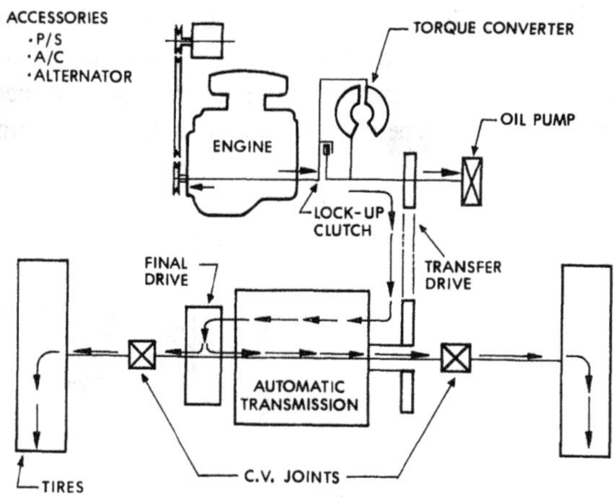

FIGURE 5. Automatic FWD transaxle.

$$\text{Speed ratio} = \frac{\text{input rpm (impeller)}}{\text{output rpm (turbine)}}$$

Figure 6 shows data for a 10.25 in. diameter converter having a stall K factor of 13.7. From the characteristic Eq. (2),

$$T_i = CN_i^2 D_i^5 \tag{2}$$

it can be seen that the capacity or "tightness" varies as input rpm^2 and fifth power of impeller toroidal flow path diameter, but is not effected by input torque level. Any given converter design of blading inlet and exit angles and proportions can be scaled up or down to obtain the desired K value.

Tighter converters tend to improve fuel economy at the sacrifice of performance, and converter sizing and/or altering characteristics through blading changes are significant factors in matching a "full-time" converter to a particular vehicle and engine combination. With the addition of a lock-up clutch, converter slip can be eliminated in all gear ratios except during low gear start-up. Sizing, therefore, becomes a less significant fuel economy factor in modern transmission designs.

The balance of losses in the transmission are divided between the oil pump

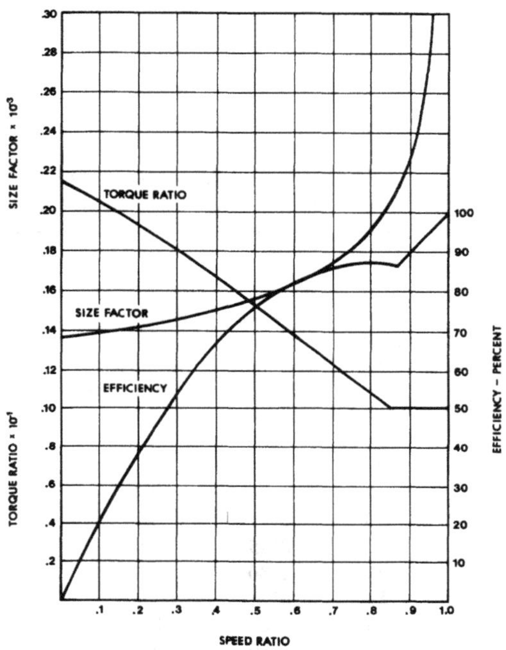

FIGURE 6. Torque converter performance.

and gear box. Common pump practice utilizes a fixed displacement internal–external gear pump, driven by the engine, to supply hydraulic pressure for actuation of shifting clutches or bands, converter charging, lubrication, and cooling. The displacement per revolution is established primarily to satisfy control system leakage and lubrication at idle rpm which is approximately 2 gpm. At higher speeds the excess pump discharge is expended as a loss. To reduce this loss the trend favors a variable displacement vane-type pump. Table 1 shows the pump drive power for a typical I-X gear pump as a function of rpm and constant vehicle speed. Included for comparison is estimated power of a V.D. pump meeting the same flow requirements.

Since pressure increases with transmission input torque, pressures during acceleration would rise above road load levels. A complete map of oil pump power over the range of rpm and pressure for a given driving cycle should be considered for fuel economy projections.

Transmission Gear Box. The selection of gear ratios for planetary gear trains found in automatic transmissions is influenced by both interrelationship of forward and reverse ratios and the complexity level of the gear arrangement. The predominant gear train in RWD vehicles consists of two sets of simple planetary gear sets with the ratios

Low:	2.46:1
Intermediate:	1.46:1
High:	1.1:1
Reverse:	−2.19:1

TABLE 1. C-4 Transmission Oil Pump Power Consumption at Road Load

		Pump horsepower		
Vehicle speed (mph)	Engine speed (rpm)	Production pump (test data)	Variable displacement pump (estimated)	Separately driven pump (estimated)
Idle	600	0.16	0.16	0.16
Idle	800	0.20	0.16	0.16
30	1242	0.33	0.21	0.17
40	1600	0.47	0.22	0.18
50	1971	0.71	0.28	0.20
60	2115	0.98	0.31	0.25
70	2736	1.31	0.44	0.28
80	3121	1.71	0.50	0.28
90	3507	2.14	0.58	0.28
100	3889	2.61	0.72	0.29

For fuel economy projections the gear box efficiency, without oil pump, at road load is important. Data for a medium size transmission are shown on Table 2. The torque converter and oil pump efficiencies are included to show the relationship to the overall transmission efficiencies. Gear box efficiencies in low and intermediate ratios are similar.

More fuel efficient three-speed automatics, such as the Ford FWD ATX, increase the forward ratio spread so that the numerical axle ratio can be lowered without sacrificing performance. Torque converter bypass is also employed. Ratios for the ATX are:

Low:	2.79:1	Open converter
Intermediate:	1.61:1	62% mechanical
High:	1:1	93% mechanical
Reverse:	−1.97:1	Open converter

TABLE 2. Transmission Components Performance

I. Road Load Horsepower Loss

Engine speed rpm	Vehicle speed mph	Engine power hp	A + B + C Transmission hp loss	A Converter hp loss	B Oil pump hp requirements	C Gear box hp loss
1242	30	9.02	2.11	1.13	0.33	0.65
1600	40	13.505	2.62	1.30	0.47	0.85
1971	50	21.52	3.53	1.54	0.71	1.28
2115	60	32.44	4.71	1.93	0.98	1.80
2736	70	47.48	6.03	2.52	1.31	2.20
3121	80	67.08	7.58	3.24	1.71	2.63
3507	90	87.20	9.33	4.03	2.14	3.16
3889	100	113.30	11.33	4.88	2.61	3.84

II. Road Load Efficiency

Vehicle speed mph	Transmission eff. (%)	Gear box with pump eff. (%)	Converter eff. (%)	Oil pump overall eff. (%)	Gear box eff. (%)
30	76.6	87.3	87.7	56.7	91.7
40	80.6	88.8	90.8	55.3	93.3
50	83.4	89.6	93.1	50.3	93.6
60	85.7	90.9	94.3	45.9	94.0
70	87.3	92.6	94.4	43.1	94.9
80	88.7	93.4	95.0	41.8	95.6
90	89.3	93.7	95.3	40.8	96.1
100	90.0	99.3	95.7	40.4	96.4

With the addition of a fourth speed, the lower three ratios from the conventional three-speed transmission can be carried over. The fourth or overdrive ratio, in conjunction with torque converter lock-up or bypass, provides improved fuel economy without effecting driveability. Ratios for the Ford AOD transmission used in RWD vehicles are:

Low:	2.40:1	Open converter
Intermediate:	1.47:1	Open converter
High:	1.00:1	50% mechanical
Overdrive:	0.67:1	100% mechanical
Reverse:	−1.97:1	Open converter

In the split power path configurations the torque converter generally operates only in the coupling range where percent slip is of concern in estimating fuel economy. It is important that the reduction in slip is a multiple benefit from split systems because the total transmission over all slip is less than converter slip. This effect can be illustrated using the Ford ATX second gear as an example, which is shown schematically in Fig. 7.

The ratio of mechanical path to fluid path is established by the ratio of the planetary ring gear pitch diameter R to the sun gear pitch diameter S, Eq. (3). In this example

$$\frac{R}{S} = \frac{\%\ \text{mechanical}}{\%\ \text{fluid}} = \frac{0.62}{0.38} = 1.63 \tag{3}$$

Percent slip in the fluid path between the converter impeller and turbine is reduced at the transmission input by the factor, Eqs. (4) and (5).

$$\left(1 + \frac{R}{S}\right) = 2.63 \tag{4}$$

$$\text{Transmission slip} = \frac{\text{converter slip}}{2.63} = 0.38 \tag{5}$$

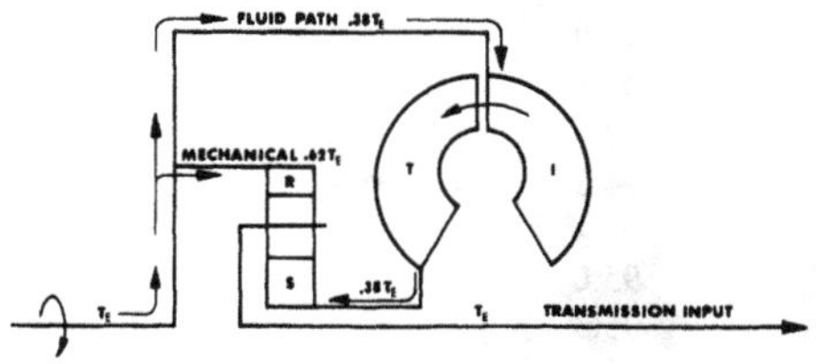

FIGURE 7. ATX second-gear split power path.

Converter slip itself is reduced because of the reduced input torque, and an effective second gear K factor can be calculated, using Eqs. (6)–(8):

$$N_0 = \text{input rpm open converter}$$

$$K_0 = \text{size factor open converter}$$

$$T_0 = \text{input torque open converter}$$

$$N_s = \text{input rpm split power}$$

$$K_s = \text{effective size factor split power}$$

$$T_s = \text{input torque split power}$$

$$N_s = K_0 T_s = K_0 0.38 = K_0(0.62) \tag{6}$$

but $N_s = N_0$ in second gear. Then

$$K_s = K_0(0.62) \tag{7}$$

Since only 38% of input torque flows through a "tighter" converter, the equivalent K_{se} as a total transmission is

$$K_{se} = K_0(0.38)(0.62) = K_0(0.235) \tag{8}$$

Preparing computer files for fuel economy projections only requires scaling the open converter K factor downward by the factor (0.235).

The upshift rpm schedule can be optimized for a given driving cycle to maximize fuel economy. There are no legislated arbitrary shift schedules that apply to automatic transmissions. A separate shift schedule would be developed to optimize performance.

3. *Transmissions—Continuously Variable*

Referring to Fig. 1, the best economy line for all power levels can only be followed with a transmission providing an infinite number of ratios. The control system for such a CVT would seek the engine speed and BMEP level corresponding to the best BSFC condition over the power range. Thus the CVT ratio and the engine throttle position or intake manifold pressure would be under full control. The control strategy would essentially optimize fuel economy for any typical driving cycle. A start-up device such as a clutch, fluid coupling, or torque converter is required since CVT mechanisms of the traction drive or friction drive type do not possess start-up capability and are also limited in ratio coverage. Hydrostatic or hydromechanical systems are often considered for automotive application, but past developments have been plagued by noise and other problems.

Promising CVT mechanisms can utilize either a rubber belt or a chain belt drive in conjunction with variable ratio pulleys. The mechanisms can then be combined in numerous configurations with start-up devices and gearing to obtain the desired overall transmission characteristics.

Improved efficiency tends to be offset by poorer ratio range and the optimum combination for best fuel economy often results in compromises. Schemes of multiple modes of operation can also be used to extend ratio coverage without sacrificing efficiency.

Many practical considerations can force operation below the best economy line, mostly attributable to the high engine load factor. These include increased noise, vibration, and possibly exhaust emissions. Recalibration of the engine for CVT application could be required to meet legislated emission levels for the driving cycle.

Efficiency characteristics and ratio coverage of the CVT mechanism are critical in achieving a fuel economy advantage over fuel efficient conventional automatic transmissions such as the Ford ATX and AOD. Since the CVT mechanism operates full time and at lower efficiency levels than geared ratios and direct drive, it starts with an efficiency disadvantage. However, the potential gain in engine efficiency can more than offset lower transmission efficiencies. Computer comparisons of these two categories, utilizing completely different computer programs, requires correlation with test results of complete transmissions installed in vehicles to be conclusive.

An efficiency map for a CVT is best constructed from individual test data on each transmission subsystem. These include the CVT mechanism, oil pump, start-up device, gearing, windage, and final drive.

Efficiency test data for an automotive variable-speed rubber belt developed by the Gates Rubber Co. are shown in Fig. 8 and 9. The first shows the effect of input speed at various input torque levels at the 1:1 ratio position. The efficiencies are high at all speeds and torques encountered during normal driving

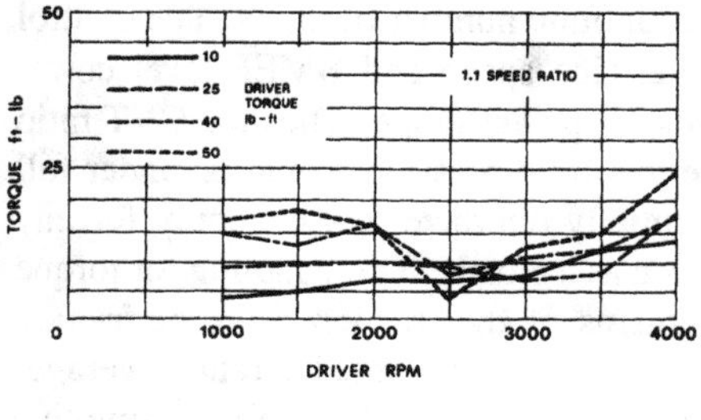

FIGURE 8. Torque loss versus speed, continuously variable ratio drive.

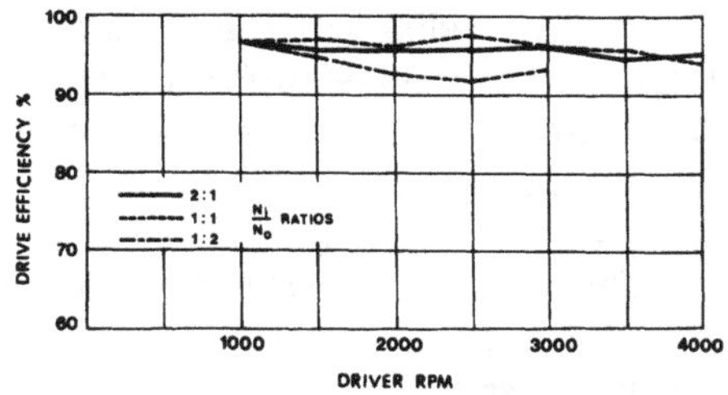

FIGURE 9. Efficiency versus speed, continuously variable ratio drive.

conditions. Figure 9 shows the effect of CVT ratio position at various input speeds, for 40 ft lb input torque. Efficiency characteristics for steel chain belts are similar to the rubber belt, but substantially higher hydraulic clamping forces are required, necessitating higher oil pumping losses.

C. Driveline Losses

Transmission of power from the transmission output to the driving wheels involves certain losses in the driveline components. Various vehicle engine–transmission installations require different driveline components. Vehicles have been produced with engines installed either longitudinally or transversely at the front or rear. Longitudinal front engine installations with RWD have been popular in the United States. However, with the trend toward smaller engines and vehicles for fuel economy, front FWD vehicles with transverse engine installations appear dominant for the future.

Several FWD drivelines which have been mass produced are shown schematically in Fig. 10. The FWDs require four constant velocity (CV) joints to transmit power and accommodate steering. Front engine RWD vehicles utilize two universal joints in the driveshaft to accommodate angular changes resulting

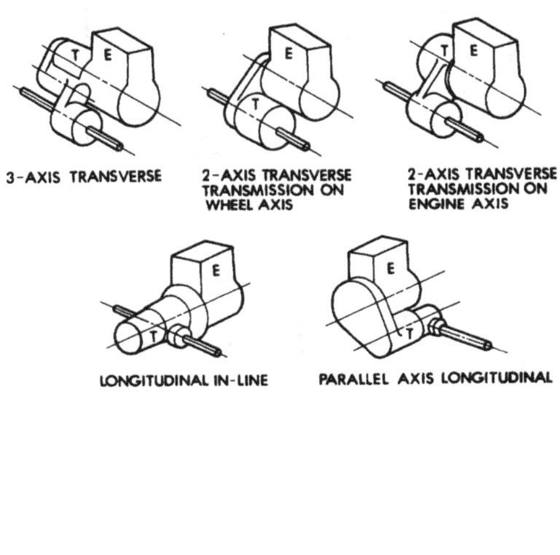

FIGURE 10. Common front-wheel drive power-train configurations.

from vehicle loading and bounce. Since the efficiency of CV and universal joints is approximately 98–99% for small angles, the total joint loss would be comparable for either drive. Essentially, the difference between efficiency in drivelines results from the number and type of transfer drives (often referred to as final drive or axle ratio).

Three common transfer drives are parallel axis gears, chains, and hypoid crossed axis gears. Figure 11 compares the efficiency of a Morse Hi-Vo chain to that of parallel axis gears. It is apparent that the chain is more efficient throughout the normal operating range with an increasing advantage at lower torques and speeds.

Figure 12 compares the efficiency of Morse Hi-Vo chain and hypoid (9.00 and 6.75 in.) final drives in a 3000-lb I.W.C. vehicle at various ratios. Plotted points represent average input torques to the final drive for the CVS-CH test cycle. The curves show the average hypoid efficiency improves as loading increases while the chain remains nearly constant. Also, the efficiency of the chain is considerably better than for the hypoid at all loads.

It may be concluded that transverse front-mounted engine FWD vehicles have an inherent driveline efficiency advantage over conventional RWD vehicles. Other advantages of transverse engine front-wheel drivelines are reduced weight and greater stiffness (discussed in detail in Sec. III).

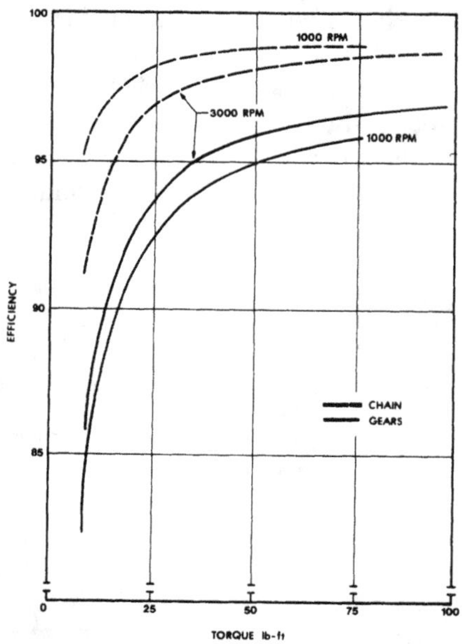

FIGURE 11. Parallel axis gears versus chain efficiency.

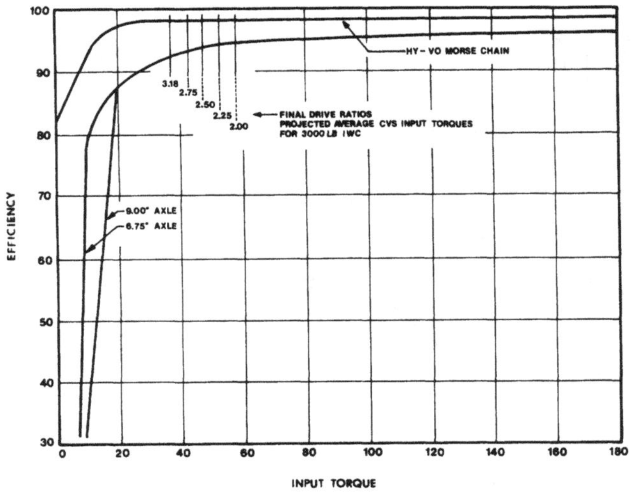

FIGURE 12. Final drive/axle efficiency.

D. Accessory Losses

According to federal procedure, vehicles tested by the EPA dynamometer driving cycle must have a properly sized alternator and cooling fan installed and operational.

Other accessories are to be included when their usage per vehicle model reaches 33%. If air conditioning is required, it is not operated during the test, but an increase in the dynamometer P.A.U. setting of 10% is mandated.

Computer projections of vehicle performance and fuel economy for power-train matching include engine accessory losses determined at each test point of the EPA driving cycle added to the required shaft power. When possible, accessory losses are determined from bench test data of the individual components. Otherwise, typical accessory power curves similar to those shown in Fig. 13 are used. The alternator load is often considered constant at the average current required for the EPA cycle. The power-steering load is assumed to be that required for straight ahead or zero steering effort.

Typical engine-driven fan power is shown in Fig. 13(c). With electric fans no power is required during the EPA cycle because it seldom energizes. Compressor power for the air-conditioning system is generally not considered in performance projections. If desired for fuel economy projections, a typical power curve is shown in Fig. 13(d).

As the size of vehicle varies, combined accessory power requirements for the alternator, power steering and air conditioning do not change substantially.

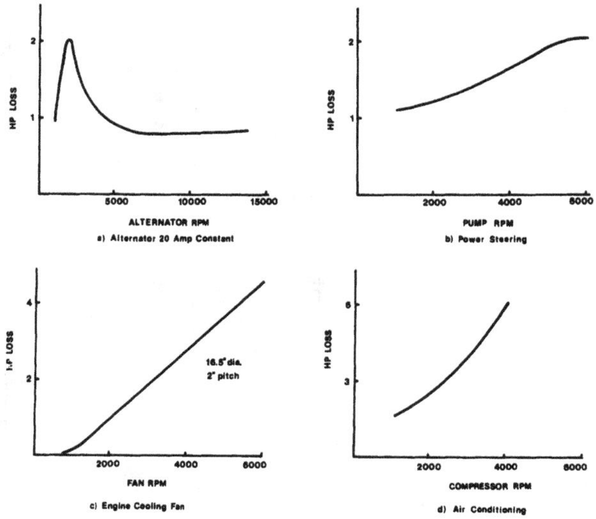

FIGURE 13. Accessory power losses.

Fan power does vary with vehicle size, particularly in small cars using electric fans.

III. VEHICLE PERFORMANCE INFLUENCES UPON FUEL ECONOMY

Fuel economy of vehicles is substantially influenced by performance targets and the driving cycle by which it is rated. Power-train optimization for fuel economy, disregarding performance or consideration of both metro and highway federal driving cycles, can result in poor customer acceptance. Trade-offs must be made between fuel economy, performance, and driving cycle ratings. Vehicle performance is rated according to

1. Acceleration
2. Gradeability
3. Maximum speed
4. Driveability

Acceleration performance of mass-produced passenger cars has changed dramatically since the oil embargo of 1974. A common measure for acceleration is 0–60 mph time. For the years 1974–1980 average 0–60 time has increased from about 10 to 15 sec. This change results from a shift toward fuel efficiency as rated on the metro driving cycle. The relationship between acceleration and fuel economy for constant N/V ratios is shown in Fig. 14.

Gradeability is related to fuel economy because it establishes limits for the overall drive-train ratio coverage. Gradeability refers to the capability of the vehicle to climb hills. A grade capability is established as a minimum on which the vehicle must accelerate upward from rest in both forward and reverse gears. This grade is typically 30% for passenger cars. Engine size, low and reverse gear ratios, final drive ratio, and torque converter selection must take into consideration gradeability specifications. Another grade consideration is 3% in high gear at both 55 and 60 mph. The purpose for these constraints is in power-train matching and is discussed further in Sec. V.B.

The desired maximum speed of the vehicle relates to fuel economy in that it will determine the engine horsepower and the final drive ratio.

Driveability is based upon a subjective jury evaluation of the vehicle, which

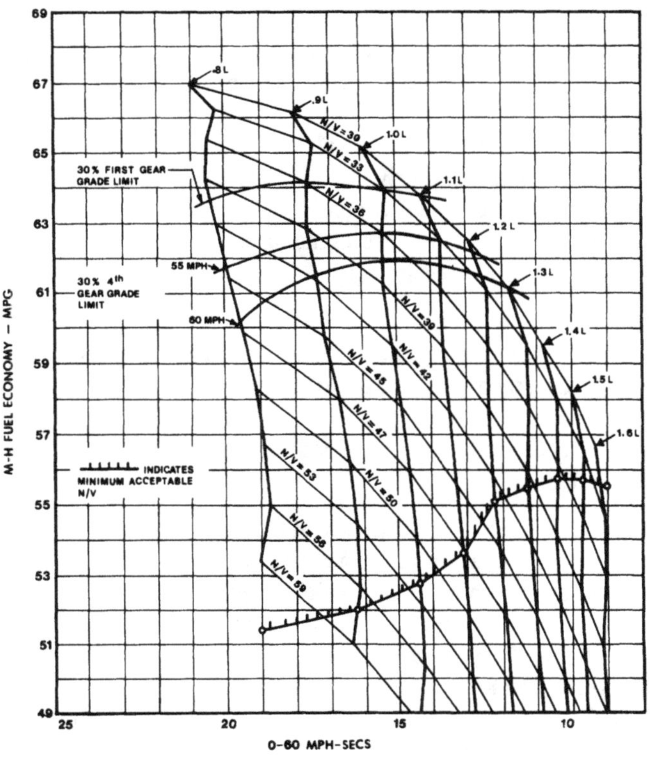

FIGURE 14. M-H fuel economy versus 0–60 mph time.

rates power-train response, noise, and vibration. Several power-train parameters affect driveability, such as engine displacement, number of cylinders, driveline N/V ratio (engine rpm/vehicle mph), and the spring constant for the type of driveline (FWD versus RWD). Larger engines and increased number of cylinders will provide larger, faster, and smoother torque response. Front-wheel drive cars have stiff drivelines because the propellor shaft, body, and tires from the RWD system have been eliminated. This also improves vehicle response. Driveline N/V ratio is an important consideration for noise and vibration. Too low a ratio will cause poor acceleration, noise, and vibration due to engine lugging. Too high a ratio will result in poor fuel economy and high-speed engine noise. Figure 14, in addition to fuel economy versus acceleration, also shows the relationship of N/V ratio and engine size. This graph is typical of information used for power-train matching and will be discussed in detail in Sec. V.

IV. COMPUTER PROJECTIONS OF FUEL ECONOMY AND PERFORMANCE

For effective power-train matching, sophisticated computer programs in conjunction with practical limits for N/V ratio must be used to analyze vehicle fuel economy and performance. A flow chart of program logic for fuel economy projections is shown in Fig. 15. The calculation starts at the drive wheels using the second by second velocity drive schedule as the basic input. In the first step of the program the drive wheel torque and rpm needed to propel the vehicle according to the drive schedule are computed. The wheel torque is calculated based on one of the following inputs:

1. Steady-state reference vehicle track test results providing the vehicle velocity, driveshaft rpm, and driveshaft torque relation.
2. Calculation incorporating the effect of aerodynamic drag and tire rolling resistance.
3. Calculation predicting road loads when the vehicle is tested on a dynamometer.

Following the calculation of the wheel torque and rpm, the driveshaft torque and rpm are determined using the drive axle ratio and efficiency.

The transmission input (engine side) torque and rpm are calculated from the driveshaft torque and rpm using the transmission gear ratio and efficiency as main parameters. The program in the first pass assumes that the transmission is in the first gear. The knowledge of transmission input torque and rpm (which is the same as the torque converter or the clutch output torque and rpm) facilitates the computation of the torque-converter (or clutch) input torque and rpm.

The main parameters in this calculation, in the case of automatic transmis-

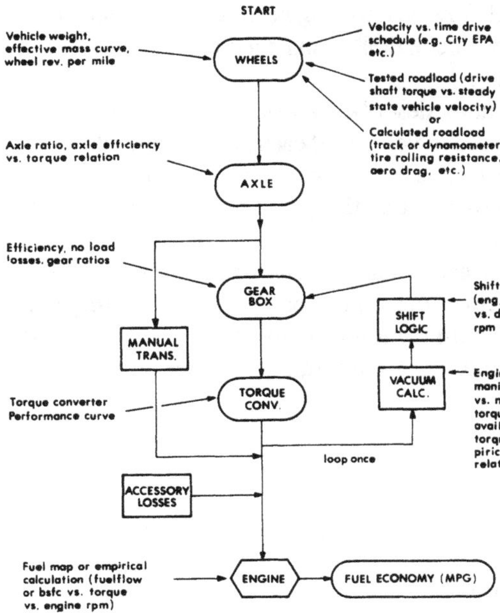

FIGURE 15. Fuel economy projection program.

sions, are the torque converter performance curves (Fig. 6). Using these performance curves the torque converter input torque and rpm are calculated.

In the case of a manual transmission it is assumed that the clutch is able to transfer the necessary torque without loss or slippage. Clutch slippage is assumed only when the vehicle velocity drops under a critical value, usually around 15 mpg, or when the clutch input rpm needed to satisfy the drive schedule is less than the idle rpm. Finally, the accessory losses are added to the torque converter input torque, resulting in the needed engine torque.

The knowledge about the available and needed engine torques allows the determination of engine vacuum by using a relation between torque and vacuum (as determined experimentally). Once the engine vacuum is known, using the automatic transmission shift schedule (gear versus vacuum versus driveshaft rpm), it is possible to determine if the automatic transmission is in the proper gear. If it is not, then the transmission changes into the proper gear and the engine torque and rpm is recalculated with this gear.

In the case of a manual transmission the proper gear is selected by comparing the vehicle velocity to the shift points (velocities at which the transmission is to be shifted). Once the proper gear and the corresponding engine torque and rpm are known, the fuel flow versus engine torque and rpm relation (determined experimentally or by calculation) is matched and the fuel flow needed to satisfy the engine recorded.

All the calculations described above are repeated second by second (the drive schedule is entered as velocity at each second) and added up for the complete drive schedule. The second by second results and/or short summary results can be printed out. A flow chart of the program logic for performance projections is shown in Fig. 16.

This calculation again starts at the engine. The "C" torque available at various engine rpm is tabulated in steps of 50 rpm (starting at idle rpm). At the beginning of the program the torque converter output (transmission side) torque and rpm are calculated using the torque converter performance curves. (In the

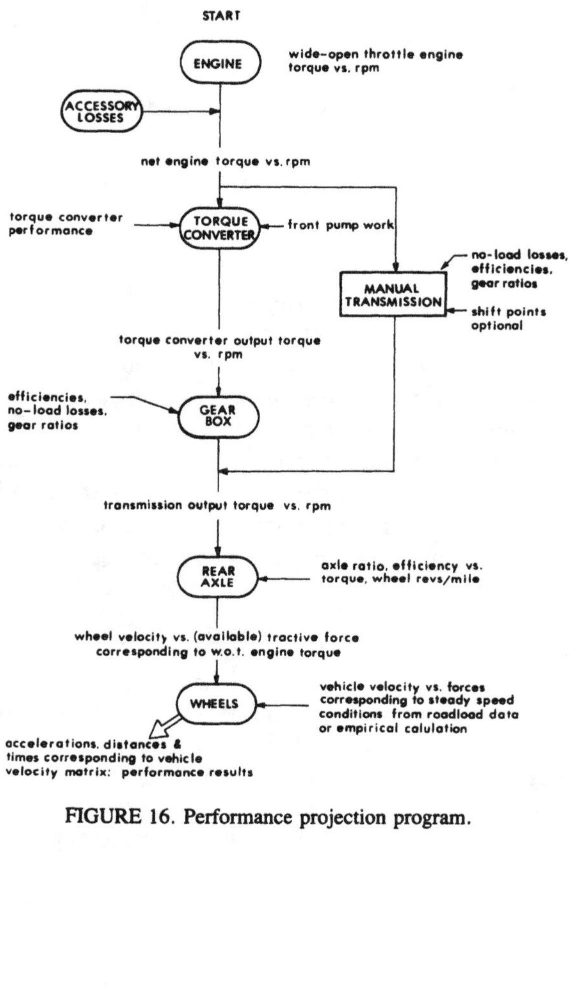

FIGURE 16. Performance projection program.

case of a manual transmission it is assumed that the clutch is capable of transmitting all the torque the engine is able to provide without any slippage.) Next, using the transmission efficiency curves and gear ratios, the driveshaft torque and rpm are calculated from the torque converter (or clutch) output torque and rpm. Using the axle ratio, axle efficiency curves, and wheel radius (calculated from the wheel revolution per mile parameter), the force available at the drive wheel for various vehicle velocities (corresponding to the driveshaft and thus to the engine rpm values used to input the "C" torque curve) are determined from the driveshaft torque and rpm.

Having calculated the force available to the vehicle at various speeds, the program then calculates the force needed at the drive wheels to propel the vehicle at constant velocities. The calculation is carried out by generating a set of vehicle velocity values and finding the corresponding steady-state driveshaft torque and rpm values. These driveshaft torque and rpm values are determined from tests (yielding the drive shaft torque versus vehicle velocity curves) or by carrying out a calculation incorporating the effect of aerodynamic drag and tire rolling resistance.

From the available force (at the drive wheel) versus vehicle velocity and the force (at the drive wheel) needed to propel the vehicle at a steady speed versus vehicle velocity, the net force available to accelerate the vehicle at a given velocity can be determined. From this net force and the equivalent mass of the vehicle, the acceleration versus velocity relation can be calculated.

Once the acceleration versus vehicle velocity relation is shown, the velocity versus time, distance versus time, acceleration versus time, etc., relations can also be calculated. From these relations the various performance parameters, (such as the time to accelerate from 0–60 mph, the distance at 10 sec, etc.) can be directly determined.

A sample computer output statement is shown in Table 3.

TABLE 3. Sample Computer Output Statement

	AMBIENT CONDITION	MPG	MPG
E	DISPL/VENT	200.0/1.	200.0/1.
N	B CURVE NO.	50201-N	50201-N
G	C CURVE NO.		
	IDLE RPM	650.	650.
	FUEL CALC	FL7A-200	FL7A-200
T	TYPE	TR7A-40C	TR7A-40C
R	GEAR RATIO 1	3.29	3.29
A	GEAR RATIO 2	1.84	1.84

(continued overleaf)

TABLE 3. (*Continued*)

N	GEAR RATIO 3	1.00	1.00
S	GEAR RATIO 4	0.81	0.81
	GEAR RATIO 5		
	CONV. CURVE	NONE	NONE
R	CALC. METHOD	/	/
O	AREA/DRAG COEF	20.7/.516	20.7/.516
A	DYNO HP		
D	TIRE SIZE	DR78-14	DR78-14
L	D TIRE PRES/CO	45./1.080	45./1.080
O	T TIRE PRES/CO	26./0.312	26./0.312
A	REVS PER MILE	827.0	827.0
D	TEST WEIGHT	3080.0	3080.0
	T DRIVE WHL WT	1423.00	1423.0
	INERTIA CLASS		
	D DRIVE WHL WT		
	AXLE RATIO	3.00	3.50
	N/V	33.494	39.076
F	CVS-CH (MPG)		
U	CVS-HWY (MPG)		
E	M-H (MPG)		
L	CITY (MPG)		
	SUBURBAN (MPG)		
E	CITY-SUB AVG		
C	ST. SPD. 30 MPH		
O	40 MPH		
N	50 MPH	27.72	26.69
	60 MPH		
	70 MPH	22.16	20.94
P	0-60 MPH (SEC)	12.90	12.85
E	25-60 MPH (SEC)	9.74	9.87
R	50-80 MPH (SEC)	16.88	16.01
F	0-10 SEC DT (FT)	469.51	480.73
	TED 50 MPH (SEC)	13.51	12.24
SHIFT SCHED		SPECIFED	SPECIFED
VEHICLE VEL.	1-2	32.34	27.72
ENGINE RPM	1-2	4400.00	4400.00
VEHICLE VEL.	2-3	57.83	49.57
ENGINE RPM	2-3	4400.00	4400.00
VEHICLE VEL.	3-4	98.00	91.21
ENGINE RPM	3-4	4052.30	4400.00
VEHICLE VEL.	4-5		
ENGINE RPM	4-5		
MAXIMUM SPEED (MPH)		99.00	97.00
DRIVESHAFT RPM		4093.65	4679.44
ENGINE RPM		3315.86	3790.35
CALCULATION CODE		SAMPLE 1	SAMPLE 2

TABLE 4. Performance and Economy Assumptions

Vehicle weight	1750 lb
Drag coefficient	0.38
Frontal area	15.4 ft^2
Tires	1032 rev/mile
Engine performance	Scaled from 2.0 liter
Engine fuel map	Scaled from 2.0 liter
Five-speed transmission ratios	3.58/2.05/1.23/.81/.69
Four-speed auto. trans. ratios	2.846/2.282/1.50/1.00
Accessory losses	Non-A/C–20 A alternator

V. POWER-TRAIN MATCHING TECHNIQUES

A. Power-Train Optimization Example

An example of matching power-train components to optimize fuel economy is shown for a small hypothetical high fuel economy car, carrying the basic assumptions shown in Table 4. The object is to determine engine displacement and final drive ratio (N/V = engine speed, rpm/vehicle speed, mph) combination providing best fuel economy but meeting all of the constraints below. Engine performance, scaled from a 2.0-liter engine, is shown in Fig. 17 and fuel flow

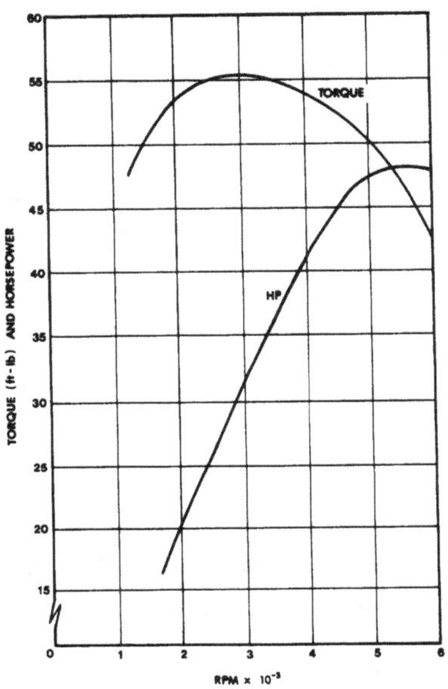

FIGURE 17. Torque and horsepower.

TABLE 5. Fuel Flow Data (1980 2.0-Liter Four-Cylinder Engine)

Speed	Torque (ft lb)										
rpm	0	10	20	30	40	50	60	70	80	90	100
500	1.3[a]	1.7	2.1	2.5	2.9	3.3	3.7	4.1	4.6	—	—
1000	1.65	2.15	2.6	3.1	3.6	4.2	4.8	5.5	6.4	7.4	—
1500	2.4	2.9	3.6	4.3	5.2	6.2	7.2	8.3	9.6	11.1	—
2000	3.0	3.9	4.8	5.8	7.0	8.2	9.5	11.0	12.7	14.5	16.9
2500	4.0	5.1	6.2	7.5	8.9	10.4	12.2	14.0	16.0	18.4	21.2
3000	5.4	6.6	8.0	9.4	11.5	13.6	15.7	17.4	20.0	22.9	26.9
3500	6.1	7.6	9.3	11.2	13.2	15.4	17.7	20.3	23.2	26.4	30.6

[a] Fuel flows listed in lb/h.

flow on Table 5. Fuel flow and torque are assumed to scale directly to engine displacement.

B. Power-Train Matching Constraints

Minimum Engine Speed after an Upshift. Selected to be 1500 rpm to minimize driveline NVH problems, especially with a three-cylinder engine.

30% First Gear Gradeability. Product acceptance standard.

3% High Gear Gradeability. Examines 3% high gear gradeability at both 55 and 60 mph. This constraint is applied to minimize "shift busyness" in freeway-type driving.

Power-Train Acceptance Chart (Fig. 18). This chart defines minimum acceptable final drive N/V as a function of the ratio of displacement/ITWC. This chart was developed by means of a regression analysis of 1979 import and domestic U.S. vehicles. All vehicles which fall to the upper right of the curve are considered to be acceptable and all those to the lower left of the curve are unacceptable.

The N/V scales are expressed in terms of five-speed versus four-speed manual rather than overdrive final drive versus direct drive, since overdrive tends to lose its meaning on front-wheel drive cars. Additionally, a somewhat lower N/V in top gear is more acceptable with a five-speed manual transmission than it is with a four-speed manual.

Included in this rationale are ratio step and shift point considerations. It also follows that an increased number of gear ratios becomes increasingly important as engine size is reduced to maintain acceptable driveability, especially with three-cylinder engines. It is, therefore, considered essential that the small car offer the five-speed manual as the base transmission.

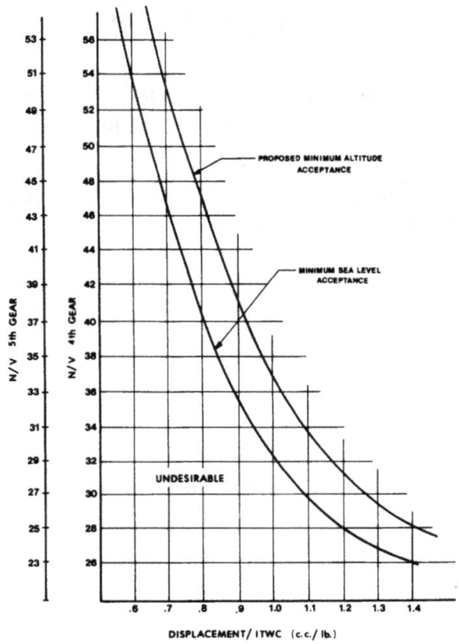

FIGURE 18. Power-train acceptance chart.

C. Power-Train Matching Results

Inputting the small car assumption data in a performance and fuel economy computer program, described in Sec. IV, generates sufficient data to construct the curves shown on Fig. 14. The EPA average fuel economy, metro–highway, is plotted against 0–60 sec vehicle maximum performance. Families of engine displacement from 0.8 to 1.5 liters and N/V ratios from 59 to 30 are superimposed on lines of constraints for driveability and gradeability. The combination meeting the 13-sec performance criteria is best met with a 1.2-liter engine and N/V ratio of 33, providing 61.2 mpg. If, however, the PAC (powertrain acceptance criteria) are met, limiting the N/V ratio to a minimum of 43 for a 1.2-liter engine, a fuel economy of 55.2 results, or a penalty of 6 mpg, defeating the prime objective of the car. Avoiding such an unacceptable penalty must be the goal for the vehicle engineer. Possible courses of action to lower minimum N/V ratios for future small cars would include:

1. Development of the power-train installation mounting system to improve harshness.
2. Modification of the engine torque characteristics to smooth out power impulses at low speeds.

3. Development of a soft spring and damper in the transmission clutch assembly.

Note that with the prevailing PAC, the optimum power-train for this vehicle example would include a 1.5-liter engine which provides a fuel economy of nearly 56 mpg along with a 10-sec 0–60 time.

The above example is useful in identifying restrictions that would seriously impair fuel economy potential.

12

Electronic Engine Control

W. B. RIBBENS

I. INTRODUCTION

Electronics in the automobile has a relatively long history beginning with the installation of AM radio receivers in automobiles. There has been essentially an orderly, continuous evolution from the earlier vacuum tube autoradios to the latest complex automotive entertainment systems employing microprocessors.

On the other hand the utilization of digital electronics for instrumentation or control application has been a major jump in automotive technology. This has resulted partly from the relatively low cost of the microprocessor and partly from its capability to implement multivariable adaptive control.

The control application of microprocessors in the automobile began in the middle 1970s, and this application is continuing to develop rapidly. By 1980 production automobiles had microprocessor-based control of the engine. In addition, there was widespread instrumentation application of microprocessors. It is clear that these applications will expand rapidly in the future, and it is likely that control of the entire power train will evolve. Moreover there is indication that electronic control of the vehicle dynamics in the form of "ride control" will be developed.

Probably the most significant control application of the microprocessor in the automobile is the electronic engine control system. This system provides much more precise control of air–fuel and spark advance than has been achievable with conventional pneumatic-mechanical control. In addition, it has the capability of optimizing engine fuel economy at any operating point subject to the exhaust emission constraints imposed by government regulation. Further, it operates the

W. B. RIBBENS • Electrical and Computer Engineering Department, University of Michigan, Ann Arbor, Michigan 48109.

evaporative emission control apparatus optimally with respect to vehicle performance and will likely control the transmission in the near future.

There is a great variety of powerplants available to the automotive designer at the present time. However, the most prevalent automotive engine at the time of this writing is the gasoline-fueled, spark-ignited internal combustion engine. This engine class is discussed exclusively in this chapter because it has been the greatest beneficiary of electronic controls to date. This engine class will be called the SI engine in this chapter.

A. Emission and Performance Constraints on the Automobile

Electronic control of the automotive engine has been motivated by the requirement to control exhaust emissions and by the requirement to meet government-mandated fuel economy standards. The latter requirement is known by the acronym CAFE (corporate average fuel economy).

Automotive exhaust emission control requirements have existed in the United States since 1966, when the earliest regulations became effective in California. Emission control limits were expanded to all states and became progressively more stringent through the late 1970s. The exhaust gases which are controlled are carbon monoxide (CO), various hydrocarbons (HC), and oxides of nitrogen (NO_x). The emission requirements and CAFE for this period are presented in Table 1.

Attempts to meet the above standards using conventional mechanical or fluidic controls have failed. Moreover, mechanical controls do not have the capability to duplicate functions with sufficient accuracy across a range of production vehicles over all operating conditions and over the life of each vehicle.

Fortunately this problem can be alleviated through incorporation of self-compensating feedback control such as is available in an electronic control system. The technology of digital control systems has advanced to the point at which a multivariable automotive engine controller is feasible. Such a system has great flexibility in that it can be reconfigurated through software changes

TABLE 1. Automotive Exhaust Emission Limits

Year	HC/CO/NO_x Federal	HC/CO/NO_x California	CAFE mpg
1978	1.5/15/2.0	0.41/9.0/1.5	18
1979	1.5/15/2.0	0.41/9.0/1.5	19
1980	0.41/7.0/2.0	0.41/9.0/1.0	20

during the development phase. Also, critical engine parameters can be changed by simply replacing data, which is stored in memory.

In addition, the cost of a multifunction digital controller has been lowered as a consequence of technological developments in the semiconductor industry. Owing to huge investments in the development of large-scale integrated (LSI) circuits and very large scale integrated (VLSI) circuits, a single-chip microcomputer is available at relatively low cost. Such a computer has the capability of providing the various functions required for a complex electronic engine control system. In fact, the cost per function is continuing to drop in absolute terms.

Of course, another significant set of constraints on the design of an automotive engine control system includes manufacturing cost, reliability, and repairability. These topics are addressed in a later section.

Nevertheless, significant progress has been made in the development of commercially feasible electronic engine control systems. It is clear that such systems will be an essential part of the future automotive internal combustion engine.

B. Motivation for Electronic Engine Controls

It is, perhaps, instructive to briefly review the motivation of the electronic control of the automotive engine, which is very well discussed in several excellent papers[1–6] from which the following discussion has been prepared.

Conventionally, the control of an automobile engine is implemented by several straightforward control systems. The timing of the spark is controlled by the distributor which is a function of engine speed, engine manifold pressure (a signal closely related to engine torque), and other operating variables such as throttle position and engine coolant temperature. The air–fuel metering system, usually a carburetor, responds to driver demand through accelerator pedal linkage to the throttle, which controls the manifold vacuum, and to the operating regime, most notably the engine speed and temperature. In the years since the enactment of emission control laws, another control element has appeared on the engine—exhaust gas recirculation or EGR. This system, which was developed to reduce the formation of oxides of nitrogen, takes exhaust gas from the engine output and mixes a small fraction of it with the fresh air and fuel at the engine intake. The amount of exhaust gas introduced is controlled in a variety of ways, but it typically is required to zero under conditions of engine idle or wide-open throttle (WOT) operation.

Finally in a discussion of drive-train control we must include the automatic transmissions, which is in use on most American cars today. The gear ratio selected is a function of vehicle speed and engine operating conditions, usually engine manifold vacuum, but also accelerator position. The transmission behavior affects the speed–torque characteristics of the load into which the engine must deliver power.

The scheduling of these control variables—spark timing, EGR, air–fuel ratio, and transmission gear ratio as a function of engine operating conditions—is called the engine calibration. In conventional automobiles each of these systems is operated as essentially independent control systems. As will be shown later, however, the control variables are not independent, and current calibrations represent considerable compromise in system design.

The traditional design of the automotive engine had three primary objectives: adequate torque for acceptable vehicle performance, good fuel economy potential as measured by brake thermal efficiency, and minimal knock. These objectives dictated partly the selection of certain parameters including displacement, number of cylinders, combustion chamber shape, connection rod and crankshaft geometry, valve area lift, geometry and timing, etc. The advent of emission control laws have placed constraints on some of these objectives and parameters.

As part of the engine design, several control variables are specified, which vary dynamically as the engine operates. In the present day engine these control variables include air–fuel ratio, ignition timing, and EGR. In addition, there is a trend toward including transmission control parameters as part of an overall drive-train control philosophy.

With the introduction of emission legislation, the design problem became considerably more conplex. In addition to the criteria mentioned above, the engineer was faced with additional constraints on emission of hydrocarbons (HC), carbon monoxide (CO), and oxides of nitrogen (NO_x). To understand the nature of these constraints, it is instructive to review the way in which they are measured.

The U.S. federal government publishes a testing procedure (Federal Register, 1972) which includes the following steps. The automobile is placed on a chassis dynamometer which has been adjusted to the proper loading and inertia requirements. The drive wheels of the vehicle rest on dynamometer rollers. One roller is connected to the loading device; the other roller is instrumented to record the vehicle speed. The vehicle is operated in accordance with a prescribed schedule of speed and load simulating an average trip. The prescribed cycle takes 1372 sec followed by a repeat of the first 505 sec to establish the effects of cold engine operation. During the operation of the vehicle in the test, a continuous sample of the exhaust components is collected, and at the end of the test, the absolute mass of each pollutant is determined. Although the legal requirements are stated in grams per mile (i.e., see Table 1) since the mileage of the test is fixed and determined by the velocity–time profile, it is convenient to think of the constraint as one of the total mass of the pollutant emitted. The driving schedule contains a representative mix of idle, acceleration, and cruise appropriate to urban driving.

Modification of the conventional mechanical-pneumatic hydraulic control methods has not been successful in meeting these emission regulations while obtaining acceptable performance. In part this results from the lack of flexibility or versatility of such control means. On the other hand electronic control theory and technology has been advanced to the point that electronic control of these engine control variables has successfully met the regulations with acceptable performance. This ability of modern electronics has motivated the electronic control of the internal combustion engine.

For an understanding of the theory of electronic engine controls, it is instructive to review the influence of engine control parameters on exhaust emissions. Of the three pollutants, CO is most sensitive to air–fuel ratios. It is formed whenever there is insufficient air in the burning mixture for complete combustion. For average gasoline approximately 14.7 g of air are required to burn 1 g of gasoline. This air–fuel ratio, 14.7:1 is referred to as the stoichiometric ratio. The emission of CO increases rapidly as the mixture is enriched (air–fuel ratio less than 14.7) and falls rapidly to low values for air–fuel ratios slightly above 14.7.

Hydrocarbons are present in the exhaust because not all of the gasoline is burned. To a great extent, this gasoline is very near the wall of the cylinder, so that the flame is extinguished by the (relatively) cold wall before the gasoline can burn. Calculations show that a rather large amount of HC is contained in this quench layer in the cylinder, but that most of it is oxidized either in the cylinder or in the hot exhaust after it leaves the cylinder. HC is the most sensitive to spark timing, i.e., the position of the piston at the time at which the spark is fired, since spark timing controls the basic thermodynamic efficiency of the engine. There is one particular spark timing for which the maximum amount of work per cycle is extracted from the cylinder (all other parameters being constant). For any other timing the amount of work decreases, and energy balance requires that the remaining energy appear as rejected heat. The more heat that is rejected to the exhaust, the more likely we are to have the combustion of those unburned hydrocarbons, and the lower the value of this particular pollutant. Unfortunately this procedure also lowers the fuel economy, since the available work per unit of gasoline expended has decreased.

Figures 1, 2, and 3 give basic sensitivity trends of each of the emission components with the control variables as well as the sensitivity of fuel consumption with the control variables. For ease in comparing values of these variables at different engine operating conditions, the production and consumption rates are normalized by dividing the value by the brake power of the engine at that condition. Variables so normalized are referred to as brake specific values, e.g., BSHC, $BSNO_x$, BSCO, and BSFC.

The effect of each control variable also depends on the other control variables. For example, the addition of EGR will cause an increase in fuel consumption at constant spark advance. This is because EGR tends to slow the

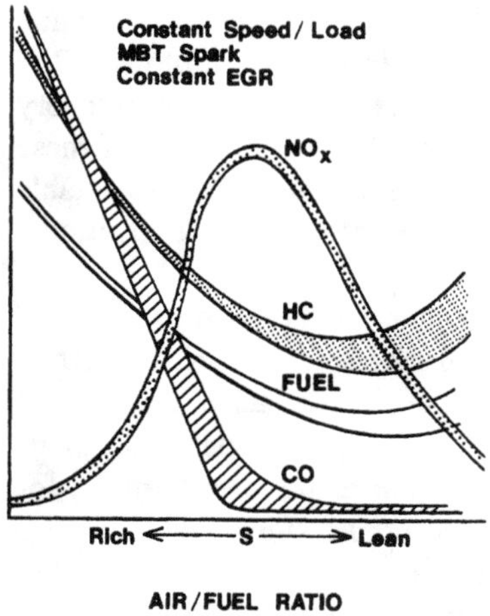

FIGURE 1. Typical emission and fuel consumption characteristics versus air–fuel ratio for the IC engine fully integrated control system. (Reprinted with permission from the Society of Automotive Engineers.)

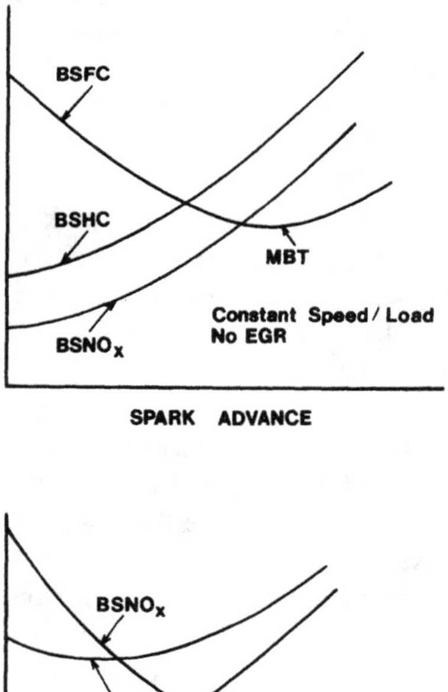

FIGURE 2. Brake-specific emissions and fuel consumption versus spark advance. (Reprinted with permission from the Society of Automotive Engineers.)

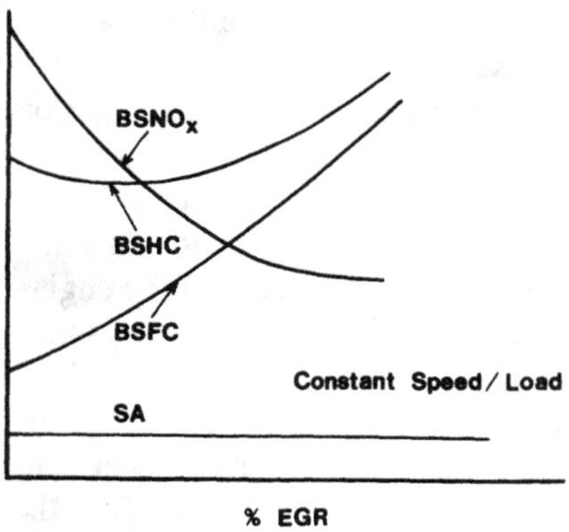

FIGURE 3. Brake-specific emissions and fuel consumption versus EGR at constant spark advance (SA). (Reprinted with permission from the Society of Automotive Engineers.)

burning rate in the cylinder, thereby introducing an effective spark retard. If, however, the spark advance is increased as EGR is added in a consistent manner, the NO_x will reduce as before, but the fuel consumption will remain nearly constant. In Fig. 4 we see the effect of EGR when spark advance is simultaneously adjusted to its optimum value (for fuel consumption). This interaction between the control variables is one of the most compelling reasons for developing a fully integrated control system.

In addition to describing the effect of the control variables on emissions, it is important to know the effects of operating conditions on emission rates. As we have already indicated NO_x is formed under conditions of high power demand. Therefore, NO_x production is greatest during accelerations, followed by cruise conditions. Carbon monoxide generation is essentially associated with fuel-metering conditions. If the metering system is adjusted sufficiently lean, steady-state values of CO will be small. In the case of the carburetor, however, because most carburetors become excessively lean during acceleration (thereby promoting misfiring), conventional practice is to use an accelerator pump which injects extra gasoline to enrich the mixture during this transient. This rich mixture condition, and the rich mixtures which are necessary during cold start, are the principal modes which generate CO. Hydrocarbon formation occurs, as we have noted, because of incomplete combustion. Because oxidation in the exhaust system is so important for the reduction of hydrocarbons, HC emission is greatest under those conditions for which there is insufficient heat or oxygen in the exhaust system. These conditions are present at idle, deceleration, and during the first few minutes after a cold start.

Having described the fundamental relationship between the control variables and the exhaust pollutants, it is possible to consider some of the control strategies and control laws which can be employed using modern control theory to meet

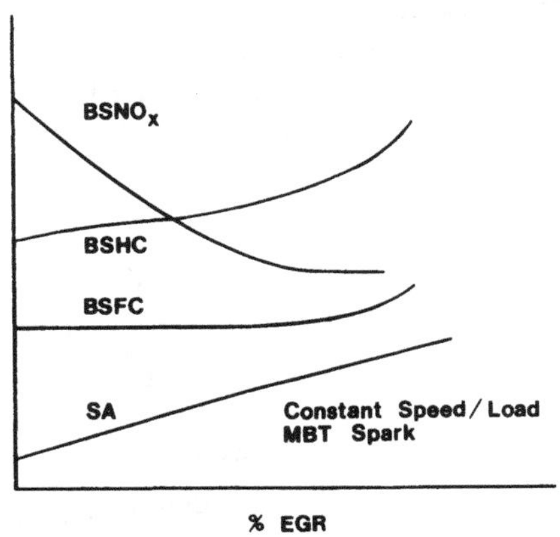

FIGURE 4. Brake-specific emissions and fuel consumption versus EGR with spark advance maintained optimum (MBT). (Reprinted with permission from the Society of Automotive Engineers.)

emission limits while satisfying performance requirements. However, it is instructive to briefly review conventional proportional control theory first.

C. Conventional Control Theory

The classical control system operated in a so-called feedback mode. Although such a system is often multidimensional (i.e., controlling several variables), it is instructive to consider a single dimensional system such as is depicted in Fig. 5.

In this figure the system output variable which is to be controlled is denoted $\xi(t)$. The control signal $r(t)$ determines the desired value of this output. A sensor s generates an output signal $v\xi(t)$ which is proportional to $\xi(t)$:

$$V_\xi(t) = k_s \xi(t)$$

An error voltage $V_e(t)$ is then generated which is proportional to the difference between the true output and the desired value:

$$V_e(t) = r(t) - V_\xi(t)$$

If the system were to operate perfectly (i.e., without error), then $V_e(t)$ would be zero at times (e.g., steady-state operation). This, of course, is impossible and a nonzero error signal exists at certain times. This error signal generates an output $V_a(t)$ from the electronic signal processing (SP). This signal operates the control actuator A which produces control input $u(t)$. This latter input changes the system in such a way as to make the error zero. In a later section a specific example of an electronic system which controls air–fuel for an IC engine is given which, perhaps, clarifies the operation of such a controller.

In modern theory there is a somewhat different approach than the simple feedback system. Many of the features of the classical control system such as the actuator and sensor are present in a modern control system. Moreover the modern control system is often implemented with a number of control signals and a variety of outputs as well.

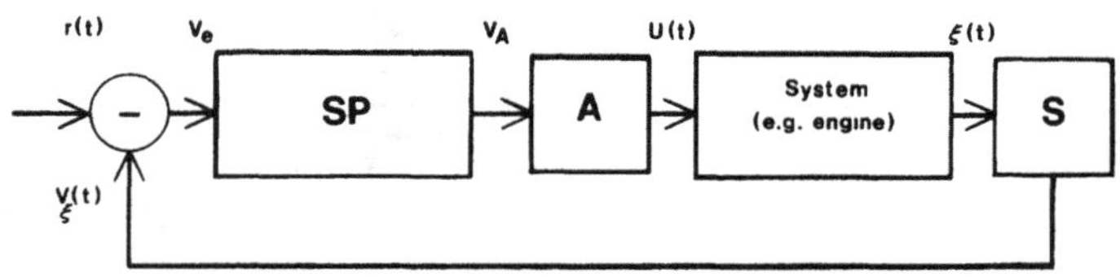

FIGURE 5. Feedback control system.

For modern control systems it is presumed that the electronic signal processing is implemented with a computer (or microcomputer). In this case a much more complex relationship than the algebraic differences between the control signal (or signals) and the system output (outputs) can be implemented. The basis of this relationship is called the control strategy. There are many control strategies which have potential in automotive engine control systems. It is instructive to review these strategies and to introduce modern control theory as it is applied to automotive electronic engine control.

II. SURVEY OF CONTROL STRATEGIES

Electronic control of the automotive engine exhaust emissions is possible because air–fuel, spark advance, and EGR strongly affect emissions and fuel economy. This fact, coupled with the ability of digital electronics to control these variables, has led to the successful implementation of electronic engine control systems, which have enabled the relatively large displacement engines to meet EPA emission limits.

One of the important issues in any digital electronic control system is the control strategy which is employed. It is instructive to survey this issue as applied to automotive control. The control strategy for the automotive engine is a set of calibration values for the control variables as a function of operating conditions. There is a large set of control strategies which are applicable to the control of the SI engine. Each has the potential of improving some engine performance measure.

It is beyond the scope of this chapter to discuss all control strategies which have been considered. Rather, we will discuss one application of optimal control theory in theoretical terms for the purpose of introducing a control strategy which optimizes fuel economy. In addition, we will discuss a control strategy and concept which has been implemented in the majority of current production cars. We begin with a theoretical discussion of the optimal control concept.

In the present discussion it is implicitly assumed that there is a digital (i.e., computer based) electronic control system which can determine the calibration variables (quasicontinuously). These variables are assigned values by an algorithm which optimizes engine performance with respect to some established set of criteria. This process of determining optimal engine calibration is a member of the class-constrained optimization problems. For an understanding of the implementation of an optimum electronic engine control, it is helpful to briefly review the theory which is applicable to this constrained optimization problem.

Theory of the Constrained Optimization Problem. As an example of the application of optimal control theory to the automotive engine, we consider the following constrained optimization problem. The goal of this constrained optimization problem is to maximize fuel economy, subject to exhaust emission constraints.[7,8] The automotive control system can be defined with respect to three vectors:*

$$\mathbf{y}' = (y_1, y_2, y_3, y_4)$$

$$\mathbf{u}' = (u_1, u_2, u_3, u_4, u_5)$$

$$\mathbf{x}' = (x_1, x_2, x_3)$$

The vector can be thought of as the system output and has components

$$y_1(\mathbf{x}(t), \mathbf{u}(t)) = \text{fuel flow rate}$$

$$y_2(\mathbf{x}(t), \mathbf{u}(t)) = \text{emission rate for HC}$$

$$y_3(\mathbf{x}(t), \mathbf{u}(t)) = \text{emission rate for CO}$$

$$y_4(\mathbf{x}(t), \mathbf{u}(y)) = \text{emission rate for NO}_x$$

The vector $\mathbf{x}(t)$ can be thought of as describing the state or operating conditions of the engine and has components

$$x_1 = \text{manifold pressure}$$

$$x_2 = \text{engine angular speed}$$

$$x_3 = \text{vehicle velocity}$$

Similarly $\mathbf{u}$(t) represents the control variables for the electronic control system and includes components

$$u_1 = \text{air–fuel ratio}$$

$$u_2 = \text{spark timing (e.g., relative to top dead center)}$$

$$u_3 = \text{EGR}$$

$$u_4 = \text{throttle position}$$

$$u_5 = \text{gear ratio for drive train}$$

In any particular control configuration there might well be additional variables introduced to any of these vector functions. However, for the purpose of the

* In these expressions the prime indicates the transpose of the vector.

present discussion, it is assumed that these are an adequate representation of the control system.

For the present constrained optimization problem, the index of performance F is the total fuel consumed during an EPA test cycle.

$$F = \int_0^T y_1(\mathbf{x}(t), u(\mathbf{t}))\, dt$$

where

$$x_3(t) = v_E(t)$$

$$= \text{EPA prescribed vehicle velocity for emission test cycle}$$

$$T = \text{period of EPA test}$$

Fuel economy is maximized if F is minimized subject to exhaust emission constraints. The exhaust emission constraints can be written

$$\int_0^T y_2(\mathbf{x}(t), \mathbf{u}(t))\, dt < G_2 = \text{allowed HC limit}$$

$$\int_0^T y_3(\mathbf{x}(t), \mathbf{u}(t))\, dt < G_3 = \text{allowed CO limit}$$

$$\int_0^T y_4(\mathbf{x}(t), \mathbf{u}(t))\, dt < G_4 = \text{allowed } NO_x \text{ limit}$$

The engine–vehicle dynamics are modeled by a set of coupled differential equations:

$$\dot{x} = \mathbf{f}(\mathbf{x}, \mathbf{u})$$

Additional constraints on this problem are imposed by engine performance requirements, driveability, and practical physical limits. These constraints can be written

$$\mathbf{x}^l < \mathbf{x}(t) < \mathbf{x}^{\mu}$$

$$\mathbf{u}^l < \mathbf{u}(t) < \mathbf{u}^l$$

$$\mathbf{y}(t) > 0$$

$$x_3(t) = V_E(t)$$

In any practical application of the constrained optimization problem, the functions $y_K(\mathbf{x}(t), \mathbf{u}(t))$ are experimentally determined. Moreover, the functions $\mathbf{f}(\mathbf{x}, \mathbf{u})$ are found approximately either from simple linear theory or from experimental measurements.

The optimal control problem, then, is to find the optimal control u^* which minimizes F subject to the emission constraints. Minimize $F(\mathbf{x}, \mathbf{u})$ subject to

$$\int_0^T V_K(x(t), \mathbf{u}(t))\, dt < G_K,\ K = 2, 3, 4$$
$$\mathbf{x}^l < \mathbf{x}(t) < \mathbf{x}^u$$
$$\mathbf{u}^l < \mathbf{u}(t) < \mathbf{x}^u$$

This formulation of the engine control problem is based upon the following assumptions:

1. The engine is assumed to be warmed up prior to the control period.
2. Driveability considerations are included by having fixed upper and lower bounds ($\mathbf{u}^l$ and $\mathbf{u}^u$) for the control vector.

It is shown in Ref. 9 that the constrained optimization problem can be formulated in terms of a set of unconstrained problems. However, a discussion of this control strategy is beyond the scope of this chapter. It is instructive to review the theory of a somewhat simpler (possibly suboptimal) electronic control system for the automotive engine. This particular control strategy is applied to an engine having a three-way catalytic converter and is employed in current production engine control systems.

The optimal control concept applied to the automotive engine is, in essence, a goal for electronic engine control system. Several experimental optimal control systems have been implemented according to this concept. Some improvements in fuel economy have been achieved. However, at the time of this writing no engine control system based upon this concept is in production. In the following section we discuss a control concept which has been implemented in production.

III. Control Concept Based upon Three-Way Catalytic Converters

One control concept currently employed places maximum emphasis upon emission control. This system is applicable to an engine–three-way catalytic (TWC) converter combination. The TWC converter changes the exhaust gas composition in such a way as to reduce the constraints on the engine-out emissions. It is, after all, the emission from the vehicle tail pipe which must be compliant with EPA regulations. The TWC has the benefit of easing the constraints on the engine exhaust such that a higher level of engine performance is possible than would be permitted if EPA limits were applied directly to these engine-out emissions. It has been shown above that the conversion efficiencies of the TWC for the three principal exhaust pollutants are dependent upon air–fuel. It is not possible to simultaneously maximize conversion efficiency for all three constituents. However, an acceptable compromise exists for air–fuel very near to the stoichiometric values (i.e., 14.7). Figure 6 illustrates the conversion efficiency of the TWC as a function of air–fuel. Note that conversion efficiencies are near stoichiometry.

This control strategy employs a feedback control system which holds the average air–fuel ratio within ±0.05 air–fuel ratios of the stoichiometric point. The other important engine control variables (e.g., ignition timing and EGR) are adjusted to meet other engine performance criteria and to assist in the control of emissions.

Catalytic control of HC and CO requires an oxidizing atmosphere whereas control of NO_x requires a reducing atmosphere. The TWC used in implementing this strategy is either a single-bed or a dual-bed converter. The dual-bed converter is more complex than the single-bed converter because it requires an extensive air management system.

Typically the first stage in a dual-bed converter provides a reducing atmosphere for reduction of NO_x. The second stage provides an oxidizing atmosphere by pumping air to this stage for control of HC and CO emissions.

The trend in automotive electronic engine control systems has been toward an integrated digitally implemented controller. In practice, a digitally implemented (i.e., microcomputer based) electronic control system, which is based upon controlling air–fuel at stoichiometry for use with a TWC, has additional functions including secondary air management control, electronic spark timing, idle-speed control, controlled canister purse, torque converter clutch control, and system self-diagnosis. These functions are discussed separately as they tend to be largely independent operations of the electronic control.

For an understanding of an electronic engine control concept which is based

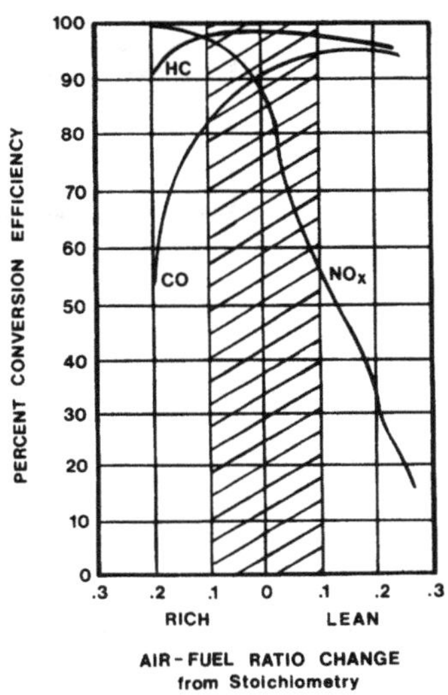

FIGURE 6. Three-way catalyst performance characteristics.

upon a TWC, it is helpful to refer to Fig. 7 which is a block diagram of a typical digital control system (e.g., GM C-4 or Ford EEC-III).

The primary purpose of this control system is to maintain air–fuel at stoichiometry. It does this in two primary modes: open loop and closed loop. The system can operate closed loop only after certain conditions are satisfied. The open-loop mode is employed whenever these conditions are not satisfied. For each mode the emissions are presumed to satisfy regulatory limits if the air–fuel is held at this correct value.

As will be seen presently, in the normal part-throttle closed-loop mode of operation the fuel control system configuration is essentially a conventional feedback controller. It employs an actuator to control engine input air–fuel and has a sensor for measuring engine output (i.e., exhaust) air–fuel. Before proceeding with a discussion of this system, it is important to understand the concept of exhaust air–fuel.

The exhaust air–fuel is defined in terms of the molar fractions of carbon, hydrogen, and oxygen. The proportions of these constituents in the exhaust can be represented in terms of corresponding input air and fuel which produced these proportions and can be expressed in air–fuel ratio.

In addition, the control system has an inhibit mode which is used during

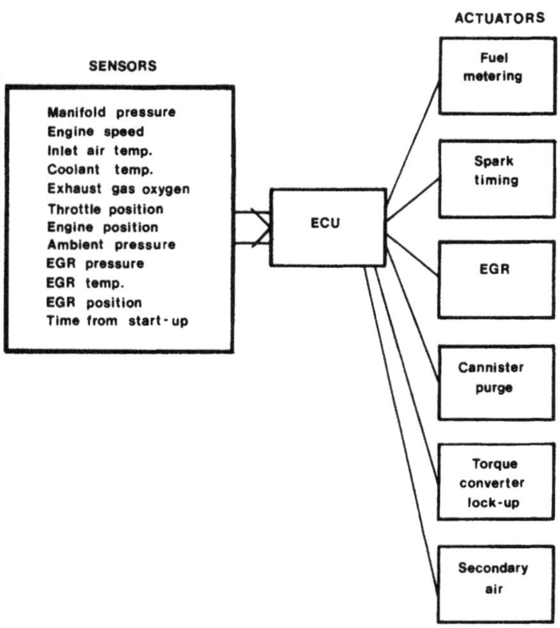

FIGURE 7. Block diagram of sensors and actuators linked through the control unit. (Reprinted with permission from the Society of Automotive Engineers.)

engine crank. This provides a rich mixture during start-up period. Still another mode of operation which occurs under very heavy load is a special enrichment mode of operation.

An important secondary function of this system is to control ignition timing. This control variable can be adjusted to optimize engine performance with respect to indicate torque or fuel economy. The secondary air control functions support the exhaust emission control subsystems.

A. Closed-Loop Operation

The normal operation of this control system from engine start begins with open loop. Then, after a set of operating conditions are satisfied (discussed later), the system is converted to closed-loop operation. It is, perhaps, simpler to begin this discussion with the closed-loop operation than with the normal sequence of events from a cold start.

Closed-loop fuel control is based upon two critically significant components: the fuel-metering actuator and the exhaust gas sensor. There are two classes of fuel-metering actuators in use at the present time. One of these is the so-called electronic carburetor which is essentially a conventional carburetor with an electrically actuated needle valve or equivalent control. In one implementation of the electronic carburetor the needle valve is stepped between a pair of limits by means of a solenoid. One limit corresponds to air–fuel which is richer than stoichiometry and the other is leaner than stoichiometry. The average air–fuel is kept near stoichiometry by varying the duty cycle of the binary-valued position of this needle valve.

The other fuel actuator is called throttle body fuel injection (TBFI). This actuator consists of one or two fuel injectors which supply fuel as determined by the control system. Normally these fuel injectors are operated in pulse mode. The air–fuel is determined by the pulse duration or equivalently the pulse duty cycle.

The other critical system component for closed loop electronic fuel is the exhaust gas oxygen (EGO) sensor which is essentially an air–fuel sensor. There are two categories of EGO sensors. One of these is a galvanic cell sensor which provides a step change in output voltage as the air–fuel varies from slightly richer to slightly leaner than stoichiometry (e.g., ZrO_2 sensor). The sensor actually responds to the oxygen partial pressure in the exhaust gas, which in turn is a function of input air–fuel. For this reason the sensor operation can be defined in terms of this air–fuel.

The output from a typical ZrO_2 sensor as a function of air–fuel into the engine is depicted in Fig. 8. Note the abruptness of the change in sensor output at an input air–fuel ratio corresponding to stoichiometric air–fuel ratio. This behavior is nonideal for the fuel control system. However, the commercially

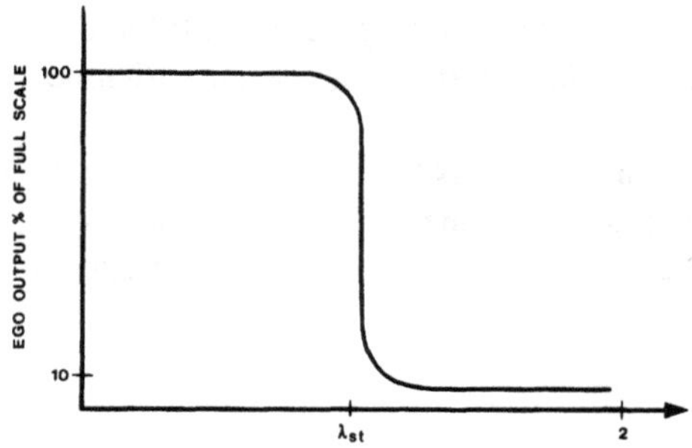

FIGURE 8. EGO output characteristics.

available ZrO_2 sensor is relatively inexpensive and permits limit cycle operation for closed-loop fuel control at the present time.

The other category of exhaust gas oxygen sensors is a variable resistance device (e.g., TiO_2). This sensor, which is not as widely used as the ZrO_2 sensor, has a similar step change in resistance as a function of air–fuel near stoichiometry.

The fuel control in closed-loop mode is a limit cycle feedback control system in which the air–fuel oscillates through stoichiometry. The oscillation frequency is usually in the range of 0.3 to about 10 Hz. The amplitude of the oscillation is normally less than 2 air–fuel units peak-to-peak.

The implementation of closed-loop control of air–fuel varies from one engine to another in detail. However, fundamental aspects of limit cycle closed-loop control are common to essentially all systems employing a TWC. The significant features of such a control system can be presented, therefore, by a representative example. Two excellent papers discussing such a system are given in Refs. 10 and 11. The following discussion of closed-loop fuel control summarizes the concept of this control methodology as described in these references. Most of the practical details of production control systems have been omitted for simplicity of explanation.

B. Practical Closed-Loop Control System

A block diagram of the conceptually simplest closed-loop control system is depicted in Fig. 9. In this system it is convenient to represent air–fuel by the equivalence ratio

$$\lambda = \frac{\text{air–fuel}}{\text{air–fuel at stoichiometry}}$$

The fuel-metering actuator regulates the fuel flow into the engine in such a way as to determine an input equivalence (i.e., λ_i). The engine exhaust gas concen-

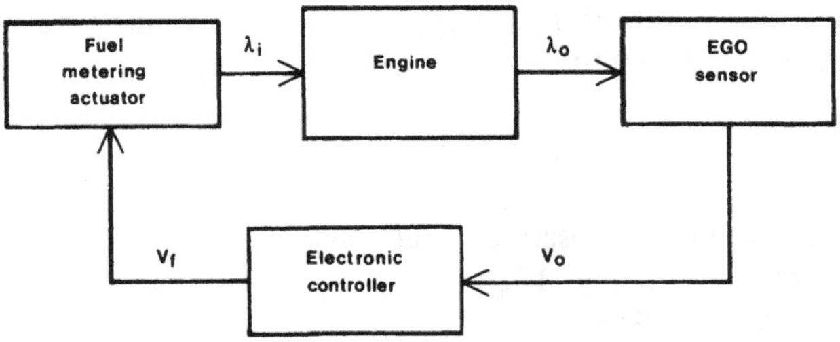

FIGURE 9. Simplified fuel control system.

tration can be thought of as defining an output equivalence ratio (i.e., λ_0) as described earlier.

The EGO sensor generates an electrical signal (V_0) which is a function of (λ_0). The controller generates an electrical output (i.e., V_f) which is a function of V_0. This signal operates the fuel-metering actuator (e.g., electronic carburetor or throttle body fuel injector) controlling fuel flow and, therefore, also λ_i. The input air–fuel is continuously adjusted by the controller in such a way as to keep the average as close to unity as possible. As we will see, the instantaneous air–fuel oscillates about the desired set point (i.e., $\lambda_i = 1$) in the limit cycle operation.

C. Closed-Loop Model

For an understanding of the operation of this fuel control system it is helpful to discuss a simplification approximate analog model to the system. In practice the fuel control system in a current production automobile is digitally implemented. For the purpose of this simplified explanation, the controller is presumed to be an integrator which is integrating the sensor output with respect to time:

$$V_f(t) = \frac{K}{\tau}\int_0^t V_0\,(t')\,dt'$$

Furthermore, the engine is presumed, reasonably, to perform a simple time delay with respect to equivalence ratio:

$$\lambda_0(t) = \lambda_i\,(t - t_D)$$

where t_D = transport delay input EGO sensor.

The fuel-metering actuator is presumed to generate input equivalence ratio which is proportional to V_f:

$$\lambda_i(t) = K_f V_f(t)$$

The EGO sensor is represented approximately as a binary-valued function of λ_0 relative to stoichiometry:

$$V_0(t) = V_0 \operatorname{sgn}(\lambda_0(t) - 1)$$

The operation of the closed-loop fuel controller can be understood qualitatively with reference to Fig. 10, which is a sketch of the various waveforms under steady-state conditions.

It is perhaps convenient to begin by considering the output equivalence ratio as it has just crossed unity (i.e., stoichiometry). At this instant the EGO

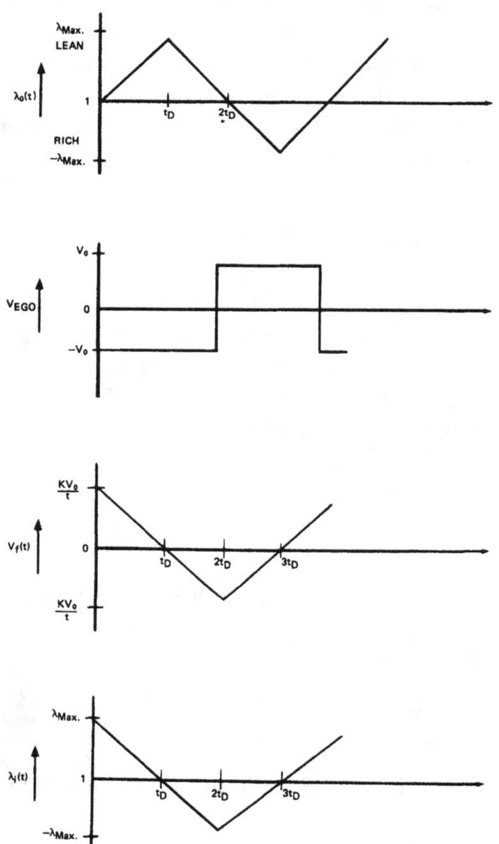

FIGURE 10. Waveforms for fuel control system.

During the time interval $(0, t_D)$, $\lambda_0\,(t)$ increases linearly with time, reaching its maximum value at t_D (i.e., the transport delay time of the engine). Then, $\lambda_0\,(t)$ begins to decrease linearly with time, crossing $\lambda_0 = 1$ at $t = 2t_D$. At this point the EGO sensor output switches polarity and the integrator output begins increasing linearly with time.

Following the transport delay this minimum occurs in the output equivalence ratio and $\lambda_0\,(t)$ increases again crossing $\lambda_0 = 1$ at $4t_D$. This cycle continues at a fixed frequency and with a peak deviation in equivalence ratio which is determined by the system parameters. It can be shown that the limit cycle frequency f_L is given by

$$f_L = \frac{1}{4t_D}, \qquad \lambda_{\max} = K_f \frac{KV_0}{\tau}\, t_D$$

The transport delay for the engine is a function of engine speed because the flow of gases through the engine depends upon engine speed. Although the exact numerical values for f_L and $\lambda_{\max}$ depend strongly upon the system parameters, the engine speed dependence for these quantities is qualitatively depicted in Fig. 11.

The preceding analysis is only an indication of the state of the art in an SI engine fuel control system. A practical system involves modifications which improve performance such as a combination of limit cycle operation and proportional control. A discussion of such systems is given in an excellent paper by Hamburg and Shulman.[10] Moreover, a typical production fuel control system is digitally implemented generating a binary-valued output which operates a "pulsed" fuel-metering actuator (e.g., throttle body fuel injection system).

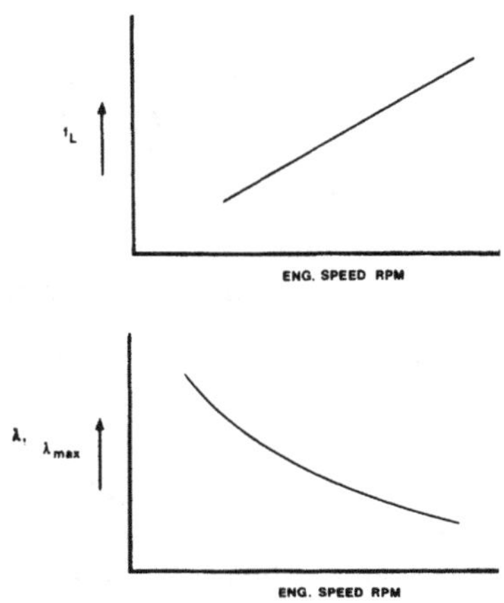

FIGURE 11. Frequency and maximum excursion of limit cycle fuel control system.

We discuss next the operation of a typical SI engine fuel control system in the open-loop mode. This class of operation can be adequately described by considering a specific example, which is taken from Ref. 10, explaining the Ford EEC-III open-loop operation. The specific system which is explained is applicable to a 5-liter V-8 engine although the concept applies to any engine.

D. Open-Loop Operation

The open-loop mode of operation for this particular system utilizes a digital computer operating in conjunction with a throttle body fuel-metering actuator and a set of sensors. This sensor set measures the following variables:

1. Manifold absolute pressure (P_m)
2. Barometric absolute pressure (P_B)
3. Engine coolant temperature (T_c)
4. Manifold charge temperature (T_m)
5. EGR position (P_e)
6. Crankshaft position (Θ_e)
7. Throttle angle position (Θ_t)
8. Crankshaft angular speed (N)

Under open-loop operation the computer processes signals from the sensors to determine the correct fuel flow rate. The fuel flow rate is fixed by the time and duration of the opening of a pair of fuel injectors. These are electromechanical valves consisting of a solenoid-operated pintle and needle valve assembly. Whenever the solenoid is activated, the valve is opened and fuel is delivered under pressure to the airstream. The fuel flow rate is determined by the fuel pressure, the duty cycle of the solenoid driver electrical signal, and the geometrical parameters of the fuel injector.

This system has five separate open-loop operating modes:

1. Cranking (engine start) rpm, $\leqslant 350$
2. Closed throttle, $\Theta_f < 7.6°$
3. Part throttle, $7.6° < \Theta_t < 61.8°$
4. Wide-open throttle, $\Theta_t > 61.8°$
5. Limited operation strategy (LOS) which permits vehicle operation with partial control system failure.

In modes 1 and 5 the system generates a fixed duration injector pulse modified as a function of T_c. In modes 2, 3, and 4 the system computes the correct injector pulse width for the desired fuel flow rate.

Moreover, the timing of each injector is also determined by the computer. In modes 2, 3, and 4 the injectors are alternately operated each 90° of the crankshaft rotation according to the following schedule:

Crankshaft position	720/0°	90°	180°	270°	360°	450°	540°	630°
Injector 1	×		×		×		×	
Injector 2		×		×		×		×

The computer determines injector pulse width τ_I so as to maintain air–fuel as close as possible to stoichiometry (i.e., 14.7). In the open-loop mode stoichiometric condition is maintained by estimating the mass flow rate of air $\dot{m}_a$ into the engine and then setting the mass flow rate of fuel $\dot{m}_f$ such that

$$\dot{m}_a \simeq 14.7\ \dot{m}_f$$

The procedure for estimating $\dot{m}_a$ begins with an estimate of the mass flow rate $\dot{m}_c$ of the combination of air and EGR into the engine. It is shown in Ref. 10 that for the Ford 5.0-liter V-8 engine this quantity is given by the following equation:

$$\dot{m}_c = \frac{0.11346 P_m N}{T_m} n_v$$

where

$$\begin{aligned} n_v &= \text{volumetric efficiency} \\ &= n_v(P_m, N) \end{aligned}$$

The values for P_m, N, and T_m are obtained from the corresponding sensors. The volumetric efficiency is obtained from a table of values which are stored in computer memory as a function of P_m and N. The numbers which are stored in memory are obtained by measuring volumetric efficiency for the particular engine class.

The mass flow rate of air $\dot{m}_a$ is found from m_c by the expression

$$\dot{m}_a = \dot{m}_c - \dot{m}_e$$

where $\dot{m}_e$ = Mass flow rate of EGR.

For the Ford 5.0-liter engine this latter quantity is given by

$$\dot{m}_e = 0.00868P - 0.0000109Pe^2$$

Each of these above computations are, of course, performed by the computer in a straightforward manner.

The desired fuel flow rate for normal operation is determined by this air flow rate and the known air–fuel at stoichiometry. For the Ford 5.0-liter engine the relationship between injector pulse width τ_I and $\dot{m}_f$ is given by

$$\dot{m}_f = A_0(\tau_I - A_1)C_0N$$

where

$$A_0 = \text{injector flow constant}$$

$$A_1 = \text{pulse width offset (for no flow)}$$

$$C_0 = \text{number of injections/revolution}$$

The Ford EEC-III electronic fuel control system has the capability of determining this injector pulse duration at a flow rate of up to 167 times/sec. This is adequate to provide fuel throughout the entire speed range of the engine.

The injector pulse is amplified suitably to have sufficient power to drive the injectors. The amplified injector pulse operates the injector solenoid thereby opening the injector valve. The quantity of fuel delivered is determined by the duration of this value opening and the fuel pressure which supplies the injector. In the Ford system, as in essentially all electronic fuel control systems, this pressure is regulated to a constant value.

The above discussion applies to normal warm-engine open-loop operation. However, air–fuel modification is desirable under certain conditions. For example, for cold-engine operation a richer than stoichiometry air–fuel is desired to compensate for reduced fuel volatility. In the Ford EEC-III fuel control system the air–fuel compensation is accomplished according to a piecewise linear function of T_c, T_e, and time start. In addition, there is a piecewise linear compensation for manifold absolute temperature and engine speed at WOT.

In open-loop modes 1 and 5 the injector pulse width is compensated for only by coolant temperature. The Ford EEC-III system utilizes a simple piecewise linear compensation for T_c.

In a conventional carburetor rapid throttle movement (e.g., step change), the air–fuel increases suddenly. To compensate for this change a single-stroke pump is coupled to the throttle. Sudden increases in throttle angle cause an extra charge of fuel to be delivered by this so-called acceleration pump.

In the Ford electronic fuel control system the equivalent of the acceleration pump is provided by additional injector pulses between the normal fuel pulses which occur each 90° of crankshaft rotation. The number and duration of such pulses depend upon throttle opening rate and initial angle. The Ford EEC-III system utilizes piecewise linear compensation algorithm for acceleration enrichment.

There are many details of this electronic fuel control system in open-loop mode which merit attention but are beyond the scope of this chapter. It is recommended that the interested reader consult Ref. 10.

The sequence of events from engine start to closed-loop operation involves a number of computer decisions based upon various sensor readings. Each individual fuel control system has its own specific set of computer operations. It is beyond the scope of this chapter to describe this operation in detail. However, as a representative example of this sequence, we present Fig. 12. This figure shows the flow diagram for the mode selection algorithm utilized in the GM C-

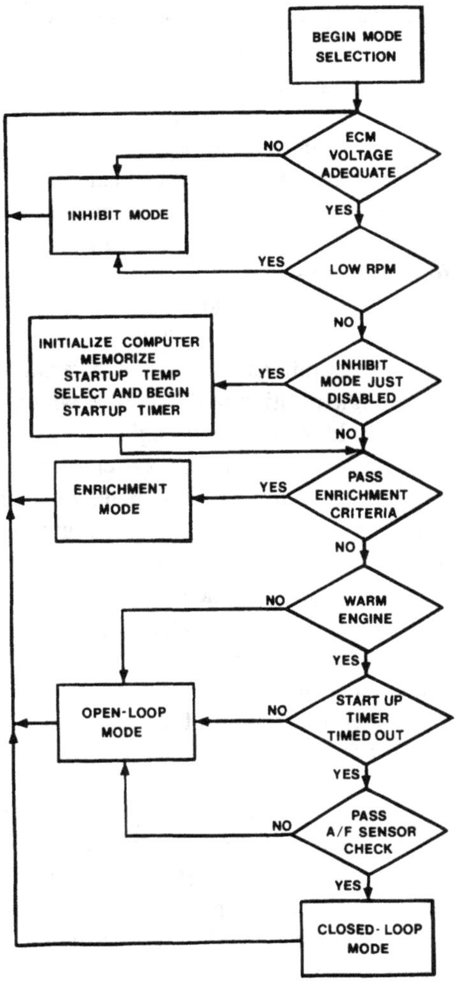

FIGURE 12. Mode selection flow chart. (Reprinted with permission from the Society of Automotive Engineers.)

4 system. For details of this procedure the reader is referred to the excellent paper.[11]

This selection flow chart indicates the requirements for each mode of operation. The inhibit mode is selected if the ECM voltage or the engine speed is too low. In the inhibit mode there is no current supplied to any output device except the engine lamp. After the inhibit mode the start-up enrichment mode is enabled for a predetermined time. The duty cycle is a function of coolant temperature.

After the end of the start-up enrichment period, the system can go to the open- or closed-loop modes. The system can operate a closed loop if the following conditions are satisfied; the exhaust oxygen sensor must have reached satisfactory operation, the start-up period must be completed, and the engine coolant temperature must be at a specified level. If these conditions are not satisfied, the system goes into the open-loop mode.

The above description of closed- and open-loop operation is only representative of current automotive engine control implementation. Each manufacturer tends to utilize a somewhat differently detailed approach. However, this discussion adequately samples present automotive fuel technology.

It has been shown earlier that ignition timing markedly affects engine performance/emissions. There is, generally speaking, an optimum ignition timing for each operating condition. This timing is normally expressed as spark advance (SA) in degrees at crankshaft rotation relative to top dead center (TDC).

The traditional method of determining ignition timing was to change the crank angle position of the point at which the ignition points opened as a function of manifold vacuum and engine speed. The ignition calibration depended upon mechanical and pneumatic system parameters.

The existence of a computer for fuel control offers the possibility of electronic ignition calibration. A configuration for such an ignition control system is shown in Fig. 13. This system, which operates in an open-loop mode, computes the value of SA from measurements of manifold absolute pressure (MAP), engine angular position relative to TDC, engine angular speed, and coolant temperature. The optimum value of spark timing is then selected using data stored in the computer memory.

As an illustration of this computation of SA, we consider this operation in the GM C-4 system.[11] The procedure for determining SA is explained by the flow chart depicted in Fig. 14. There are two modes of operation. First, is "bypass mode." This mode of operation is for cranking the engine and for backup should a computer malfunction occur. In this mode the SA is a fixed predetermined value which is relatively retarded and which will permit the vehicle to be driven at reduced performance levels from computer controlled SA. The bypass mode test is used to assure the conditions are satisfied to allow computer control of spark advance.

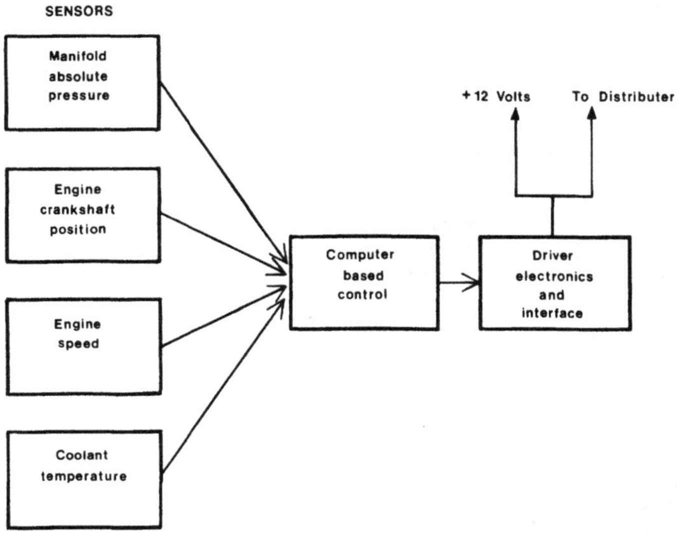

FIGURE 13. Electronic spark advance block diagram. (Reprinted with permission from the Society of Automotive Engineers.)

When the conditions are satisfied, the computer will calculate the proper time in the engine cycle for ignition to occur. Spark advance is the algebraic sum of advances from five tables. These are:

1. *Main Advance*—A table of spark advance based on LOAD and rpm.
2. *Start-up Advance*—A table of spark advance based on change in coolant temperature. This temperature change is the current coolant temperature minus the start-up coolant temperature. This factor is deleted when a calibrated coolant temperature is exceeded.
3. *Supplemental Spark Advance*—A table of spark advance based on LOAD. To enable this advance mode, a set of steady-state conditions (rpm, LOAD, and temperature) must be satisfied for a period of time. If any of the conditions are not satisfied, the mode is disabled without delay.
4. *Coolant Advance*—A table of spark advance based on LOAD and COOLANT TEMPERATURE.
5. *Altitude Advance*—A table of spark advance based on LOAD and BAROMETRIC PRESSURE.

The values for SA for these tables can be obtained empirically for any given engine–vehicle combination to provide desired performance.

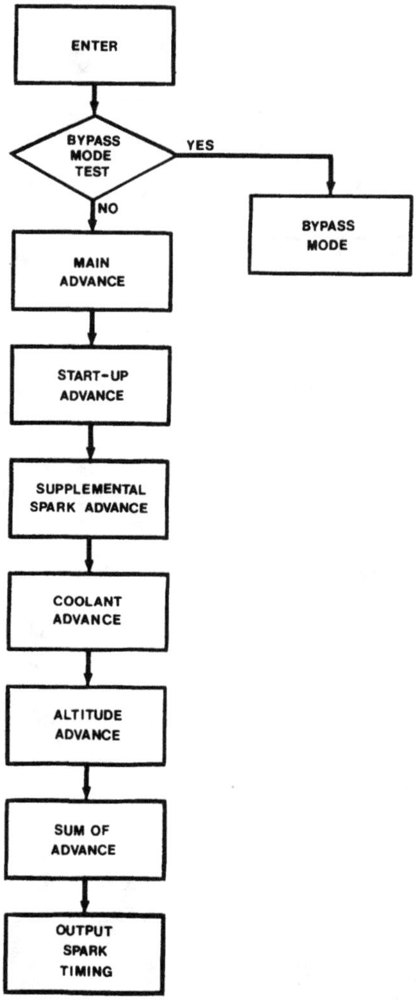

FIGURE 14. Flow chart for spark advance. (Reprinted with premission from the Society of Automotive Engineers.)

E. Closed-Loop Spark Advance

The ignition timing for maximum torque (corresponding to best fuel economy) is a function of engine speed, throttle position, air–fuel, and EGR. This is the optimum spark timing for the engine and can be obtained using a closed-loop control system discussed in Ref. 12.

A measurement of torque can be obtained from a measurement of cylinder instantaneous pressure and from a measurement of engine position. The closed-loop system of Ref. 12 uses an inexpensive piezoelectric sensor in the form of a ring which is located under the spark plug.

This sensor generates an output signal which closely estimates instantaneous cylinder pressure. It can also detect the relatively high-frequency pressure oscillations which are associated with knock.

The concept for the closed-loop spark-timing control system is based upon the trace of cylinder pressure as a function of crankshaft angle. This pressure begins to rise after ignition, reaching a maximum value at a crankshaft angle Θ_p. Engine performance is related directly to the value Θ_p. There is, in fact, an optimum value Θ_p^* for this angle. The SA closed-loop control system varies ignition timing so as to maintain Θ_p at or near Θ_p^*.

This system uses the piezoelectric pressure sensor to measure pressure waveform $p(t)$. The signal processing electronics generates an output which is proportional to Θ_p. The feedback control system varies SA in such a way as to maintain Θ_p at or very near Θ_p^*.

In any practical implementation this SA control system must detect the occurrence of knock. At certain operating conditions the SA required to maintain $\Theta_p = \Theta_p^*$ might cause excessive knock. In this case the ignition timing should be retarded until knock is reduced to an acceptable level.

At the time of this writing the above system is experimental. It is not known when or whether such a system will be employed in commercially available vehicles. However, some sort of closed-loop electrical spark is likely to be implemented on automobiles in the near future.

F. Electronic Engine Control System Performance

The performance of production electronic automotive engine control has amply demonstrated the ability of production automobiles to meet EPA exhaust emission standards. In addition, the fuel economy improvements of recent years can at least partly be attributed to the performance of digital electronic control.

It is instructive to illustrate the actual performance of a typical engine feedback control system such as is described above. In Ref. 13 the measured performance of a proportional integral engine control system is given. Figure 15 is taken from this reference and displays the actuator command signal, the sensor output voltage (ZrO_2) and the throttle position in percent of WOT.

Beginning with an initial (0–10 sec) steady-state period, the throttle is abruptly advanced from 8 to 20% WOT. In open-loop operation (i.e., no feedback signal from the sensor) the air–fuel changes by 2 from 15:1 to 13:1. In the normal closed-loop operation, the sensor output is momentarily changed causing the actuator command signal to change from about zero to near -5 V. This causes the air–fuel to return to near stoichiometry (i.e., about 14.7:1).

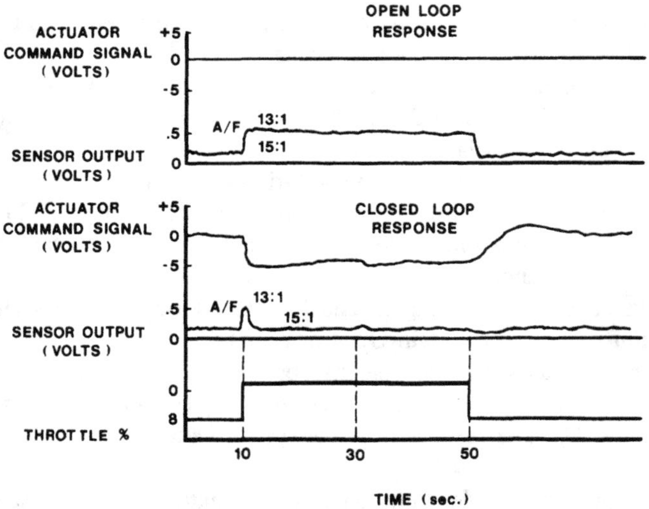

FIGURE 15. Waveforms illustrating performance of fuel control system. (Reprinted with permission from General Motors Corporation.)

At 50 sec the throttle is abruptly returned to the initial position. The command signal returns to near zero potential during a relatively short transient response interval.

This illustration of the performance of an integral proportional controller has purposely been simplified. There are many aspects of the engine model (e.g., dynamics) which have been simplified in our discussion of the theoretical system performance as well. These two discussions are given at about the same level of complexity. The interested reader can learn more about either the theory or the actual performance by consulting the references cited.

G. Other Computer Control Functions

The digital engine control system has a capacity for functions other than fuel and spark. For example, it can enable or disable EGR. It can operate the evaporative emission control system canister purge and can enable a clutch shunting the torque converter, effectively "locking" the torque converter. These functions are described in detail in Ref. 11.

H. Future of Electronic Drive-Train Control

It is not possible at the time of this writing to predict the exact future course of electronic drive-train control. However, it is perhaps worthwhile to speculate on certain potential future control concepts.

It is likely that the trend toward an integrated multifunction digital engine–transmission control system will continue. However, it is not clear that the control concept based upon TWC will continue indefinitely. There are several concepts which have been evaluated experimentally and which offer great promise for improved performance while meeting government standards. These involve the use of very lean mixtures (gasoline fueled), the use of alternate fuels, closed-loop SA, turbocharging, variable cycle engines, etc.

Just which subset or superset of potential automotive technologies will emerge in production vehicles is very difficult to predict. It is hoped that this chapter has provided adequate background for the interested reader to follow the future developments in this exciting field.

REFERENCES

1. Rivard, J. G. "Closed Loop Electronic Fuel Injection Control of the Internal Combustion Engine," SAE paper 730005 (1973).
2. Toelle, A. D. "Microprocessor Control of the Automobile Engine," SAE paper 77008 (1977).
3. Leshner, M. D., Stuart, J. W., Jr., and Leshner, E. "Closed-Loop Control for Adaptive Lean-Limit Operation," SAE paper 780039 (1979).
4. Hubbard, M. "Closed-Loop Control of Lean Fuel/Air Ratios Using a Temperature Compensated Zirconia Oxygen Sensor," SAE paper 760287 (1976).
5. Reddy, J. N. "Application of Automotive Sensors to Engine Control," SAE paper 780210 (1978).
6. Cassidy, J. F., Jr. "Electronic Engine Control: A Technological Revolution for the Automotive Industry," General Motors Research Publication (1980).
7. Rao, H. S., and Cohen, A. I. "Engine Control Optimization Via Nonlinear Programming," SAE paper 790177 (1979).
8. Athans, M. "The Role of Modern Control Theory for Engine Control," SAE paper 780852 (1978).
9. Cassidy, J. F., Jr. "A Computerized On-Line Approach to Calculating Optimum Engine Calibrations," General Motors Research Publication (1976).
10. Reid, R. A., and Czadzeck, G. H. "Ford's 1980 Central Fuel Injection System," SAE paper 790742 (1979).
11. Bremmer, R. J., Grimm, R. A., and Stonestreet, S. P. "GM Microcomputer Engine Control System," SAE paper 800053 (1980).
12. Hosey, R. J., and Powell, J. D. "Closed-Loop Knock Adaptive Spark Timing Based on Cylinder Pressure," Paper presented at 1978 Winter ASME, December (1978).
13. Cassidy, J. F., Jr. "Electronic Closed-Loop Control for the Automobile," General Motors Research Publication (1974).

Index

Printed in the USA
CPSIA information can be obtained
at www.ICGtesting.com
CBHW062145021024
15299CB00003B/39